DATA SERIES 67

HANDBOOK OF COMPARATIVE WORLD STEEL STANDARDS

U.S.A. | UNITED KINGDOM | GERMANY | FRANCE | RUSSIA | JAPAN

CANADA | AUSTRALIA | INTERNATIONAL

Albert S. Melilli, editor

ASTM Publication Code Number (PCN)
28-067000-02

ASTM
100 Barr Harbor Drive
West Conshohocken, PA 19428-2959

Library of Congress Cataloging-in-Publication Data

Handbook of comparative world steel standards = Sekai tekkō zairyō
Kikaku hikaku taishō sōran : U.S.A., United Kingdom, Germany, France
Russia, Japan, Canada, Australia, international / Albert S. Melilli, editor,
(Data series ; 67)
"ASTM publication code number (PCN): 28-067000-02
Includes bibliographical references and index.
ISBN 0-8031-1825-2
1. Steel--Standards--Handbooks, manuals, etc. 2. Steel alloys-Standards
--Handbooks, manuals, etc. I. Melilli, Albert S. II. Series: ASTM data series
publication ; DS 67.
TA472.H24 1996 96-9914
672'.0218--dc20 CIP

Printed in Philadelphia, PA
Aug. 1996

Preface

The steel standards of a country are generally established on the basis of chemical composition, physical properties and applications. However, they often reflect and incorporate the country's conditions and needs, and so they vary from one country to another, even for the same grade of steel.

Such differences are a constant source of inconvenience, especially to those who are engaged in foreign trade and who deal with steel, steel products and machines and plants that use all kinds and types of steel. It is a time consuming process to identify and compare the steel standards of one country to those of others for the same grade of steel or to compare specifications such as chemical composition.

And so in response to the requests and needs of business and industry, The Handbook of Comparative World Steel Standards was first published by The International Technical Information Institute (ITI) in 1974, and revised by ITI in 1980, 1982, 1985 and 1990. The handbook has been accepted globally by various fields of business such as the import and export of steel materials, electrical equipment, machinery, tools, parts, tubing materials, transportation materials such as for shipbuilding and railway, as well as for construction, steel bridges and plant engineering.

This 1996 edition, published by ASTM, contains numerous revisions from the previous edition, including the most current country standards as well as references to the latest European Standards (EN), as applicable. The book has been published in English, but Japanese headings have been retained for the sake of continuity and additional convenience.

EXPLANATORY NOTES ON HANDBOOK

1. Composition

Chapter I **Steel Plates for General Structure**

Chapter II **Steel Plates for Boiler and Pressure Vessels**

Chapter III **Steel Pipes and Tubes**

Chapter IV **Steels for Machine Strucural Use**

Chapter V **Steels for Special Purposes**

2. Standards

Japan .... JIS

U.S.A. .... ASTM, AISI, API

U.K. .... BS

West Germany ...... DIN

France .... NF

U.S.S.R. .... ГОСТ

Canada .... CSA

Australia .... AS

International .... ISO

3. Comparative Tables of Standards

The standards are tabulated in a convenient, comparative style.
For example, the tables in Chapter I show to what numbers and designations of the standards of U.S.A., U.K., West Germany, France, Canada and Australia, correspond to the JIS G3101-87, "Rolled Steel for General Structure," and permit ready comparison of specifications included in the standards. These specifications are individually listed in the following tables.

4. Items Included in Standards Tables

(1) Designation

(2) Number

(3) Year

(4) Grade and Type

(5) Material Number

(6) Manufacturing Process

(7) Chemical Composition

(8) Mechanical Properties

  a. Heat treatment

  b. Thickness

  c. Yield point

  d. Tensile Strength

  e. Notch toughness (Test piece, Temperature, Thickness)

(9) Index Number

5. Index

Each chapter contains index of grades, types and designations by country and standard. The indexes are so arranged that search is made not by the page number but by the index number assigned to each steel grade and type, assuring rapid and accurate retrieval.

6. Language

The standards and specifications are written in English and Japanese for additional convenience in the use of the Handbook.

7. Updating

Efforts have been made to maintain the accuracy of the handbook by updating. The standards and specifications found in the handbook are those available as of 1990.

Though the steel standards and comparative tables in this handbook have been prepared most carefully and accurately to facilitate the convenience of use and transaction of steel materials and products, the standards of each country have their own background of development and incorporation, thus making, sometimes, it practically impossible to compare in the most desirable format. Therefore, care should be taken to the choice and use of the steel materials and products. We shall not be held responsible for any legal and other problems arising from the use of this handbook.

CONTENTS 目　次

Chapter I　Steel Plates for General Structure 一般構造用鋼板

Chapter II Steel Plates for Boiler and Pressure Vessels
ボイラおよび圧力容器用鋼板

Chapter III Steel Pipes and Tubes 鋼 管

Chapter Ⅳ Steels for Machine Structural Use 機械構造用鋼

Chapter Ⅴ Steels for Special Purposes 特殊用途鋼

CHAPTER I

STEEL PLATES
FOR
GENERAL STRUCTURE

第 1 部

一般構造用鋼板

Abbreviations
略　語　表

(1) Heat Treatment　熱処理

A: Anneal　焼なまし

N: Normalize　焼ならし

Q: Quench　焼入れ

S: Solution Heat Treatment　固溶化処理

T: Temper　焼もどし

(2) Impact Test Piece　衝撃試験片

2V: 2mm V Notch　2mmVノッチ

2U: 2mm U Notch　2mmUノッチ

5U: 5mm U Notch　5mmUノッチ

DVM: DIN, DVM

(3) Designation　種類番号

Gr.: Grade　グレード

Cl.: Class　クラス

Typ.: Type　タイプ

(4) Country　国名記号

JP: Japan　日　本

US: The United States of America (U.S.A.)　アメリカ

UK: The United Kingdom　イギリス

GR: West Germany　西ドイツ

FR: France　フランス

CA: Canada　カナダ

AU: Australia　オーストラリア

1. STANDARDS LIST 収録各国規格リスト

1-1 JIS（Japan 日 本）

G3101-87 Rolled Steel for General Structure
一般構造用圧延鋼材

G3106-92 Rolled Steel for Welded Structure
溶接用構造圧延鋼材

G3114-88 Hot-rolled Atmospheric Corrosion Resisting Steels for Welded Structure
溶接構造用耐候性熱間圧延鋼材

G3125-87 Superior Atmospheric Corrosion Resisting Steels
高耐候性圧延鋼材

G3128-87 High Yield Strength Steel Plates for Welded Structure
溶接構造用高降伏点鋼板

1-2 ASTM（U. S. A. アメリカ）

A36M-94 Structural Steel
構造用鋼材

A242M-93a High-Strength Low-Alloy Structural Steel
構造用高張力低合金鋼鋼材

A283M-93a Low and Intermediate Tensile Strength Carbon Steel Plates, Shapes and Bars
構造用低・中抗張力炭素鋼鋼材

A284M-88 Low and Intermediate Tensile Strength Carbon-Silicon Steel Plates for Machine Parts and General Construction (Discontinued 1990; Replaced by A283)
機械部品及び一般構造用低・中抗張力C-Si鋼鋼材

A514M-94a High-Yield-Strength, Quenched and Tempered Alloy Steel Plate, Suitable for Welding
溶接性調質高降伏点合金鋼鋼材

A529M-94 Structural Steel with 290MPa Minimum Yield Point（13mm Maximum Thickness）
構造用最低降伏点290MPa鋼鋼材

A572M-94c High-Strength Low-Alloy Columbium-Vanadium Steels of Structural Quality
構造用低合金Nb-V高張力鋼鋼材

A573M-93a Structural Carbon Steel Plates of Improved Toughness
構造用靭性改良型炭素鋼鋼材

A588M-94 High-Strength Low-Alloy Structural Steel with 345MPa Minimum Yield Point 100mm Thickness
構造用最低降伏点低合金鋼鋼材

A633M-94a Normalized High-Strength Low-Alloy Structural Steel
構造用低合金焼ならし高張力鋼鋼材

A656M-93a Hot-Rolled StructuralSteel, High Strength Low Alloy Plates with Improved Formability
易加工性低合金熱間圧延高張力鋼鋼材

A678M-94a　Quenched and Tempered Carbon Steel Plates for Structural Application
構造用焼入れ焼戻し炭素鋼鋼材

A710M-94　Low-Carbon Age-Hardening Nickel-Copper-Chromium-Molybdenum-Columbium and Nickel-Copper-Columbium Alloy Steels
低炭素・時効硬化Ni-Cu-Cr-Mo-Nb及びNi-Cu-Nb合金鋼鋼材

1-3 BS（U.K.　イギリス）

4360-86　Weldable Structural Steels (Superseded, in part, by European Standard EN 10113-1, -2 & - 3: 1993 and EN 10155: 1993)
溶接性構造用鋼鋼材

1-4 DIN（West Germany 西ドイツ）

17100-80　Steels for General Structural Purposes (Superseded by European Standard EN 10025: 90)
一般構造用鋼鋼材

17102-83　Weldable Fine Grain Steels (Superseded by European Standard EN 10113-1 & -2: 1993)
溶接細粒鋼鋼材

1-5 NF（France　フランス）

A35-501-87　Structural Steels
構造用鋼鋼材

A35-502-84　Structural Steels with Improved Atmospheric Corrosion Resistance
耐候性構造用鋼

1-6 CSA（Canada カナダ）

CAN/CSA -G40.21-92　Structural Quality Steels
構造用鋼鋼材

1-7 AS（Australia オーストラリア）

1204-80　Structural Steels — Ordinary Weldable Grades
一般構造用鋼鋼材

1205-80　Structural Steels — Weather-Resistant Weldable Grades
溶接耐候性構造用鋼鋼材

2. Steel Plates for General Structure　一般構造用鋼板

2-1 Tensile Strength 300 N／mm² or MPa (30 kgf／mm²) Class

Country 国	Standard No. 規格番号及び年号	Designation 種類記号	Chemical Composition 化学成分 (%) Thickness(t) mm板厚	C	Si	Mn	P	S	Cu	Cr	Ni
JP日	JIS 63101-87	SS330 (SS34)	–	–	–	–	≦0.050	≦0.050	–	–	–
US米	ASTM A283M-93a	Gr.A	t ≦40	≦0.14	≦0.40	≦0.90	≦0.04	≦0.04	When Specified ≧0.20	–	–
			40< t		0.15〜0.40						
		Gr.B	t ≦40	≦0.17	≦0.40	≦0.90	≦0.04	≦0.04	When Specified ≧0.20	–	–
			40< t		0.15〜0.40						
		Gr.C	t ≦40	≦0.17	≦0.40	≦0.90	≦0.04	≦0.04	When Specified ≧0.20	–	–
			40< t		0.15〜0.40						
UK英	–	–	–	–	–	–	–	–	–	–	–
GR独	DIN 17102-83	STE255	–	≦0.18	≦0.40	≦0.50	≦0.035	≦0.030	≦0.20	≦0.30	≦0.30
		STE285	–	≦0.18	≦0.40	0.60〜1.00	≦0.035	≦0.030	≦0.20	≦0.30	≦0.30
		Superseded by European Standard EN 10113-1: 1993, Grade S275N, Material No. 1.0490									
FR仏	NF A35-501-87	A33	–	–	–	–	–	–	–	–	–
		Superseded by European Standard EN 10025: 1993, Grade S185, Material No. 1.0035									

引張り強さ 300N／mm² または MPa（30キロ）級

Mo	V	Others	Mechanical Properties 機械的性質 Heat Treatment 熱処理	Thickness (t)mm 板厚	Yield Point min. N/mm² or MPa (kgf/mm²) 最小降伏点	Tensile Strength N/mm² or MPa (kgf/mm²) 引張強さ	Notch Toughness 衝撃値 Test Piece 試験片	Test Temp.(℃) 試験温度	Absorbed Energy J (min. kg-m) 吸収ｴﾈﾙｷﾞｰ	Index No. 索引番号
–	–	–	–	t ≦15 16<t≦40 40< t	205 (21) 195 (20) 175 (18)	330～430 (34～44)	–	–	–	A001
–	–	–	–	–	165	310～415	–	–	–	
–	–	–	–	–	185	345～450	–	–	–	
–	–	–	–	–	205	380～515	–	–	–	
–	–	–	–	–	–	–	–	–	–	
≦0.08	–	N≦0.020 Al≦0.020 Nb≦0.03		t ≦16 16<t≦35 35<t≦50 50<t≦60 60<t≦70 70<t≦85 85<t≦100 100<t≦125 125<t≦150	255 255 245 235 235 225 215 205 195	t ≦70 360～480 70< t ≦85 350～470 85<t≦100 340～460 100< t ≦125 330～450 125< t ≦150 320～440	–	–	–	
≦0.08	–	N≦0.020 Al≦0.020 Nb≦0.03 Nb+Ti+V ≦0.05		t ≦16 16<t≦35 35<t≦50 50<t≦60 60<t≦70 70<t≦85 85<t≦100 100<t≦125 125<t≦15	285 285 275 265 265 255 245 235 225	t ≦70 390～510 70< t ≦85 380～500 85<t≦100 370～490 100< t ≦125 360～480 125< t ≦150 350～470	–	–	–	
–	–	–	–	t≦3 3<t≦30 30<t≦50	155 175 175	t ≦3 320～560 3<t≦150	–	–	–	

Country 国	Standard No. 規格番号 及び年号	Desig-nation 種類記号	Chemical Composition 化 学 成 分 (%)								
			Thick-ness(t) mm板厚	C	Si	Mn	P	S	Cu	Cr	Ni
FR仏	NF A35-501-87	E24-2	Superseded by EN 10025, Grade S235JR, Material No. 1.0037 –	≦0.17	–	–	≦0.045	≦0.045	–	–	–
		E24-2E	–	≦0.17	–	–	≦0.045	≦0.045	–	–	–
		E24-2NE	–	≦0.17	–	–	≦0.045	≦0.045	–	–	–
		E24-3NE	–	≦0.17	–	–	≦0.045	≦0.045	–	–	–
		E24-4CS	–	≦0.16	–	–	≦0.035	≦0.035	–	–	–
	NF A35-502-87	E24W-2	–	≦0.13	0.10～0.40	0.20～0.60	≦0.040	≦0.035	0.20～0.50	0.40～0.80	≦0.65
		E24W-3	Superseded by European Standard EN 10155: 1993, Grade S235 JOW, Material No. 1.8958 –	≦0.13	0.10～0.40	0.20～0.60	≦0.040	≦0.035	0.20～0.50	0.40～0.80	≦0.65
		E24W-4	Superseded by European Standard EN 10155: 1993, Grade S235 J2W, Material No. 1.8961								
CA加	–	–	–	–	–	–	–	–	–	–	–
AU濠	–	–	–	–	–	–	–	–	–	–	–

$1N/mm^2$ or $MPa = 1.01972 \times 10^{-1} kgf/mm^2$
$1J = 1.01972 \times 10^{-1} kg\text{-}m$

			Mechanical Properties 機械的性質							Index No. 索引番号
Mo	V	Others	Heat Treatment 熱処理	Thickness (t)mm 板厚	Yield Point min. N/mm^2 or MPa (kgf/mm^2) 最小降伏点	Tensile Strength N/mm^2 or MPa (kgf/mm^2) 引張強さ	Notctch Toughness 衝撃値 Test Piece 試験片	Test Temp.(℃) 試験温度	Absorbed Energy J (min. kg-m) 吸収ｴﾈﾙｷﾞｰ	
－ －	－ －	N≦0.008 N≦0.007	－	t≦3 3<t≦30 30<t≦50	215 235 215	t ≦3 360～480	2 V	20	35	A001
－	－	N≦0.008	－	50<t≦80 80<t≦110 110<t≦150	205 195 185	3<t≦150 340～460	2 V	20	35	
－	－	－	－	t≦3 3<t≦30 30<t≦50 50<t≦80 80<t≦110 110<t≦150	215 235 215 205 195 185	t ≦3 360～480 3<t≦150 340～460	2 V	0	35	
－	－	AL≧0.02	－	t≦3 3<t≦30 30<t≦50 50<t≦80 80<t≦110 110<t≦150	215 235 215 205 195 185	t≦3 360～480 3<t≦150 340～460	2 V	-20	35	
－	－	－	－	t≦3 3<t≦30 30<t≦50 50<t≦80 80<t≦110	215 235 215 205 195	t≦50 360～460 t>50 ≧360	－	－	－	
－	－	－	－	t≦3 3<t≦30 30<t≦50 50<t≦80 80<t≦110	215 235 215 205 195	t≦50 360～460 t>50 ≧360	2 V	0	35	
－	－	－	－	t≦3 3<t≦30 30<t≦50 50<t≦80 80<t≦110	215 235 215 205 195	t≦50 360～460 t>50 ≧360	2 V	-20	35	
－	－	－	－	－	－	－	－	－	－	
－	－	－	－	－	－	－	－	－	－	

2-2 Tensile Strength 400 N/mm² or MPa (40 kgf/mm²) Class

Country 国	Standard No. 規格番号及び年号	Designation 種類記号	Thickness (t) mm 板厚	C	Si	Mn	P	S	Cu	Cr	Ni
			Chemical Composition 化学成分 (%)								
JP 日	JIS G3101-87	SS400 (SS41)	–	–	–	–	≦0.050	≦0.050	–	–	–
	JIS G3106-92	SM400 (SM41) A	t ≦50	≦0.23	–	≧2.5C	≦0.035	≦0.035	Alloy elements when necessary. 必要に応じて合		
		A	50<t ≦200	≦0.25							
		B	t ≦50	≦0.20	≦0.35	0.16~1.40					
		B	50<t ≦200	≦0.22							
		C		≦0.18	≦0.35	≦1.40					
	JIS G3114-88	SMA 400 (SMA 41) W A, B, C	–	≦0.18	0.15~0.65	≦1.25	≦0.035	≦0.035	0.30~0.50	0.45~0.75	0.05~0.30
		P A, B, C			≦0.55				0.20~0.35	0.30~0.55	–
US 米	ASTM A36M-94	–	t ≦20	≦0.25	–	–	≦0.04	≦0.05	When specified ≧0.20	–	–
			20<t ≦40	≦0.25	–	0.80~1.20					
			40<t ≦65	≦0.26	≦0.40	0.80~1.20					
			65<t ≦100	≦0.27	≦0.40	0.80~1.20					
			100< t	≦0.29	≦0.40	0.80~1.20					
	ASTM A283M-93a	Gr. D	t ≦40	≦0.27	<0.40	≦0.09	≦0.04	≦0.04	When specified ≧0.20	–	–
			20 < t		0.15~0.40						
	ASTM A284M-89	Gr. C	t ≦25	≦0.24	0.15~0.40	≦0.09	≦0.04	≦0.05		–	–
			25<t ≦50	≦0.27							
			50<t ≦100	≦0.29							
			100<t ≦200	≦0.33							
			200<t ≦300	≦0.36							
		Gr. D	t ≦25	≦0.27	0.15~0.40	≦0.09	≦0.04	≦0.05	–	–	–
			25<t ≦50	≦0.29							
			50<t ≦100	≦0.31							
			100<t ≦200	≦0.35							
	ASTM A529M-94	Gr. 42 (290)	t ≦12.5	≦0.27	–	≦1.20	≦0.04	≦0.05	When specified ≧0.20	–	–
		Gr. 50 (345)	t ≦25	≦0.27	–	≦1.35	≦0.04	≦0.05		–	–
	ASTM A572M-94b	Gr. 42 (290)	t ≦150	≦0.21	(t ≦40) ≦0.40	≦1.35	≦0.04	≦0.05	When specified ≧0.20	–	–
		Gr. 50 (345)	t ≦100	≦0.23	(40<t) 0.1~0.40						
	ASTM A573M-93a	Gr. 58 (400)	t ≦40	≦0.23	0.10~0.35	0.60~0.90	≦0.035	≦0.04	–	–	–

ASTM A284 was discontinued in 1990 and replaced by ASTM A283

引張り強さ 400 N／mm² または MPa（40キロ）級

Mo	V	Others	Heat Treatment 熱処理	Thickness (t)mm 板厚	Yield Point min. N/mm²or MPa (kgf/mm²) 最小降伏点	Tensile Strength N/mm²or MPa (kgf/mm²) 引張強さ	Notch Toughness 衝撃値 Test Piece 試験片	Test Temp.(℃) 試験温度	Absorbed Energy J (min. kg-m) 吸収ｴﾈﾙｷﾞｰ	Index No. 索引番号
–	–	–	–	t ≦16	245 (25)	400～510 (41～52)	–	–	–	A002
				16<t≦40	235 (24)					
				40<t	215 (22)					
may be added, 金元素の添加可。			–	t ≦16	245 (25)	400～510 (41～52)	–	–	–	
				16<t≦40	235 (24)					
				40<t≦100	215 (22)		2 V	0	27(2.8)	
				100<t≦160	205 (21)					
				160<t≦200	195 (20)		2 V	0	47(4.8)	
One or more of Mo, Nb, Ti, V and/or Zr may be added. (≦0.15%) Mo, Nb, Ti, V, Zr のうち1種以上の0.15%以下の添加可。			–	t ≦16	245 (25)	400～540 (41～55)	–	–	–	
				16<t≦40	235 (24)		2 V	0	27(2.8)	
				40< t	215 (22)		2 V	0	47(4.8)	
							–	–	–	
							2 V	0	27(2.8)	
							2 V	0	47(4.8)	
–	–	–	–	–	250	400～550	–	–	–	
–	–	–	–	–	230	415～550	–	–	–	
–	–	–	–	–	205	≧415	–	–	–	
–	–	–	–	–	230	≧415	–	–	–	
–	–	–	–	–	290	415～585	–	–	–	
–	–	–	–	–	345	485～690	–	–	–	
–	–	–	–	–	290	≧415	–	–	–	
					345	≧450				
–	–	–	–	–	220	400～490	–	–	–	

Country 国	Standard No. 規格番号及び年号	Designation 種類記号	Chemical Composition 化 学 成 分 （%）								
			Thickness(t) mm板厚	C	Si	Mn	P	S	Cu	Cr	Ni
US米	ASTM A573M-93a	Gr.65 (450)	t ≦13	≦0.24	0.15～0.40	0.85～1.20	≦0.035	≦0.04	–	–	–
			13<t ≦40	≦0.26							
	ASTM A633M-94a	Gr.A	≦100	≦0.18	0.15～0.50	1.00～1.35	≦0.035	≦0.04	–	–	–
	ASTM A656M-93a	Type 3 Gr.50 (345)	–	≦0.18	≦0.60	≦1.65	≦0.025	≦0.035	–	–	–
		Type 7 Gr.50 (345)	–	≦0.18	≦0.60	≦1.65	≦0.025	≦0.035	–	–	–
UK英	BS 4360-86	Gr.40A	–	≦0.22	≦0.50	≦1.60	≦0.050	≦0.050	–	–	–
		Gr.40B	–	≦0.20	≦0.50	≦1.50	≦0.050	≦0.050	–	–	–
		Gr.40C	–	≦0.18	≦0.50	≦1.50	≦0.040	≦0.040	–	–	–
		Gr.40D	–	≦0.16	≦0.50	≦1.50	≦0.040	≦0.040	–	–	–
		Gr.40E	–	≦0.16	0.10～0.50	≦1.50	≦0.040	≦0.040	–	–	–
		Superseded by European Standard EN 10113-1: 1993, Grade S275 NL, Material No. 1.0491									
		Gr.43A	–	≦0.25	≦0.50	≦1.60	≦0.050	≦0.050	–	–	–

$1N/mm^2$ or $MPa = 1.01972 \times 10^{-1} kgf/mm^2$
$1J = 1.01972 \times 10^{-1} kg\text{-}m$

Mo	V	Others	Mechanical Properties 機械的性質							Index No. 索引番号
			Heat Treatment 熱処理	Thickness (t)mm 板厚	Yield Point min. N/mm^2 or MPa (kgf/mm^2) 最小降伏点	Tensile Strength N/mm^2 or MPa (kgf/mm^2) 引張強さ	Notctch Toughness 衝撃値			
							Test Piece 試験片	Test Temp.(℃) 試験温度	Absorbed Energy J (min. kg-m) 吸収エネルギー	
–	–	–	–	–	240	450～510	–	–	–	A002
–	–	Nb≦0.05	N	≦100	290	430～570	–	–	–	
–	≦0.08	Nb 0.005~0.15	–	–	345	≧415	–	–	–	
–	(≦0.005)	Nb 0.005~0.10 (V+Nb≦0.02)	–	–	345	≧415	–	–	–	
–	–	–	–	t≦16 16<t≦40 40<t≦63 63<t≦100 100<t	235 225 215 205 185	340～500	–	–	–	
–	–	–	–	t≦16 16<t≦40 40<t≦63 63<t≦100 100<t	235 225 215 205 185	340～450	2V	20	27	
–	–	–	–	t≦16 16<t≦40 40<t≦63 63<t≦100 100<t	235 225 215 210 185	340～550	2V	0	27	
–	–	–	–	t≦16 16<t≦40 40<t≦63 63<t≦100 100<t	235 225 215 215 205	340～500	2V	−20	27	
–	–	–	–	t≦16 16<t≦40 40<t≦63 63<t≦100 100<t	260 245 240 225 215	340～500	2V	−50	27	
–	–	–	–	t≦16 16<t≦40 40<t≦63 63<t≦100 100<t	275 275 255 245 225	430～580	–	–	–	

Country 国	Standard No. 規格番号及び年号	Desig-nation 種類記号	Chemical Composition 化学成分（%）								
			Thick-ness(t) mm板厚	C	Si	Mn	P	S	Cu	Cr	Ni
UK英	BS 4360-86	Gr.43B	−	≦0.21	≦0.50	≦1.50	≦0.050	≦0.050	−	−	−
		Gr.43C	−	≦0.18	≦0.50	≦1.50	≦0.040	≦0.040	−	−	−
		Gr.43D	−	≦0.16	≦0.50	≦1.50	≦0.040	≦0.040	−	−	−
		Gr.43EE	−	≦0.16	≦0.50	≦1.50	≦0.040	≦0.040	−	−	−
GR独	DIN 17100-80	ST44-2	t≦40	≦0.21	−	−	≦0.050	≦0.050	−	−	−
			t>40	≦0.22	−	−	≦0.050	≦0.050	−	−	−
		ST44-3	−	≦0.20	−	−	≦0.040	≦0.040	−	−	−

1N/mm² or MPa＝1.01972×10⁻¹kgf/mm²
1J＝1.01972×10⁻¹kg-m

Mo	V	Others	Mechanical Properties 機械的性質							Index No. 索引番号
			Heat Treatment 熱処理	Thickness (t)mm 板厚	Yield Point min. N/mm² or MPa (kgf/mm²) 最小降伏点	Tensile Strength N/mm² or MPa (kgf/mm²) 引張強さ	Notctch Toughness 衝撃値 Test Piece 試験片	Notctch Toughness 衝撃値 Test Temp.(℃) 試験温度	Notctch Toughness 衝撃値 Absorbed Energy J (min.kg-m) 吸収ｴﾈﾙｷﾞｰ	
				t ≦16	275					
				16<t≦40	265					
−	−	−	−	40<t≦63	255	430～580	2 V	20	27	A002
				63<t≦100	245					
				100<t	225					
				t ≦16	275					
				16<t≦40	265					
−	−	−	−	40<t≦63	255	430～580	2 V	0	27	
				63<t≦100	245					
				100<t	225					
				t ≦16	275					
				16<t≦40	265					
−	−	−	−	40<t≦63	255	430～580	2 V	−20	27	
				63<t≦100	245					
				100<t	225					
				t ≦16	275					
				16<t≦40	265					
−	−	−	−	40<t≦63	255	430～580	2 V	−50	27	
				63<t≦100	245					
				100<t	225					
				t ≦16	275	t ≦16				
−	−	≦0.009	−	16<t≦40	265	430～580	−	−	−	
				40<t≦63	255	3≦t≦100				
				63<t≦80	245	410～540				
−	−	≦0.009	−	80<t≦100	235	100< t	−	−	−	
				100<t	by agreement	by agreement				
		Additional		t ≦16	275	t ≦16				
		N−		16<t≦40	265	430～580	−	−	−	
−	−	Combining	−	40<t≦63	255	3≦t≦100				
		elements		63<t≦80	245	410～540				
		(yes)		80<t≦100	245	100< t	−	−	−	
				100<t	by agreement	by agreement				

Country 国	Standard No. 規格番号及び年号	Desig-nation 種類記号	Chemical Composition 化 学 成 分 （%）								
			Thick-ness(t) mm板厚	C	Si	Mn	P	S	Cu	Cr	Ni
ＧＲ独	ＤＩＮ 17102-83	STE-315	–	≦0.18	≦0.45	0.70～1.50	≦0.035	≦0.030	≦0.20	≦0.30	≦0.30
ＦＲ仏	ＮＦ A35-501-87	E28-2NE	–	≦0.20	–	–	≦0.045	≦0.045	–	–	–
		E28-3NE	–	≦0.16	–	–	≦0.040	≦0.040	–	–	–
		E28-4CS	–	≦0.18	–	–	≦0.035	≦0.035	–	–	–
ＣＡ加	CAN/CSA G40.21-92	33G	–	≦0.26	≦0.40	≦1.20	≦0.05	≦0.05	–	–	–
		38W	–	≦0.20	≦0.40	0.50～1.50	≦0.04	≦0.05	–	–	–
		44W	–	≦0.22	≦0.40	0.50～1.50	≦0.04	≦0.05	–	–	–
		50W	–	≦0.23	≦0.40	0.50～1.50	≦0.04	≦0.05	–	–	–
		38WT	–	≦0.20	0.15～0.40	0.80～1.50	≦0.03	≦0.04	–	–	–

1N/mm² or MPa=1.01972×10⁻¹kgf/mm²
1J=1.01972×10⁻¹kg-m

Mo	V	Others	Heat Treatment 熱処理	Thickness (t)mm 板厚	Yield Point min. N/mm² or MPa (kgf/mm²) 最小降伏点	Tensile Strength N/mm² or MPa (kgf/mm²) 引張強さ	Notch Toughness 衝撃値: Test Piece 試験片	Notch Toughness 衝撃値: Test Temp.(℃) 試験温度	Notch Toughness 衝撃値: Absorbed Energy J (min. kg-m) 吸収ｴﾈﾙｷﾞｰ	Index No. 索引番号
≤0.08	–	N≤0.020 Al≤0.020 Nb≤0.03 Nb+Ti+V ≤0.05	–	t≤16	315	t≤70	–	–	–	A002
				16<t≤35	315	440~560				
				35<t≤50	305	70<t≤85				
				50<t≤60	295	430~550				
				60<t≤70	295	85<t≤100				
				70<t≤85	285	420~540				
				85<t≤100	275	100<t≤125 410~530				
				100<t≤125	265					
				125<t≤150	255	125<t≤150 400~520				
–	–	N≤0.008	–	t<3	255		2V	20	35	
				3≤t≤30	275	t<3				
				30<t≤50	255	420~560				
				50<t≤80	245	3≤t≤150				
				80<t≤110	235	400~540				
				110<t≤150	225					
–	–	–	–	3≤t≤30	275	t<3	2V	0	35	
				30<t≤50	255	420~560				
				50<t≤80	245	3≤t≤150				
				80<t≤110	235	400~540				
				110<t≤150	225					
–	–	Al≥0.02	–	t<3	255		2V	-20	35	
				3≤t≤30	275	t<3				
				30<t≤50	255	420~560				
				50<t≤80	245	3≤t≤150				
				80<t≤110	235	400~540				
				110<t≤150	225					
–	–	Al or Cb or V≤0.10	–	–	230	380~495	–	–	–	
–	–	Al or Cb or V≤0.10	–	≤64	260	415~585	–	–	–	
				>64	250					
–	–	Al or Cb or V≤0.10	–	≤64	305	450~620	–	–	–	
				>64	275					
–	–	Al or Cb or V≤0.10	–	≤64	345	450~655	–	–	–	
				>64	315					
–	–	Al or Cb or V≤0.10	–	≤64	260	415~585	–	–	–	
				>64	250					

Country 国	Standard No. 規格番号及び年号	Designation 種類記号	Chemical Composition 化学成分 (%)								
			Thickness(t) mm板厚	C	Si	Mn	P	S	Cu	Cr	Ni
CA加	CAN/CSA G40, 21-92	44WT	–	≦0.22	0.15～0.40	0.80～1.50	≦0.03	≦0.04	–	–	–
AU豪	AS1204-80	200	–	≦0.15	–	(C+Mn/6 ≦0.25)	≦0.030	≦0.030	–	–	–
		250	–	≦0.25	≦0.40	(C+Mn/6 ≦0.43)	≦0.040	≦0.040	–	–	–
		250L0 250L15	–	≦0.20	≦0.40	≦1.50 (C+Mn/6 ≦0.43)	≦0.040	≦0.040	–	–	–

$1N/mm^2$ or $MPa = 1.01972 \times 10^{-1} kgf/mm^2$
$1J = 1.01972 \times 10^{-1} kg\text{-}m$

			Mechanical Properties 機械的性質							Index No. 索引番号
			Heat Treatment 熱処理	Thickness (t)mm 板厚	Yield Point min. N/mm²or MPa (kgf/mm²) 最小降伏点	Tensile Strength N/mm²or MPa (kgf/mm²) 引張強さ	Notch Toughness 衝撃値			
Mo	V	Others					Test Piece 試験片	Test Temp.(℃) 試験温度	Absorbed Energy J (min. kg-m) 吸収ｴﾈﾙｷﾞｰ	
–	–	Al or Cb or V≦0.10	–	≦64 >64	305 275	450～620	–	–	–	A002
–	–	–	–	t≦8 8≦ t ≦12	200 200	≧300	–	–	–	
–	–	–	–	t ≦50 50<t≦100 100 < t	260 250 230	≧410	2V	0 -15	27 27	
–	–	–	–	t≦12 12<t≦40 40<t	260 250 230	≧410	2V	0 -15	27 27	

2-3 Tensile Strength 500 N/mm² or MPa (50 kgf/mm²) Class

Country 国	Standard No. 規格番号及び年号	Designation 種類記号		Thickness (t) mm板厚	C	Si	Mn	P	S	Cu	Cr	Ni
					Chemical Composition 化学成分 (%)							
JP日	JIS G3101-87	SS490 (SS50)		–	–	–	–	≦0.050	≦0.050	–	–	–
		SS540 (SS55)		–	≦0.30	–	≦1.60	≦0.040	≦0.040	–	–	–
	JIS G3110-88	SM 490 (SM50)	A	t ≦50	≦0.20	≦0.55	≦1.60	≦0.035	≦0.035	Alloy elements when necessary. 必要に応じて		
				50<t ≦2000	≦0.22							
			B	t ≦50	≦0.18							
				50<t ≦2000	≦0.20							
			C	t ≦100	≦0.18							
		SM 490Y (SM50Y)	A, B	–	≦0.20	≦0.55	≦1.60	≦0.035	≦0.035			
		SM 520 (SM53)	A, B	–	≦0.20	≦0.55	≦1.60	≦0.035	≦0.035			
	JIS G3114-88	SMA 490 (SMA 50)	W A, B, C	–	≦0.18	0.15〜0.65	≦1.40	≦0.035	≦0.035	0.30〜0.50	0.45〜0.75	–
			P A, B, C			≦0.55				0.20〜0.35	0.30〜0.55	
	JIS G3125-87	SPA-H		–	≦0.12	0.25〜0.75	0.20〜0.50	0.070〜 0.150	≦0.040	0.25〜0.60	0.30〜1.25	≦0.65
US米	ASTM A242M-93a	Type 1		–	≦0.15	–	≦1.00	≦0.15	≦0.05	≧0.20	–	–
	ASTM A588M-94	Gr. A		–	≦0.19	0.30〜0.65	0.80〜1.25	≦0.04	≦0.05	0.25〜0.40	0.40〜0.65	≦0.40
		Gr. B		–	≦0.20	0.15〜0.50	0.75〜1.35	≦0.04	≦0.05	0.20〜0.40	0.40〜0.70	≦0.50
		Gr. C		–	≦0.15	0.15〜0.30	0.80〜1.35	≦0.04	≦0.05	0.20〜0.50	0.30〜0.50	0.25〜0.50
		Gr. K		–	≦0.17	0.25〜0.50	0.50〜1.20	≦0.04	≦0.05	0.30〜0.50	0.40〜0.70	≦0.40

引張り強さ500 N／㎟ または MPa（50キロ）級

Mo	V	Others	Heat Treatment 熱処理	Thickness (t)mm 板厚	Yield Point min. N/mm² or MPa (kgf/mm²) 最小降伏点	Tensile Strength N/mm² or MPa (kgf/mm²) 引張強さ	Notch Toughness 衝撃値: Test Piece 試験片	Test Temp. (°C) 試験温度	Absorbed Energy J (min. kg-m) 吸収エネルギー	Index No. 索引番号
–	–	–	–	t≦16	285(29)	490～610 (50～62)	–	–	–	A003
				16<t≦40	275(28)					
				40<t	255(26)					
–	–	–	–	t≦16	400(41)	≧540 (≧55)	–	–	–	
				16<t≦40	390(40)					
may be added. 合金元素の添加可。			–	t≦16	225(33)	490～610 (50～62)	–	–	–	
				16<t≦40	315(32)					
				40<t≦75	295(30)		2V	0	27(2.8)	
				75<t≦100	295(30)					
				100<t≦160	285(29)					
				160<t≦200	275(28)		2V	0	47(4.8)	
			–	t≦16	365(37)	490～610 (50～62)	–	–	–	
				16<t≦40	355(36)					
				40<t≦75	335(34)		2V	0	27(2.8)	
				75<t≦100	325(33)					
			–	t≦16	365(37)	520～640 (53～65)	2V	0	27(2.8)	
				16<t≦40	355(36)					
				40<t≦75	335(34)		2V	0	47(4.8)	
				[illegible]t≦100	[illegible](33)					
[illegible]	[illegible]	[illegible]	[illegible]	[illegible]	[illegible]55(37)	490～610 (50～62)	–	–	–	
				[illegible]	[illegible]55(36)		2V	0	47(4.8)	
				[illegible]	335(34)		–	–	–	
							2V	0	27(2.8)	
							2V	0	27(2.8)	
							2V	0	27(2.8)	
[illegible]	[illegible]	[illegible]	[illegible]	[illegible]	345(35)	≧480(≧49)	–	–	–	
				[illegible]	355(36)	≧490(≧50)				
[illegible]	[illegible]	[illegible]	[illegible]	[illegible]	345	≧480	–	–	–	
				[illegible]	315	≧460				
				[illegible]0	290	≧435				
≦0.10	–	Cb 0.005～0.05	[illegible]	[illegible]	345	≧485	–	–	–	
				[illegible]125	315	≧460				
				[illegible]≦200	290	≧435				

Country 国	Standard No. 規格番号及び年号	Designation 種類記号	Chemical Composition 化学成分 (%)								
			Thickness(t) mm板厚	C	Si	Mn	P	S	Cu	Cr	Ni
US米	ASTM A572M-94b	Gr. 60 (415)	≦32	≦0.26	≦0.40	≦1.35	≦0.04	≦0.05	When specified ≧2.00	–	–
		Gr. 65 (450)	13<t≦32	≦0.23	≦0.40	≦1.65	≦0.04	≦0.05		–	–
			≦13	≦0.26							
	ASTM A573M-93a	Gr. 70 (485)	t ≦13	≦0.27	0.15~0.40	0.85~1.26	≦0.035	≦0.04	–	–	–
			13<t ≦40	≦0.28							
	ASTM A633M-94a	Gr. C	t ≦40	≦0.20	0.15~0.50	1.15~1.50	≦0.035	≦0.04	–	–	–
			40<t ≦100								
		Gr. D	t ≦40	≦0.20	0.15~0.50	–	≦0.035	≦0.04	≦0.35	≦0.25	≦0.25
			40<t ≦100								
		Gr. E	≦1.50	≦0.22	0.15~0.50	1.15~1.50	≦0.035	≦0.04	–	–	–
	ASTM A678M-94a	Gr. A	≦40	≦0.16	0.15~0.50	0.90~1.50	≦0.035	≦0.04	When specified ≧0.20	–	–
		Gr. B	≦40	≦0.20	0.15~0.50	0.70~1.35	≦0.035	≦0.04		–	–
			40<t ≦65	≦0.20	0.15~0.50	1.00~1.60					
	ASTM A710M-94	Gr. A Cl. 2	–	≦0.07	≦0.40	0.40~0.70	≦0.025	≦0.025	1.00~1.30	0.60~0.90	0.70~1.00
	ASTM A656M-93a	Type 3	–	≦0.18	≦0.60	≦1.65	≦0.025	≦0.035	–	–	–
		Type 7	–	≦0.18	≦0.60	≦1.65	≦0.025	≦0.035	–	–	–
UK英	BS 4360-86	Gr. 50A	–	≦0.23	≦0.50	≦1.60	≦0.050	≦0.050	–	–	–

$1N/mm^2$ or $MPa = 1.01972 \times 10^{-1} kgf/mm^2$
$1J = 1.01972 \times 10^{-1} kg\text{-}m$

Mo	V	Others	Mechanical Properties 機械的性質				Notch Toughness 衝撃値			Index No. 索引番号
			Heat Treatment 熱処理	Thickness (t)mm 板厚	Yield Point min. N/mm^2 or MPa (kgf/mm^2) 最小降伏点	Tensile Strength N/mm^2 or MPa (kgf/mm^2) 引張強さ	Test Piece 試験片	Test Temp. (℃) 試験温度	Absorbed Energy J (min. kg-m) 吸収エネルギー	
–	–	–	–	–	415	≧520	–	–	–	A003
–	–	–	–	–	450	≧550	–	–	–	
–	–	–	–	–	290	485~620	–	–	–	
–	–	Cb0.01~0.05	N	t≦65	345	485~620	–	–	–	
				65<t≦100	315	450~590				
≦0.08	–	–	N	t≦65	290	485~620	–	–	–	
				65<t≦100	315	450~590				
–	0.04~0.11	N0.01~0.03 Nb(0.01 ~0.05)	N	≦65	415	550~690	–	–	–	
				65<t≦100	415	550~690				
				100<t≦150	380	515~655				
–	–	B may be added by agreement. 協議	QT	≦40	345	485~620	–	–	–	
–	–		QT	–	415	550~690	–	–	–	
0.15~0.25	–	Cb≧0.02	N and P	t≦25	450	495	–	–	–	
				25<t≦50	415					
				50<t≦100	380	450				
				100<t	345	415				
–	–	N≦0.020 Nb 0.005~0.15	–	–	Gr.60 415	Gr.60 485	–	–	–	
					Gr.70 485	Gr.70 550				
–	0.005~0.15	N≦0.020 Nb 0.005~0.15	–	–	Gr.60 415	Gr.60 485	–	–	–	
					Gr.70 485	Gr.70 550				
–	0.003~0.10	Nb 0.003~0.10	–	t≦16	355	490~640	–	–	–	
				16<t≦40	345					
				40<t≦63	340					
				63<t≦100	325					
				100<t	305					

Country 国	Standard No. 規格番号及び年号	Designation 種類記号	Chemical Composition 化学成分 (%)								
			Thickness(t) mm板厚	C	Si	Mn	P	S	Cu	Cr	Ni
UK英	BS 4360-86	Gr. 50B	–	≦0.20	≦0.50	≦1.50	≦0.040	≦0.040	–	–	–
		Gr. 50C	–	≦0.20	≦0.50	≦1.50	≦0.040	≦0.040	–	–	–
		Gr. 50D	–	≦0.18	0.10～0.50	≦1.50	≦0.040	≦0.040	–	–	–
		Gr. 50DD	–	≦0.18	0.10～0.50	≦1.50	≦0.040	≦0.040	–	–	–
		Gr. 50EE	–	≦0.18	0.10～0.50	≦1.50	≦0.040	≦0.040	–	–	–
			Superseded by European Standard EN 10113-1: 1993, Grade S355Nl, Material No. 1.0546								
		Gr. 50F	–	≦0.16	0.10～0.50	≦1.50	≦0.025	≦0.025	–	–	–
		Gr. 55C	–	≦0.22	≦0.60	≦1.60	≦0.040	≦0.040	–	–	–
		Gr. 55EE	–	≦0.22	0.10～0.50	≦1.60	≦0.040	≦0.030	–	–	–
			Superseded by European Standard EN 10113-1: 1993, Grade S460Nl, Material No. 1.8903								
		Gr. 55F	–	≦0.16	0.10～0.50	≦1.50	≦0.025	≦0.025	–	–	–

1N/mm² or MPa=1.01972×10⁻¹kgf/mm²
1J=1.01972×10⁻¹kg-m

Mo	V	Others	Mechanical Properties 機械的性質							Index No. 索引番号
			Heat Treatment 熱処理	Thickness (t)mm 板厚	Yield Point min. N/mm²or MPa (kgf/mm²) 最小降伏点	Tensile Strength N/mm²or MPa (kgf/mm²) 引張強さ	Notctch Toughness 衝撃値			
							Test Piece 試験片	Test Temp.(℃) 試験温度	Absorbed Energy J (min. kg-m) 吸収エネルギー	
–	0.003~0.10	Nb 0.003~0.10	t≧12.5mm N	t≦16	355	490～640	2V	20	27	A003
				16<t≦40	345					
				40<t≦63	340					
				63<t≦100	325					
				100<t	305					
–	0.003~0.10	Nb 0.003~0.10	t≧12.5mm N	t≦16	355	490～640	2V	0	27	
				16<t≦40	345					
				40<t≦63	340					
				63<t≦100	325					
				100<t	305					
–	0.003~0.10	Nb 0.003~0.10	N	t≦16	355	490～640	2V	-20	27	
				16<t≦40	345					
				40<t≦63	340					
				63<t≦100	325					
				100<t	305					
–	0.003~0.10	Nb 0.003~0.10	N	t≦16	355	490～640	2V	-50	27	
				16<t≦40	345					
				40<t≦63	340					
				63<t≦100	325					
				100<t	305					
–	0.003~0.10	Nb 0.003~0.10	N	t≦16	355	490～640	2V	-50	27	
				16<t≦40	345					
				40<t≦63	340					
				63<t≦100	325					
				100<t	305					
–	0.003~0.10	Nb 0.003~0.08	QT	t≦16	355	490～640	2V	-60	27	
				16<t≦40	345					
				40<t≦63	340					
				63<t≦100	325					
				100<t	305					
–	0.003~0.20	Nb 0.003~0.10	t≧16mm N	t≦16	450	550～700	2V	0	27	
				16<t≦40	430					
				40<t	410					
–	0.003~0.10	Nb 0.003~0.08	N	t≦16	450	550～700	2V	-50	27	
				16<t≦40	430					
				40<t≦63	415					
				63<t	400					
–	0.003~0.10	Nb 0.003~0.08	N	t≦16	450	550～700	2V	-60	27	
				16<t≦40	430					
				40<t	410					

Country 国	Standard No. 規格番号及び年号	Designation 種類記号	Chemical Composition 化 学 成 分 (%)								
			Thickness(t) mm板厚	C	Si	Mn	P	S	Cu	Cr	Ni
GR独	DIN 17100-80	ST52-3	t ≦30	≦0.20	≦0.55	≦1.60	≦0.040	≦0.040	–	–	–
			t <30	≦0.20	≦0.55	≦1.60	≦0.040	≦0.040	–	–	–
		ST50-2	–	about 0.30	–	–	≦0.050	≦0.045	–	–	–
	DIN 17102-83	StE355	–	≦0.20	0.10～0.50	0.90～1.65	≦0.035	≦0.030	≦0.20	≦0.30	≦0.30
		Superseded by European Standard EN 10113-1: 1993, Grade S355N, Material No. 10545									
		STE-380	–	≦0.20	0.10～0.60	1.00～1.70	≦0.035	≦0.030	≦0.20	≦0.30	≦1.00

1N/mm² or MPa=1.01972×10⁻¹kgf/mm²
1J=1.01972×10⁻¹kg-m

Mo	V	Others	Heat Treatment 熱処理	Thickness (t)mm 板厚	Yield Point min. N/mm²or MPa (kgf/mm²) 最小降伏点	Tensile Strength N/mm²or MPa (kgf/mm²) 引張強さ	Notctch Toughness 衝撃値 Test Piece 試験片	Test Temp.(℃) 試験温度	Absorbed Energy J (min.kg-m) 吸収ｴﾈﾙｷﾞｰ	Index No. 索引番号
–	–	–	–	t≦16	335	t≦3	–	–	–	A003
				16<t≦40	345	510～680				
				40<t≦63	335	3<t≦100				
–	–	–	–	63<t≦80	325	490～630				
				80<t≦100	315	100<t				
				100<t	by agreement	by agreement				
–	–	N≦0.009	–	t≦16	295	t<3	–	–	–	
				16<t≦40	285	490～660				
				40<t≦63	275	3<t≦100				
				63<t≦80	265	470～610				
				80<t≦100	255	100<t				
				100<t	by agreement	by agreement				
≦0.08	≦0.10	Nb≦0.05 Nb+Ti+V ≦0.12 Al≦0.30 N≦0.020	–			t≦70	–	–	–	
				t≦16	355	490～630				
				16<t≦35	355	70<t≦85				
				35<t≦50	345	480～620				
				50<t≦60	335	85<t≦100				
				60<t≦70	335	470～610				
				70<t≦85	325	100<t≦125				
				85<t≦100	315	400～600				
				100<t≦125	305	125<t≦150				
				125<t≦150	295	450～590				
≦0.08	≦0.20	Nb≦0.05 Nb+Ti+V ≦0.12 Al≦0.30 N≦0.020	–			t≦70	–	–	–	
				t≦16	380	500～650				
				16<t≦35	375	70<t≦85				
				35<t≦50	365	490～640				
				50<t≦60	355	85<t≦100				
				60<t≦70	345	480～630				
				70<t≦85	335	100<t≦125				
				85<t≦100	325	470～620				
				100<t≦125	315	125<t≦150				
				125<t≦150	305	460～610				

Country 国	Standard No. 規格番号及び年号	Designation 種類記号	Chemical Composition 化学成分 (%)								
			Thickness(t) mm板厚	C	Si	Mn	P	S	Cu	Cr	Ni
GR独	DIN 17102-83	StE420	–	≦0.20	0.10~0.60	1.00~1.70	≦0.035	≦0.030	≦0.20	≦0.30	≦1.00
		Superseded by European Standard EN 10113-1: 1993, Grade S420N, Material No. 1.8902									
		StE460	–	≦0.20	0.10~0.60	1.00~1.70	≦0.035	≦0.030	≦0.20	≦0.30	≦1.00
		Superseded by European Standard EN 10113-1: 1993, Grade S460N, Material No. 1.8901									
FR仏	NF A35-501-87	E36-2NE	–	≦0.24	–	–	≦0.045	≦0.045	–	–	–
		E36-3NE	–	≦0.20	–	–	≦0.040	≦0.040	–	–	–
		E36-4CS	–	≦0.20	–	–	≦0.045	≦0.045	–	–	–
		A50-2NE	–	–	–	–	≦0.045	≦0.045	–	–	–

1N/mm² or MPa=1.01972×10⁻¹kgf/mm²
1J=1.01972×10⁻¹kg-m

Mo	V	Others	Mechanical Properties 機械的性質 Heat Treatment 熱処理	Thickness (t)mm 板厚	Yield Point min. N/mm² or MPa (kgf/mm²) 最小降伏点	Tensile Strength N/mm² or MPa (kgf/mm²) 引張強さ	Notctch Toughness 衝撃値 Test Piece 試験片	Test Temp.(℃) 試験温度	Absorbed Energy J (min. kg-m) 吸収ｴﾈﾙｷﾞｰ	Index No. 索引番号
≦0.10	≦0.20	Nb≦0.05 Nb+Ti+V ≦0.12 Al≦0.30 N≦0.020	–	t≦16 16<t≦35 35<t≦50 50<t≦60 60<t≦70 70<t≦85 85<t≦100 100<t≦125 125<t≦150	420 410 400 390 385 375 365 355 345	t≦70 530～680 70<t≦85 520～670 85<t≦100 510～660 100<t≦125 500～650 125<t≦150 490～640	–	–	–	A003
≦0.10	≦0.20	Nb≦0.05 Nb+Ti+V ≦0.12 Al≦0.30 N≦0.020	–	t≦16 16<t≦35 35<t≦50 50<t≦60 60<t≦70 70<t≦85 85<t≦100 100<t≦125 125<t≦150	460 450 440 430 420 410 400 390 380	t≦70 560～730 70<t≦85 550～720 85<t≦100 540～710 100<t≦125 530～700 125<t≦150 570～740	–	–	–	
–	–	–	–	–	–	–	–	–	–	
–	–	–	–	t<3 3<t≦30 30<t≦50 50<t≦80 80<t≦110 110<t≦150	325 355 335 335 315 305	t<3 570～650 3<t≦150 490～630	2V	0	35	
–	–	Al ≧0.02	–	t<3 3<t≦30 30<t≦50 50<t≦80 80<t≦110 110<t≦150	325 355 335 335 315 305	t<3 510～650 3<t≦150 510～650	2V	0	50	
–	–	–	–	t<3 3<t≦30 30<t≦50 50<t≦80 80<t≦110 110<t≦150	275 295 275 275 265 255	t<3 490～630 3<t≦150 490～610	–	–	–	

Country 国	Standard No. 規格番号及び年号	Desig-nation 種類記号	Chemical Composition 化学成分 (%)								
			Thick-ness(t) mm板厚	C	Si	Mn	P	S	Cu	Cr	Ni
FR仏	NF A35-502-84	A 2	–	≦0.12	0.20～0.75	≦1.00	0.07~0.15	≦0.04	0.25～0.55	0.30～1.25	≦0.65
		A 3	–	≦0.12	0.20～0.75	≦1.00	0.07~0.15	≦0.040	0.25～0.55	0.30～1.25	≦0.65
		A 4	–	≦0.12	0.20～0.75	≦1.00	0.07~0.15	≦0.040	0.25～0.55	0.30～1.25	≦0.65
		B 3	–	≦0.19	≦0.50	≦1.50	≦0.040	≦0.040	0.20～0.55	0.40～0.80	≦0.65
		B 4	–	≦0.19	≦0.50	≦1.50	≦0.040	≦0.040	0.20～0.55	0.40～0.80	≦0.65
CA加	CAN/CSA G40.21~92	50G	–	≦0.28	≦0.04	≦1.65	≦0.04	≦0.05	–	–	–
		60G	–	≦0.28	≦0.04	≦1.65	≦0.04	≦0.05	–	–	–
		60W	–	≦0.23	≦0.04	0.50～1.50	≦0.04	≦0.05	–	–	–
		50WT	–	≦0.22	0.15～0.40	0.80～1.50	≦0.03	≦0.04	–	–	–
		60WT	–	≦0.22	0.15～0.40	0.80～1.50	≦0.03	≦0.04	–	–	–
		50R	–	≦0.16	≦0.75	≦0.75	0.05~0.15	≦0.04	0.20～0.60	0.30～1.25	≦0.90
		50A	–	≦0.20	0.15～0.40	0.75～1.35	≦0.03	≦0.04	0.20～0.60	≦0.70	≦0.90
		60A	–	≦0.20	0.15～0.40	0.75～1.35	≦0.03	≦0.03	0.20～0.60	≦0.70	≦0.90
		50AT	–	≦0.20	0.15～0.40	0.75～1.35	≦0.03	≦0.04	0.20～0.60	≦0.70	≦0.90
		60AT	–	≦0.20	0.15～0.40	0.75～1.35	≦0.03	≦0.04	0.20～0.60	≦0.70	≦0.90

1N/mm² or MPa$=1.01972\times10^{-1}$kgf/mm²
1J$=1.01972\times10^{-1}$kg-m

Mo	V	Others	Heat Treatment 熱処理	Thickness (t)mm 板 厚	Yield Point min. N/mm² or MPa (kgf/mm²) 最小降伏点	Tensile Strength N/mm² or MPa (kgf/mm²) 引張強さ	Notch Toughness 衝撃値 Test Piece 試験片	Test Temp.(℃) 試験温度	Absorbed Energy J (min. kg-m) 吸収ｴﾈﾙｷﾞｰ	Index No. 索引番号
–	–	–	–	t<3 3<t≦30	315 355	t≦50 480～580 t>50 ≧480	–	–	–	A003
–	–	–	–	t<3 3<t≦30	315 355	t≦50 480～580 t>50 ≧480	2 V	0	35	
–	–	–	–	t<3 3<t≦30	315 355	t≦50 480～580 t>50 ≧480	2 V	-20	35	
≦0.30	–	Zr≦0.15	–	3<t≦30 30<t≦50 50<t≦80 80<t≦110	355 335 335 315	t≦50 480～580 t>50 ≧480	2 V	0	35	
≦0.30	–	Zr≦0.15	–	3<t≦30 30<t≦50 50<t≦80 80<t≦110	355 335 335 315	t≦50 480～580 t>50 ≧480	2 V	-20	50	
–	–	Al or Cb or V≦0.10	–	≦64	345	485～690	–	–	–	
–	–	Al or Cb or V≦0.10	–	≦64	415	550～725	–	–	–	
–	–	Al or Cb or V≦0.10	–	≦64	415	515～690	–	–	–	
–	–	Al or Cb or V≦0.10	–	≦64 64～150	345 330	485～655	–	–	–	
–	–	Al or Cb or V≦0.10	–	≦64	415	515～690	–	–	–	
–	–	Al or Cb or V≦0.10	–	≦64	345	485～655	–	–	–	
–	–	Al or Cb or V≦0.10	–	≦100	345	485～655	–	–	–	
–	–	Al or Cb or V≦0.10	–	≦64	415	515～690	–	–	–	
–	–	Al or Cb or V≦0.10	–	≦100	345	485～655	–	–	–	
–	–	Al or Cb or V≦0.10	–	≦64	415	515～690	–	–	–	

Country 国	Standard No. 規格番号及び年号	Designation 種類記号	Chemical Composition 化学成分 (%)								
			Thickness(t) mm板厚	C	Si	Mn	P	S	Cu	Cr	Ni
AU豪	AS 1204-80	350	–	≦0.22	≦0.50	≦1.50 (C+Mn/6 ≦0.45)	≦0.040	≦0.040	–	–	–
		350L0 350L15	–	≦0.22	≦0.50	≦1.50 (C+Mn/6 ≦0.45)	≦0.040	≦0.040	–	–	–
	AS 1205-80	W350/1 W350/1L0	≦20	≦0.12	0.20～0.70	≦1.00	0.06～0.15	≦0.040	0.15～0.45	0.40～1.00	≦0.50
		W350/2 W350/2L0 W350/2L15	≦20	≦0.19	0.20～0.60	≦1.00	≦1.35	≦0.040	0.20～0.45	0.25～0.70	≦0.50

1N/mm² or MPa = 1.01972 × 10^{-1} kgf/mm²
1J = 1.01972 × 10^{-1} kg-m

Mo	V	Others	Mechanical Properties 機械的性質							Index No. 索引番号
			Heat Treatment 熱処理	Thickness (t)mm 板厚	Yield Point min. N/mm² or MPa (kgf/mm²) 最小降伏点	Tensile Strength N/mm² or MPa (kgf/mm²) 引張強さ	Notctch Toughness 衝撃値			
							Test Piece 試験片	Test Temp. (℃) 試験温度	Absorbed Energy J (min. kg-m) 吸収ｴﾈﾙｷﾞｰ	
–	–	–	–	t ≦ 50 50 < t ≦ 100 100 < t	360 340 330	≧480	2 V	0 -15	27 27	A003
–	–	–	–	t ≦ 50 50 < t ≦ 100 100 < t	360 340 330	≧480	2 V	0 -15	27 27	
–	–	–	–	–	340	≧480	2 V	0 -15	27 27	
–	–	–	–	–	340	≧480	2 V	0 -15	27 27	

2-4 Tensile Strength 600 N／mm² or MPa (60 kgf／mm²) Class

Country 国	Standard No. 規格番号及び年号	Designation 種類記号		Chemical Composition 化学成分（%）								
			Thickness(t) mm板厚	C	Si	Mn	P	S	Cu	Cr	Ni
JP日	JIS G3106-92	SM570 (SM58)		–	≦0.18	≦0.55	≦1.60	≦0.035	≦0.035	–	–	–
	JIS G3114-88	SM570 (SM58)	W	–	≦0.18	0.15～0.65	≦1.40	≦0.035	≦0.035	0.30～0.50	0.45～0.75	0.05～0.30
			P			≦0.55				0.20～0.35	0.30～0.55	–
US米	ASTM A656M-93a	Type 3 Gr. 80		–	≦0.18	≦0.60	≦1.65	≦0.025	≦0.035	–	–	–
		Type 7 Gr. 80		–	≦0.18	≦0.60	≦1.65	≦0.025	≦0.035	–	–	–
	ASTM A678M-94a	Gr. C		–	≦0.22	0.20～0.50	1.00～1.60	≦0.035	≦0.04	When specified ≧0.20	–	–
	ASTM A710M-94	Gr. A	Cl. 1	–	≦0.07	≦0.40	0.40～0.70	≦0.025	≦0.025	1.00～1.30	0.60～0.90	0.70～1.00
		Gr. A	Cl. 3	–	≦0.07	≦0.40	0.40～0.70	≦0.025	≦0.025	1.00～1.30	0.60～0.90	0.70～1.00
		Gr. B		–	≦0.06	0.15～0.40	0.40～0.65	≦0.025	≦0.025	1.00～1.30	–	1.20～1.50
UK英	–	–		–	–	–	–	–	–	–	–	–
GR独	DIN 17100-80	ST60-2		–	about 0.4	–	–	≦0.050	≦0.050	–	–	–

引張り強さ 600 N／mm² または MPa（60キロ）級

Mo	V	Others	Mechanical Properties 機械的性質 Heat Treatment 熱処理	Thickness (t)mm 板厚	Yield Point min. N/mm²or MPa (kgf/mm²) 最小降伏点	Tensile Strength N/mm²or MPa (kgf/mm²) 引張強さ	Notch Toughness 衝撃値 Test Piece 試験片	Test Temp.(℃) 試験温度	Absorbed Energy J (min. kg-m) 吸収エネルギー	Index No. 索引番号
–	–	* t≦50 Ceq≦0.44 Pcm≦0.28 50<t≦100 Ceq≦0.47 Pcm≦0.30	N QT	≦16 16<t≦40 40<t≦75 75<t≦100	460(47) 450(46) 430(44) 420(43)	570～720 (58～73)	2V	-5	47 (4.8)	A004
One or more of Mo, Ni, Nb, Ti, V and/ or Zr may be added. Mo, Ni, Nb, Ti V, Zr のいずれか1種以上添加可。			–	≦16 16<t≦40 >40	460(47) 450(46) 430(44)	570～720 (58～73)	2V	-5	47 (4.8)	
–	≦0.08	N≦0.020 Nb0.005~0.15	–	–	550	≧620	–	–	–	
–	0.005~0.15	N≦0.020 Cb0.005~0.15	–	–	550	≧620	–	–	–	
–	–	Cb(0.01～0.09) B may be added by agreemant 協議	QT	t≦20 20<t≦40 40<t≦50	515 485 450	655～790 620～760 585～720	–	–	–	
0.15～0.25	–	Nb≧0.02	P	t≦8 8<t≦20	585 550	≧620	2V	-45	27J	
0.15～0.25	–	–	Q and P	t≦50 50<t≦100 100<t	515 450 415	≧585 ≧515 ≧485	2V	-45	27J	
–	–	Nb≧0.02	P	t≦6.5 6.5<t≦12.5 12.5<t≦20 20<t≦25	585 565 550 515	≧620 ≧620 ≧615 –	2V	-45	27J	
–	–	–	–	–	–	–	–	–	–	
–	–	N≧0.009	–	t≦16 16<t≦40 40<t≦63 63<t≦80 80<t≦100 100<t	335 325 315 305 295 by agreement	t<3 590～770 3≦t≦100 570～710 100<t by agreement	–	–	–	

$$*\ Ceq = C + \frac{Mn}{6} + \frac{Si}{24} + \frac{Ni}{40} + \frac{Cr}{5} + \frac{Mo}{4} + \frac{V}{14},\quad Pcm = C + \frac{Si}{30} + \frac{Mn}{20} + \frac{Cu}{20} + \frac{Cr}{20} + \frac{Mo}{15} + \frac{10}{V} + 5B$$

Country 国	Standard No. 規格番号及び年号	Designation 種類記号	Chemical Composition 化学成分 (%)								
			Thickness(t) mm板厚	C	Si	Mn	P	S	Cu	Cr	Ni
GR独	DIN 17102-83	STE-500	–	≦0.20	0.10〜0.60	1.00〜1.70	≦0.035	≦0.030	≦0.20	≦0.30	≦1.00
FR仏	NF A35-501-87	A60-2NE	Superseded by European Standard EN 10025: 1993, Grade E 335, Material No. 1.0060 –	–	–	–	≦0.045	≦0.045	–	–	–
CA加	CAN/CSA G40.21-92	70W	–	≦0.26	0.15〜0.40	0.50〜1.50	≦0.04	≦0.05	–	–	–
		70WT	–	≦0.26	0.15〜0.40	0.80〜1.50	≦0.03	≦0.04	–	–	–
		70A	–	≦0.20	0.15〜0.40	1.00〜1.60	≦0.025	≦0.035	0.20〜0.60	≦0.70	0.25〜0.50
		70AT	–	≦0.20	0.15〜0.40	1.00〜1.60	≦0.025	≦0.035	0.20〜0.60	≦0.70	0.25〜0.50
AU豪	–	–	–	–	–	–	–	–	–	–	–

$1N/mm^2$ or $MPa = 1.01972 \times 10^{-1} kgf/mm^2$
$1J = 1.01972 \times 10^{-1} kg\text{-}m$

Mo	V	Others	Mechanical Properties 機械的性質							Index No. 索引番号
			Heat Treatment 熱処理	Thickness (t)mm 板厚	Yield Point min. N/mm²or MPa (kgf/mm²) 最小降伏点	Tensile Strength N/mm²or MPa (kgf/mm²) 引張強さ	Notch Toughness 衝撃値			
							Test Piece 試験片	Test Temp.(℃) 試験温度	Absorbed Energy J (min.kg-m) 吸収エネルギー	
≦0.10	≦0.22	Nb≦0.05 Nb+Ti+V ≦0.22 N≦0.020 Al≦0.030	–	t ≦16 16<t≦35 35<t≦50 50<t≦60 60<t≦70 70<t≦85 85<t≦100 100<t≦125 125<t≦150	500 480 470 460 450 440 430 420 410	t≦70 610～780 70<t≦85 600～700 85≦t<100 590～860 100<t≦125 580～750 125<t≦150 570～740	–	–	–	A004
–	–	–	–	3≦ t ≦30 30<t≦50 50<t≦80 80<t≦110 110<t≦150	335 315 315 305 295	3≦t≦150 590～710	–	–	–	
–	–	Al or Cb or V≦0.10	–	≦64	485	585～795	–	–	–	
–	–	Al or Cb or V≦0.10	–	≦64	485	585～795	–	–	–	
–	–	Al or Cb or V≦0.10	–	≦64	485	585～795	–	–	–	
–	–	Al or Cb or V≦0.10	–	≦64	485	585～795	–	–	–	
–	–	–	–	–	–	–	–	–	–	

2-5 Tensile Strength≧700 N／mm² or MPa (70 蓑kgf／mm²) Class

Country 国	Standard No. 規格番号 及び年号	Designation 種類記号	Chemical Composition 化学成分 (%) Thickness(t) mm板厚	C	Si	Mn	P	S	Cu	Cr	Ni
JP日	JIS G3128-87	SHY685 (SHY70)	–	≦0.18	≦0.55	≦1.50	≦0.030	≦0.025	≦0.50	≦1.20	–
		SHY685N (SHY70N)	–	≦0.18	≦0.55	≦1.50	≦0.030	≦0.025	≦0.50	≦0.80	0.30～1.50
		SHY685NS (SHY70NS)	–	≦0.14	≦0.55	≦1.50	≦0.015	≦0.015	≦0.50	≦0.80	0.30～1.50
US米	ASTM A514M-94a	Gr. A	t ≦32	0.15～0.21	0.40～0.80	0.80～1.10	≦0.035	≦0.035	–	0.50～0.80	–
		Gr. B	t ≦32	0.12～0.21	0.20～0.35	0.70～1.00	≦0.035	≦0.035	–	0.40～0.65	–
		Gr. C	t ≦32	0.10～0.20	0.15～0.30	0.10～1.50	≦0.035	≦0.035	–	–	–
		Gr. E	t ≦150	0.12～0.20	0.20～0.40	0.40～0.70	≦0.035	≦0.035	–	1.42～2.10	–
		Gr. F	t ≦65	0.10～0.20	0.15～0.35	0.60～1.00	≦0.035	≦0.035	0.15～0.50	0.40～0.65	0.75～1.00
		Gr. H	t ≦50	0.12～0.21	0.20～0.35	0.95～1.30	≦0.035	≦0.035	–	0.40～0.65	0.30～0.70
		Gr. J	t ≦32	0.12～0.21	0.20～0.35	0.45～0.70	≦0.035	≦0.035	–	–	–
		Gr. K	t ≦50	0.10～0.20	0.15～0.30	1.10～1.50	≦0.035	≦0.035	–	–	–
		Gr. M	t ≦50	0.12～0.21	0.20～0.35	0.45～0.70	≦0.035	≦0.035	–	–	1.20～1.50
		Gr. P	t ≦150	0.12～0.21	0.20～0.35	0.45～0.70	≦0.035	≦0.035	–	0.85～1.20	1.20～1.50
		Gr. Q	t ≦150	0.14～0.21	0.15～0.35	0.95～1.30	≦0.035	≦0.035	–	1.00～1.50	1.20～1.50
		Gr. R	t ≦50	0.15～0.20	0.20～0.35	1.10～1.50	≦0.035	≦0.035	–	0.35～0.65	0.90～1.10
		Gr. S	t ≦50	0.11～0.21	0.15～0.45	1.10～1.50	≦0.035	≦0.020	–	–	–
		Gr. T	t ≦50	0.08～0.14	0.40～0.60	1.20～1.50	≦0.035	≦0.035	–	–	–
UK英	–	–	–	–	–	–	–	–	–	–	–
GR独	DIN 17100-80	ST70-2	about 0.50	–	–	≦0.050	≦0.050	–	–	–	–

引張り強さ 700 N／mm² または MPa（70キロ）級

Mo	V	Others	Mechanical Properties 機械的性質 Heat Treatment 熱処理	Thickness (t)mm 板厚	Yield Point min. N/mm² or MPa (kgf/mm²) 最小降伏点	Tensile Strength N/mm² or MPa (kgf/mm²) 引張強さ	Notch Toughness 衝撃値 Test Piece 試験片	Test Temp.(℃) 試験温度	Absorbed Energy J (min. kg-m) 吸収エネルギー	Index No. 索引番号
≦0.60	≦0.10	B≦0.005	QT	t≦50	685	780~930	2V	-20	47	
				50<t≦100	665	760~910				
≦0.60	≦0.10	B≦0.005	QT	t≦50	685	780~930	2V	-20	47	A005
				50<t≦100	665	760~910				
≦0.60	≦0.05	B≦0.005	QT	t≦50	685	780~930	–	–	–	
				50<t≦100	665	760~910				
0.18~0.28	–	Zr0.05~0.15 B≦0.0025								
0.15~0.25	0.03~0.08	Ti0.01~0.03 B0.0005~0.005								
0.20~0.30	–	B0.001~0.005								
0.40~0.60	–	Ti0.04~0.10 B0.001~0.005								
0.40~0.60	0.03~0.08	B0.0005~0.006								
0.20~0.30	0.03~0.08	B0.0005~0.005	QT	t≦20 20<t≦65 65<t≦150	690 690 620	760~895 760~895 690~895	–	–	–	
0.50~0.65	–	B0.001~0.005								
0.45~0.55		B0.001~0.005								
0.45~0.60	–	B0.001~0.005								
0.45~0.60	–	B0.001~0.005								
0.40~0.60	0.03~0.08	–								
0.15~0.25	0.03~0.08	–								
0.10~0.60	≦0.06	B0.001~0.005 Cb≦0.06								
0.10~0.35	0.03~0.08	B0.001~0.005								
–	–	–	–	–	–	–	–	–	–	
–	–	N≦0.009	–	t≦16 16<t≦40 40<t≦63 63<t≦80 80<t≦100 100<t	365 355 345 335 325 by agreement	t<3 690~895 3≦t≦100 670~825 100<t by agreement	–	–	–	

Country 国	Standard No. 規格番号及び年号	Desig-nation 種類記号	Chemical Composition 化学成分 (%)								
			Thick-ness(t) mm板厚	C	Si	Mn	P	S	Cu	Cr	Ni
NF仏	NF A35-501-87	A70-2NE	Superseded by European Standard EN 10025: 1993, Grade E 360, Material No. 1.0070 –	–	–	–	≦0.045	≦0.45	–	–	–
CA加	CAN/CSA G40.21-92	100Q	–	≦0.20	0.15～0.40	≦1.50	≦0.03	≦0.04	–	–	–
		100QT	–	≦0.20	0.15～0.40	≦1.50	≦0.03	≦0.04	–	–	–
AU豪	–	–	–	–	–	–	–	–	–	–	–

1N/mm² or MPa＝1.01972×10^{-1}kgf/mm²
1J＝1.01972×10^{-1}kg-m

Mo	V	Others	Heat Treatment 熱処理	Thickness (t)mm 板 厚	Yield Point min. N/mm²or MPa (kgf/mm²) 最小降伏点	Tensile Strength N/mm²or MPa (kgf/mm²) 引張強さ	Notctch Toughness 衝撃値 Test Piece 試験片	Test Temp.(℃) 試験温度	Absorbed Energy J (min.kg-m) 吸収ｴﾈﾙｷﾞｰ	Index No. 索 引 番 号
—	—	—	—	3≦ t ≦30 30<t≦50 50<t≦80 80<t≦110 110<t≦150	365 345 345 335 325	3≦t≦150 690～830	—	—	—	A005
—	—	B 0.0005 ～0.005	—	≦64	690	795～930	—	—	—	
—	—	B 0.0005 ～0.005	—	≦64	690	795～930	—	—	—	
—	—	—	—	—	—	—	—	—	—	

INDEX OF CHAPTER I

STEEL PLATES FOR GENERAL STRUCTURE

第1部　一般構造用鋼板

索　　引

"Index No."stands for the "Index No." in the last column of each page.

「索引番号」は本文各頁の右端の「索引番号」の欄の数字をあらわす。

Standard No. 規格番号	Grade グレード	Index No. 索引番号
〔JIS〕		
JIS-G3101	SS330(SS34)	A001
	SS400(SS41)	A002
	SS490(SS50)	A003
	SS540(SS55)	A003
JIS-G3106	SM490(SM50)	A003
	SM490Y(SM50Y)	A003
	SM520(SM53)	A003
	SM570(SM58)	A004
JIS-G3114	SM400(SM41)	A002
	SMA490(SMA50)	A003
	SMA570(SMA58)	A004
JIS-G3125	SPA-H	A003
JIS-G3128	SHY685(SHY70)	A005
	SHYN685(SHYN70)	A005
	SHYNS685(SHYNS70)	A005
〔ASTM〕		
ASTM-A36M	—	A002
ASTM-A242M	Type1	A003
ASTM-A283M	Gr.A	A001
	Gr.B	A001
	Gr.C	A001
	Gr.D	A002
ASTM-A284M	Gr.C	A002
	Gr.D	A002
ASTM-A514M	Gr.A	A005
	Gr.B	A005
	Gr.C	A005
	Gr.E	A005
	Gr.F	A005
	Gr.H	A005
	Gr.J	A005
	Gr.M	A005
	Gr.P	A005
	Gr.Q	A005
	Gr.R	A005
	Gr.S	A005
	Gr.T	A005
ASTM-A529M	—	A002
ASTM-A572M	Gr.42(290)	A002
	Gr.50(345)	A002
ASTM-A573M	Gr.58(400)	A002
	65(450)	A002
ASTM-A588M	Gr.A	A003
	Gr.B	A003
	Gr.C	A003
	Gr.D	A003
	Gr.K	A003
ASTM-A633M	Gr.A	A002
ASTM-A656M	Gr.50(345)	A002
	Gr.60(415)	A003
	Gr.70(485)	A003
	Gr.80(550)	A004
ASTM-A678M	Gr.A	A003
	Gr.B	A003
ASTM-A710M	Gr.A,C1,1	A004
	Gr.A,C1,2	A003
	Gr.A,C1,3	A004
	Gr.B	A004
〔BS〕		
BS-4360	Gr.40A	A002
	Gr.40B	A002
	Gr.40C	A002
	Gr.40D	A002
	Gr.40EE	A002
	Gr.43A	A002
	Gr.43B	A002
	Gr.43C	A002
	Gr.43D	A002
	Gr.43EE	A002
	Gr.50A	A003
	Gr.50B	A003
	Gr.50C	A003
	Gr.50D	A003
	Gr.50EE	A003
	Gr.50F	A003
	Gr.55C	A003
	Gr.55EE	A003
	Gr.55F	A003
〔DIN〕		
DIN-17100	St44-2	A002
	St44-3	A002
	St52-3	A003
	St50-2	A003

Standard No. 規 格 番 号	Grade グレード	Index No. 索引番号
	St60-2	A004
	St70-2	A005
DIN-17102	StE255	A001
	StE285	A001
	StE315	A002
	StE355	A003
	StE380	A003
	StE420	A003
	StE460	A003
	StE500	A004
〔N F〕		
NFA35-501	A33	A001
	A50-2NE	A003
	A60-2NE	A004
	A70-2NE	A005
	E24-2	A001
	E24-2E	A001
	E24-2NE	A001
	E24-3NE	A001
	E24-4CS	A001
	E28-2NE	A002
	E28-3NE	A002
	E28-4CS	A002
	E36-2NE	A003
	E36-3NE	A003
	E36-4CS	A003
NFA35-502	E24W2	A001
	E24W3	A001
	E24W4	A001
	E36W-A2	A003
	E36W-A3	A003
	E36W-A4	A003
	E36W-B3	A003
	E36W-B4	A003
〔C S A〕		
CAN/CSA G.40.21	33G	A002
	38W	A002
	38WT	A002
	44W	A002
	44WT	A002
	50A	A003
	50AT	A003
	50R	A003
	50W	A002
	50WT	A003
	60A	A003
	60AT	A003
	60G	A003
	60W	A003
	60WT	A003
	70A	A004
	70W	A004
	70WT	A004
	100Q	A005
	100QT	A005
〔A S〕		
AS-1204	200	A002
	250	A002
	250Lo	A002
	250L15	A002
	350	A003
	350L15	A003
AS-1205	W350/1	A005
	W350/1Lo	A005
	W350/2	A005
	W350/2Lo	A005
	W350/2L15	A005

CHAPTER II

STEEL PLATES
FOR
BOILER AND PRESSURE VESSELS

第 2 部

ボイラおよび圧力容器用鋼板

Abbreviations

略 語 表

(1) Heat Treatment 熱処理

A: Anneal 焼なまし

N: Normalize 焼ならし

Q: Quench 焼入れ

S: Solution Heat Treatment 固溶化処理

T: Temper 焼もどし

(2) Impact Test Piece 衝撃試験片

2V: 2mm V Notch 2mmVノッチ

2U: 2mm U Notch 2mmUノッチ

5U: 5mm U Notch 5mmUノッチ

DVM: DIN, DVM

(3) Designation 種類番号

Gr.: Grade グレード

Cl.: Class クラス

Typ.: Type タイプ

(4) Country 国名記号

JP: Japan 日 本

US: The United States of America(U.S.A.) アメリカ

UK: The United Kingdom イギリス

GR: West Germany 西ドイツ

FR: France フランス

UR: U.S.S.R. ソ 連

CHAPTER II CONTENTS　第2部　目次

1. STANDARDS LIST 収録各国規格リスト

1-1 JIS (Japan 日 本)

G3103-87 Carbon Steel and Molybdenum Alloy Steel Plates for Boilers and Other Pressure Vessels
ボイラ及び圧力容器用炭素鋼及びモリブデン鋼鋼材

G3115-90 Steel Plates for Pressure Vessels for Intermediate Temperature Service
圧力容器用鋼板

G3116-90 Steel Sheets, Plates and Strip for Gas Cylinders
圧力ガス容器用鋼板及び鋼帯

G3118-87 Carbon Steel Plates for Pressure Vessels for Intermediate and Moderate Temprature Service
中・常温圧力容器用炭素鋼鋼材

G3119-87 Manganese-Molybdenum and Manganese-Molybdenum-Nickel Alloy Steel Plates for Boilers and Other Pressure Vessels
ボイラ及び圧力容器用マンガンモリブデン鋼及びマンガンモリブデンニッケル鋼鋼板

G3120-87 Manganese-Molybdenum and Manganese-Molybdenum-Nickel Alloy Steel Plates Quenched and Tempered for Pressure Vessels
圧力容器用調質型マンガンモリブデン鋼及びマンガンモリブデンニッケル鋼鋼板

G3124-87 High Strength Steel Plates for Pressure Vessel for Intermediate and Moderate Temperature Service
中・常温圧力容器用高強度鋼鋼板

G3126-90 Carbon Steel Plates for Pressure Vessels for Low Temperature Service
低温圧力容器用炭素鋼鋼板

G3127-90 Nickel Steel Plates for Pressure Vessels for Low Temperature Service
低温圧力容器用ニッケル鋼鋼板

G4109-87 Chromium-Molybdenum Alloy Steel Plates for Pressure Vessels
ボイラ及び圧力容器用クロムモリブデン鋼鋼板

G4304-91 Hot Rolled Stainless Steel Sheets and Plates
熱間圧延ステンレス鋼板

G4312-91 Heat-Resisting Steel Sheets and Plates
耐熱鋼板

1-2 ASTM (U.S.A. アメリカ)

A167-94a Stainless and Heat-Resisting Chromium-Nickel Steel Plate, Sheet and Strip
ステンレス及び耐熱Cr-Ni 鋼板, 鋼帯

A176-94 Stainless and Heat-Resisting Chromium Steel Plates, Sheet and Strip
ステンレス及び耐熱Cr鋼板, 鋼帯

A202M-93　Pressure Vessel Plates, Alloy Steel, Chromium-Manganese-Silicon
圧力容器用Cr-Mn-Si合金鋼鋼板

A203M-93　Pressure Vessel Plates, Alloy Steel, Nickel
圧力容器用Ni合金鋼鋼板

A204M-93　Pressure Vessel Plates, Alloy Steel, Molybdenum
圧力容器用Mo合金鋼鋼板

A225M-93　Pressure Vessel Plates, Alloy Steel, Manganese-Vanadium-Nickel
圧力容器用Mn-V-Ni 合金鋼鋼板

A240-95　Heat-Resisting Chromium and Chromium-Nickel Stainless Steel Plate, Sheet and Strip for Fusion-Welded Unfired Pressure Vessels
溶融溶接火なし圧力容器用耐熱Cr及びCr-Ni ステンレス鋼板, 鋼帯

A285M-90　Pressure Vessel Plates, Carbon Steel, Low and Intermediate Tensile Strength
圧力容器用低・中抗張力炭素鋼鋼板

A299M-90　Pressure Vessel Plates, Carbon Steel Manganese-Silicon
圧力容器用C-Mn-Si 鋼鋼板

A302M-93　Pressure Vessel Plates, Alloy Steel, Manganese-Molybdenum, and Manganese-Molybdenum-Nickel
圧力容器用Mn-Mo 及びMn-Mo-Ni鋼鋼板

A353M-93　Pressure Vessel Plates, Alloy Steel, 9 Percent Nickel Double-Normalized and Tempered
圧力容器用 9 %Ni 2 回焼ならし焼もどし合金鋼鋼板

A387-92　Pressure Vessel Plates, Alloy Steel, Chromium-Molybdenum
圧力容器Cr-Mo 合金鋼鋼板

A442M-90　Pressure Vessel Plates, Carbon Steel, Improved Transition Properties
圧力容器用遷移特性改良型炭素鋼鋼板

A455M-90　Pressure Vessel Plates, Carbon Steel, High Strength Manganese
圧力容器用高強度炭素Mn鋼鋼板

A515M-92　Pressure Vessel Plates, Carbon Steel, for Intermediate and Higher-Temperature Service
圧力容器用中高温用炭素鋼鋼板

A516M-90　Pressure Vessel Plates, Carbon Steel, for Moderate and Lower-Temperature Service
圧力容器用中低温用炭素鋼鋼板

A517M-93　Pressure Vessel Plates, Alloy Steel, High-Strength, Quenched and Tempered
圧力容器用高強度焼入れ焼もどし合金鋼鋼板

A533M-93　Pressure Vessel Plates, Alloy Steel, Quenched and Tempered, Manganese-Molybdenum and Manganese-Molybdenum-Nickel
圧力容器用焼入れ焼もどしMn-Mo 及びMn-Mo-Ni合金鋼鋼板

A537M-91　Pressure Vessel Plates, Heat-Treated, Carbon-Manganese-Silicon Steel
圧力容器用熱処理C-Mn-Si 鋼鋼板

A543-93　Pressure Vessel Plates, Alloy Steel, Quenched and Tempered Nickel-Chromium-Molybdenum
圧力容器用焼入れ焼もどしNi-Cr-Mo合金鋼鋼板

A562M-90 Pressure Vessel Plates, Carbon Steel, Manganese-Titanium for Glass or Diffused Metallic Coatings
圧力容器用ガラスまたは金属拡散コーティング用Mn-Ti 入り炭素鋼鋼板

A612M-90 Pressure Vessel Plates, Carbon Steel, High Strength for Moderate and Lower Temperature Service
圧力容器用常・低温用炭素鋼高張力鋼板

A645M-87 Pressure Vessel Plates, 5 Percent Nickel Alloy Steel, Specially Heat Treated
圧力容器用特殊熱処理５%Ni合金鋼鋼板

A662M-93 Pressure Vessel Plates, Carbon-Manganese, for Moderate and Lower Temperature Service
圧力容器用常・低温用C-Mn鋼鋼板

A724M-90 Pressure Vessel Plates, Carbon Steel, Quenched and Tempered, for Welded Layered Pressure Vessels
溶接積層圧力容器用焼入れ焼もどし鋼鋼板

A734M-87a Pressure Vessel Plates, Alloy Steel and High-Strength Low-Alloy Steel, Quenched Tempered
圧力容器用焼入れ焼もどし合金鋼及び高張力低合金鋼鋼板

A735M-90 Pressure Vessel Plates, Low-Carbon Manganese-Molybdenum-Columbium Alloy Steel, for Moderate and Lower Temperature Service
常・低温圧力容器用C-Mn-Mo-Nb鋼鋼板

A736M-88 Pressure Vessel Plates, Low-Carbon Age-Hardening Nickel-Chromium-Molybdenum-Clumbium-Alloy Steel
圧力容器用時効硬化処理低C-Ni-Cu-Cr-Mo-Nb鋼鋼板

A737M-87 Pressure Vessel Plates, High-Strength Low Alloy Steel
圧力容器用高張力低合金鋼鋼板

A738M-90 Pressure Vessel Plates, Heat-Treated, Carbon-Manganese-Silicon Steel, for Moderate and Lower Temperature Service
常・低温圧力容器用熱処理C-Mn-si 鋼鋼板

A841M-94 Steel Plates for Pressure Vessels, produced by the Thermo-Mechanical control Process (TMCP)
圧力容器用熱機構制御法製鋼（TMCP鋼）鋼板

A844M-93 Steel Plates, 9% Nickel Alloy, for Pressure Vessels, produced by Direct-Quenching Process
圧力容器用直接急冷法製９%Ni合金鋼鋼板

1-3 AISI（アメリカ鉄鋼協会）

MN06/222-74 Stainless and Heat Resisting Steels
ステンレス及び耐熱鋼

1-4 BS (U.K.　イギリス)

1501　Steel for Use in The Chemical, Petroleum and Allied Industries
圧力容器用化学・石油工業用鋼鋼材

1501 Part 1-80　Steels for Fired and Unfired Pressure Vessels, Plates: Carbon and Carbon Manganese Steels (Superseded by European Standard EN 10028 -1 + -3)
ボイラ及び圧力容器用C, C-Mn 鋼鋼板

1501 Part 2-70　Steels for Fired and Unfired Pressure Vessels, Plates: Alloy Steels
ボイラ及び圧力容器用合金鋼鋼板 (Superseded by European Standard EN 10028-3, in part.)

1501 Part 3-90　Steels for Pressure Purposes: Plates; Sheet and Strip Corrosion and Heat-Resisting Steels
ボイラ及び圧力容器用耐蝕耐熱鋼鋼板

1-5 DIN (West Germany 西ドイツ)

17102-83　Weldable Fine Grain Steels (Superseded by European Standard EN 10113, Parts 1&2 and European Standard EN 10028, Parts 1&3)
溶接細粒鋼鋼材

17155-83　Plate and Strip for Boilers (Superseded by European Standard EN 10028, Parts 1&2)
ボイラー用鋼板

17280-85　Steels with Low Temperature Toughness
低温用鋼材

17440-85　Stainless Steels
ステンレス鋼鋼材

17441-85　Stainless Steels
ステンレス鋼鋼材

1-6 NF (France　フランス)

A35-573-81　Stainless Steel for General Purposes
一般ステンレス鋼鋼材

A36-205-82　Steel Plates for Boilers and Pressure Vessels, Carbon and Carbon Manganese Steels
ボイラ及び圧力容器用非合金鋼鋼板

A36-206-83　Mo, Mn-Mo and Cr-Mo Alloy Steel Plates for Boilers and Pressure Vessels
ボイラ及び圧力容器用Mo, Mn-Mo 及びCr-Mo 鋼鋼板

A36-207-82　High Yield Strength Steel Plates for Pressure Vessels
圧力容器用高張力鋼鋼板

A36-208-82　Nickel Alloy Steel Plates for Low Temprature Boilers and Pressure Vessels
低温ボイラ・圧力容器用Ni合金鋼鋼板

A36-209-82　Austenitic Stainless Steels for Boilers and Pressure Vessels
ボイラ・圧力容器用オーステナイト系ステンレス鋼鋼材

1-7 ГОСТ (U.S.S.R.: ソ　連)

5520-79　Carbon and Alloy Steel Plates for Boilers and Pressure Vessels
ボイラ・圧力容器用炭素鋼及び合金鋼鋼板

5532-72　Highly Alloyed Steels and Corrosion-Proof, Heat-Resisting and Heat Treated Alloy
耐蝕耐熱鋼及び高合金鋼鋼材

2. Carbon Steel Plates 炭素鋼板

Country 国	Standard No. 規格番号及び年号	Designation 種類記号	Chemical Composition 化学成分（%） Thickness(t) mm板厚	C	Si	Mn	P	S	Cu	Cr	Ni
Tensile Strength 310, 340, 350 N/mm^2 or MPa (32, 35, 36 kgf/mm^2) Class											
JP日	–	–	–	–	–	–	–	–	–	–	–
US米	ASTM A285M-90	Gr. A	≦50	≦0.17	–	≦0.90	≦0.035	≦0.035	–	–	–
		Gr. B	≦50	≦0.22	–	≦0.90	≦0.035	≦0.035	–	–	–
UK英	BS1501 Part1-80	141 Gr. 360	≦19 No Longer Active	≦0.16	–	≦0.50	≦0.050	≦0.050	≦0.30	≦0.25	≦0.30
		151 Gr. 360 No Longer Active	≦100	≦0.17	≦0.35	0.40～1.20	≦0.030	≦0.045	≦0.30	≦0.25	≦0.30
		154 Gr. 360	≦9.5 No Longer Active	≦0.20	–	0.30～1.20	≦0.050	≦0.050	≦0.30	≦0.25	≦0.30
		161 Gr. 360 Active in UK only	≦150	≦0.17	0.10～0.35	0.40～1.20	≦0.030	≦0.030	≦0.30	≦0.25	≦0.30
		164 Gr. 360 Superseded by European Standard EN 10028-2, Grade P235GH, Material No. 1.0345	≦150	≦0.20	0.10～0.35	0.40～1.20	≦0.030	≦0.030	≦0.30	≦0.25	≦0.30
GR独	DIN 17155-83	H1 Superseded by European Standard EN 10028-2, Grade P235GH; Material No. 1.0345	–	≦0.16	≦0.35	0.40～1.20	≦0.035	≦0.30	≦0.030	≦0.25	≦0.30
FR仏	–	–	–	–	–	–	–	–	–	–	–
URソ	ГОСТ 5520-79	12K	–	0.08～0.16	0.17～0.37	0.40～0.70	≦0.040	≦0.040	–	–	–
Tensile Strength 360, 370, 380 N/mm^2 or MPa (37, 38, 39 kgf/mm^2) Class											
JP日	–	–	–	–	–	–	–	–	–	–	–
US米	ASTM A285M-90	Gr. C	≦50	≦0.28	–	≦0.90	≦0.035	≦0.035	–	–	–

$1N/mm^2$ or $MPa = 1.01972 \times 10^{-1} kgf/mm^2$
$1J = 1.01972 \times 10^{-1} kg\text{-}m$

Mo	V	Others	Mechanical Properties 機械的性質 Heat Treatment 熱処理	Thickness (t)mm 板厚	Yield Point min. N/mm²or MPa (kgf/mm²) 最小降伏点	Tensile Strength N/mm²or MPa (kgf/mm²) 引張強さ	Notctch Toughness 衝撃値 Test Piece 試験片	Test Temp.(℃) 試験温度	Absorbed Energy J (min. kg-m) 吸収ｴﾈﾙｷﾞｰ	Index No. 索引番号
引張り強さ 310, 340, 350 N または MPa (32, 35, 36キロ) 級										
−	−	−	−	−	−	−	−	−	−	B001
−	−	−	−	−	165	310～450	−	−	−	
−	−	−	−	−	185	345～485	−	−	−	
≦0.10	−	Cu+Cr+Ni+ Mo≦0.70	−	3<t≦16 16<t≦19	200 200	360～480	−	−	−	
≦0.10	−	Cu+Cr+Ni+ Mo≦0.70	N	3<t≦16 16<t≦40 40<t≦63 63<t≦100	200 195 185 175	350～480	−	−	−	
≦0.10	−	Cu+Cr+Ni+ Mo≦0.70	−	≦(9.5)	200	350～480	−	−	−	
≦0.10	−	Cu+Cr+Ni+ Mo≦0.70	N	3<t≦16 16<t≦40 40<t≦63 63<t≦100 100<t≦150	200 200 190 170 170	350～480	−	−	−	
≦0.10	Al≧0.015	Cu+Cr+Ni+ Mo≦0.70	N	3<t≦16 16<t≦40 40<t≦63 63<t≦100 100<t≦150	265 235 220 − −	350～480	2V	RT 0 -20	32 24.5 21.5	
−	≦0.03	Al≧0.020 Ti≧0.03	N	t≦16 16<t≦40 40<t≦60 60<t≦100 100<t≦150	235 225 215 205 185	t≦100 350～470 100<t≦150 350～480	2V	0	21	
−	−	−	−	−	−	−	−	−	−	
−	−	−	N	t≦50 21<t≦40 41<t≦60	(23) (22) (21)	(36～45)	2U	20	(8) (8) (8)	
引張り強さ 360, 370, 380 N または MPa (37, 38, 39キロ) 級										
−	−	−	−	−	−	−	−	−	−	B002
−	−	−	−	−	205	380～515	−	−	−	

Country 国	Standard No. 規格番号及び年号	Designation 種類記号	Thickness(t) mm板厚	C	Si	Mn	P	S	Cu	Cr	Ni
			Chemical Composition 化学成分 (%)								
US米											
	ASTM A516M-90	Gr.55 (Gr.380)	t≦12.5 12.5<t≦50 50<t≦100 100<t≦200 200<t	≦0.18 ≦0.20 ≦0.22 ≦0.24 ≦0.26	0.15～0.40	t≦12.5 0.60～0.90 12.5<t 0.60～1.20	≦0.035	≦0.035	–	–	–
	ASTM A562M-90	–	–	≦0.12	0.15～0.50	≦12.0	≦0.035	≦0.035	≦0.15	–	–
UK英	–	–	–	–	–	–	–	–	–	–	–
GR独	–	–	–	–	–	–	–	–	–	–	–
FR仏	NF A36-205-82	A37CP	–	≦0.16	≦0.30	≧0.40	≦0.035	≦0.030	–	–	–
		A37AP	–	≦0.16	≦0.30	≧0.40	≦0.035	≦0.030	–	–	–
		A37FP	–	≦0.15	≦0.20	≧0.40	≦0.030	≦0.020	–	–	–
		A37P2	≦110	≦0.15	≦0.20	≧0.40	≦0.035	≦0.035	≦0.25	–	–
URソ	ГОСТ 5520-79	15K	–	0.12～0.20	0.15～0.30	0.35～0.65	≦0.040	≦0.040	–	–	–
Tensile Strength 390, 400, 410 N／mm² or MPa (40, 41, 42 kgf／mm²) Class											
JP日	JIS G3116-90	SG255 (SG26)	2.3～6.	≦0.20	–	≧0.30	≦0.040	≦0.040	–	–	–

1N/mm² or MPa = 1.01972 × 10^{-1} kgf/mm²
1J = 1.01972 × 10^{-1} kg-m

Mo	V	Others	Heat Treatment 熱処理	Thickness (t)mm 板 厚	Yield Point min. N/mm² or MPa (kgf/mm²) 最小降伏点	Tensile Strength N/mm² or MPa (kgf/mm²) 引張強さ	Notch Toughness 衝撃値 Test Piece 試験片	Notch Toughness 衝撃値 Test Temp. (℃) 試験温度	Notch Toughness 衝撃値 Absorbed Energy J (min. kg-m) 吸収エネルギー	Index No. 索引番号
–	–	–	t>40mm N	–	205	380～515	–	–	–	
–	–	Ti ≧ 4×C	N	–	205	380～515	–	–	–	
–	–	–	–	–	–	–	–	–	–	
–	–	–	–	–	–	–	–	–	–	
–	–	–	N	t ≦ 30 30<t ≦ 50 50<t ≦ 80 80<t	235 215 205 195	360～430	2 V	20 0	32 28	
–	–	–	N	t ≦ 30 30<t ≦ 50 50<t ≦ 80 80<t	235 215 205 195	360～430	2 V	0 -20	32 28	
–	–	–	N	t ≦ 30 30<t ≦ 50 50<t ≦ 80 80<t	235 215 205 195	360～430	2 V	0 -20 -40	40 32 28	
–	–	–	N	t ≦ 30 30<t ≦ 50 50<t ≦ 80 80<t	235 215 205 185	360～430	2 V	-20	28	
–	–	–	–	t ≦ 20 21<t ≦ 40 41<t ≦ 60	(23) (22) (21)	(36～49)	2 U	0	(7) (6.5) (6)	
引張り強さ 390, 400, 410 N または MPa（40, 41, 42キロ）級										
–	–	–	–	–	255 (26)	≧ 402 (≧ 41)	–	–	–	B003

Country 国	Standard No. 規格番号及び年号	Designation 種類記号	Chemical Composition 化学成分 (%)								
			Thickness(t) mm板厚	C	Si	Mn	P	S	Cu	Cr	Ni
JP日	JIS G3103-87	SB410 (SB42)	t ≦25 25<t ≦50 50<t ≦200	≦0.24 ≦0.27 ≦0.30	0.15～0.30	≦0.90	≦0.040	≦0.040	–	–	–
	JIS G3118-87	SGV410 (SGV42)	t≦12.5 12.5<t ≦50 50<t ≦100 100<t ≦200	≦0.21 ≦0.23 ≦0.25 ≦0.27	0.15～0.30	0.85～1.20	≦0.035	≦0.040	–	–	–
	JIS G3115-90	SPV235 (SPV24)	t ≦100 100< t	≦0.18 ≦0.20	0.15～0.35	≦1.40	≦0.030	≦0.030	–	–	–
US米	ASTM A515M-92	Gr.60 (Gr.415)	t ≦25 25<t ≦50 50<t ≦100 100<t ≦200 200< t	≦0.24 ≦0.27 ≦0.29 ≦0.31 ≦0.31	0.15～0.40	≦0.90	≦0.035	≦0.035	–	–	–
	ASTM A516M-90	Gr.60 (Gr.415)	t ≦12.5 12.5<t ≦50 50<t ≦100 100<t ≦200 200< t	≦0.21 ≦0.23 ≦0.25 ≦0.27 ≦0.27	0.15～0.40	t≦12.5 0.60～0.90 12.5<t 0.85～1.20	≦0.035	≦0.035	–	–	–
UK英	BS1501 Part1-80	151 Gr.400 No Longer Active	≦100	≦0.20	≦0.35	0.50～1.30	≦0.030	≦0.045	≦0.30	≦0.25	≦0.30
		154 Gr.400	≦9.5 No Longer Active	≦0.24	–	0.40～1.20	≦0.050	≦0.050	≦0.30	≦0.25	≦0.30
		161 Gr.400 Active in UK only	≦150	≦0.20	0.10～0.35	0.50～1.30	≦0.030	≦0.030	≦0.30	≦0.25	≦0.30
		164 Gr.400 No Longer Active	≦150	≦0.23	0.10～0.35	0.50～1.30	≦0.030	≦0.030	≦0.30	≦0.25	≦0.30

$1N/mm^2$ or $MPa = 1.01972 \times 10^{-1} kgf/mm^2$
$1J = 1.01972 \times 10^{-1} kg\text{-}m$

			Mechanical Properties 機械的性質							Index No. 索引番号
			Heat Treatment 熱処理	Thickness (t)mm 板厚	Yield Point min. N/mm^2 or MPa (kgf/mm^2) 最小降伏点	Tensile Strength N/mm^2 or MPa (kgf/mm^2) 引張強さ	Notch Toughness 衝撃値			
Mo	V	Others					Test Piece 試験片	Test Temp.(℃) 試験温度	Absorbed Energy J (min. kg-m) 吸収ｴﾈﾙｷﾞｰ	
–	–	–	t>50mm N	–	225 (23)	410～550 (42～56)	–	–	–	B003
–	–	–	t>38mm N	–	225 (23)	410～490 (42～50)	–	–	–	
–	–	–	–	6< t ≦50 50<t≦100 100<t≦200	235(24) 215(22) 195(20)	400～510 (41～52)	2 V	0	47 (4.8)	
–	–	–	t>50mm N	–	220	415～550	–	–	–	
–	–	–	t>40mm N	–	220	415～550	–	–	–	
≦0.10	–	Cu+Cr+Ni+ Mo ≦0.70	N	3<t≦16 16<t≦40 40<t≦63 63<t≦100	225 215 205 195	400～520	–	–	–	
≦0.10	–	Cu+Cr+Ni+ Mo ≦0.70	–	≦9.5	235	400～520	–	–	–	
≦0.10	–	Cu+Cr+Ni+ Mo ≦0.70	N	3<t≦16 16<t≦40 40<t≦63 63<t≦100 100<t≦150	225 215 205 195 195	400～520	–	–	–	
≦0.10	Al ≦0.015	Cu+Cr+Ni+ Mo ≦0.70	N	3<t≦16 16<t≦40 40<t≦63 63<t≦150	275 265 245 –	460～520	2 V	RT 0 -20	32 24.5 21.5	

Country 国	Standard No. 規格番号及び年号	Designation 種類記号	Chemical Composition 化学成分（%）								
			Thickness(t) mm板厚	C	Si	Mn	P	S	Cu	Cr	Ni
UK英	BS1501 Part1-80	224 Gr. 400	≦150	≦0.18	0.10～0.35	0.90～1.50	≦0.030	≦0.030	≦0.30	≦0.25	≦0.30
		Superseded by European Standard EN 10028-2, Grade P265GH, Material No. 1.0425									
GR独	DIN 17155-83	HⅡ	−	≦0.20	≦0.35	0.50～1.30	≦0.035	≦0.030	≦0.30	≦0.25	−
		Superseded by European Standard EN 10028-2, Grade P265GH, Material No. 1.0425									
FR仏	NF A36-205-82	A42CP	≦110	≦0.18	≦0.35	≧0.50	≦0.04	≦0.04	−	−	−
		A42AP	≦110	≦0.18	≦0.35	≧0.50	≦0.04	≦0.04	−	−	−
		A42FP	≦110	≦0.18	≦0.25	≧0.60	≦0.035	≦0.035	≦0.25	−	−
URソ	ГOCT 5520-79	16K	−	0.12～0.20	0.17～0.37	0.45～0.75	≦0.040	≦0.040	−	−	−
		20K	−	0.16～0.24	0.15～0.30	t<20mm 0.35～0.65 t ≧20 0.35～0.80	≦0.040	≦0.040	≦0.30	≦0.30	≦0.30
Tensile Strength 430, 440, 450 N／mm² or MPa (44, 45, 46 kgf／mm²) Class											
JP日	JIS G3116-90	SG295 (SG30)	2.3~6.0	≦0.20	≦0.35	≦1.00	≦0.040	≦0.040	−	−	−
	JIS G3103-87	SB450 (SB46)	≦25 25<t ≦50 25<t ≦200	≦0.28 ≦0.31 ≦0.33	0.15～0.30	≦0.90	≦0.035	≦0.040	−	−	−

$1N/mm^2$ or $MPa = 1.01972 \times 10^{-1} kgf/mm^2$
$1J = 1.01972 \times 10^{-1} kg\text{-}m$

Mo	V	Others	Mechanical Properties 機械的性質 Heat Treatment 熱処理	Thickness (t)mm 板厚	Yield Point min. N/mm²or MPa (kgf/mm²) 最小降伏点	Tensile Strength N/mm²or MPa (kgf/mm²) 引張強さ	Notctch Toughness 衝撃値 Test Piece 試験片	Test Temp.(℃) 試験温度	Absorbed Energy J (min. kg-m) 吸収エネルギー	Index No. 索引番号
≦0.10	Al≦0.015	Cu+Cr+Ni+ Mo ≦0.70	N	3<t≦16 16<t≦40 40<t≦63 63<t≦150	275 265 245 –	400～520	2V	RT 0 -20 -30 -40 -50	49 44 32 32 24.5 21.5	B003
≦0.10	≦0.03	Al≧0.020 Nb≦0.10 Ti≦0.03	N	t≦16 16<t≦40 40<t≦60 60<t≦100 100<t≦150	265 255 245 215 205	t≦100 410～530 100<t≦150 400～530	DVM	RT 常温	55	
–	–	–	N	t≦30 30<t≦50 50<t≦80 80<t	255 235 225 215	410～490	2V	0	32	
–	–	–	N	t≦30 30<t≦50 50<t≦80 80<t	255 235 225 215	410～490	2V	0 20	32 28	
–	–	–	N	t≦30 30<t≦50 50<t≦80 80<t	255 235 235 225	410～490	2V	0 -20 -40	40 32 28	
–	–	–	–	t≦20 21<t≦40 41<t≦60	255 245 235	(41)～(50)	2U	20	(7)	
–	As≦0.08	N≦0.003	–	t≦20 21≦t≦40 41≦t≦60	(25) (24) (23)	(41)～(50)	2U	20	(6) (5.5) (5)	
引張り強さ 430, 440, 450 N または MPa (44, 45, 46キロ) 級										
–	–	–	–	–	295 (30)	≧440 (≧45)	–	–	–	B004
–	–	–	t>50mm N	–	245 (25)	450～590 (46～60)	–	–	–	

Country 国	Standard No. 規格番号及び年号	Designation 種類記号	Chemical Composition 化学成分 (%) Thickness(t) mm板厚	C	Si	Mn	P	S	Cu	Cr	Ni
JP日	JIS G3118-87	SGV450 (SGV46)	t ≦12.5 12.5<t ≦50 50<t ≦100 100<t ≦200	≦0.24 ≦0.26 ≦0.28 ≦0.29	0.15~0.30	0.85~1.20	≦0.035	≦0.040	–	–	–
US米	ASTM A515M-92	Gr. 65 (Gr. 450)	t ≦25 25<t ≦50 50<t	≦0.28 ≦0.31 ≦0.33	0.15~0.40	≦0.90	≦0.035	≦0.035	–	–	–
	ASTM A516M-90	Gr. 65 (Gr. 450)	t ≦12.5 12.5<t ≦50 50<t ≦100 100<t ≦200 200< t	≦0.24 ≦0.26 ≦0.28 ≦0.29 ≦0.29	0.15~0.40	0.85~1.20	≦0.035	≦0.035	–	–	–
UK英	BS1501 Part1-80	151 Gr. 430 No Longer Active	≦100	≦0.25	≦0.35	0.60~1.40	≦0.030	≦0.045	≦0.30	≦0.25	≦0.30
		154 Gr. 443 No Longer Active	≦9.5	≦0.25	–	0.40~1.20	≦0.050	≦0.050	≦0.30	≦0.25	≦0.30
		161 Gr. 430 Active in UK Only	≦100	≦0.25	0.10~0.35	0.60~1.40	≦0.030	≦0.030	≦0.30	≦0.25	≦0.30
		224 Gr. 430 Superseded by European Standard EN 10028-2, Grade P295GH, Material No. 1.0481	≦150	≦0.20	0.10~0.40	0.90~1.50	≦0.030	≦0.030	≦0.30	≦0.25	≦0.30
GR独	–	–	–	–	–	–	–	–	–	–	–
FR仏	–	–	–	–	–	–	–	–	–	–	–
URソ	ΓOCT 5520-79	18K	–	0.14~0.22	0.17~0.37	0.55~0.85	≦0.040	≦0.040	–	–	–
Tensile Strength 460, 470 N／mm² or MPa (47, 48 kgf／mm²) Class											
JP日	–	–	–	–	–	–	–	–	–	–	–

1N/mm² or MPa＝1.01972×10⁻¹kgf/mm²
1J＝1.01972×10⁻¹kg-m

Mo	V	Others	Mechanical Properties 機械的性質				Notch Toughness 衝撃値			Index No. 索引番号
			Heat Treatment 熱処理	Thickness (t)mm 板厚	Yield Point min. N/mm² or MPa (kgf/mm²) 最小降伏点	Tensile Strength N/mm² or MPa (kgf/mm²) 引張強さ	Test Piece 試験片	Test Temp.(℃) 試験温度	Absorbed Energy J (min. kg-m) 吸収エネルギー	
–	–	–	t>38mm N	–	240 (25)	450～540 (46～55)	–	–	–	B004
–	–	–	t>50mm N	–	240	450～585	–	–	–	
–	–	–	t>40mm N	–	240	450～585	–	–	–	
≦0.10	–	Cu+Cr+Ni+ Mo≦0.70	N	3<t≦16 16<t≦40 40<t≦63 63<t≦100	255 235 225 215	430～550	–	–	–	
≦0.10	–	Cu+Cr+Ni+ Mo≦0.70	–	≦0.95	225	430～550	–	–	–	
≦0.10	–	Cu+Cr+Ni+ Mo≦0.70	–	3<t≦16 16<t≦40 40<t≦63 63<t≦100 100<t≦150	255 235 225 215 205	430～550	–	–	–	
≦0.10	Al≦0.015	Cu+Cr+Ni+ Mo≦0.70	–	3<t≦16 16<t≦40 40<t≦63 63<t≦150	305 285 – –	430～550	2V	RT 0 -20 -30 -40 -50	49 44 37 32 24.5 21.5	
–	–	–	–	–	–	–	–	–	–	
–	–	–	–	–	–	–	–	–	–	
–	–	–	N	t≦20 21≦t≦40 41≦t≦60	(28) (27) (26)	(44～53)	2U	20	(6) (6) (6)	
引張り強さ 460,470 N または MPa（47,48キロ）級										
–	–	–	–	–	–	–	–	–	–	B005

Country 国	Standard No. 規格番号及び年号	Desig-nation 種類記号	Chemical Composition 化 学 成 分 （ % ）								
			Thick-ness(t) mm板厚	C	Si	Mn	P	S	Cu	Cr	Ni
US米	–	–	–	–	–	–	–	–	–	–	–
UK英	BS1501 Patr1-80	223 Gr.460	≦150	≦0.20	0.10～0.40	0.80～1.60	≦0.030	≦0.030	≦0.30	≦0.25	≦0.30
		No Longer Active									
		224 Gr.460	≦150	≦0.22	0.10～0.40	0.90～1.60	≦0.030	≦0.030	≦0.30	≦0.25	≦0.30
		No Longer Active									
GR独	DIN 17155-83	17Mn4	–	0.14～0.40	≦0.40	0.90～1.40	≦0.035	≦0.030	≧0.30	≦0.25	≦0.30
		Superseded by European Standard EN 10028-2, Grade P295GH, Material No. 1.0481									
FR仏	NF A36-205-82	A48CP	–	≦0.20	≦0.35	0.80～1.50	≦0.035	≦0.030	–	–	–
		A48AP	–	≦0.20	≦0.35	0.80～1.50	≦0.035	≦0.030	–	–	–
		A48FP	–	≦0.20	≦0.35	0.80～1.50	≦0.030	≦0.020	–	≦0.25	≦0.30
		A48CPR	–	≦0.20	≦0.50	1.0～1.60	≦0.035	≦0.030	–	–	–
		A48APR	–	≦0.20	≦0.50	1.0～1.60	≦0.035	≦0.030	–	–	–
		A48FPR	–	≦0.20	≦0.50	1.0～1.60	≦0.030	≦0.020	–	–	–

1N/mm² or MPa = 1.01972 × 10^{-1} kgf/mm²
1J = 1.01972 × 10^{-1} kg-m

Mo	V	Others	Mechanical Properties 機械的性質							Index No. 索引番号
			Heat Treatment 熱処理	Thickness (t)mm 板厚	Yield Point min. N/mm² or MPa (kgf/mm²) 最小降伏点	Tensile Strength N/mm² or MPa (kgf/mm²) 引張強さ	Notctch Toughness 衝撃値			
							Test Piece 試験片	Test Temp.(℃) 試験温度	Absorbed Energy J (min. kg-m) 吸収ｴﾈﾙｷﾞｰ	
–	–	–	–	–	–	–	–	–	–	
≦0.10	Nb 0.010~0.060	Cu+Cr+Ni+ Mo≦0.70	N	3<t≦16 16<t≦40 40<t≦63 63<t≦150	350 340 340 –	460～580	2 V	RT 0 -15 -30	49 44 32.5 21.5	B005
≦0.10	Al 0.015	Cu+Cr+Ni+ Mo≦0.70	N	3<t≦16 16<t≦40 40<t≦63 63<t≦150	320 320 305 –	460～580	2 V	RT 0 -20 -30 -40 -50	49 44 37 32.5 24.5 21.5	
≦0.10	≦0.03	Al≧0.020 Nb≧0.01 Ti≦0.03	N	t≦16 16<t≦40 40<t≦60 60<t≦100 100<t≦150	295 285 285 255 235	t≦60 460～580 60<t≦100 450～570 100<t≦150 440～570	2 V	0	31.5	
–	–	–	N	t≦30 30<t≦50 50<t≦80 80<t	295 275 265 255	470～560	–	20 0	47 40	
–	–	–	N	t≦30 30<t≦50 50<t≦80 80<t	295 275 265 255	470～560	–	0 -20	43 40	
≦0.10	≦0.05	–	N	t≦30 30<t≦50 50<t≦80 80<t	295 275 265 255	470～560	–	0 -20 -40	56 47 40	
–	–	–	N	t≦30 30<t≦50 50<t≦80 80<t	295 275 265 255	470～560	–	–	–	
–	–	–	N	t≦30 30<t≦50 50<t≦80 80<t	295 275 265 255	470～560	–	–	–	
–	–	–	N	t≦30 30<t≦50 50<t≦80 80<t	295 275 265 255	470～560	–	–	–	

Country 国	Standard No. 規格番号及び年号	Designation 種類記号	Chemical Composition 化学成分 (%)								
			Thickness(t) mm板厚	C	Si	Mn	P	S	Cu	Cr	Ni
URソ	–	–	–	–	–	–	–	–	–	–	–
Tensile Strength 480, 490, 500 N／mm² or MPa (49, 50, 51 kgf／mm²) Class											
JP日	JIS G3116-90	SG325 (SG33)	2.3~6.0	≦0.20	≦0.55	≦1.50	≦0.040	≦0.040	–	–	–
	JIS G3103-87	SB480 (SB49)	≦25 25<t ≦50 25<t ≦50	≦0.31 ≦0.33 ≦0.35	0.15~0.30	≦0.90	≦0.035	≦0.040	–	–	–
	JIS G3118-87	SGV480 (SGV49)	t≦12.5 12.5<t ≦50 50<t ≦100 100<t ≦200	≦0.27 ≦0.28 ≦0.30 ≦0.31	0.15~0.30	0.85~1.20	≦0.035	≦0.040	–	–	–
	JIS G3115-90	SPV315 (SPV32)	–	0.18	0.15~0.55	≦1.50	≦0.030	≦0.040	–	–	–
US米	ASTM A515M-92	Gr.70 (Gr.415)	t ≦25 25<t ≦50 50<t ≦200	≦0.31 ≦0.33 ≦0.35	0.15~0.40	≦1.20	≦0.035	≦0.035	–	–	–
	ASTM A516M-90	Gr.485 (Gr.70)	t≦12.5 12.5<t ≦50 50<t ≦100 100<t ≦200 200< t	≦0.27 ≦0.28 ≦0.30 ≦0.31 ≦0.31	0.15~0.40	0.85~1.20	≦0.035	≦0.035	–	–	–
	ASTM A537M-91	Cl.1	≦100	≦0.24	0.15~0.50	t≦38 0.70~1.35 38<t 1.00~1.65	≦0.035	≦0.035	≦0.35	≦0.25	≦0.25
	ASTM A737M-87	Gr.B	–	≦0.20	0.15~0.50	1.15~1.50	≦0.035	≦0.030	–	–	–
UK英	BS1501 Part1-80	223 Gr.490 No Longer Active	≦150	≦0.20	0.10~0.50	0.90~1.60	≦0.030	≦0.030	≦0.30	≦0.25	≦0.30
		224 Gr.490 Superseded by European Standard EN 10028-2, Grade P355GH, Material No. 1.0473	≦150	≦0.22	0.10~0.40	0.90~1.60	≦0.030	≦0.030	≦0.30	≦0.25	≦0.30

$1N/mm^2$ or $MPa = 1.01972 \times 10^{-1} kgf/mm^2$
$1J = 1.01972 \times 10^{-1} kg\text{-}m$

Mo	V	Others	Mechanical Properties 機械的性質				Notch Toughness 衝撃値			Index No. 索引番号
			Heat Treatment 熱処理	Thickness (t)mm 板厚	Yield Point min. N/mm^2 or MPa (kgf/mm^2) 最小降伏点	Tensile Strength N/mm^2 or MPa (kgf/mm^2) 引張強さ	Test Piece 試験片	Test Temp.(℃) 試験温度	Absorbed Energy J (min. kg-m) 吸収ｴﾈﾙｷﾞｰ	
–	–	–	–	–	–	–	–	–	–	B005
引張り強さ 480, 490, 500 N または MPa（49, 50, 51キロ）級										
–	–	–	–	–	325 (33)	≧490 (≧50)	–	–	–	B006
–	–	–	t>50mm N	–	265 (27)	480～620 (49～63)	–	–	–	
–	–	–	t>38mm N	–	265 (27)	480～620 (49～50)	–	–	–	
–	–	–	–	6<t≦50 50<t≦100	315(32) 295(30)	490～610 (50～62)	2V	0	47 (4.8)	
–	–	–	t>50mm N		260	485～620				
–	–	–	t>40mm N	–	260	485～620	–	–	–	
≦0.08	–	–	N	t≦65 65<t≦100	345 310	485～620 450～585	–	–	–	
–	–	Nb≦0.05	N	–	345	485～625 450～585	–	–	–	
≦0.10	Nb 0.010~0.006	Cu+Cr+Ni+ Mo≦0.70	N	3<t≦16 16<t≦40 40<t≦63 63<t≦150	350 345 345 –	490～610	2V	RT 0 -15 -30	49 44 37 21.5	
≦0.10	Al≦0.015	Cu+Cr+Ni+ Mo≦0.70	N	3<t≦16 16<t≦40 40<t≦63 63<t≦150	325 325 305 –	490～610	2V	RT 0 -20 -30 -40 -50	51 44 37 32.5 24.5 21.5	

Country 国	Standard No. 規格番号及び年号	Designation 種類記号	Chemical Composition 化学成分（%）								
			Thickness(t) mm板厚	C	Si	Mn	P	S	Cu	Cr	Ni
UK英	BS1501 Part1-80	225 Gr.490	≦150	≦0.20	0.10~0.50	0.90~1.60	≦0.030	≦0.030	≦0.30	≦0.25	≦0.30
GR独	–	–	–	–	–	–	–	–	–	–	–
FR仏	–	–	–	–	–	–	–	–	–	–	–
URソ	–	–	–	–	–	–	–	–	–	–	–
Tensile Strength 510, 520, 530 N／mm² or MPa (52, 53, 54 kgf／mm²) Class											
JP日	JIS G3115-90	SPV355 (SPV36)	–	≦0.20	0.15~0.55	≦1.60	≦0.030	≦0.030	–	–	–
US米	ASTM A455M-90	–	–	≦0.33	≦0.10	0.85~1.20	≦0.035	≦0.04	–	–	–
US米	ASTM A455M-87	–	t ≦20	≦0.33	≦0.10	0.85~1.20	≦0.035	≦0.035	–	–	–
UK英	–	–	–	–	–	–	–	–	–	–	–
GR独	DIN 17155-83	19Mn6	–	0.15~0.22	0.30~0.60	1.00~1.60	≦0.035	≦0.030	≦0.30	≦0.25	≦0.30
		Superseded by European Standard EN 10028-2, Grade P355GH, Material No. 1.0473									
FR仏	NF A36-205-82	A52CP	–	≦0.20	≦0.50	1.0~1.60	≦0.035	≦0.030	≦0.30	≦0.25	≦0.30
		A52AP	–	≦0.20	≦0.50	1.0~1.60	≦0.035	≦0.030	≦0.30	≦0.25	≦0.30
		A52FP	–	≦0.20	≦0.50	1.0~1.60	≦0.030	≦0.020	≦0.30	≦0.25	≦0.30
		A52CPR	–	≦0.20	≦0.50	1.0~1.60	≦0.035	≦0.030	≦0.30	≦0.25	≦0.30
		A52APR	–	≦0.20	≦0.50	1.0~1.60	≦0.035	≦0.030	≦0.30	t > 25mm ≦0.25	t > 25mm ≦0.30

$1N/mm^2$ or $MPa = 1.01972 \times 10^{-1} kgf/mm^2$
$1J = 1.01972 \times 10^{-1} kg\text{-}m$

Mo	V	Others	Mechanical Properties 機械的性質 Heat Treatment 熱処理	Thickness (t)mm 板厚	Yield Point min. N/mm²or MPa (kgf/mm²) 最小降伏点	Tensile Strength N/mm²or MPa (kgf/mm²) 引張強さ	Notch Toughness 衝撃値 Test Piece 試験片	Test Temp.(℃) 試験温度	Absorbed Energy J (min. kg-m) 吸収ｴﾈﾙｷﾞｰ	Index No. 索引番号
≦0.10	Nb 0.010~0.006 Al≦0.015	Cu+Cr+Ni+ Mo≦0.70	N	3<t≦16 16<t≦40 40<t≦63 63<t≦150	–	490～610	2V	-20 -30 -40	49 37 21.5	B006
–	–	–	–	–	–	–	–	–	–	
–	–	–	–	–	–	–	–	–	–	
–	–	–	–	–	–	–	–	–	–	
引張り強さ 510, 520, 530 N または MPa (52, 53, 54キロ) 級										
–	–	–	–	6<t≦50 50<t≦100	355(36) 335(34)	520～640 (53～65)	2V	0	47 (4.8)	B007
–	–	–	–	t≦9.5 9.5<t≦15 15<t≦20	260 255 240	515～655 505～640 485～620	–	–	–	
–	–	–	–	t≦9.5 9.5<t≦15 15<t≦20	260 255 240	515～655 505～640 485～620	–	–	–	
–	–	–	–	–	–	–	–	–	–	
≦0.10	≦0.03	Al≧0.020 Nb≦0.01 Ti≦0.03	–	t≦26 26<t≦40 40<t≦60 60<t≦100 100<t≦150	350 345 335 315 295	t≦60 (52～66) 60<t≦100 (50～64) 100<t≦150 (49～64)	2V	0	23	
≦0.15	≦0.050	Nb≦0.040	N	t≦30 30<t≦80 80<t	350 335 325	510～610	2V	20 0	48 40	
≦0.15	≦0.050	Nb≦0.040	N	t≦30 30<t≦80 80<t	355 335 325	510～610	2V	20 0	48 40	
≦0.15	≦0.050	Nb≦0.040	N and SR*	t≦30 30<t≦80 80<t	350 335 325	510～610	2V	0 -20 -40	57 48 40	
≦0.15	≦0.050	Nb ≦0.040	N and SR*	t≦30 30<t≦80 80<t	350 335 325	510～610	–	–	–	
≦0.15	≦0.050	Nb ≦0.040	N	t≦30 30<t≦80 80<t	350 335 325	510～610	–	–	–	

* SR : Stress Relieving

Country 国	Standard No. 規格番号及び年号	Designation 種類記号	Thickness(t) mm板厚	C	Si	Mn	P	S	Cu	Cr	Ni
				Chemical Composition 化学成分 (%)							
FR仏	NF A36-205-82	A52FPR	−	≦0.20	≦0.50	1.0～1.60	≦0.030	≦0.020	≦0.30	≦0.25	≦0.40
	NF A36-207-82	A510	t ≦35	≦0.18	≦0.50	≦1.60	−	−	≦0.30	≦0.20	≦0.20
			35< t	≦0.20							
		A 530 Type 1	t ≦35	≦0.20	≦0.50	≦1.60	−	−	≦0.30	≦0.20	≦0.20
			35< t	≦0.20							
		A 530 Type 2	t ≦35	≦0.18							
			35< t	≦0.20							
URソ	−	−	−	−	−	−	−	−	−	−	−

Tensile Strength 540 N／mm² or MPa (55 kgf／mm²) or More Class

Country 国	Standard No. 規格番号及び年号	Designation 種類記号	Thickness(t) mm板厚	C	Si	Mn	P	S	Cu	Cr	Ni
JP日	JIS G3116-90	SG365 (SG37)	2.3～6.0	≦0.20	≦0.55	≦1.50	≦0.040	≦0.040	−	−	−
	JIS G3115-90	SPV450 (SPV46)	6<t≦50	≦0.18	0.15～0.75	≦1.60	≦0.030	≦0.030	−	t≦50mm	* Ceq≦0.44
			50<t ≦100						−	50<t≦75mm	Ceq≦0.46
		SPV490 (SPV50)	t ≦50	≦0.18	0.15～0.75	≦1.60	≦0.030	≦0.030	−	t≦50mm	Ceq≦0.45
			50<t ≦100						−	50<t≦75mm	Ceq≦0.47
US米	ASTM A737M-87	Gr. C	−	≦0.22	0.15～0.50	1.15～1.50	≦0.035	≦0.030	−	−	−
UK英	−	−	−	−	−	−	−	−	−	−	−
GR独	−	−	−	−	−	−	−	−	−	−	−
FR仏	NF A36-207-82	A 550 Type1	−	≦0.20	≦0.50	≦1.60	−	−	≦0.30	≦0.20	≦0.20
		A 550 Type2	−	≦0.22	≦0.55	≦1.60			≦0.35	≦0.25	0.20～0.70
		A 590 Type1	−	≦0.20	≦0.50	≦1.70	−	−	≦0.30	≦0.20	≦0.20
		A 590 Type2	−	≦0.18	≦0.40	≦1.70	−	−	≦0.60	≦0.40	0.20～0.70
URソ	−	−	−	−	−	−	−	−	−	−	−

$1N/mm^2$ or $MPa = 1.01972 \times 10^{-1} kgf/mm^2$
$1J = 1.01972 \times 10^{-1} kg\text{-}m$

Mo	V	Others	Mechanical Properties 機械的性質 Heat Treatment 熱処理	Thickness (t)mm 板厚	Yield Point min. N/mm^2 or MPa (kgf/mm^2) 最小降伏点	Tensile Strength N/mm^2 or MPa (kgf/mm^2) 引張強さ	Notch Toughness 衝撃値 Test Piece 試験片	Test Temp.(℃) 試験温度	Absorbed Energy J (min. kg-m) 吸収ｴﾈﾙｷﾞｰ	Index No. 索引番号
≦0.15	≦0.050	Nb ≦0.040	N	t ≦30 30<t≦80 80<t	350 335 325	510～610	–	–	–	B007
≦0.10	–	Nb 0.010～0.060	–	t ≦16 16<t≦35 35<t≦50 50<t≦80 80<t≦100	335 335 315 315 305	t≦35 550～670 35<t≦80 540～660	–	–	–	
≦0.10	–	Nb 0.010～0.060	–	t ≦16 16<t≦35 35<t≦50 50<t≦80	355 345 335 325	t≦35 590～670 35<t≦80 570～700	–	–	–	
–	–	–	–	–	–	–	–	–	–	
引張り強さ 540 N または MPa（55キロ）以上級										
–	–	–	–	–	365 (37)	≧540 (≧55)	–	–	–	B008
–	–	–	QT	6< t ≦50	450(46)	570～770 (58～70)	2 V	-10	47 (4.8)	
–	–	–		50<t≦70	430(44)					
–	–	–		6< t ≦50	490(50)	610～740 (62～75)	2 V	-10	47 (4.8)	
–	–	–		50<t≦75	470(48)					
–	0.04～0.11	N≦0.03	N	–	415	550～690	–	–	–	
–	–	–	–	–	–	–	–	–	–	
–	–	–	–	–	–	–	–	–	–	
≦0.15	t≦35 0.02～0.12 35<t≦80 0.02～0.15	Nb 0.010～0.060	–	t ≦16 16<t≦35 35<t≦50 50<t≦80	400 390 380 360	t ≦35 550～670 35<t≦80 540～660	–	–	–	
–	0.02～0.15	Nb 0.010～0.060	–	t ≦16 16<t≦35 35<t≦50 50<t≦80	400 390 380 360	t ≦35 550～670 35<t≦80 540～660	–	–	–	
≦0.15	t≦35 0.02～0.12 35<t≦80 0.02～0.18									
–	–	–	–	–	–	–	–	–	–	

$$*Ceq = C + \frac{Mn}{6} + \frac{Si}{24} + \frac{Ni}{40} + \frac{Cr}{5} + \frac{Mo}{4} + \frac{V}{14}$$

Country 国	Standard No. 規格番号及び年号	Designation 種類記号	Chemical Composition 化学成分（%）								
			Thick-ness(t) mm板厚	C	Si	Mn	P	S	Cu	Cr	Ni
1/4 Mo High Stregth Steel　1/4 Mo 高張度鋼（高温強度保証鋼）											
ＪＰ日	ＪＩＳ G3124-87	SEV245 (SEV25)	−	≦0.20	0.15〜0.60	0.80〜1.60	≦0.035	≦0.035	≦0.35	−	−
		SEV295 (SEV30)	−	≦0.19	0.15〜0.60	0.80〜1.60	≦0.035	≦0.035	≦0.35	−	−
		SEV345 (SEV35)	−	≦0.19	0.15〜0.60	0.80〜1.70	≦0.035	≦0.035	≦0.35	−	−
ＵＳ米	−	−	−	−	−	−	−	−	−	−	−
ＵＫ英	−	−	−	−	−	−	−	−	−	−	−
ＧＲ独	−	−	−	−	−	−	−	−	−	−	−
ＦＲ仏	−	−	−	−	−	−	−	−	−	−	−
ＵＲソ	−	−	−	−	−	−	−	−	−	−	−

$1N/mm^2$ or $MPa = 1.01972 \times 10^{-1} kgf/mm^2$
$1J = 1.01972 \times 10^{-1} kg\text{-}m$

			Mechanical Properties 機械的性質							Index No. 索引番号
			Heat Treatment 熱処理	Thickness (t)mm 板厚	Yield Point min. N/mm²or MPa (kgf/mm²) 最小降伏点	Tensile Strength N/mm²or MPa (kgf/mm²) 引張強さ	Notctch Toughness 衝撃値			
Mo	V	Others					Test Piece 試験片	Test Temp.(℃) 試験温度	Absorbed Energy J (min.kg-m) 吸収ｴﾈﾙｷﾞｰ	
≦0.35	≦0.10	≦0.05	−	t ≦50 50<t≦100 100<t≦125 125<t≦150	370(38) 355(36) 345(35) 335(34)	510〜650 (52〜66) 510〜650 (52〜66) 490〜630 (50〜64) 490〜630 (50〜64)	−	−	−	B009
0.10〜0.40	≦0.10	≦0.05	−	t ≦50 50<t≦100 100<t≦125 125<t≦150	420(43) 400(41) 390(40) 380(39)	540〜690 (55〜70) 530〜690 (55〜70) 530〜680 (54〜69) 520〜670 (53〜68)	−	−	−	
0.15〜0.50	≦0.10	≦0.05	−	t ≦50 50<t≦100 100<t≦125 125<t≦150	430(44) 430(44) 420(43) 410(42)	590〜740 (60〜75) 590〜740 (60〜75) 580〜730 (59〜74) 570〜720 (58〜73)	−	−	−	
−	−	−	−	−	−	−	−	−	−	
−	−	−	−	−	−	−	−	−	−	
−	−	−	−	−	−	−	−	−	−	
−	−	−	−	−	−	−	−	−	−	
−	−	−	−	−	−	−	−	−	−	

3. Low Alloy Steel Plates 低合金鋼鋼板

Country 国	Standard No. 規格番号及び年号	Designation 種類記号	Chemical Composition 化学成分 (%) Thickness(t) mm板厚	C	Si	Mn	P	S	Cu	Cr	Ni
1/4 Mo Steel											
JP日	–	–	–	–	–	–	–	–	–	–	–
US米	–	–	–	–	–	–	–	–	–	–	–
UK英	–	–	–	–	–	–	–	–	–	–	–
GR独	DIN 17155-83	15Mo3	–	0.15~0.20	0.10~0.35	0.40~0.90	≦0.035	≦0.030	≦0.30	≦0.25	–
		Superseded by European Standard EN 10028-2, Grade 16 Mo3, Material No. 1.5415									
FR仏	NF A36-206-83	15D3	≦80	≦0.18	0.15~0.30	0.50~0.80	≦0.035	≦0.030	–	–	≦0.30
URソ	–	–	–	–	–	–	–	–	–	–	–
1/2 Mo Steel, 1/2 Mo-B Steel											
JP日	JIS G3103-87	SB450M (SB46M)	≦25	≦0.18	0.15~0.30	≦0.90	≦0.035	≦0.040	–	–	–
			25<t ≦50	≦0.21							
			50<t ≦100	≦0.23							
			100<t ≦150	≦0.25							
		SB480M (SB49M)	≦25	≦0.20	0.15~0.30	≦0.90	≦0.035	≦0.040	–	–	–
			25<t ≦50	≦0.23							
			50<t ≦100	≦0.25							
			100<t ≦150	≦0.27							
US米	ASTM A204M-93	Gr. A	t≦25	≦0.18	0.15~0.40	≦0.90	≦0.035	≦0.035	–	–	–
			25<t ≦50	≦0.21							
			50<t ≦100	≦0.23							
			100< t	≦0.25							
		Gr. B	t≦25	≦0.20	0.15~0.40	≦0.90	≦0.035	≦0.035	–	–	–
			25<t ≦50	≦0.23							
			50<t ≦100	≦0.25							
			100< t	≦0.27							
		Gr. C	t≦25	≦0.23	0.15~0.40	≦0.90	≦0.035	≦0.035	–	–	–
			25<t ≦50	≦0.26							
			50<t ≦100	≦0.28							
			100< t	≦0.28							
UK英	BS1501 Part2-70	240	≦100	0.12~0.23	0.15~0.30	0.50~0.90	≦0.035	–	≦0.30	–	–
		No Longer Active									

1N/mm² or MPa = 1.01972 × 10^{-1} kgf/mm²
1J = 1.01972 × 10^{-1} kg-m

Mo	V	Others	Heat Treatment 熱処理	Thickness (t)mm 板厚	Yield Point min. N/mm² or MPa (kgf/mm²) 最小降伏点	Tensile Strength N/mm² or MPa (kgf/mm²) 引張強さ	Notch Toughness 衝撃値: Test Piece 試験片	Notch Toughness 衝撃値: Test Temp. (°C) 試験温度	Notch Toughness 衝撃値: Absorbed Energy J (min. kg-m) 吸収エネルギー	Index No. 索引番号
–	–	–	–	–	–	–	–	–	–	B010
–	–	–	–	–	–	–	–	–	–	
–	–	–	–	–	–	–	–	–	–	
0.25～0.35	–	–	N	t ≦16 16<t ≦40 40<t ≦60	275 265 265	430～520	DVM	RT 常温	47	
0.25～0.35	≦0.04	–	N+T	t ≦16 16<t ≦40 40<t ≦60 60<t ≦100 100<t ≦150	275 275 265 245 225	t ≦16 440～590 60<t ≦100 430～580 100<t ≦150 420～570	2V	20	t ≦16 31.5 60<t ≦150 27.5	
–	–	–	–	–	–	–	–	–	–	
0.45～0.60	–	–	–	–	255 (26)	450～590 (46～60)	–	–	–	B011
0.45～0.60	–	–	–	–	275 (28)	480～620 (49～63)	–	–	–	
0.45～0.60	–	–	t ≧40mm N	–	255	450～585	–	–	–	
0.45～0.60	–	–	t ≧40mm N	–	275	485～620	–	–	–	
0.45～0.60	–	–	t ≧40mm N	–	295	515～655	–	–	–	
0.45～0.60	–	–	N	t ≦16 16<t ≦50 t >63	285 275 -1% per 6mm	450～590	5U	RT	15	

Country 国	Standard No. 規格番号及び年号	Desig-nation 種類記号	Chemical Composition 化学成分（%）								
			Thick-ness(t) mm板厚	C	Si	Mn	P	S	Cu	Cr	Ni
UK英	BS1501 Part2-70	261	≦87.5	0.10～0.17	0.10～0.40	0.40～0.80	≦0.040	≦0.040	≦0.30	≦0.25	≦0.30
GR独	–	–	–	–	–	–	–	–	–	–	–
FR仏	–	–	–	–	–	–	–	–	–	–	–
URソ	–	–	–	–	–	–	–	–	–	–	–

Mn-1/2 Mo, Mn-1/2 Mo-V Steel

Country 国	Standard No. 規格番号及び年号	Desig-nation 種類記号	Thick-ness(t) mm板厚	C	Si	Mn	P	S	Cu	Cr	Ni
JP日	JIS G3119-87	SBV 1A	t≦25 25<t≦50 50<t≦100	≦0.20 ≦0.25 ≦0.25	0.15～0.30	0.95～1.30	≦0.035	≦0.035	–	–	–
		SBV 1B	t≦25 25<t≦50 50<t≦100	≦0.20 ≦0.25 ≦0.25	0.15～0.30	1.15～1.50	≦0.035	≦0.040	–	–	–
	JIS G3120-87	SQV 1A	–	≦0.25	0.15～0.30	1.15～1.50	≦0.035	≦0.040	–	–	–
		SQV 1B	–	≦0.25	0.15～0.30	1.15～1.50	≦0.035	≦0.040	–	–	–
US米	ASTM A302M-93	Gr. A	t≦25 25<t≦50 50<t	≦0.20 ≦0.23 ≦0.25	0.15～0.40	0.95～1.30	≦0.035	≦0.035	–	–	–
		Gr. B	t≦25 25<t≦50 50<t	≦0.20 ≦0.23 ≦0.25	0.15～0.40	1.15～1.50	≦0.035	≦0.035	–	–	–
	A533M-93	TypeA-1 TypeA-3	–	≦0.25	0.15～0.30	1.15～1.50	≦0.035	≦0.035	–	–	–
UK英	–	–	–	–	–	–	–	–	–	–	–
GR独	–	–	–	–	–	–	–	–	–	–	–
FR仏	NF A36-206-83	18MD 4.05	–	≦0.20	0.15～0.35	0.15～0.35	≦0.030	≦0.030	–	≦0.30	–
		15MDV 4.05	–	≦0.18	0.15～0.35	0.15～0.35	≦0.030	≦0.030	–	≦0.30	–
URソ	–	–	–	–	–	–	–	–	–	–	–

Mn-1/2 Mo-1/4 Ni Steel

Country 国	Standard No. 規格番号及び年号	Desig-nation 種類記号	Thick-ness(t) mm板厚	C	Si	Mn	P	S	Cu	Cr	Ni
JP日	–	–	–	–	–	–	–	–	–	–	–

$1N/mm^2$ or $MPa = 1.01972 \times 10^{-1} kgf/mm^2$
$1J = 1.01972 \times 10^{-1} kg\text{-}m$

Mo	V	Others	Mechanical Properties 機械的性質 Heat Treatment 熱処理	Thickness (t)mm 板厚	Yield Point min. N/mm²or MPa (kgf/mm²) 最小降伏点	Tensile Strength N/mm²or MPa (kgf/mm²) 引張強さ	Notch Toughness 衝撃値 Test Piece 試験片	Test Temp.(℃) 試験温度	Absorbed Energy J (min.kg-m) 吸収ｴﾈﾙｷﾞｰ	Index No. 索引番号
0.45~0.60	–	B 0.001~0.005	N	t ≦25 25<t ≦50 50<t	46 46 43	560~670	2V	RT 常温	39 27.5 20.5	B011
–	–	–	–	–	–	–	–	–	–	
–	–	–	–	–	–	–	–	–	–	
–	–	–	–	–	–	–	–	–	–	
0.45~0.60	–	–	t >50mm N	–	315 (32)	520~615 (53~67)	–	–	–	B012
0.45~0.60	–	–	t >50mm N	–	345 (35)	550~690 (56~70)	–	–	–	
0.45~0.60	–	–	QT	–	345 (35)	550~690 (56~70)	2V	by agreement 協議	40 (4.1)	
0.45~0.60	–	–	QT	–	480 (49)	620~790 (63~81)	2V	by agreement 協議	47 (4.8)	
0.45~0.60	–	–	t >50mm N	–	310	515~655	–	–	–	
0.45~0.60	–	–	t >50mm N	–	345	550~690	–	–	–	
0.45~0.60	–	–	QT	–	345 570	550~690 690~860	–	–	–	
–	–	–	–	–	–	–	–	–	–	
–	–	–	–	–	–	–	–	–	–	
0.40~0.60	≦0.04	–	N+T	t ≦60 60<t ≦80 80<t	345 345 345	510~650	–	–	–	
0.40~0.60	0.04~0.08	–	N+T	t ≦60 60<t ≦80 80<t	345 345 325	510~650	–	–	–	
–	–	–	–	–	–	–	–	–	–	
–	–	–	–	–	–	–	–	–	–	B013

Country 国	Standard No. 規格番号及び年号	Designation 種類記号	Chemical Composition 化学成分 (%) Thickness(t) mm板厚	C	Si	Mn	P	S	Cu	Cr	Ni
JP日	ASTM A533M-87	TypeD-1 TypeD-2 TypeD-3	–	≦0.25	0.15～0.40	1.15～1.50	≦0.035	≦0.040	–	–	0.20～0.40
UK英	–	–	–	–	–	–	–	–	–	–	–
GR独	–	–	–	–	–	–	–	–	–	–	–
FR仏	–	–	–	–	–	–	–	–	–	–	–
URソ	–	–	–	–	–	–	–	–	–	–	–
Mn-1/2 Mo-1/2 Ni Steel											
JP日	JIS G3119-87	SBV2	t ≦25 25<t ≦50 50<t ≦100	≦0.20 ≦0.23 ≦0.25	0.15～0.35	1.15～1.50	≦0.035	≦0.040	–	–	0.40～0.70
	JIS G3120-87	SQV2A	–	≦0.25	0.15～0.30	1.15～1.50	≦0.035	≦0.040	–	–	0.40～0.70
		SQV2B	–	≦0.25	0.15～0.30	1.15～1.50	≦0.035	≦0.040	–	–	0.40～0.70
US米	ASTM A302M-93	Gr. C	t ≦25 25<t ≦50 50<t	≦0.20 ≦0.23 ≦0.25	0.15～0.40	1.15～1.50	≦0.035	≦0.035	–	–	0.40～0.70
	ASTM A533M-93	Type B-1 Type B-2 Type B-3	–	≦0.25	0.15～0.40	1.15～1.50	≦0.035	≦0.035	–	–	0.40～0.70
UK英	–	–	–	–	–	–	–	–	–	–	–
GR独	–	–	–	–	–	–	–	–	–	–	–
FR仏	–	–	–	–	–	–	–	–	–	–	–
URソ	–	–	–	–	–	–	–	–	–	–	–
Mn-1/2 Mo-3/4 Ni Steel											
JP日	JIS G3119-87	SBV3	t ≦25 25<t ≦50 50<t ≦150	≦0.20 ≦0.23 ≦0.25	0.15～0.30	1.15～1.50	≦0.035	≦0.040	–	–	0.70～1.00
	JIS G3120-87	SQV3A	–	≦0.25	0.15～0.30	1.15～1.50	≦0.035	≦0.040	–	–	0.70～1.00
		SQV3B	–	≦0.25	0.15～0.30	1.15～1.50	≦0.035	≦0.040	–	–	0.70～1.00
US米	ASTM A302M-93	Gr. D	t ≦25 25<t ≦50 50<t	≦0.20 ≦0.23 ≦0.25	0.15～0.40	1.15～1.50	≦0.035	≦0.035	–	–	0.70～1.00

1N/mm² or MPa$=1.01972\times10^{-1}$kgf/mm²
1J$=1.01972\times10^{-1}$kg-m

Mo	V	Others	Heat Treatment 熱処理	Thickness (t)mm 板厚	Yield Point min. N/mm² or MPa (kgf/mm²) 最小降伏点	Tensile Strength N/mm² or MPa (kgf/mm²) 引張強さ	Notctch Toughness 衝撃値 Test Piece 試験片	Test Temp.(℃) 試験温度	Absorbed Energy J (min. kg-m) 吸収ｴﾈﾙｷﾞｰ	Index No. 索引番号
0.45～0.60	–	–	QT	345	550～690	–	–	–	–	B013
				485	620～795					
				570	690～850					
–	–	–	–	–	–	–	–	–	–	
–	–	–	–	–	–	–	–	–	–	
–	–	–	–	–	–	–	–	–	–	
–	–	–	–	–	–	–	–	–	–	
0.45～0.60	–	–	t >50mm N	–	345 (35)	550～690 (56～70)	–	–	–	B014
0.45～0.60	–	–	QT	–	345 (35)	550～690 (56～70)	–	–	–	
0.45～0.60	–	–	QT	–	480 (49)	620～790 (63～81)	–	–	–	
0.45～0.60	–	–	t >50mm N	–	345	550～690	–	–	–	
0.45～0.60	–	–	QT	–	345	550～690	–	–	–	
					485	620～795				
					570	690～860				
–	–	–	–	–	–	–	–	–	–	
–	–	–	–	–	–	–	–	–	–	
–	–	–	–	–	–	–	–	–	–	
–	–	–	–	–	–	–	–	–	–	
0.45～0.60	–	–	t >50mm N	–	345 (35)	550～690 (56～70)	–	–	–	B015
0.45～0.60	–	–	QT	–	345 (35)	550～690 (56～70)	2 V	by agreement 協議	40 (4.1)	
0.45～0.60	–	–	QT	–	480 (49)	620～790 (63～81)	2 V	by agreement 協議	47 (4.8)	
0.45～0.60	–	–	t >50mm N	–	345	550～690	–	–	–	

Country 国	Standard No. 規格番号 及び年号	Designation 種類記号	Chemical Composition 化学成分 (%)								
			Thick-ness(t) mm板厚	C	Si	Mn	P	S	Cu	Cr	Ni
US米	ASTM A533M-93	Type C-1 Type C-2 Type C-3	–	≦0.25	0.15~0.40	1.15~1.50	≦0.035	≦0.035	–	–	0.70~1.00
UK英	–	–	–	–	–	–	–	–	–	–	–
GR独	–	–	–	–	–	–	–	–	–	–	–
FR仏	–	–	–	–	–	–	–	–	–	–	–
URソ	–	–	–	–	–	–	–	–	–	–	–
3/4 Cr-1/2 Mo Steel											
JP日	JIS G4109-87	SCMV1-1 SCMV1-2	6~200	≦0.21	≦0.40	0.55~0.80	≦0.030	≦0.030	–	0.50~0.80	–
US米	ASTM A387M-92	Gr.2CL.1 Gr.2CL.2	–	0.05~0.21	0.15~0.40	0.55~0.80	≦0.035	≦0.040	–	0.50~0.60	–
UK英	–	–	–	–	–	–	–	–	–	–	–
GR独	–	–	–	–	–	–	–	–	–	–	–
FR仏	NF A36-206-83	15CD205	–	≦0.18	0.15~0.30	0.50~0.90	≦0.030	≦0.030	–	0.40~0.60	–
URソ	–	–	–	–	–	–	–	–	–	–	–
1 Cr-1/2 Mo Steel											
JP日	JIS G4109-87	SCMV2-1 SCMV2-2	6~200	≦0.17	≦0.040	0.40~0.65	≦0.030	≦0.030	–	0.80~1.15 0.80~1.15	–
US米	ASTM A387-92	Gr.12 CL.1 Gr.12 CL.2	–	0.05~0.17	0.15~0.40	0.40~0.65	≦0.035	≦0.035	–	0.80~1.15	–
UK英	BS1501 Part2-70	620Gr. 20 Superseded by European Standard EN 10028-2, Grade 13 CrMo 4-5, Material No.1.7335	≦150	0.09~0.15	0.10~0.40	0.40~0.70	≦0.040	≦0.040	≦0.30	0.70~1.20	≦0.30
		620Gr. 31 No longer active	≦150	0.12~0.18	0.10~0.40	0.40~0.70	≦0.040	≦0.040	≦0.30	0.70~1.20	≦0.30
GR独	DIN 17155-83	13CrMo44 Superseded by European Standard EN 10028-2, Grade 13 CrMo 4-5, Material No.1.7335	–	0.08~0.18	0.10~0.35	0.40~1.00	≦0.035	≦0.030	≦0.30	0.70~1.20	–
FR仏	NF 17155-83	A36-206	–	≦0.18	0.15~0.35	0.40~0.80	≦0.030	≦0.030	–	0.80~1.20	–

$1N/mm^2$ or $MPa = 1.01972 \times 10^{-1} kgf/mm^2$
$1J = 1.01972 \times 10^{-1} kg\text{-}m$

			Mechanical Properties 機械的性質							Index No. 索引番号
			Heat Treatment 熱処理	Thickness (t)mm 板厚	Yield Point min. N/mm²or MPa (kgf/mm²) 最小降伏点	Tensile Strength N/mm²or MPa (kgf/mm²) 引張強さ	Notch Toughness 衝撃値			
Mo	V	Others					Test Piece 試験片	Test Temp.(℃) 試験温度	Absorbed Energy J (min. kg-m) 吸収ｴﾈﾙｷﾞｰ	
0.45~0.60	–	–	QT	–	345	550~690	–	–	–	B015
					485	620~795				
					570	690~860				
–	–	–	–	–	–	–	–	–	–	
–	–	–	–	–	–	–	–	–	–	
–	–	–	–	–	–	–	–	–	–	
–	–	–	–	–	–	–	–	–	–	
0.45~0.60	–	–	A N+T		225(22)	380~550 (39~56)	–	–	–	B016
			N+T		315(32)	480~620 (49~63)				
0.45~0.60	–	–	A N+T	–	230 310	380~550 485~620	–	–	–	
–	–	–	–	–	–	–	–	–	–	
–	–	–	–	–	–	–	–	–	–	
0.40~0.60	≦0.04	N+T	–	≦60 60<t≦80 ≦80	275 265 255		–	–	–	
–	–	–	–	–	–	–	–	–	–	
0.45~0.60	–	–	A N+T	–	225(23)	380~550 (39~56)	–	–	–	B017
			N+T		275(28)	480~620 (46~60)				
0.45~0.60	–	–	A N+T	–	230 275	380~550 450~585	–	–	–	
0.45~0.65	–	Sn≦0.03	N+T	t<75 t≧75	345 315	(49~61) (46~58)	–	–	–	
0.45~0.65	–	Sn≦0.03	N+T	t<75 t≧75	345 315	(49~61) (46~58)	–	–	–	
0.40~0.65	–	–	N+T	t≦16 16<t≦40 40<t≦60	305 295 295	t≦60 440~590 60<t≦100 430~580 100<t≦150 420~570	2V	20	t≦60 31.5 60<t≦150 27.5	
0.40~0.60	≦0.04	–	N+T	t≦60 60<t≦80 80<t	295 285 275	470~600	–	–	–	

Country 国	Standard No. 規格番号及び年号	Designation 種類記号	Chemical Composition 化学成分 (%) Thickness(t) mm板厚	C	Si	Mn	P	S	Cu	Cr	Ni
URソ	–	–	–	–	–	–	–	–	–	–	–

$1\frac{1}{4}$ Cr-1/2 Mo Steel

Country	Standard	Designation	Thickness	C	Si	Mn	P	S	Cu	Cr	Ni
JP日	JIS G4109-87	SCMV3-1 SCMV3-2	6~200	≦0.17	0.50~0.80	0.40~0.65	≦0.030	≦0.030	–	1.00~1.50	–
US米	ASTM A387M-92	Gr.11 CL.1 Gr.11,CL.2	–	0.05~0.17	0.50~0.80	0.40~0.65	≦0.035	≦0.035	–	1.00~1.50	–
UK英	BS1501 Part2-70	621 Active in UK only	≦1.50	0.19~0.15	0.15~0.35	0.40~0.70	≦0.040	≦0.040	≦0.30	1.00~1.50	≦0.30
GR独	–	–	–	–	–	–	–	–	–	–	–
FR仏	–	–	–	–	–	–	–	–	–	–	–
URソ	–	–	–	–	–	–	–	–	–	–	–

$2\frac{1}{4}$ Cr-1 Mo Steel

Country	Standard	Designation	Thickness	C	Si	Mn	P	S	Cu	Cr	Ni
JP日	JIS G4109-87	SCMV4-1 SCMV4-2	6~300	≦0.17	≦0.50	0.30~0.60	≦0.030	≦0.030	–	2.00~2.50	–
US米	ASTM A387-92	Gr.22CL.1 Gr.22CL.2	–	0.05~0.15	≦0.50	0.30~0.60	≦0.035	≦0.035	–	2.00~2.50	–
		Gr.22LCL.1	–	≦0.10	≦0.50	0.30~0.60	≦0.035	≦0.035	–	2.00~2.50	–
UK英	BS1501 Part2-70	622Gr.31	≦1.50	0.10~0.15	0.20~0.50	0.40~0.80	≦0.040	≦0.040	≦0.30	2.00~2.50	≦0.30
	Superseded by European Standard EN 10028-2 Grade 11 Cr Mo 9-10, Material No. 1.7383										
		622Gr.45	≦1.50	0.10~0.15	0.20~0.50	0.40~0.80	≦0.040	≦0.040	≦0.30	2.00~2.50	≦0.30
	Superseded by European Standard EN 10028-2 Grade 10 Cr Mo 9-10, Material No. 1.7380										
GR独	–	–	–	–	–	–	–	–	–	–	–
FR仏	NF A36-206-83	10CD9.10	–	≦0.15	0.15~0.35	0.40~0.80	≦0.030	≦0.030	–	2.00~2.50	–
URソ	–	–	–	–	–	–	–	–	–	–	–

3 Cr-1 Mo Steel

Country	Standard	Designation	Thickness	C	Si	Mn	P	S	Cu	Cr	Ni
JP日	JIS G4109-87	SCMV5-1 SCMV5-2	6~300	≦0.17	≦0.50	0.30~0.60	≦0.030	≦0.030	–	2.75~3.25	–
US米	ASTM A387M-92	Gr.21CL.1 Gr.21CL.2	–	0.05~0.15	≦0.50	0.30~0.60	≦0.035	≦0.035	–	2.75~3.25	–
	ASTM A387-92	Gr.21L CL.1		≦0.10	≦0.50	0.30~0.60	≦0.035	≦0.035	–	2.00~2.50	–
UK英	–	–	–	–	–	–	–	–	–	–	–
GR独			–	–	–	–	–	–	–	–	–
FR仏	–	–	–	–	–	–	–	–	–	–	–
URソ	–	–	–	–	–	–	–	–	–	–	–

$1 N/mm^2$ or $MPa = 1.01972 \times 10^{-1} kgf/mm^2$
$1 J = 1.01972 \times 10^{-1} kg\text{-}m$

			Mechanical Properties 機械的性質							Index No.
			Heat Treatment 熱処理	Thickness (t)mm 板厚	Yield Point min. N/mm²or MPa (kgf/mm²) 最小降伏点	Tensile Strength N/mm²or MPa (kgf/mm²) 引張強さ	Notch Toughness 衝撃値			
Mo	V	Others					Test Piece 試験片	Test Temp.(℃) 試験温度	Absorbed Energy J (min. kg-m) 吸収ｴﾈﾙｷﾞｰ	索引番号
–	–	–	–	–	–	–	–	–	–	B024
0.45～0.65	–	–	A N+T	–	235(24)	410～590 (42～60)	–	–	–	B018
			N+T		315(32)	520～690 (53～70)				
0.45～0.65	–	–	A N+T	–	240 310	415～585 515～690	–	–	–	
0.45～0.65	–	–	N+T	≦75 >75	345 315	480～600 450～570	–	–	–	
–	–	–	–	–	–	–	–	–	–	
–	–	–	–	–	–	–	–	–	–	
–	–	–	–	–	–	–	–	–	–	
0.90～1.10	–	–	A N+T	–	205(21)	410～590 (42～60)	–	–	–	B019
			N+T		315(32)	520～690 (53～70)				
0.90～1.10	–	–	A N+T	–	205 310	415～585 515～690	–	–	–	
0.90～1.10	–	–	A N+T	–	205	415～485	–	–	–	
0.90～1.20	–	Sn≧0.03	N+T	–	280	480～600	–	–	–	
0.90～1.20	–	Sn≧0.03	N+T	–	555	700～825	–	–	–	
–	–	–	–	–	–	–	–	–	–	
0.90～1.10	≦0.04	–	N+T	305	315	520～670	–	–	–	
–	–	–	–	–	–	–	–	–	–	
0.90～1.10	–	–	A N+T	–	205(21)	410～590 (42～60)	–	–	–	B020
			N+T		315(32)	520～690 (53～70)				
0.90～1.10	–	–	A N+T	–	205 310	415～585 515～690	–	–	–	
0.90～1.10	–	–	A N+T	–	205	415～485	–	–	–	
–	–	–	–	–	–	–	–	–	–	
–	–	–	–	–	–	–	–	–	–	
–	–	–	–	–	–	–	–	–	–	
–	–	–	–	–	–	–	–	–	–	

Country 国	Standard No. 規格番号及び年号	Designation 種類記号	Chemical Composition 化学成分（%）								
			Thickness(t) mm板厚	C	Si	Mn	P	S	Cu	Cr	Ni
5 Cr-1/2 Mo Steel											
JP日	JIS G4109-87	SCMV6-1	6~300	≦0.15	≦0.50	0.30～0.60	≦0.030	≦0.030	–	4.00～6.00	–
		SCMV6-2									
US米	ASTM A387M-92	GR. 5	–	≦0.15	≦0.50	0.30～0.60	≦0.035	≦0.030	–	4.00～6.00	–
UK英	BS1501 Part2-70	625 No longer active	–	≦1.50	≦0.50	0.30～0.70	≦0.045	≦0.045	≦0.40	4.00～6.00	≦0.40
GR独	–	–	–	–	–	–	–	–	–	–	–
FR仏	NF A36-206-83	Z10CD505	–	≦0.15	≦0.50	0.30～0.60	≦0.030	≦0.030	–	4.0～6.0	–
URソ	ГОСТ 5632-72	15X5	–	≦0.15	≦0.5	≦0.5	≦0.030	≦0.025	–	4.5～6.0	–
		15X5M	–	≦0.15	≦0.5	≦0.5	≦0.030	≦0.025	–	4.5～6.0	–
		15X5Bφ	–	≦0.15	0.3～0.6	≦0.5	≦0.030	≦0.025	–	4.5～6.0	–
9 Cr-1 Mo Steel											
JP日	–	–	–	–	–	–	–	–	–	–	–
US米	ASTM A387M-92	Gr. 9	–	0.05～0.15	≦1.00	0.30～0.60	≦0.030	≦0.030	–	8.00～10.00	–
		Gr.91 CL.2	–	0.08～0.12	0.20～0.50	0.30～0.60	≦0.020	≦0.010	–	8.00～9.50	≦0.40
UK英	–	–	–	–	–	–	–	–	–	–	–
GR独	–	–	–	–	–	–	–	–	–	–	–
FR仏	–	–	–	–	–	–	–	–	–	–	–
URソ	–	–	–	–	–	–	–	–	–	–	–
Ni-Cr-Mo Steel											
JP日	–	–	–	–	–	–	–	–	–	–	–

1N/mm² or MPa = 1.01972 × 10⁻¹kgf/mm²
1J = 1.01972 × 10⁻¹kg-m

			Mechanical Properties 機械的性質							Index No. 索引番号
			Heat Treatment 熱処理	Thickness (t)mm 板厚	Yield Point min. N/mm²or MPa (kgf/mm²) 最小降伏点	Tensile Strength N/mm²or MPa (kgf/mm²) 引張強さ	Notch Toughness 衝撃値			
Mo	V	Others					Test Piece 試験片	Test Temp.(℃) 試験温度	Absorbed Energy J (min. kg-m) 吸収ｴﾈﾙｷﾞｰ	
0.45~0.65	–	–	A N+T	–	205(21)	410~590 (42~60)	–	–	–	B021
			N+T		315(32)	520~690 (53~70)				
0.45~0.65	–	–	A N+T	–	310	515~690	–	–	–	
0.45~0.65	–	–	A	–	(22)	≧(43)	–	–	–	
–	–	–	–	–	–	–	–	–	–	
0.45~0.65	≦0.04	–	N+T	Class 1	305	520~680	–	–	–	
				Class 2	335	590~745				
–	–	–	–	–	–	–	–	–	–	
0.45~0.65	–	–	–	–	–	–	–	–	–	
–	0.4~0.6	W0.4~0.7	–	–	–	–	–	–	–	
–	–	–	–	–	–	–	–	–	–	B023
0.90~1.10	–	–	A N+T	–	310	515~690	–	–	–	
0.85~1.05	0.18~0.25	Cb0.06~0.10 N0.03~0.070 Al≦0.04	A N+T	–	415	585~760	–	–	–	
–	–	–	–	–	–	–	–	–	–	
–	–	–	–	–	–	–	–	–	–	
–	–	–	–	–	–	–	–	–	–	
–	–	–	–	–	–	–	–	–	–	
–	–	–	–	–	–	–	–	–	–	B024

Country 国	Standard No. 規格番号及び年号	Designation 種類記号	Thickness(t) mm板厚	C	Si	Mn	P	S	Cu	Cr	Ni
				Chemical Composition 化学成分 (%)							
US米	ASTM A543M-93	Tp.B-Cl.1 Tp.B-Cl.2 Tp.B-Cl.3	≧50	≦0.23	0.20~0.40	≦0.40	≦0.020	≦0.020	–	1.50~2.00	t ≦100 2.60~3.25 100<t 3.00~4.00
		Tp.C-Cl.1 Tp.C-Cl.2 Tp.C-Cl.3	≧50	≦0.23	0.20~0.40	≦0.40	≦0.020	≦0.020	–	1.20~1.80	t ≦100 2.60~3.25
UK英	BS1501 Part2-70	Gr.281 Active in U.K. only	≦150	0.90~0.15	≦0.40	0.90~1.30	≦0.040	≦0.040	≦0.30	0.40~0.70	0.70~1.00
		Gr.282 No longer active	≦150	0.12~0.17	≦0.40	0.90~1.30	≦0.040	≦0.040	≦0.30	0.30~0.70	1.40~1.50
GR独	–	–	–	–	–	–	–	–	–	–	–
FR仏	–	–	–	–	–	–	–	–	–	–	–
URソ	–	–	–	–	–	–	–	–	–	–	–

Mn-Cr-Mo Steel

Country	Standard No.	Designation	Thickness	C	Si	Mn	P	S	Cu	Cr	Ni
JP日	–	–	–	–	–	–	–	–	–	–	–
US米	–	–	–	–	–	–	–	–	–	–	–
UK英	BS1501 Part2-70	Gr.271	≦150	0.11~0.27	≦0.40	1.00~1.50	≦0.040	≦0.040	≦0.30	0.40~0.70	≦0.7
GR独	–	–	–	–	–	–	–	–	–	–	–
FR仏	–	–	–	–	–	–	–	–	–	–	–
URソ	–	–	–	–	–	–	–	–	–	–	–

Mn-Cr-Si Steel

Country	Standard No.	Designation	Thickness	C	Si	Mn	P	S	Cu	Cr	Ni
JP日	–	–	–	–	–	–	–	–	–	–	–
US米	ASTM A202M-93	Gr.A	–	≦0.17	0.60~0.90	1.05~1.40	≦0.035	≦0.035	–	0.35~0.60	–
		Gr.B		≦0.25							
UK英	–	–	–	–	–	–	–	–	–	–	–
GR独	–	–	–	–	–	–	–	–	–	–	–
FR仏	–	–	–	–	–	–	–	–	–	–	–
URソ	–	–	–	–	–	–	–	–	–	–	–

Mn-V Steel

Country	Standard No.	Designation	Thickness	C	Si	Mn	P	S	Cu	Cr	Ni
JP日	–	–	–	–	–	–	–	–	–	–	–

$1N/mm^2$ or $MPa = 1.01972 \times 10^{-1} kgf/mm^2$
$1J = 1.01972 \times 10^{-1} kg\text{-}m$

Mo	V	Others	Heat Treatment 熱処理	Thickness (t)mm 板 厚	Yield Point min. N/mm² or MPa (kgf/mm²) 最小降伏点	Tensile Strength N/mm² or MPa (kgf/mm²) 引張強さ	Notctch Toughness 衝撃値: Test Piece 試験片	Test Temp. (℃) 試験温度	Absorbed Energy J (min. kg-m) 吸収エネルギー	Index No. 索引番号
					585	725～860				
0.45～0.60	≦0.03	－	QT	－	690	795～930	－	－	－	B024
					485	620～795				
					585	725～860				
0.45～0.60	≦0.03	－	QT	－	690	795～930	－	－	－	
					485	620～795				
		Nb≦0.10		t≦9.4	460	590～700				
0.20～0.28	0.40～0.12		NT	9.4<t≦25	460	590～680	5U	RT	44	
		N≦0.015		25<t≦75	420	560～680	2V	-10	68	
				75<t	380	560～680				
				t≦9.4	490	590～710				
0.30～0.40	0.08～0.12	－	NT	9.4<t≦25	480	590～710	5U	RT	34	
				25<t≦75	450	590～710	2V	-10	68	
				75<t	420	570～700				
－	－	－	－	－	－	－	－	－	－	
－	－	－	－	－	－	－	－	－	－	
－	－	－	－	－	－	－	－	－	－	
－	－	－	－	－	－	－	－	－	－	B025
－	－	－	－	－	－	－	－	－	－	
				t≦9.4	460	590～700				
0.20～0.28	0.04～0.12	－	NT	9.4<t≦25	460	590～680	5U	RT	24.5	
				25<t≦75	420	560～680	2V	0	24.5	
				75<t	380	560～680				
－	－	－	－	－	－	－	－	－	－	
－	－	－	－	－	－	－	－	－	－	
－	－	－	－	－	－	－	－	－	－	
－	－	－	－	－	－	－	－	－	－	B026
－	－	－	－	－	310	515～655	－	－	－	
					325	585～760				
－	－	－	－	－	－	－	－	－	－	
－	－	－	－	－	－	－	－	－	－	
－	－	－	－	－	－	－	－	－	－	
－	－	－	－	－	－	－	－	－	－	
－	－	－	－	－	－	－	－	－	－	B027

Country 国	Standard No. 規格番号及び年号	Designation 種類記号	Chemical Composition 化学成分 (%)								
			Thickness(t) mm板厚	C	Si	Mn	P	S	Cu	Cr	Ni
US米	ASTM A225M-93	Gr. C	–	≦0.25	0.15～0.40	≦1.60	≦0.035	≦0.040	–	–	–
		Gr. D	15～50	≦0.20	0.10～0.50	≦1.70	≦0.035	≦0.035	–	–	–
UK英	–	–	–	–	–	–	–	–	–	–	–
GR独	–	–	–	–	–	–	–	–	–	–	–
FR仏	–	–	–	–	–	–	–	–	–	–	–
URソ	–	–	–	–	–	–	–	–	–	–	–

1N/mm² or MPa = 1.01972×10^{-1} kgf/mm²
1J = 1.01972×10^{-1} kg-m

Mo	V	Others	Mechanical Properties 機械的性質							Index No. 索引番号
			Heat Treatment 熱処理	Thickness (t)mm 板厚	Yield Point min. N/mm² or MPa (kgf/mm²) 最小降伏点	Tensile Strength N/mm² or MPa (kgf/mm²) 引張強さ	Notctch Toughness 衝撃値			
							Test Piece 試験片	Test Temp. (℃) 試験温度	Absorbed Energy J (min. kg-m) 吸収ｴﾈﾙｷﾞｰ	
–	0.13～0.18	–	–	–	485	725～935	–	–	–	B027
–	0.10～0.18	–	–	t≦75	380	525～725	–	–	–	
				75<t	415	515～690				
–	–	–	–	–	–	–	–	–	–	
–	–	–	–	–	–	–	–	–	–	
–	–	–	–	–	–	–	–	–	–	
–	–	–	–	–	–	–	–	–	–	

4. Quenched and Tempered High Strength Plates 調質高張力鋼板

Country 国	Standard No. 規格番号及び年号	Designation 種類記号	Chemical Composition 化学成分（%） Thickness(t) mm板厚	C	Si	Mn	P	S	Cu	Cr	Ni
JP日	JIS G3115-90	SPV450 (SPV46)	6～75	≦0.18	0.15～0.75	≦1.60	≦0.030	≦0.030	t≦50mm *Ceq≦0.44		
		SPV490 (SPV50)	6～75	≦0.18	0.15～0.75	≦1.60	≦0.030	≦0.030	t≦50mm *Ceq≦0.45		
US米	ASTM A517M-93a	Gr. A	≦32	0.15～0.21	0.40～0.80	0.80～1.10	P≦0.035 S≦0.035	Zr0.05～0.15	−	0.50～0.80	−
		Gr. B	≦32	0.15～0.21	0.20～0.35	0.70～1.10		Ti0.01～0.03	−	0.50～0.65	−
		Gr. C	≦32	0.10～0.20	0.15～0.30	0.80～1.10		−	−	−	−
		Gr. D	≦32	0.13～0.20	0.20～0.35	0.40～0.70		Ti0.04～0.10	0.20～0.40	0.85～1.20	−
		Gr. E	≦150	0.12～0.20	0.20～0.35	0.40～0.70		Ti0.04～0.10	0.20～0.40	1.40～2.00	−
		Gr. F	≦65	0.10～0.20	0.15～0.35	0.60～1.00		−	0.15～0.50	0.40～0.65	0.70～1.00
		Gr. H	≦50	0.12～0.21	0.20～0.35	0.95～1.30		−	−	0.40～0.65	0.30～0.70
		Gr. J	≦32	0.12～0.21	0.20～0.35	0.45～0.70		−	−	−	−
		Gr. K	≦50	0.10～0.20	0.15～0.30	1.10～1.50		−	−	−	−
		Gr. M	≦50	0.12～0.21	0.20～0.35	0.45～0.70		−	−	−	1.20～1.50
		Gr. P	≦100	0.12～0.21	0.20～0.35	0.45～0.70		−	−	0.85～1.20	1.20～1.50
		Gr. Q	≦150	0.14～0.21	0.15～0.35	0.45～1.30		−	−	1.00～1.50	1.20～1.50
		Gr. S	≦50	≦0.10	≦0.15	1.10～1.50		Ti≦0.06	−	−	−
		Gr. T	≦50	0.08～0.14	0.40～0.60	1.20～1.50		−	−	−	−
	ASTM A537M-91	Cl.2	≦150	≦0.24	0.15～0.50	t≦40 0.70～1.35 40<t 1.00～1.60	≦0.035	≦0.035	≦0.35	≦0.25	≦0.25
		Cl.3	≦150								
	ASTM A724M-90	Gr. A	≦22	≦0.18	≦0.55	1.00～1.60	≦0.035	≦0.035	≦0.35	≦0.25	≦0.25
		Gr. B	≦22	≦0.20	≦0.50	1.00～1.60					
		Gr. C	≦50	≦0.22	0.20～0.60	1.10～1.60					
	ASTM A734-87a	Type A	−	≦0.12	≦0.40	0.45～0.75	≦0.035	≦0.015	−	0.90～1.20	0.90～1.20
		Type B	−	≦0.17	≦0.35	≦1.60	≦0.035	≦0.015	≦0.35	≦0.25	−
UK英	−	−	−	−	−	−	−	−	−	−	−
GR独	−	−	−	−	−	−	−	−	−	−	−
FR仏	−	−	−	−	−	−	−	−	−	−	−
URソ	−	−	−	−	−	−	−	−	−	−	−

1N/mm² or MPa = 1.01972 × 10⁻¹ kgf/mm²
1J = 1.01972 × 10⁻¹ kg-m

Mo	V	Others	Heat Treatment 熱処理	Thickness (t)mm 板厚	Yield Point min. N/mm² or MPa (kgf/mm²) 最小降伏点	Tensile Strength N/mm² or MPa (kgf/mm²) 引張強さ	Notch Toughness 衝撃値: Test Piece 試験片	Notch Toughness 衝撃値: Test Temp. (℃) 試験温度	Notch Toughness 衝撃値: Absorbed Energy J (min. kg-m) 吸収エネルギー	Index No. 索引番号
50 < t ≦ 75mm * Ceq ≦ 0.46			QT	6<t≦50	450(46)	570~700	2V	-10	47(4.8)	B028
				50<t≦75	430(44)	(58~71)				
50 < t ≦ 75mm * Ceq ≦ 0.47				6<t≦50	490(50)	610~740	2V	-10	47(4.8)	
				50<t≦75	470(48)	(62~75)				
0.18~0.28	–	≦0.025	QT	t ≦65 65<t≦150	690 620	795~930 725~930	–	–	–	
0.15~0.25	0.03~0.08	0.0005~0.005								
0.20~0.30	–	0.001~0.005								
0.15~0.25	–	0.0015~0.005								
0.40~0.60	–	0.0015~0.005								
0.40~0.60	–	0.0005~0.006								
0.20~0.30	0.03~0.08	≦0.0005								
0.50~0.65	–	0.001~0.005								
0.45~0.55	–	–								
0.45~0.60	–	0.001~0.005								
0.45~0.60	–	0.001~0.005								
0.45~0.60	0.03~0.08	–								
0.10~0.35	–	–								
0.45~0.60	0.03~0.08	0.01~0.005								
≦0.08	–	–	QT	t ≦65	415	550~690	–	–	–	
				65<t≦100	380	515~655				
				100<t≦150	315	485~620				
			QT	t ≦65	380	550~690				
				65<t≦100	345	515~655				
				100<t≦150	275	485~620				
≦0.08	≦0.08	–	QT	–	485	620~760	–	–	–	
		–			515	655~795				
		B ≦0.005			485	620~760				
0.25~0.40	–	Al ≦0.06	QT	–	450	530~670	–	–	–	
Al ≦0.06	V ≦0.011	N ≦0.030	QT	–	450	530~670	–	–	–	
–	–	–	–	–	–	–	–	–	–	
–	–	–	–	–	–	–	–	–	–	
–	–	–	–	–	–	–	–	–	–	
–	–	–	–	–	–	–	–	–	–	

$$*\,Ceq = C + \frac{Mn}{6} + \frac{Si}{24} + \frac{Ni}{40} + \frac{Cr}{5} + \frac{Mo}{4} + \frac{V}{14}$$

5. Steel Plates for Low Temperature Service 低温用鋼板

Country 国	Standard No. 規格番号及び年号	Designation 種類記号	Chemical Composition 化学成分 (%) Thickness(t) mm板厚	C	Si	Mn	P	S	Cu	Cr	Ni
JP日	JIS G3126-90	SLA235A (SLA24A)	6～50	≦0.15	0.15～0.30	0.70～1.50	≦0.030	≦0.025	–	–	–
		SLA235B (SLA24B)	6～50	≦0.15	0.15～0.30	0.70～1.50	≦0.030	≦0.025	–	–	–
		SLA325A (SLA33A)	6～32	≦0.16	0.15～0.55	0.80～1.60	≦0.030	≦0.025	–	–	–
		SLA325B (SLA33B)	6～32	≦0.16	0.15～0.55	0.80～1.60	≦0.030	≦0.025	–	–	–
		SLA360 SLA(37)	6～32	≦0.18	0.15～0.55	≦0.030	≦0.030	≦0.025	–	–	–
US※	ASTM A612M-90	–	t ≦20	≦0.25	0.15～0.40	1.00～1.35	≦0.035	≦0.035	≦0.35	≦0.25	≦0.25
			20<t ≦20	≦0.25	0.15～0.50	1.00～1.50					

$1N/mm^2$ or $MPa = 1.01972 \times 10^{-1} kgf/mm^2$
$1J = 1.01972 \times 10^{-1} kg\text{-}m$

Mo	V	Others	Mechanical Properties 機械的性質 Heat Treatment 熱処理	Thickness (t)mm 板厚	Yield Point min. N/mm²or MPa (kgf/mm²) 最小降伏点	Tensile Strength N/mm²or MPa (kgf/mm²) 引張強さ	Notch Toughness 衝撃値 Test Piece 試験片	Test Temp.(℃) 試験温度	Absorbed Energy J (min. kg-m) 吸収エネルギー	Index No. 索引番号
–	–	–	N	t ≦40mm	235(24)	400～510 (41～52)	2 V	6≦t≦20 -5 20<t -10	½Emax	B 029
				40mm<t	215(22)					
–	–	–	N	t ≦40mm	235(24)	400～510 (41～52)	2 V	6≦t<8.5 -30 8.5≦t<12 -20 12≦t<20 -15 20<t -30	½Emax	
				40mm<t	215(22)					
–	–	–	N	–	325(33)	440～560 (45～57)	2 V	6≦t<8.5 -40 8.5≦t<12 -30 12≦t<20 -25 20<t -35	½Emax	
–	–	–	QT	–	325(33)	440～560 (45～57)	2 V	6≦t<8.5 -60 8.5≦t<12 -50 12≦t<20 -45 20<t -55	½Emax	
–	–	–	QT	–	360(37)	490～610 (50～62)	2 V	6≦t<8.5 -60 8.5≦t<12 -50 12≦t<20 -45 20<t -55	½Emax	
≦0.08	≦0.08	–	–	t ≦12.5	345	570～725	–	–	–	
				12.5<t ≦25	345	560～695				

Country 国	Standard No. 規格番号及び年号	Designation 種類記号	Chemical Composition 化学成分 (%)								
			Thickness(t) mm板厚	C	Si	Mn	P	S	Cu	Cr	Ni
US米	ASTM A662M-93	Gr. A	≦50	≦0.14	0.15～0.40	0.90～1.35	≦0.035	≦0.035	–	–	–
		Gr. B	≦50	≦0.19	0.15～0.40	0.85～1.50					
		Gr. C	≦50	≦0.20	0.15～0.50	1.00～1.60					
	ASTM A735M-90	Cl. 1 Cl. 2 Cl. 3 Cl. 4	–	≦0.06	≦0.40	t≦16 1.20～1.90 16<t 1.50～2.20	≦0.035	≦0.025	When Specified 0.20～0.35	–	–
	ASTM A736M-88	Gr. A Cl. 1 Cl. 2 Cl. 3	–	≦0.07	≦0.40	0.40～0.70	≦0.025	≦0.025	1.00～1.30	0.60～0.70	0.70～1.00
		Gr. C Cl. 1 Cl. 2	–	≦0.07	≦0.40	1.30～1.65	≦0.025	≦0.025	1.00～1.30	0.60～0.90	0.70～1.00
	ASTM A738-90a	Gr. A	–	≦0.24	0.15～0.50	t≦65mm ≦1.50 t>65mm ≦1.00	≦0.035	≦0.035	≦0.35	≦0.25	≦0.50
		Gr. B	≦65	≦0.20	0.15～0.50	t≦65mm 0.90～1.50 t>65mm –	≦0.030	≦0.040	≦0.35	≦0.30	≦0.60
		Gr. C	≦150	≦0.20	0.15～0.50	t≦65mm ≦1.50 t>65mm	≦0.020	≦0.025	≦0.35	≦0.25	≦0.50
	ASTM A841M-94	–	≦100	≦0.20	0.15～0.50	t≦40mm 0.75～1.35 t>40mm 1.00～1.60	≦0.030	≦0.030	≦0.35	≦0.25	≦0.25
UK英	–	–	–	–	–	–	–	–	–	–	–

$1 N/mm^2$ or $MPa = 1.01972 \times 10^{-1} kgf/mm^2$
$1 J = 1.01972 \times 10^{-1} kg\text{-}m$

Mo	V	Others	Mechanical Properties 機械的性質							Index No. 索引番号
			Heat Treatment 熱処理	Thickness (t)mm 板厚	Yield Point min. N/mm^2 or MPa (kgf/mm^2) 最小降伏点	Tensile Strength N/mm^2 or MPa (kgf/mm^2) 引張強さ	Notch Toughness 衝撃値			
							Test Piece 試験片	Test Temp. (℃) 試験温度	Absorbed Energy J (min. kg-m) 吸収エネルギー	
–	–	–	N	–	275	400～540	–	–	–	B029
				–	275	450～585				
				–	295	485～620				
0.23～0.47	–	Cb 0.03～0.09	QT	–	450	550～690	–	–	–	
			QT	–	485	585～725				
			QT	–	515	620～760				
			QT or P*	–	550	655～790				
0.15～0.25	–	Cb≧0.02	P*	–	550	620～760	–	–	–	
			N	t≦20	450	495～635				
				20<t≦25	450	495～635				
				25<t≦50	415	495～635				
				50<t≦100	380	450～585				
				100<t	345	415～550				
			Q	t≦20	515	585～725				
				20<t≦25	515	585～725				
				25<t≦50	515	585～725				
				50<t≦100	450	515～655				
				100<t	415	485～625				
0.15～0.25	–	Cb≧0.02	P*	–	620	690～820	–	–	–	
			N	t≦20	585	695～795				
				20<t≦25	550	620～760				
				25<t≦50	550	620～760				
≦0.08	≦0.07	Cb≦0.04 N+V≦0.08	t≦65 N	95<t≦100	310	515～655	–	–	–	
			QT t>65				–	–	–	
							–	–	–	
≦0.20	≦0.07	Nb≦0.04 Nb+V≦0.08	QT		415	585～705	–	–	–	
≦0.08	≦0.05	–	QT	t≦65	415	550～690	–	–	–	
				65<t≦100	380	515～655	–	–	–	
				100<t≦150	315	485～620	–	–	–	
≦0.08	≦0.06	Cb≦0.03 Al≧0.02 or ≧0.015 acidsoluble	–	t≦65	345	485～620	–	–	–	
				t>65	310	450～585				
–	–	–	–	–	–	–	–	–	–	

* Precipitation Heat-Treatment

Country 国	Standard No. 規格番号及び年号	Designation 種類記号	Chemical Composition 化学成分 (%)								
			Thickness(t) mm板厚	C	Si	Mn	P	S	Cu	Cr	Ni
GR独	DIN 17102-83	TSTE 255	−	≦0.16	≦0.40	0.50～1.30	≦0.030	≦0.025	≦0.20	≦0.30	≦0.30
		ESTE 255	−	≦0.16	≦0.40	0.50～1.30	≦0.025	≦0.015	≦0.20	≦0.30	≦0.30
		TSTE 285	−	≦0.16	≦0.40	0.60～1.40	≦0.030	≦0.025	≦0.20	≦0.30	≦0.30
			Superseded by European Standard EN 10028-1 &-3, Grade P275NL1, Material No. 1.0488								
		ESTE 285	−	≦0.16	≦0.40	0.60～1.40	≦0.025	≦0.015	≦0.20	≦0.30	≦0.30
			Superseded by European Standard EN 10028-1 &-3, Grade P275NL2, Material No. 1.0486								
		TSTE 315	−	≦0.18	≦0.40	0.70～1.50	≦0.030	≦0.025	≦0.20	≦0.30	≦0.30
		ESTE 315	−	≦0.18	≦0.40	0.70～1.50	≦0.025	≦0.015	≦0.20	≦0.30	≦0.30
		TSTE 355	−	≦0.18	0.10～0.50	0.90～1.65	≦0.030	0.025	≦0.20	≦0.30	≦0.30
			Superseded by European Standard EN 10028-1 & -3, Grade P355NL1, Material No. 1.0566								
		ESTE 355	−	≦0.18	0.10～0.50	0.90～1.65	≦0.025	0.015	≦0.20	≦0.30	≦0.30
			Superseded by European Standard EN 10028-1 & -3, Grade P355NL2, Material No. 1.0566								

$1N/mm^2$ or $MPa = 1.01972 \times 10^{-1} kgf/mm^2$
$1J = 1.01972 \times 10^{-1} kg\text{-}m$

Mo	V	Others	Mechanical Properties 機械的性質							Index No. 索引番号
			Heat Treatment 熱処理	Thickness (t)mm 板厚	Yield Point min. N/mm^2 or MPa (kgf/mm^2) 最小降伏点	Tensile Strength N/mm^2 or MPa (kgf/mm^2) 引張強さ	Notctch Toughness 衝撃値			
							Test Piece 試験片	Test Temp.(℃) 試験温度	Absorbed Energy J (min. kg-m) 吸収ｴﾈﾙｷﾞｰ	
≦0.08	–	N≦0.020 Nb≦0.03 Al≦0.020	–	t ≦16 16<t≦35 35<t≦50 50<t≦60 60<t≦70 70<t≦85 85<t≦100 100<t≦125 125<t≦150	255 255 245 235 235 225 215 205 195	t≦70 360～480 70<t≦85 350～470 85<t≦100 340～460 100<t≦125 330～450 125<t≦150 320～440	–	–	–	B029
≦0.08	–	N≦0.020 Nb≦0.03 Al≦0.020								
≦0.08	Ni+Ti+V ≦0.05	N≦0.020 Nb≦0.03 Al≦0.020	–	t ≦16 16<t≦35 35<t≦50 50<t≦60 60<t≦70 70<t≦85 85<t≦100 100<t≦125 125<t≦150	285 255 245 235 235 225 215 205 195	t≦70 390～510 70<t≦85 380～500 85<t≦100 370～490 100<t≦125 360～480 125<t≦150 320～440	–	–	–	
≦0.08	Ni+Ti+V ≦0.05	N≦0.020 Nb≦0.03 Al≦0.020								
≦0.08	Ni+Ti+V ≦0.05	N≦0.020 Nb≦0.03 Al≦0.020	–	t ≦16 16<t≦35 35<t≦50 50<t≦60 60<t≦70 70<t≦85 85<t≦100 100<t≦125 125<t≦150	315 315 305 295 295 285 275 265 255	t≦70 440～560 70<t≦85 430～550 85<t≦100 420～540 100<t≦125 410～530 125<t≦150 400～520	–	–	–	
≦0.08	Ni+Ti+V ≦0.05	N≦0.020 Nb≦0.03 Al≦0.020								
≦0.08	V≦0.10 Ni+Ti+V ≦0.12	N≦0.020 Nb≦0.03 Al≦0.020	–	t ≦16 16<t≦35 35<t≦50 50<t≦60 60<t≦70 70<t≦85 85<t≦100 100<t≦125 125<t≦150	355 355 345 335 335 325 315 305 295	t≦70 490～630 70<t≦85 480～620 85<t≦100 470～610 100<t≦125 460～600 125<t≦150 450～590	–	–	–	
≦0.08	V≦0.10 Ni+Ti+V ≦0.12	N≦0.020 Nb≦0.03 Al≦0.020								

Country 国	Standard No. 規格番号及び年号	Designation 種類記号	Chemical Composition 化学成分 (%)							
			Thickness(t) mm板厚	C	Si	Mn	P	S	Cu	Cr	Ni
GR独	DIN 17102-83	TSTE 380	–	≦0.20	0.10〜0.60	1.00〜1.70	≦0.030	≦0.025	≦0.20	≦0.30	≦1.00
		ESTE 380	–	≦0.20	0.10〜0.60	1.00〜1.70	≦0.025	≦0.015	≦0.20	≦0.30	≦1.00
		TSTE 420	–	≦0.20	0.10〜0.60	1.00〜1.70	≦0.035	≦0.030	≦0.20	≦0.30	≦1.00
		ESTE 420	–	≦0.20	0.10〜0.60	1.00〜1.70	≦0.025	≦0.015	≦0.20	≦0.30	≦1.00
		TSTE 460	–	≦0.20	0.10〜0.60	1.00〜1.70	≦0.035	≦0.030	≦0.20	≦0.30	≦1.00
		Superseded by European Standard EN 10 028 -1 &-3, Grade P460 NL1, Material No. 1.8915									
		ESTE 460	–	≦0.20	0.10〜0.60	1.00〜1.70	≦0.025	≦0.015	≦0.20	≦0.30	≦1.00
		Superseded by European Standard EN 10 028 -1 & -3, Grade P460 NL2, Material No. 1.8918									
		TSTE 500	–	≦0.21	0.10〜0.60	1.00〜1.70	≦0.030	≦0.030	≦0.20	≦0.30	≦1.00
		ESTE 500	–	≦0.21	0.10〜0.60	1.00〜1.70	≦0.025	≦0.015	≦0.20	≦0.30	≦1.00

1N/mm² or MPa=1.01972×10⁻¹kgf/mm²
1J=1.01972×10⁻¹kg-m

Mo	V	Others	Heat Treatment 熱処理	Thickness (t)mm 板厚	Yield Point min. N/mm² or MPa (kgf/mm²) 最小降伏点	Tensile Strength N/mm² or MPa (kgf/mm²) 引張強さ	Notctch Toughness 衝撃値: Test Piece 試験片	Notctch Toughness 衝撃値: Test Temp. (℃) 試験温度	Notctch Toughness 衝撃値: Absorbed Energy J (min. kg-m) 吸収ｴﾈﾙｷﾞｰ	Index No. 索引番号
≦0.08	V≦0.20 Ni+Ti+V ≦0.22	N≦0.020 Nb≦0.03 Al≦0.020	–	t≦16 16<t≦35 35<t≦50 50<t≦60 60<t≦70 70<t≦85 85<t≦100 100<t≦125 125<t≦150	380 375 365 355 345 335 325 315 305	t≦70 500～650 70<t≦85 490～640 85<t≦100 480～630 100<t≦125 470～620 125<t≦150 460～610	–	–	–	B029
≦0.08	V≦0.20 Ni+Ti+V ≦0.22	N≦0.020 Nb≦0.03 Al≦0.020								
≦0.10	V≦0.20 Ni+Ti+V ≦0.22	N≦0.020 Nb≦0.03 Al≦0.020	–	t≦16 16<t≦35 35<t≦50 50<t≦60 60<t≦70 70<t≦85 85<t≦100 100<t≦125 125<t≦150	420 410 400 390 385 375 365 355 345	t≦70 530～680 70<t≦85 520～670 85<t≦100 510～660 100<t≦125 500～650 125<t≦150 490～640	–	–	–	
≦0.10	V≦0.20 Ni+Ti+V ≦0.22	N≦0.020 Nb≦0.03 Al≦0.020								
≦0.10	V≦0.22 Ni+Ti+V ≦0.22	N≦0.020 Nb≦0.03 Al≦0.020	–	t≦16 16<t≦35 35<t≦50 50<t≦60 60<t≦70 70<t≦85 85<t≦100 100<t≦125 125<t≦150	460 450 440 430 420 410 400 390 380	t≦70 560～730 70<t≦85 550～720 85<t≦100 540～710 100<t≦125 530～700 125<t≦150 520～690	–	–	–	
≦0.10	V≦0.22 Ni+Ti+V ≦0.22	N≦0.020 Nb≦0.03 Al≦0.020								
≦0.10	V≦0.22 Ni+Ti+V ≦0.22	N≦0.020 Nb≦0.03 Al≦0.020	–	t≦16 16<t≦35 35<t≦50 50<t≦60 60<t≦70 70<t≦85 85<t≦100 100<t≦125 125<t≦150	500 480 470 460 450 440 430 420 410	t≦70 610～780 70<t≦85 600～770 85<t≦100 590～760 100<t≦125 580～750 125<t≦150 570～740	–	–	–	
≦0.08	V≦0.22 Ni+Ti+V ≦0.22	N≦0.020 Nb≦0.03 Al≦0.020								

Country 国	Standard No. 規格番号及び年号	Desig-nation 種類記号	Chemical Composition 化学成分（%） Thick-ness(t) mm板厚	C	Si	Mn	P	S	Cu	Cr	Ni
FR仏	−	−	−	−	−	−	−	−	−	−	−
URソ	−	−	−	−	−	−	−	−	−	−	−
1/2 % Ni Steel											
JP日	−	−	−	−	−	−	−	−	−	−	−
US米	−	−	−	−	−	−	−	−	−	−	−
UK英	−	−	−	−	−	−	−	−	−	−	−
GR独	−	−	−	−	−	−	−	−	−	−	−
FR仏	NF A36-208-82	0.5Ni285	−	≦0.14	≦0.35	0.70〜1.50	≦0.030	≦0.025	≦0.30	≦0.25	0.40〜0.75
		0.5Ni355	−	≦0.16	≦0.35	0.85〜1.65	≦0.030	≦0.025	≦0.30	≦0.25	0.40〜0.75
URソ	−	−	−	−	−	−	−	−	−	−	−
1 1/2 % Ni Steel											
JP日	−	−	−	−	−	−	−	−	−	−	−
US米	−	−	−	−	−	−	−	−	−	−	−
UK英	−	−	−	−	−	−	−	−	−	−	−
GR独	DIN 17280-85	14NiMn6	−	≦0.18	≦0.35	0.80〜1.50	≦0.025	≦0.020	−	−	1.30〜1.70
FR仏	NF A36-208-82	1.5Ni285	−	≦0.18	≦0.35	0.30〜0.70	≦0.025	≦0.020	≦0.30	≦0.25	1.30〜1.70
		1.5Ni335	−	≦0.18	≦0.35	0.80〜1.50	≦0.025	≦0.020	≦0.30	≦0.25	1.30〜1.70
URソ	−	−	−	−	−	−	−	−	−	−	−
2 1/4 % Ni Steel											
JP日	JIS G3127-90	SL2N255 (SL2N26)	6〜50	≦0.17	0.15〜0.30	≦0.70	≦0.025	≦0.025	−	−	2.10〜2.50
US米	ASTM A203M-93	Gr. A	t ≦50 50<t ≦100 100<t	≦0.17 ≦0.20 ≦0.23	0.15〜0.40	≦0.70 ≦0.80 ≦0.80	≦0.035	≦0.040	−	−	2.10〜2.50
		Gr. B	t ≦50 50<t ≦100 100<t	≦0.21 ≦0.24 ≦0.25	0.15〜0.40	≦0.70 ≦0.80 ≦0.80	≦0.035	≦0.035	−	−	2.10〜2.50

1N/mm² or MPa=1.01972×10^{-1}kgf/mm²
1J=1.01972×10^{-1}kg-m

			Mechanical Properties 機械的性質							Index No.
			Heat Treatment	Thickness (t)mm	Yield Point min. N/mm²or MPa (kgf/mm²)	Tensile Strength N/mm²or MPa (kgf/mm²)	Notctch Toughness 衝撃値			
Mo	V	Others	熱処理	板厚	最小降伏点	引張強さ	Test Piece 試験片	Test Temp.(℃) 試験温度	Absorbed Energy J (min.kg-m) 吸収ｴﾈﾙｷﾞｰ	索引番号
–	–	–	–	–	–	–	–	–	–	B029
–	–	–	–	–	–	–	–	–	–	
–	–	–	–	–	–	–	–	–	–	B030
–	–	–	–	–	–	–	–	–	–	
–	–	–	–	–	–	–	–	–	–	
–	–	–	–	–	–	–	–	–	–	
–	V≦0.08 Nb≦0.06	Al≧0.015	N	t≦30 30<t≦50 50<t	285 275 265	420～510	2V	-60	t≦50mm 40	
–	V≦0.08 Nb≦0.06	Al≧0.015	N	t≦30 30<t≦50 50<t	355 345 335	490～590	2V	-60	t≦50mm 40	
–	–	–	–	–	–	–	–	–	–	
–	–	–	–	–	–	–	–	–	–	B031
–	–	–	–	–	–	–	–	–	–	
–	–	–	–	–	–	–	–	–	–	
–	–	–	–	t≦30 30<t≦50 50<t≦70	355 345 –	470～640	–	–	–	
≦0.10	≦0.04	Al≦0.015	N, NT or QT	t≦30 30<t≦50 50<t	285 275 265	490～580	2V	-80	t≦50mm 40	
≦0.10	≦0.04	Al≦0.015	N, NT or QT	t≦30 30<t≦50 50<t	355 345 335	500～610	2V	-80	t≦50mm 40	
–	–	–	–	–	–	–	–	–	–	
–	–	–	N	–	255 (26)	450～590 (46～60)	2V	-70	11(1.1) 17(1.7) 21(2.1)	B032
–	–	–	N	–	255	450～585	–	–	–	
–	–	–	N	–	275	485～620	–	–	–	

Country 国	Standard No. 規格番号及び年号	Designation 種類記号	Thickness(t) mm板厚	C	Si	Mn	P	S	Cu	Cr	Ni
			Chemical Composition 化学成分 (%)								
UK英	–	–	–	–	–	–	–	–	–	–	–
GR独	–	–	–	–	–	–	–	–	–	–	–
FR仏	–	–	–	–	–	–	–	–	–	–	–
URソ	–	–	–	–	–	–	–	–	–	–	–
3 1/2 % Ni Steel											
JP日	JIS G3127-90	SL3N255 (SL3N26)	6～50	≦0.15	≦0.30	≦0.70	≦0.025	≦0.025	–	–	3.25～3.75
		SL3N275 (SL3N28)	6～50	≦0.17	≦0.30	≦0.70	≦0.025	≦0.025	–	–	3.25～3.75
		SL3N440 (SL3N45)	6～50	≦0.15	≦0.30	≦0.70	≦0.025	≦0.025	–	–	3.25～3.75
US米	ASTM A203M-93	Gr. D	≦50	≦0.17	0.15～0.40	≦0.070	≦0.035	≦0.035	–	–	3.25～3.75
			50<t ≦100	≦0.20		≦0.080					
			100< t	≦0.20		≦0.080					
		Gr. E	≦50	≦0.20	0.15～0.40	≦0.070	≦0.035	≦0.035	–	–	3.25～3.75
			50<t ≦100	≦0.23		≦0.080					
			100< t	–		≦0.080					
		Gr. F	≦50	≦0.20	0.15～0.40	≦0.070	≦0.035	≦0.035	–	–	3.25～3.75
			50<t ≦100	≦0.23		≦0.080					
			100< t	–		≦0.080					
UK英	BS1501 Part2-70	503	≦ 1 1/2 in	≦0.15 No longer active	0.10～0.35	0.30～0.80	≦0.025	≦0.030	≦0.30	≦0.30	3.25～3.75
GR独	DIN 17280-85	10Ni14	–	≦0.15	≦0.35	0.30～0.80	≦0.025	≦0.020	–	–	3.25～3.75
FR仏	NF A36-208-82	3.5Ni285	–	≦0.15	≦0.35	0.30～0.80	≦0.025	≦0.020	≦0.30	≦0.25	3.25～3.75
		3.5Ni355	–	≦0.15	≦0.35	0.30～0.80	≦0.025	≦0.020	≦0.30	≦0.25	3.25～3.75
URソ	–	–	–	–	–	–	–	–	–	–	–

1N/mm² or MPa = 1.01972 × 10^{-1} kgf/mm²
1J = 1.01972 × 10^{-1} kg-m

Mo	V	Others	Mechanical Properties 機械的性質							Index No. 索引番号
			Heat Treatment 熱処理	Thickness (t)mm 板厚	Yield Point min. N/mm² or MPa (kgf/mm²) 最小降伏点	Tensile Strength N/mm² or MPa (kgf/mm²) 引張強さ	Notch Toughness 衝撃値 Test Piece 試験片	Test Temp. (℃) 試験温度	Absorbed Energy J (min. kg-m) 吸収エネルギー	
–	–	–	–	–	–	–	–	–	–	B032
–	–	–	–	–	–	–	–	–	–	
–	–	–	–	–	–	–	–	–	–	
–	–	–	–	–	–	–	–	–	–	
–	–	–	N	–	255 (26)	450～590 (46～60)	2 V	-101	21(2.1)	B033
–	–	–	N	–	275 (28)	480～620 (49～63)	2 V	-101	21(2.1)	
–	–	–	QT	–	440 (45)	540～690 (55～70)	2 V	-110	21(2.1)	
–	–	–	N	–	255	450～585	–	–	–	
–	–	–	N	–	275	485～620	–	–	–	
–	–	–	QT	t ≦50	380	550～690	–	–	–	
				50 < t	345	515～655				
≦0.10	–	Total Al ≧0.020	NT or QT	–	260	(≧450)	2 V	-80 -100	34.5 18	
–	≦0.05	–	–	t ≦30	365	470～640	–	–	–	
				30<t≦50	345					
				50<t≦70	335					
≦0.10	–	–	N, NT or QT	t ≦30	285	460～570	2 V	-100	t≦50mm 40	
				30<t≦50	275					
				50<t	265					
≦0.10	≦0.04	Al ≧0.015	N, NT or QT	t ≦30	355	460～570	2 V	-100	t≦50mm 40	
				30<t≦50	345					
				50<t	335					
–	–	–	–	–	–	–	–	–	–	

Country 国	Standard No. 規格番号 及び年号	Designation 種類記号	Thickness(t) mm板厚	C	Si	Mn	P	S	Cu	Cr	Ni
			Chemical Composition 化学成分（%）								
5% Ni Steel											
JP日	JIS G3127-90	SL5N590 (SL5N60)	6～50	≦0.13	≦0.30	≦1.50	≦0.025	≦0.025	–	–	4.75～6.00
US米	ASTM A645M-87	–	–	≦0.13	0.20～0.40	0.30～0.60	≦0.025	≦0.025	–	–	4.75～5.25
UK英	–	–	–	–	–	–	–	–	–	–	–
GR独	DIN 17280-85	12Ni19	–	≦0.15	≦0.35	0.30～0.80	≦0.025	≦0.020	–	–	4.50～5.30
FR仏	NF A36-208-82	5Ni300	–	≦0.12	≦0.35	0.30～0.80	≦0.025	≦0.020	≦0.30	≦0.25	4.75～5.25
URソ	–	–	–	–	–	–	–	–	–	–	–
9% Ni Steel											
JP日	JIS G3127-90	SL9N520 (SL9N53)	6～50	≦0.12	0.15～0.30	≦0.90	≦0.025	≦0.025	–	–	8.50～9.50
		SL9N590 (SL9N60)	6～50	≦0.12	0.15～0.30	≦0.90	≦0.025	≦0.025	–	–	8.50～9.50
US米	ASTM A353M-93	–	≦50	≦0.13	0.15～0.40	≦0.90	≦0.035	≦0.035	–	–	8.50～9.50
	ASTM A844M-93	–	–	≦0.13	0.15～0.40	≦0.90	≦0.020	≦0.020	–	–	8.50～9.50
UK英	BS1501 Part2-70	509	≦50	≦0.10	0.10～0.30	0.30～0.80	≦0.025	≦0.030	≦0.30	≦0.30	8.75～9.75
		510	≦50	≦0.10	0.10～0.30	0.30～0.80	≦0.025	≦0.030	≦0.30	≦0.30	8.75～9.75
GR独	DIN 17280-85	X8Ni9	–	≦0.10	≦0.35	0.30～0.80	≦0.025	≦0.020	–	–	8.00～10.00
FR仏	NF A36-208-82	9Ni490	–	≦0.10	≦0.35	0.30～0.80	≦0.025	≦0.020	≦0.30	≦0.025	8.5～10
		9Ni585	–	≦0.10	≦0.35	0.30～0.80	≦0.025	≦0.020	≦0.30	≦0.25	8.5～10
URソ	–	–	–	–	–	–	–	–	–	–	–

1N/mm² or MPa = 1.01972×10⁻¹kgf/mm²
1J = 1.01972×10⁻¹kg-m

Mo	V	Others	Mechanical Properties 機械的性質				Notctch Toughness 衝撃値			Index No. 索引番号
			Heat Treatment 熱処理	Thickness (t)mm 板厚	Yield Point min. N/mm²or MPa (kgf/mm²) 最小降伏点	Tensile Strength N/mm²or MPa (kgf/mm²) 引張強さ	Test Piece 試験片	Test Temp.(℃) 試験温度	Absorbed Energy J (min.kg-m) 吸収ｴﾈﾙｷﾞｰ	
–	–	–	QT	–	590 (60)	690～830 (70～85)	2V	-130	41(4.2)	B034
0.20～0.35	N≦0.020	Al0.02~0.12	QT	–	450 (46)	655～795 (67～81)	–	–	–	
–	–	–	–	–	–	–	–	–	–	
–	–	–	–	t≦30 30<t≦50 50<t≦70	390 380 –	50～570	–	–	–	
≦0.10	≦0.05	Al≦0.015	N, NT or QT	t≦30 30<t≦50 50<t≦70	390 380 370	540～740	2V	-120	t≦50mm 41	
–	–	–	–	–	–	–	–	–	–	
–	–	–	N, NT	–	520 (53)	690～830 (70～85)	2V	-196	34(3.5)	B035
–	–	–	QT	–	590 (60)	690～830 (70～85)	2V	-196	41(4.2)	
–	–	–	N, NT	–	515	690～825	–	–	–	
≦0.020	≦0.020	–	–	–	585	690～825	–	–	–	
≦0.020	–	Total Al ≧0.020	N, NT or QT	–	530	≧710	2V	-100 -160 -196	(6.9) (4.8) (3.5)	
≦0.020	–	Total Al ≧0.020	QT	–	590	≧710	2V	-100 -160 -196	(6.9) (4.8) (3.5)	
≦0.10	–	–	–	t≦30 30<t≦50 50<t≦70	490 480 470	640～840	–	–	–	
≦0.10	≦0.04	Al≦0.015	N, NT or QT	t≦30 30<t≦50	450 480	640～850	2V	-196	t≦50mm (4.1)	
≦0.10	≦0.04	Al≦0.015	QT	t≦30 30<t≦50	590 580	690～845	2V	-196	t≦50mm (4.1)	
–	–	–	–	–	–	–	–	–	–	

6. Stainless Steel Plates　　ステンレス鋼板

Type of Steel 鋼種	Country 国	Standard No. 規格番号及び年号	Designation 種類記号	Chemical Composition 化学成分 (%)							
				C	Si	Mn	P	S	Cu	Cr	Ni
Cr-Mn-Ni Steel											
201	JP日	JIS G4304-91	SUS201	≦0.15	≦1.00	5.50～7.50	≦0.060	≦0.030	−	16.00～18.00	3.50～5.50
	米	ASTM A240-94a	TYP. 201 UNS S20100	≦0.15	≦1.00	5.50～7.50	≦0.060	≦0.030	−	16.00～18.00	3.50～5.50
	UK英	−	−	−	−	−	−	−	−	−	−
	GR独	−	−	−	−	−	−	−	−	−	−
	FR仏	−	−	−	−	−	−	−	−	−	−
	URソ	−	−	−	−	−	−	−	−	−	−
202	JP日	JIS G4304-91	SUS202	≦0.15	≦1.00	7.50～10.00	≦0.060	≦0.030	−	17.00～19.00	4.00～6.00
	US米	ASTM A240-94a	TYP. 202 UNS S20200	≦0.15	≦1.00	7.50～10.00	≦0.060	≦0.030	−	17.00～19.00	4.00～6.00
	UK英	−	−	−	−	−	−	−	−	−	−
	GR独	−	−	−	−	−	−	−	−	−	−
	FR仏	−	−	−	−	−	−	−	−	−	−
	URソ	ΓOCT 5632-72	12X17 Γ9AH4	≦0.12	≦0.8	8.00～10.5	≦0.035	≦0.020	−	16.0～18.0	3.5～4.5
209	JP日	−	−	−	−	−	−	−	−	−	−
	US米	ASTM A240M-94a	TYP. XM19 UNS S20910	≦0.06	≦0.75	4.00～4.60	≦0.040	≦0.030	−	20.50～23.50	11.50～13.50
	UK英	−	−	−	−	−	−	−	−	−	−
	GR独	−	−	−	−	−	−	−	−	−	−
	FR仏	−	−	−	−	−	−	−	−	−	−
	URソ	−	−	−	−	−	−	−	−	−	−
214	JP日	−	−	−	−	−	−	−	−	−	−
	US米	ASTM A240M-94a	TYP.XM31 UNS S21400	≦0.12	0.30～1.00	14.00～16.00	≦0.045	≦0.030	−	17.00～18.00	≦1.00
	UK英	−	−	−	−	−	−	−	−	−	−
	GR独	−	−	−	−	−	−	−	−	−	−
	FR仏	−	−	−	−	−	−	−	−	−	−
	URソ	ΓOCT 5632-72	15X17A Γ14	≦0.15	≦0.8	13.5～15.5	≦0.03	≦0.020	−	16.0～18.0	≦0.6
			10X14A Γ15	≦0.10	≦0.8	14.5～16.5	≦0.035	≦0.030	−	13.0～15.0	−

$1N/mm^2 = 1.01972 \times 10^{-1} kgf/mm^2$

			Mechanical Properties 機械的性質					Index No. 索引番号
			Heat Treatment 熱処理	Thickness (t)mm 板厚	Min. Yield Strength N/mm^2 or MPa(kgf/mm^2) 耐力		Tensile Strength N/mm^2 or MPa (kgf/mm^2)	
Mo	V	Others			0.2% Proof Stress	1% Proof Stress		
–	–	N ≦0.25	–	–	245 (25)	–	≧640 (≧65)	B036
–	–	N ≦0.25	–	–	260	–	655	
–	–	–	–	–	–	–	–	
–	–	–	–	–	–	–	–	
–	–	–	–	–	–	–	–	
–	–	–	–	–	–	–	–	
–	–	N ≦0.25	S	–	245 (25)	–	≧590 (≧60)	B037
–	–	N ≦0.25	–	–	260	–	620	
–	–	–	–	–	–	–	–	
–	–	–	–	–	–	–	–	
–	–	–	–	–	–	–	–	
–	–	N 0.15~0.25	–	–	–	–	–	
–	–	–	–	–	–	–	–	B038
1.50~3.00	V0.10~0.30 Cb0.10~0.30	N 0.20~0.40	–	Sheet	415	–	≧725	
				Strip	380		≧690	
–	–	–	–	–	–	–	–	
–	–	–	–	–	–	–	–	
–	–	–	–	–	–	–	–	
–	–	–	–	–	–	–	–	
–	–	–	–	–	–	–	–	B039
–	–	N ≦0.35	–	Sheet	485	–	≧860	
				Plate	380	–	≧725	
–	–	–	–	–	–	–	–	
–	–	–	–	–	–	–	–	
–	–	–	–	–	–	–	–	
–	–	N 0.25~0.37	–	–	–	–	–	
–	–	N 0.15~0.25	–	–	–	–	–	

Type of Steel 鋼種	Country 国	Standard No. 規格番号及び年号	Designation 種類記号	Chemical Composition 化学成分（%） C	Si	Mn	P	S	Cu	Cr	Ni
216	JP日	–	–	–	–	–	–	–	–	–	–
	US米	ASTM A240-94a	TYP.XM17 UNS S21600	≦0.08	≦0.75	7.50～9.00	≦0.045	≦0.030	–	17.50～22.00	5.00～7.00
			TYP.XM18 UNS S21603	≦0.03	≦0.75	7.50～9.00	≦0.045	≦0.030	–	17.50～22.00	5.00～7.00
	UK英	–	–	–	–	–	–	–	–	–	–
	GR独	–	–	–	–	–	–	–	–	–	–
	FR仏	–	–	–	–	–	–	–	–	–	–
	URソ	–	–	–	–	–	–	–	–	–	–
240	JP日	–	–	–	–	–	–	–	–	–	–
	US米	ASTM A240-94a	TYP.XM29 UNS S24000	≦0.08	≦0.75	11.50～14.50	≦0.060	≦0.030	–	17.00～19.00	2.25～3.75
	UK英	–	–	–	–	–	–	–	–	–	–
	GR独	–	–	–	–	–	–	–	–	–	–
	FR仏	–	–	–	–	–	–	–	–	–	–
	URソ	–	–	–	–	–	–	–	–	–	–
Others	FR仏	NF-A36-209-82	Z3CMN18-08-07AZ	≦0.040	≦1.0	6.5～8.5	≦0.040	–	17～19	6～8	–
	URソ	ΓOCT 5632-72	07X21Γ7AH	≦0.07	≦0.7	6.0～7.5	≦0.030	–	19.5～21.0	5.0～6.0	–

Cr-Ni Steels

Type of Steel 鋼種	Country 国	Standard No. 規格番号及び年号	Designation 種類記号	C	Si	Mn	P	S	Cu	Cr	Ni
301	JP日	G4304-91	SUS301	≦0.15	≦1.00	≦2.00	≦0.045	≦0.030	–	16.00～18.00	6.00～8.00
			SUS301J1	0.08～0.12	≦1.00	≦2.00	≦0.045	≦0.030	–	16.00～18.00	7.00～9.00
	US米	ASTM A240-94a	TYP.301 UNS S30100	≦0.15	≦0.75	≦2.00	≦0.045	≦0.030	–	16.00～18.00	6.00～8.00
	UK英	–	–	–	–	–	–	–	–	–	–
	GR独	–	–	–	–	–	–	–	–	–	–
	NF仏	NF-A35-573-90	Z12CN17 90-07	0.08～0.15	≦1.0	–	≦0.040	≦0.030	–	16.0～18.0	6.0～8.0
	URソ	ΓOCT 5632-72	07X16H6	0.05～0.09	≦0.8	≦0.8	≦0.035	≦0.020		1.55～1.75	5.0～8.0

$1N/mm^2 = 1.01972 \times 10^{-1} kgf/mm^2$

Mo	V	Others	Heat Treatment 熱処理	Thickness (t)mm 板厚	Min. Yield Strength N/mm² or MPa(kgf/mm²) 耐力: 0.2% Proof Stress	1% Proof Stress	Tensile Strength N/mm² or MPa (kgf/mm²)	Index No. 索引番号
–	–	–	–	–	–	–	–	B040
2.00～3.00	–	N 0.25～0.50	–	Sheet	415	–	≧690	
				Plate	345	–	≧620	
2.00～3.00	–	N 0.25～0.50	–	Sheet	415	–	≧690	
				Plate	345	–	≧620	
–	–	–	–	–	–	–	–	
–	–	–	–	–	–	–	–	
–	–	–	–	–	–	–	–	
–	–	–	–	–	–	–	–	
–	–	–	–	–	–	–	–	B041
–	–	N 0.20～0.40	–	Sheet	415	–	≧690	
				Plate	380		≧690	
–	–	–	–	–	–	–	–	
–	–	–	–	–	–	–	–	
–	–	–	–	–	–	–	–	
–	–	–	–	–	–	–	–	
–	–	N 0.15～0.25	–	t < 5	345	380	t < 20 590～805	B042
				5 ≦ t ≦75	440	370	20≦ t ≦75 580～785	
–	–	N 0.15～0.25	–	–	–	–	–	
–	–	–	–	–	205 (21)	–	≧520 (≧53)	B043
–	–	–	–	–	205 (21)	–	≧570 (≧58)	
		–	–	–	205	–	≧515	
–	–	–	–	–	–	–	–	
–	–	–	–	–	–	–	–	
–	–	–	–	t < 5	265	305	610～855	
				5 ≦ t ≦50	255	295		
–	–	–	–	–	–	–	–	

Type of Steel 鋼種	Country 国	Standard No. 規格番号及び年号	Designation 種類記号	Chemical Composition 化学成分 (%) C	Si	Mn	P	S	Cu	Cr	Ni
302	JP日	JIS G4304-91	SUS302	≦0.15	≦1.00	≦2.00	≦0.045	≦0.030	–	17.00 ~ 19.00	–
	US米										
		ASTM A240-94a	TYP.302 UNS S30200	≦0.15	≦0.75	≦2.00	≦0.045	≦0.030	–	17.00~ 19.00	8.00~ 10.00
	UK英	–	–	–	–	–	–	–	–	–	–
	GR独	DIN 17440-85	X10CrS 189	≦0.12	–	–	≦0.060	0.15~ 0.35	–	17.0 ~ 19.0	8.0~10.0
	FR仏	NF-A35-573-90	Z12CN 18-09	≦0.12	≦0.75	1.00~1.50 ≦2.0	≦0.040	≦0.030	–	17.0 ~ 19.0	8.0~9.0
	URソ	ГОСТ 5632-72	12X18H9	≦0.12	≦0.8	≦2.0	≦0.035	≦0.020	–	17.0 ~ 19.0	7.5~9.5
			17X18H9	0.13~0.21	≦0.8	≦2.0	≦0.035	≦0.020	–	17.0 ~ 19.0	8.0~10.0
302B	JP日	JIS G4304-91	SUS302B	≦0.15	2.00~3.00	≦2.00	≦0.045	≦0.030	–	17.00 ~ 19.00	8.00~ 10.00
	US米										
		ASTM A167-94a	TYP.302B UNS S30215	≦0.15	2.00~3.00	≦2.00	≦0.045	≦0.030	–	17.00 ~ 19.00	8.00~ 10.00
	UK英	–	–	–	–	–	–	–	–	–	–
	GR独	–	–	–	–	–	–	–	–	–	–
	FR仏	–	–	–	–	–	–	–	–	–	–
	URソ	–	–	–	–	–	–	–	–	–	–
304	JP日	JIS G4304-91	SUS304	≦0.08	≦1.00	≦2.00	≦0.045	≦0.030	–	18.00 ~ 20.00	8.00~ 15.00
	US米										
		ASTM A240-94a	TYP.304 UNS S30400	≦0.08	≦0.75	≦2.00	≦0.045	≦0.030	–	18.00 ~ 20.00	8.00~ 10.50
			TYP.304H UNS S30409	0.04~0.10	≦0.75	≦2.00	≦0.045	≦0.030	–	18.00 ~ 20.00	8.00~ 10.50
	UK英	BS1501 Part3-90	304S31	≦0.07	≦-1.00	≦-2.00	≦0.045	≦0.025	–	17.00~ 19.00	8.0 ~11.0

$1N/mm^2 = 1.01972 \times 10^{-1} kgf/mm^2$

Mo	V	Others	Mechanical Properties 機械的性質					Index No. 索引番号
			Heat Treatment 熱処理	Thickness (t)mm 板厚	Min. Yield Strength N/mm^2 or MPa(kgf/mm^2) 耐力		Tensile Strength N/mm^2 or MPa (kgf/mm^2)	
					0.2% Proof Stress	1% Proof Stress		
–	–	–	S	–	205 (21)	–	≥520 (≥53)	B044
–	–	N ≤0.10	–	–	205	–	≥515	
–	–	–	–	–	–	–	–	
–	–	–	–	–	195	230	500-700	
–	–	–	–	t < 5 5 ≤ t ≤50	255 245	295 285	560~75.5	
–	–	–	–	–	–	–	–	
–	–	–	–	–	–	–	–	
–	–	–	S	–	205 (21)	–	≥520 (≥53)	B045
–	–	–	–	–	205	–	≥515	
–	–	–	–	–	–	–	–	
–	–	–	–	–	–	–	–	
–	–	–	–	–	–	–	–	
–	–	–	–	–	–	–	–	
–	–	–	S	–	205 (21)	–	≥520 (≥53)	B046
–	–	N ≤0.10	–	–	205	–	≥515	
–	–	–	–	–	205	–	≥515	
–	–	–	S	–	195~230	245	500~700	

Type of Steel 鋼種	Country 国	Standard No. 規格番号及び年号	Desig-nation 種類記号	Chemical Composition 化学成分 (%)							
				C	Si	Mn	P	S	Cu	Cr	Ni
304	GR独	DIN 17440	X5CrNi 1810	≦0.07	–	–	–	–	–	17.0~19.0	8.5 ~10.5
	FR仏	NF-A35-573-81	Z 6 CN 18-09	≦0.07	≦1.0	≦2.0	≦0.040	≦0.030	–	17 ~20	8.0~11
		NF-A-36-209-91	Z7 CN 18-09	≦0.07	≦0.75	≦2.0	≦0.040	≦0.015	–	17.0~19.0	8.0~10.0
	URソ	ГОСТ 5632-72	08X18H10	≦0.08	≦0.8	≦2.0	≦0.035	≦0.020	–	17.0~19.0	9.0~11.0
304L	JP日	JIS G4304-91	SUS304L	≦0.030	≦1.00	≦2.00	≦0.045	≦0.030	–	18.00 ~20.00	9.00~13.00
	US米										
		ASTM A240-94a	TYP.304L UNS S30403	≦0.030	≦0.75	≦2.00	≦0.045	≦0.030	–	18.00 ~20.00	8.00~12.00
	UK英	BS1501 Part3-90	304S11	≦0.03	≦ 1.00	≦ 2.00	≦0.045	≦0.025	–	17.00~19.00	9.00~12.00
	GR独	17440-85	X2CrNi 1911	≦0.030	–	–	–	–	–	18.0~20.0	10.0~12.
	NF仏	NF-A35-573-81	Z 2 CN 18-10	≦0.030	≦1.0	≦2.0	≦0.040	≦0.030	–	17.0 ~19.0	9.0~11.0
		NFA36-209-91	Z3 CN 18-10	≦0.03	≦0.75	≦2.0	≦0.040	≦0.015	–	17.0~19.0	9.0~11.0
	URソ	ГОСТ 5632-72	04X18H10	≦0.04	≦0.8	≦2.0	≦0.035	≦0.020	–	17.0~19.0	9.0~11.0
304N	JP日	JIS G4304-91	SUS304N1	≦0.08	≦1.00	≦2.50	≦0.045	≦0.030	–	18.00 ~20.00	7.00~10.50
	US米										
		ASTM A240-94a	TYP.304N UNS S30451	≦0.080	≦0.75	≦2.00	≦0.045	≦0.030	–	18.00 ~20.00	8.00~10.50
	UK英	–	–	–	–	–	–	–	–	–	–
			–	–	–	–	–	–	–	–	–

1N/mm² = 1.01972 ×10⁻¹kgf/mm²

			Mechanical Properties 機械的性質					Index No. 索引番号
			Heat Treatment 熱処理	Thickness (t)mm 板厚	Min. Yield Strength N/mm² or MPa(kgf/mm²) 耐力		Tensile Strength N/mm² or MPa (kgf/mm²)	
Mo	V	Others			0.2% Proof Stress	1% Proof Stress		
–	–	–	–	–	195	230	500 ~700	B046
–	–	–	–	t < 5 5 ≦ t ≦50	215 205	255 245	(55~75)	
–	–	–	–	t < 5 5 ≦ t ≦20 20≦ t ≦75	215 215 215	255 245 245	t ≦20 530~740 20 < t ≦75 520~730	
–	–	–	–	–	–	–	–	
–	–	–	S	–	175 (18)	–	≧480 (≧49)	B047
–	–	N ≦0.10	–	–	170	–	≧485	
–	–	–	S	–	180	–	480 ~680	
–	–	–	–	–	180	205	450 700	
–	–	–	–	t < 5 5 ≦ t ≦50	205 195	245 235	490 ~690	
–	–	–	S	t < 5 5 ≦ t ≦20 20≦ t ≦75	205 195 185	245 235 225	t ≦20 500 ~700 20 < t ≦75 490 ~700	
–	–	–	–	–	–	–	–	
–	–	N 0.10~0.25	–	–	275 (28)		≧550 (≧56)	B048
–	–	N 0.10~0.16	–	–	240	–	≧550	
–	–	–	–	–	–	–	–	
–	–	–	–	–	–	–	–	

Type of Steel 鋼種	Country 国	Standard No. 規格番号及び年号	Designation 種類記号	Chemical Composition 化学成分 (%)							
				C	Si	Mn	P	S	Cu	Cr	Ni
304N	GR独	17440-85	X2CrNiN 1810	≦0.03	≦1.0	≦2.00	≦0.045	≦0.030	–	17.0～19.0	8.0～11.5
	FR仏	NF-A36-69 209-91	Z3CN 18-10AZ	≦0.03	≦0.75	≦2.0	≦0.040	≦0.015	–	17.0～19.0	9.0～11.0
			Z6 CN 18-09AZ	≦0.06	≦0.75	≦2.0	≦0.040	≦0.015	–	18.0～20.0	8.0～10.0
	URソ	–	–	–	–	–	–	–	–	–	–
304 Modity	JP日	JIS G4304-91	SUS304N2	≦0.08	≦1.00	≦2.50	≦0.040	≦0.030	–	18.00～20.00	7.50～11.50
			SUS304LN	≦0.03	≦1.00	≦2.00	≦0.045	≦0.030	–	17.00～19.00	8.50～11.00
	US	ASTM A240-94a	TYP.304LN UNS S30453	≦0.03	≦0.75	≦2.00	≦0.045	≦0.030	–	18.00～20.00	8.00～12.00
	UK英	BS1501 Part3-90	304S61	≦0.03	≦1.00	≦2.00	≦0.045	≦0.025	–	17.00～19.00	8.5～11.5
	GR独	–	–	–	–	–	–	–	–	–	–
	FR仏	–	–	–	–	–	–	–	–	–	–
	URソ	–	–	–	–	–	–	–	–	–	–
305	JP日	JIS G4304-91	SUS305	≦0.12	≦1.00	≦2.00	≦0.045	≦0.030	–	17.00～19.00	10.50～13.00
	US米										
		ASTM A240-94a	TYP.305 UNS S30500	≦0.12	≦0.75	≦2.00	≦0.045	≦0.030	–	17.00～19.00	10.50～13.00
	UK英	–	–	–	–	–	–	–	–	–	–
	GR独	17440-85	X5CrNi 1911	≦0.07	–	–	–	–	–	17.0～20.0	10.5～12.5
	FR仏	–	–	–	–	–	–	–	–	–	–
	URソ	ГОСТ 5632-72	03X18H11	≦0.03	≦0.8	≦2.0	≦0.035	≦0.020	–	17.0～19.0	10.5～12.5
			06X18H11	≦0.06	≦0.8	≦2.0	≦0.035	≦0.020	–	17.0～19.0	10.0～12.0
			03X18H12	≦0.03	≦0.4	≦2.0	≦0.035	≦0.020	–	17.0～19.0	11.5～13.0

1N/mm² = 1.01972 ×10⁻¹kgf/mm²

Mo	V	Others	Heat Treatment 熱処理	Thickness (t)mm 板厚	Mechanical Properties 機械的性質 Min. Yield Strength N/mm² or MPa(kgf/mm²) 耐力 0.2% Proof Stress	 1% Proof Stress	 Tensile Strength N/mm² or MPa (kgf/mm²)	Index No. 索引番号
–	–	N 0.12~0.20	Q	–	200	235	500 ~730	B048
–	–	N 0.12~0.20	S	t < 5 5 ≦ t ≦20 20≦ t ≦75	285 275 265	325 315 305	t ≦20 570 ~775 20 < t ≦75 580 ~785	
–	–	N 0.12~0.20	S	t < 5 5 ≦ t ≦20 20≦ t ≦75	305 295 285	345 335 325	t ≦20 590 ~795 20 < t ≦75 580 ~7858	
–	–	–	–	–	–	–	–	
–	N 0.15~0.30	Nb ≦0.15	–	–	345 (35)	–	≧690 (≧70)	B049
–	–	N 0.12~0.22	S	–	245 (25)	–	≧550 (≧56)	
–	–	0.10~0.16	–	–	205	–	≦515	
–	–	No.12~0.22	S	–	270		550~750	
–	–	–	–	–	–	–	–	
–	–	–	–	–	–	–	–	
–	–	–	–	–	–	–	–	
–	–	–	S	–	175 (18)	–	≧480 (≧49)	B050
–	–	–	–	–	205	–	≧515	
–	–	–	–	–	–	–	–	
–	–	–	Q	–	180	215	500 ~700	
–	–	–	–	–	–	–	–	
–	–	–	–	–	–	–	–	
–	–	–	–	–	–	–	–	
–	–	Ti ≦0.005	–	–	–	–	–	

Type of Steel 鋼種	Country 国	Standard No. 規格番号及び年号	Designation 種類記号	Chemical Composition 化学成分 (%)							
				C	Si	Mn	P	S	Cu	Cr	Ni
308	JP日	–	–	–	–	–	–	–	–	–	–
	US米										
		ASTM A167-94a	TYP.308 UNS S30800	≦0.08	≦0.75	≦2.00	≦0.045	≦0.030	–	19.00 ～ 21.00	10.00 ～ 12.00
	UK英	–	–	–	–	–	–	–	–	–	–
	GR独	–	–	–	–	–	–	–	–	–	–
	FR仏	–	–	–	–	–	–	–	–	–	–
	URソ	–	–	–	–	–	–	–	–	–	–
309	JP日	JIS G4312-91	SUH309	≦0.20	≦1.00	≦2.00	≦0.040	≦0.030	–	22.00 ～ 24.00	12.00 ～ 15.00
	US米										
		ASTM A167-94a	TYP.309 UNS S30900	≦0.20	≦0.75	≦2.00	≦0.05	≦0.030	–	22.00 ～ 24.00	12.00 ～ 15.00
	UK英	–	–	–	–	–	–	–	–	–	–
	GR独	–	–	–	–	–	–	–	–	–	–
	FR仏	–	–	–	–	–	–	–	–	–	–
	URソ	ГОСТ 5632-72	20X22H13	≦0.20	≦1.0	≦2.0	≦0.035	≦0.025	–	22.0～ 25.0	12.0～ 15.0
309S	JP日	JIS G4312-91	SUS309S	≦0.08	≦1.00	≦2.00	≦0.045	≦0.030	–	22.00 ～ 24.00	12.00 ～ 15.00
	US米										
		ASTM A240-94a	TYP.309S UNS S30908	≦0.08	≦0.75	≦2.00	≦0.045	≦0.030	–	22.00 ～ 24.00	12.00 ～ 15.00
	UK英	BS1501 Part 3-90	309S16	≦0.08	≦1.00	≦2.00	≦0.045	≦0.025	–	22.00 ～ 25.00	13.00 ～ 16.00
	GR独	–	–	–	–	–	–	–	–	–	–
	FR仏	–	–	–	–	–	–	–	–	–	–
	URソ	–	–	–	–	–	–	–	–	–	–
310	JP日	JIS G4312-91	SUH310	≦0.25	≦1.50	≦2.00	≦0.040	≦0.030	–	24.00 ～ 26.00	19.00 ～ 22.00
	US米										
		ASTM A167-94a	TYP.310 UNS S31000	≦0.25	≦1.50	≦2.00	≦0.045	≦0.030	–	24.00 ～ 26.00	19.00 ～ 22.00

1N/mm² = 1.01972 ×10⁻¹kgf/mm²

Mo	V	Others	Mechanical Properties 機械的性質					Index No. 索引番号
			Heat Treatment 熱処理	Thickness (t)mm 板厚	Min. Yield Strength N/mm² or MPa(kgf/mm²) 耐力		Tensile Strength N/mm² or MPa (kgf/mm²)	
					0.2% Proof Stress	1% Proof Stress		
–	–	–	–	–	–	–	–	B051
–	–	–	–	–	205	–	≧515	
–	–	–	–	–	–	–	–	
–	–	–	–	–	–	–	–	
–	–	–	–	–	–	–	–	
–	–	–	–	–	–	–	–	
–	–	–	S	–	205 (21)	–	≧560 (≧57)	B052
–	–	–	–	–	205	–	≧515	
–	–	–	–	–	–	–	–	
–	–	–	–	–	–	–	–	
–	–	–	–	–	–	–	–	
–	–	–	–	–	–	–	–	
–	–	–	–	–	205 (2)	–	≧520 (≧53)	B053
–	–	–	–	–	205	–	≧515	
–	–	–	–	–	205	–	510~710	
–	–	–	–	–	–	–	–	
–	–	–	–	–	–	–	–	
–	–	–	–	–	–	–	–	
–	–	–	S	–	205 (21)	–	≧590 (≧60)	B054
–	–	–	–	–	205	–	≧515	

Type of Steel 鋼種	Country 国	Standard No. 規格番号及び年号	Designation 種類記号	Chemical Composition 化学成分（%）							
				C	Si	Mn	P	S	Cu	Cr	Ni
310	UK英	BS1501 Part 3-90	310S16	≦0.08	≦1.00	≦2.00	≦0.045	≦0.025	–	23.00 ～ 26.00	19.00 ～ 22.00
	GR独	–	–	–	–	–	–	–	–	–	–
	FR仏	–	–	–	–	–	–	–	–	–	–
	URソ	ГОСТ 5632-72	20X23H18	≦0.20	≦1.0	≦2.0	≦0.035	≦0.020	–	22.0～ 25.0	17.0～ 20.0
310S	JP日	JIS G4304-91	SUH310S	≦0.08	≦1.50	≦2.00	≦0.040	≦0.030	–	24.00 ～ 26.00	19.00 ～ 22.00
	US米	ASTM A240-94a	TYP.310S UNS S31008	≦0.08	≦1.50	≦2.00	≦0.045	≦0.030	–	24.00 ～ 26.00	19.00 ～ 22.00
	UK英	–	–	–	–	–	–	–	–	–	–
	GR独	–	–	–	–	–	–	–	–	–	–
	FR仏	–	–	–	–	–	–	–	–	–	–
	URソ	ГОСТ 5632-72	10X23H18	≦0.10	≦1.0	≦2.0	≦0.035	≦0.020	–	22.0～ 25.0	17.0～ 20.0
316	JP日	JIS G4304-87	SUS316	≦0.08	≦1.00	≦2.00	≦0.045	≦0.030	–	16.00 ～ 18.00	10.00 ～ 14.00
	US米	ASTM A240-94a	TYP.316 UNS S31600	≦0.08	≦0.75	≦2.00	≦0.045	≦0.030	–	16.00 ～ 18.00	10.00 ～ 14.00
			TYP.316H UNS S31609	0.04～0.10	≦1.00	0.50～2.00	≦0.045	≦0.030	–	16.00 ～ 18.00	10.00 ～ 14.00
	UK英	BS1501 Part3-90	316S31	≦0.07	≦1.00	≦2.00	≦0.045	≦0.030	–	16.5～ 18.5	10.5～ 13.50
			316S33	≦0.07	≦1.00	≦2.00	≦0.045	≦0.025	–	16.5～ 18.5	11.0～ 14.00
			316S51	0.04～0.10	≦1.00	≦2.00	≦0.045	≦0.025	–	16.50～ 18.50	10.00～ 13.00
			316S53	0.04～0.10	≦1.00	≦2.00	≦0.045	≦0.025	–	16.50～ 18.50	10.50～ 13.50
	GR独	DIN 17440-85	X5CrNiMo 17122	≦0.07	–	–	–	–	–	16.5～ 18.5	10.5～ 13.5
	NF仏	NF-A35-573-90	Z 7 CND 17-11	≦0.07	≦1.0	≦2.0	≦0.040	≦0.030	–	16.0 ～ 18.0	10.0～12.0
	URソ	–	–	–	–	–	–	–	–	–	–
316L	JP日	JIS G4304-91	SUS316L	≦0.030	≦1.00	≦2.00	≦0.045	≦0.030	–	16.00 ～ 18.00	12.00 ～ 15.00

$1N/mm^2 = 1.01972 \times 10^{-1} kgf/mm^2$

Mo	V	Others	Mechanical Properties 機械的性質 Heat Treatment 熱処理	Thickness (t)mm 板厚	Min. Yield Strength N/mm² or MPa(kgf/mm²) 耐力 0.2% Proof Stress	1% Proof Stress	Tensile Strength N/mm² or MPa (kgf/mm²)	Index No. 索引番号
–	–	–	–	–	205	–	510~710	B054
–	–	–	–	–	–	–	–	
–	–	–	–	–	–	–	–	
–	–	–	–	–	–	–	–	
–	–	–	S	–	205 (21)	–	≥520 (≥53)	B055
–	–	–	–	–	205	–	≥515	
–	–	–	–	–	–	–	–	
–	–	–	–	–	–	–	–	
–	–	–	–	–	–	–	–	
–	–	–	–	–	–	–	–	
2.00~3.00	–	–	–	–	205 (21)	–	≥520 (≥53)	B056
2.00~3.00	–	N ≤0.10	–	–	205	–	≥515	
2.00~3.00	–	–	–	–	205	–	≥515	
2.00~2.50	–	–	S	–	205	–	510~710	
2.50~3.00	–	–	S	–	205	–	510~710	
2.00~2.50	–	–	–	–	205	–	510~710	
2.50~3.00	–	–	–	–	205	–	510~710	
2.0~2.5	–	–	–	–	205	240	510~710	
2.0~2.5	–	–	–	t < 5 5 ≤ t ≤50	225 215	265 255	560~760	
–	–	–	–	–	–	–	–	
2.00~3.00	–	–	S	–	175 (18)	–	≥480 (≥49)	B057

Type of Steel 鋼 種	Country 国	Standard No. 規格番号 及び年号	Designation 種類記号	Chemical Composition 化 学 成 分 （%）							
				C	Si	Mn	P	S	Cu	Cr	Ni
316L	US米										
		ASTM A240-94a	TYP.316L UNS S31603	≦0.030	≦0.75	≦2.00	≦0.045	≦0.030	–	16.00 ～ 18.00	10.00 ～ 14.00
	UK英	BS1501 Part3-90	316S11	≦0.030	≦1.00	≦2.00	≦0.045	≦0.025	–	16.50～ 18.50	11.0～ 14.0
			316S13	≦0.030	≦1.00	≦2.00	≦0.045	≦0.025	–	16.50～ 18.50	11.50～ 14.50
	GR独	DIN 17440-85	X2CrNiMo 17132	≦0.030	–	–	–	–	–	16.5～ 18.5	11.0～ 14.0
	NF仏	NF-A35-573-90	Z3NCND 17-11-02	≦0.03	≦1.0	≦2.0	≦0.040	≦0.030	–	16.0 ～ 18.0	10.0～ 12.0
			Z3NCND 17-12-03	≦0.03	≦1.0	≦2.0	≦0.040	≦0.015	–	16.5 ～ 18.5	11.0～ 13.0
	URソ	ГОСТ 5632-72	03X17H 14M2	≦0.03	≦0.8	1.0 ～2.0	≦0.035	≦0.020	–	16.0～ 18.0	13.0～ 15.0
316N	JP日	JIS G4304-91	SUS316N	≦0.08	≦1.00	≦2.00	≦0.045	≦0.030	–	16.00 ～ 18.00	10.00 ～ 14.00
			SUS316LN	≦0.03	≦1.00	≦2.00	≦0.045	≦0.030	–	16.50 ～ 18.50	10.50 ～ 14.50
	US米										
		ASTM A240-94a	TYP.316N UNS S31651	≦0.08	≦0.75	≦2.00	≦0.045	≦0.030	–	16.00 ～ 18.00	10.00 ～ 14.00
	UK英	BS1501 Part3-90	316S61	≦0.03	≦1.00	≦2.00	≦0.045	≦0.025	–	16.50 ～ 18.50	10.50 ～ 13.50
			316S63	≦0.03	≦1.00	≦2.00	≦0.045	≦0.025	–	16.50 ～ 18.50	11.50～ 14.50
	GR独	–	–	–	–	–	–	–	–	–	–
	FR仏	–	–	–	–	–	–	–	–	–	–
	URソ	–	–	–	–	–	–	–	–	–	–
	JP日	–	–	–	–	–	–	–	–	–	–
316Ti	US米	ASTM A240-94a	TYP.316Ti UNS S31635	≦0.08	≦0.75	≦2.00	≦0.045	≦0.030	–	16.00 ～ 18.00	10.00 ～ 14.00
	UK英	BS1501 Part3-90	320S31	≦0.08	≦1.00	≦2.00	≦0.045	≦0.025	–	16.50 ～ 18.50	11.00 ～ 14.00

1N/mm² = 1.01972 × 10⁻¹ kgf/mm²

Mo	V	Others	Heat Treatment 熱処理	Thickness (t)mm 板厚	Min. Yield Strength N/mm² or MPa(kgf/mm²) 耐力: 0.2% Proof Stress	Min. Yield Strength 耐力: 1% Proof Stress	Tensile Strength N/mm² or MPa (kgf/mm²)	Index No. 索引番号
								B057
2.00~3.00	–	N ≦0.10	–	–	170	–	≧485	
2.00~2.50	–	–	S	–	190	245	490 ~690	
2.50~3.00	–	–	S	–	190	245	490 ~690	
2.0 ~2.5	–	–	–	–	190	225	490 ~690	
2.0~2.5	–	–	–	t < 5 5 ≦ t ≦50	215 205	255 245	510 ~710	
2.5~3.0	–	–	–	t < 5 5 ≦ t ≦50	215 205	255 245	510 ~710	
2.0~2.3	–	–	–	–	–	–	–	
2.00~3.00	–	N 0.12~0.22	–	–	275 (28)	–	≧550 (≧56)	B058
–	–	N 0.12~0.22	–	–	245 (25)	–	≧550 (≧56)	
2.00~3.00	–	N 0.10~0.16	–	–	240	–	≧550	
2.00~2.50	–	N 0.12~0.22	S	–	280	345	580 ~ 780	
2.50~3.00	–	N 0.12~0.22	S	–	280	345	580 ~ 780	
2.0~3.0	–	N ≦0.10	–	–	–	–	–	
–	–	–	–	–	–	–	–	
–	–	–	–	–	–	–	–	
–	–	–	–	–	–	–	–	B059
–	N ≦0.10	Ti 5C~0.70	–	–	205	–	≧515	
2.00~2.50	–	Ti 5XC ≦ 0.80	S	–	210	–	510 ~710	

Type of Steel 鋼種	Country 国	Standard No. 規格番号及び年号	Designation 種類記号	Chemical Composition 化学成分（%） C	Si	Mn	P	S	Cu	Cr	Ni
316Ti	GR独	DIN 17440-85	X6CrNiMo Ti17122	≦0.08	–	–	–	–	–	16.5～18.5	10.5～13.5
	NF仏	NF-A35-573-81	Z6CNDT 17-12	≦0.08	≦1.0	≦2.0	≦0.040	≦0.030	–	16.0～18.0	10.5～13.0
	URソ	ГОСТ 5632-72	08X17H 13M2T	≦0.08	≦0.8	≦2.0	≦0.035	≦0.020	–	16.0～18.0	12.0～14.0
			10X17H 13M2T	≦0.08	≦1.0	≦2.0	≦0.035	≦0.020	–	16.0～18.0	12.0～14.0
			08X17H 15M3T	≦0.08	≦0.8	≦2.0	≦0.035	≦0.020	–	16.0～18.0	14.0～16.0
316Nb	JP日	–	–	–	–	–	–	–	–	–	–
	US米	ASTM A240-94a	TYP.316cb UNS S31640	≦0.08	≦0.75	≦2.00	≦0.045	≦0.030	–	16.00～18.00	10.00～14.00
	UK英	–	–	–	–	–	–	–	–	–	–
	GR独	DIN 17440-85	X6CrNiMo Ti17122	≦0.08	–	–	–	–	–	16.5～18.5	10.5～13.5
	NF仏	NF-A35-573-90	Z6CNDNb 17-12	≦0.06	≦0.75	≦2.0	≦0.040	≦0.015	–	16.0～18.0	10.5～12.5
	URソ	ГОСТ 5632-72	08X16H 13M2Б	0.06～0.12	≦0.8	≦1.0	≦0.035	≦0.020	–	15.0～17.0	12.5～14.5
			09X16H 15M2Б	≦0.08	≦0.8	≦0.8	≦0.035	≦0.020	–	15.0～17.0	14.0～16.0
316Cu	JP日	JIS G4304-91	SUS316JI	≦0.08	≦1.00	≦2.00	≦0.045	≦0.030	1.00～2.50	17.00～19.00	10.00～14.00
			SUS316 JIL	≦0.03	≦1.00	≦2.00	≦0.045	≦0.030	1.00～2.50	17.00～19.00	12.00～16.00
	US米	–	–	–	–	–	–	–	–	–	–
	UK英	–	–	–	–	–	–	–	–	–	–
	GR独	–	–	–	–	–	–	–	–	–	–
	FR仏	–	–	–	–	–	–	–	–	–	–
	URソ	–	–	–	–	–	–	–	–	–	–
317	JP日	JIS G4304-91	SUS317	≦0.08	≦1.00	≦2.00	≦0.045	≦0.030	–	18.00～20.00	11.00～15.00
	US米										
		ASTM A240-94a	TYP.3M UNS S31700	≦0.08	≦0.75	≦2.00	≦0.045	≦0.030	–	18.00～20.00	11.00～15.00

$1N/mm^2 = 1.01972 \times 10^{-1} kgf/mm^2$

Mo	V	Others	Mechanical Properties 機械的性質 Heat Treatment 熱処理	Thickness (t)mm 板厚	Min. Yield Strength N/mm² or MPa(kgf/mm²) 耐力 0.2% Proof Stress	1% Proof Stress	Tensile Strength N/mm² or MPa (kgf/mm²)	Index No. 索引番号
2.0 ~2.5	–	Ti 4X%~0.80	–	–	210	245	500 ~730	B059
2.0~2.5	–	Ti 5C ~0.6	–	t < 5 5 ≦ t ≦50	235 225	275 265	550 ~750	
2.5~3.0	–	Ti 5C ~0.7	–	–	–	–	–	
2.0~2.3	–	Ti 5C ~0.6	–	–	–	–	–	
3.0~4.3	–	Ti 0.3~0.6	–	–	–	–	–	
–	–	–	–	–	–	–	–	B060
2.0 ~3.0	N ≦0.10	Cb 10C ~1.10	–	–	205	–	≧515	
–	–	–	–	–	–	–	–	
2.0 ~2.5	–	Nb 10C ~1.0	Q	–	215	250	510 ~740	
2.0~2.5	–	Nb + Ta 10C ~1.0	–	t < 5 5 ≦ t ≦50	235 225	275 265	550 ~750	
2.0~2.5	–	Nb 0.9~1.3	–	–	–	–	–	
2.5~3.0	–	Nb 0.9~1.3	–	–	–	–	–	
1.20~2.75	–	–	–	–	205 (21)	–	≧520 (≧53)	B061
1.20~2.75	–	–	–	–	175 (18)	–	≧480 (≧49)	
–	–	–	–	–	–	–	–	
–	–	–	–	–	–	–	–	
–	–	–	–	–	–	–	–	
–	–	–	–	–	–	–	–	
–	–	–	–	–	–	–	–	
3.00~4.00	–	–	–	–	205 (21)	–	≧520 (≧53)	B062
3.00~4.00	–	N ≦0.10	–	–	205	–	≧515	

Type of Steel 鋼種	Country 国	Standard No. 規格番号及び年号	Designation 種類記号	Chemical Composition 化学成分 (%) C	Si	Mn	P	S	Cu	Cr	Ni
317	UK英	−	−	−	−	−	−	−	−	−	−
	GR独	−	−	−	−	−	−	−	−	−	−
	FR仏	−	−	−	−	−	−	−	−	−	−
	URソ	−	−	−	−	−	−	−	−	−	−
317L	JP日	JIS G4304-91	SUS317L	≦0.03	≦1.00	≦2.00	≦0.045	≦0.030	−	18.00～20.00	11.00～15.00
	US米										
		ASTM A240-94a	TYP.317L UNS S31703	≦0.30	≦0.75	≦2.00	≦0.045	≦0.030	−	18.00～20.00	11.00～15.00
	UK英	−	−	−	−	−	−	−	−	−	−
	GR独	−	−	−	−	−	−	−	−	−	−
	FR仏	NF-A35-573-90	Z3CND 19-15-04	≦0.03	≦0.75	≦2.0	≦0.035	≦0.010	−	17.5～19.5	14.0～16.0
	URソ	−	−	−	−	−	−	−	−	−	−
321	JP日	JIS G4304-91	SUS321	≦0.08	≦1.00	≦2.00	≦0.040	≦0.030	−	17.00～19.00	9.00～13.00
	US米										
		ASTM A240-94a	TYP.321 UNS S32100	≦0.08	≦0.75	≦2.00	≦0.045	≦0.030	−	17.00～19.00	9.00～12.00
			TYP.321H UNS S32109	0.04～0.10	≦0.75	≦2.00	≦0.045	≦0.030	−	17.00～19.00	9.00～12.00
	UK英	BS1501 Part3-90	321S31	≦0.08	≦1.00	≦2.00	≦0.045	≦0.025	−	17.0～19.0	9.00～12.00
			321S51	0.04～0.10	≦1.00	≦2.00	≦0.045	≦0.025	−	17.0～19.0	9.00～12.00

$1N/mm^2 = 1.01972 \times 10^{-1} kgf/mm^2$

			Mechanical Properties 機械的性質					Index No. 索引番号
			Heat Treatment 熱処理	Thickness (t)mm 板厚	Min. Yield Strength N/mm² or MPa(kgf/mm²) 耐力		Tensile Strength N/mm² or MPa (kgf/mm²)	
Mo	V	Others			0.2% Proof Stress	1% Proof Stress		
–	–	–	–	–	–	–	–	B062
–	–	–	–	–	–	–	–	
–	–	–	–	–	–	–	–	
–	–	–	–	–	–	–	–	
3.00～4.00	–	–	–	–	175	–	≧480	B063
3.00～4.00	–	N ≦0.10	–	–	205	–	≧515	
–	–	–	–	–	–	–	–	
–	–	–	–	–	–	–	–	
3.0～4.0	–	–	Q	–	195	235	500～700	
–	–	–	–	–	–	–	–	
–	–	Ti ≧5C	–	–	205 (21)	–	≧520 (≧53)	B064
–	N ≦0.10	Ti 5(C+N)～0.70	–	–	205	–	≧515	
–	N ≦0.10	Ti 4(C+N)～0.70	–	–	205	–	≧515	
–	–	Ti 5C～0.80	S	–	200	–	510～710	
–	–	Ti 5C～0.80	S	–	175	–	490～690	

Type of Steel 鋼種	Country 国	Standard No. 規格番号及び年号	Designation 種類記号	Chemical Composition 化学成分（%）							
				C	Si	Mn	P	S	Cu	Cr	Ni
321	GR独	DIN 17440-85	X6CrNi Ti1818	≦0.08	–	–	–	–	–	17.0～19.0	9.00～12.00
	FR仏	NF-A35-573-81	Z6CNT 18-10	≦0.08	≦1.0	≦2.0	≦0.040	≦0.030	–	17.0～19.0	9.00～12
		NF-A36-209-91	Z6CNT 18-10	≦0.06	≦0.75	≦2.0	≦0.040	≦0.015	–	17.0～19.0	9.00～11.0
	URソ	ГОСТ 5632-72	12X18H	0.06～0.12	≦0.8	≦1.0	≦0.035	≦0.020	–	15.0～17.0	12.5～14.5
			08X18H 10T	≦0.08	≦0.8	≦0.8	≦0.035	≦0.020	–	15.0～17.0	14.0～16.0
			12X18H 10T	≦0.08	≦0.8	≦0.8	≦0.035	≦0.020	–	15.0～17.0	14.0～16.0
		ГОСТ 5632-72	08X18H 12T	≦0.08	≦0.8	≦2.0	≦0.035	≦0.020	–	17.0～19.0	11.0～13.0
			12X18H 12T	≦0.12	≦0.8	≦2.0	≦0.035	≦0.020	–	17.0～19.0	11.0～13.0
329	JP日	JIS G4304-91	SUS329J1	≦0.08	≦1.00	≦1.50	≦0.040	≦0.030	–	23.00～28.00	3.00～6.00
			SUS329 J3L	≦0.030	≦1.00	≦1.50	≦0.040	≦0.030	–	22.00～26.00	4.50～6.50
	US米	ASTM A240-94a	TYP.329 UNS S32900	≦0.08	≦0.75	≦1.00	≦0.040	≦0.030	–	23.00～28.00	2.50～5.00
	UK英	–	–	–	–	–	–	–	–	–	–
	GR独	–	–	–	–	–	–	–	–	–	–
	FR仏	–	–	–	–	–	–	–	–	–	–
	URソ	–	–	–	–	–	–	–	–	–	–
330	JP日	JIS G4312-91	SUH330	≦0.15	≦1.50	≦2.00	≦0.040	≦0.030	–	14.00～17.00	33.00～37.00
	US米	AISI MN06-222	330	≦0.08	0.75～1.50	≦2.00	≦0.040	≦0.030	–	17.00～20.00	34.00～37.00
	UK英	–	–	–	–	–	–	–	–	–	–
	GR独	–	–	–	–	–	–	–	–	–	–
	FR仏	–	–	–	–	–	–	–	–	–	–
	URソ	–	–	–	–	–	–	–	–	–	–
347	JP日	JIS G4312-87	SUS347	≦0.08	≦1.00	≦2.00	≦0.045	≦0.030	–	17.00～18.00	9.00～13.00

$1N/mm^2 = 1.01972 \times 10^{-1} kgf/mm^2$

Mo	V	Others	Mechanical Properties 機械的性質 Heat Treatment 熱処理	Thickness (t)mm 板厚	Min. Yield Strength N/mm² or MPa(kgf/mm²) 耐力 0.2% Proof Stress	1% Proof Stress	Tensile Strength N/mm² or MPa (kgf/mm²)	Index No. 索引番号
–	–	Ti 5C~0.80	Q	–	205	245	510 ~740	B064
–	–	Ti 5C ~0.6	–	t < 5	–	225	530 ~730	
				5≦ t ≦25	–	205		
–	–	N ≦ 0.050 Ti 5(C+N) ≦ 0.70	–	t < 5	225	265	t ≦20 530 ~730	
				5 ≦ t ≦75	215	255	20 < t ≦75 570 ~710	
2.0~2.5	–	Nb 0.9~1.3	–	–	–	–	–	
2.5~3.0	–	Nb 0.9~1.3	–	–	–	–	–	
2.5~3.0	–	Nb 0.9~1.3	–	–	–	–	–	
–	–	Ti 5C~0.8	–	–	–	–	–	
–	–	Ti 5C~0.8	–	–	–	–	–	
1.00~3.00	–	–	S	–	390 (40)	–	≧590 (≧60)	B065
2.50~3.50	–	N 0.08~0.20	S	–	450 (46)	–	≧620 (≧63)	
1.00~2.00	–	–	–	–	485	–	≧620	
–	–	–	–	–	–	–	–	
–	–	–	–	–	–	–	–	
–	–	–	–	–	–	–	–	
–	–	–	–	–	–	–	–	
–	–	–	S	–	205 (21)	–	≧560 (≧57)	B066
–	–	–	–	–	–	–	–	
–	–	–	–	–	–	–	–	
–	–	–	–	–	–	–	–	
–	–	–	–	–	–	–	–	
–	–	–	–	–	–	–	–	
–	–	Nb+Ta ≧10C	S	–	205 (21)	–	≧520 (≧53)	B067

Type of Steel 鋼種	Country 国	Standard No. 規格番号及び年号	Desig-nation 種類記号	Chemical Composition 化学成分 (%)							
				C	Si	Mn	P	S	Cu	Cr	Ni
347	US米										
		ASTM A240-94a	TYP.347 UNS S34700	≦0.08	≦0.75	≦2.00	≦0.045	≦0.030	–	17.00 ～ 19.00	9.00 ～ 13.00
			TYP.347H UNS S34709	0.04～0.10	≦0.75	≦2.00	≦0.045	≦0.030	–	17.00 ～ 19.00	9.00 ～ 13.00
	UK英	BS1501 Part3-90	347S31	≦0.08	≦1.00	≦2.00	≦0.045	≦0.025	–	17.0～ 19.0	9.0～ 12.0
			347S51	0.04～0.10	≦1.00	≦2.00	≦0.045	≦0.025	–	17.0～ 19.0	9.0～ 12.0
	GR独	DIN 17440-85	X6CrNiNb 1810	≦0.08	–	–	–	–	–	17.0～ 19.0	9.0～ 12.0
	FR仏	NF-A36-209-91	Z 6 CNNb 18-10	≦0.06	≦0.75	≦0.75	≦0.040	≦0.015	–	17.0 ～ 19.0	9.00 ～ 11.0
	URソ	ГОСТ 5632-72	08X18H 12Б	≦0.08	≦0.8	≦2.0	≦0.035	≦0.020	–	17.0～ 19.0	11.0～ 13.0
348	JP日	–	–	–	–	–	–	–	–	–	–
	US米										
		ASTM A240-94a	TYP. 348 UNS S34800	≦0.08	≦0.75	≦2.00	≦0.045	≦0.030	Co ≦0.20	17.00 ～ 19.00	9.00 ～ 13.00
			TYP.348H UNS S34809	0.04～0.10	≦0.75	≦2.00	≦0.045	≦0.030	Co ≦0.20	17.00 ～ 19.00	9.00 ～ 13.00
	UK英	–	–	–	–	–	–	–	–	–	–
	GR独	–	–	–	–	–	–	–	–	–	–
	FR仏	–	–	–	–	–	–	–	–	–	–
	URソ	–	–	–	–	–	–	–	–	–	–
381	JP日	JIS G4312-91	SUSXM 15J1	≦0.08	3.00～5.00	≦2.00	≦0.045	≦0.035	–	15.00 ～ 20.00	11.50 ～ 15.00

$1N/mm^2 = 1.01972 \times 10^{-1} kgf/mm^2$

Mo	V	Others	Mechanical Properties 機械的性質 Heat Treatment 熱処理	Thickness (t)mm 板厚	Min. Yield Strength N/mm^2 or $MPa(kgf/mm^2)$ 耐力 0.2% Proof Stress	 1% Proof Stress	Tensile Strength N/mm^2 or MPa (kgf/mm^2)	Index No. 索引番号
								B067
–	–	Cb+Ta 10C ~1.00	–	–	205	–	≧515	
–	–	Cb+Ta 8C ~1.00	–	–	205	–	≧515	
–	–	Nb 10xC≦1.0	S	–	205	–	510~710	
–	–	Nb 10xC≦1.2	S	–	205	–	510~710	
–	–	Nb 10C~1.00	S	–	205	240	510 ~710	
–	–	Nb+Ta 10C~1.0	–	t < 5 5 < t ≦75	225 215	265 255	t≦20 (56-57) 20< t ≦75 (54-75)	
–	–	Nb 10C~1.1	–	–	–	–	–	
–	–	–	–	–	–	–	–	B068
Co ≦0.20	Ta ≦0.10	Cb+Ta 10C ~1.00	–	–	205	–	≧515	
Co ≦0.20	Ta ≦0.10	Cb+Ta 8C ~1.00	–	–	205	–	≧515	
–	–	–	–	–	–	–	–	
–	–	–	–	–	–	–	–	
–	–	–	–	–	–	–	–	
–	–	–	–	–	–	–	–	
–	–	–	S	–	205 (21)	–	≧520 (≧53)	B069

Type of Steel 鋼種	Country 国	Standard No. 規格番号及び年号	Designation 種類記号	C	Si	Mn	P	S	Cu	Cr	Ni
				Chemical Composition 化学成分 (%)							
381	US米										
		ASTM A240-94a	TYP.XM-15 UNS S38100	≦0.08	1.50～2.50	≦2.00	≦0.030	≦0.030	－	17.00 ～ 19.00	17.50 ～ 18.50
	UK英	－	－	－	－	－	－	－	－	－	－
	GR独	－	－	－	－	－	－	－	－	－	－
	FR仏	－	－	－	－	－	－	－	－	－	－
	URソ	－	－	－	－	－	－	－	－	－	－
Others	JP日	JIS G4304-91	SUS631	≦0.09	≦1.00	≦1.00	≦0.040	≦0.030	－	16.00 ～ 18.00	6.50 ～ 7.55
	UK英	BS1501 Part3-90	318S13	≦0.03	≦1.00	≦2.00	≦0.025	≦0.020		21.00～ 23.00	4.50 ～ 6.50
			904S13	≦0.03	≦1.00	≦2.00	≦0.040	≦0.025	1.00～2.00	19.00～ 22.00	24～27
	URソ	ГОСТ 5632-72	09X17HЮ	≦0.09	≦0.8	≦0.8	≦0.030	≦0.020	－	16.0～ 17.5	7.0～8.0
		ГОСТ 5632-72	09X17H Ю1	≦0.09	≦0.8	≦0.8	≦0.035	≦0.025	－	16.5～ 18.5	6.5～7.5
Cr-Steel											
403	JP日	JIS G4304-91	SUS403	≦0.15	≦0.50	≦1.00	≦0.040	≦0.030	－	11.50 ～ 13.00	≦0.60
	US米										
		ASTM A176-94	TYP.403 UNS A40300	≦0.15	≦0.50	≦1.00	≦0.040	≦0.030	－	11.50 ～ 13.00	≦0.60
	UK英	BS1501 Part3-90	403S17	≦0.08	≦0.80	≦1.0	≦0.040	≦0.030	－	12.0～ 14.0	≦0.50
	GR独	DIN 17440	X6Cr13	≦0.08	－	－	－	－	－	12.0～ 14.0	－
	FR仏	－	－	－	－	－	－	－	－	－	－
	URソ	ГОСТ 5632-72	12X13	0.09～0.15	≦0.8	≦0.8	≦0.030	≦0.025	－	12.0～ 14.0	－
			08X13	≦0.08	≦0.8	≦0.8	≦0.030	≦0.025	－	12.0～ 13.0	－
405	JP日	JIS G4304-91	SUS405	≦0.08	≦1.00	≦1.00	≦0.040	≦0.030	－	11.50 ～ 14.50	≦0.60

1N/mm² =1.01972 ×10⁻¹kgf/mm²

			Mechanical Properties 機械的性質					Index No. 索引番号
			Heat Treatment 熱処理	Thickness (t)mm 板厚	Min. Yield Strength N/mm² or MPa(kgf/mm²) 耐力		Tensile Strength N/mm² or MPa (kgf/mm²)	
Mo	V	Others			0.2% Proof Stress	1% Proof Stress		
								B069
–	–	–	–	–	205	–	≧515	
–	–	–	–	–	–	–	–	
–	–	–	–	–	–	–	–	
–	–	–	–	–	–	–	–	
–	–	–	–	–	–	–	–	
–	–	AL 0.75～1.50	S	–	380 (39)	–	≧1030 (≧105)	B070
2.50～3.50	–	N 0.08～0.20	S	> 20 80<t<100	450 480	–	680～880 640～840	
4.00～5.00	–	–	S	–	220	–	520～720	
–	–	AL 0.5～0.8	–	–	–	–	–	
–	–	Al 0.7～1.1	–	–	–	–	–	
–	–	–	A	–	205 (21)	–	≧440 (≧45)	B071
–	–	–	–	–	205	–	≧485	
–	–	–	A	–	205	265	≧420	
–	–	–	–	–	250	–	400～600	
–	–	–	–	–	–	–	–	
–	–	–	–	–	–	–	–	
–	–	–	–	–	–	–	–	
–	–	Al 0.10～0.30	A	–	175 (18)	–	≧410 (≧42)	B072

Type of Steel 鋼 種	Country 国	Standard No. 規格番号及び年号	Designation 種類記号	Chemical Composition 化 学 成 分 (%)							
				C	Si	Mn	P	S	Cu	Cr	Ni
4 0 5	US米										
		ASTM A240-94a	TYP.405 UNS S40500	≦0.08	≦1.00	≦1.00	≦0.040	≦0.030	–	11.50 ～ 14.50	≦0.60
	UK英	BS1501 Part3-73	405S17	≦0.08	≦0.80	≦1.0	≦0.040	≦0.030	–	12.0～ 14.0	≦0.50
	GR独	DIN 17440	X6CrA113	≦0.08	–	–	–	–	–	12.0～ 14.0	–
	FR仏	–	–	–	–	–	–	–	–	–	–
	URソ	–	–	–	–	–	–	–	–	–	–
4 0 9	JP日										
	US米										
		ASTM A240-94a	TYP.409 UNS S40900	≦0.08	≦1.00	≦1.00	≦0.45	≦0.45	–	10.50 ～ 11.75	≦0.50
	UK英	–	–	–	–	–	–	–	–	–	–
	GR独	–	–	–	–	–	–	–	–	–	–
	FR仏	–	–	–	–	–	–	–	–	–	–
	URソ	–	–	–	–	–	–	–	–	–	–
4 1 0	JP日	JIS 4304-91	SUS410	≦0.15	≦1.00	≦1.00	≦0.040	≦0.030	–	11.50 ～ 13.50	≦0.60
			SUS410L	≦0.03	≦1.00	≦1.00	≦0.040	≦0.030	–	11.00 ～ 13.50	–
	US米										
		ASTM A240-94a	TYP.410 UNS S41000	≦0.15	≦1.00	≦1.00	≦0.040	≦0.030	–	11.50 ～ 13.50	≦0.75
	UK英	–	–	–	–	–	–	–	–	–	–
	GR独	DIN 17440-85	X6Cr13	≦0.08	–	–	–	–	–	12.0～ 14.0	–
			X10Cr13	0.08～0.12	–	–	–	–	–	12.0～ 14.0	–
	FR仏	–	–	–	–	–	–	–	–	–	–
	URソ	–	–	–	–	–	–	–	–	–	–
410S	JP日	JIS G4304-91	SUS410S	≦0.08	≦1.00	≦1.00	≦0.040	≦0.030	–	11.50 ～ 13.50	≦0.60

$1N/mm^2 = 1.01972 \times 10^{-1} kgf/mm^2$

			Mechanical Properties 機械的性質					Index No. 索引番号
			Heat Treatment 熱処理	Thickness (t)mm 板厚	Min. Yield Strength N/mm^2 or MPa(kgf/mm^2) 耐力		Tensile Strength N/mm^2 or MPa (kgf/mm^2)	
Mo	V	Others			0.2% Proof Stress	1% Proof Stress		
								B072
–	–	Al 0.10~0.30	–	–	170	–	≧415	
–	–	Al 0.10~0.30	A	–	205	265	≧420	
–	–	Al 0.10~0.30	–	–	250	–	400 ~600	
–	–	–	–	–	–	–	–	
–	–	–	–	–	–	–	–	
								B073
–	–	Ti 6C~0.75	–	–	205	–	≧380	
–	–	–	–	–	–	–	–	
–	–	–	–	–	–	–	–	
–	–	–	–	–	–	–	–	
–	–	–	–	–	–	–	–	
–	–	–	A	–	205 (21)	–	≧440 (≧45)	B074
–	–	–	A	–	195 (20)	–	≧360 (≧37)	
–	–	–	–	–	205	–	≧450	
–	–	–	–	–	–	–	–	
–	–	–	–	–	250	–	400~600	
–	–	–	–	–	250	–	450~650	
–	–	–	–	–	–	–	–	
–	–	–	–	–	–	–	–	
–	–	–	–	–	205 (21)	–	≧440 (≧45)	B075

Type of Steel 鋼 種	Country 国	Standard No. 規格番号 及び年号	Designation 種類記号	Chemical Composition 化 学 成 分 (%)							
				C	Si	Mn	P	S	Cu	Cr	Ni
410S	US米										
		ASTM A240-94a	TYP.410S UNS S41008	≦0.08	≦1.00	≦1.00	≦0.040	≦0.030	–	11.50 ～ 13.50	≦0.60
	UK英	–	–	–	–	–	–	–	–	–	–
	GR独	DIN 17440-85	X6Cr13	≦0.08	–	–	–	–	–	12.0～ 14.0	–
	FR仏	–	–	–	–	–	–	–	–	–	–
	URソ	–	–	–	–	–	–	–	–	–	–
4 2 0	JP日	–	–	–	–	–	–	–	–	–	–
	US米	ASTM A176-94	TYP.420 UNS S42000	0.30～0.40	≦1.00	≦1.00	≦0.040	≦0.030	–	12.00 ～ 14.00	≦0.75
	UK英	–	–	–	–	–	–	–	–	–	–
	GR独	DIN 17440-85	X20Cr13	0.17～0.25	–	–	–	–	–	12.0～ 14.0	–
	FR仏	–	–	–	–	–	–	–	–	–	–
	URソ	ГОСТ 5632-72	20X13	0.16～0.25	≦0.8	≦0.8	≦0.030	≦0.025	–	12.0～ 14.0	≦0.60
4 2 9	JP日	JIS G4304-91	SUS429	≦0.12	≦1.00	≦1.00	≦0.040	≦0.030	–	14.00 ～ 16.00	≦0.60
			SUS429J1	0.25～0.40	≦1.00	≦1.00	≦0.040	≦0.030	–	15.00 ～ 17.00	≦0.60
	US米										
		ASTM A240-94a	TYP.429 UNS S42900	≦0.12	≦1.00	≦1.00	≦0.040	≦0.030	–	14.00 ～ 16.00	≦0.75
	UK英	–	–	–	–	–	–	–	–	–	–
	GR独	–	–	–	–	–	–	–	–	–	–
	FR仏	–	–	–	–	–	–	–	–	–	–
	URソ	–	–	–	–	–	–	–	–	–	–

1N/mm² =1.01972 ×10⁻¹kgf/mm²

Mo	V	Others	Heat Treatment 熱処理	Thickness (t)mm 板厚	Mechanical Properties 機械的性質 Min. Yield Strength N/mm² or MPa(kgf/mm²) 耐力 0.2% Proof Stress	1% Proof Stress	Tensile Strength N/mm² or MPa (kgf/mm²)	Index No. 索引番号
								B075
–	–	–	–	–	205 (21)	–	≧415 (≧42)	
–	–	–	–	–	–	–	–	
–	–	–	–	–	250	–	400~600	
–	–	–	–	–	–	–	–	
–	–	–	–	–	–	–	–	
								B076
–	–	–	–	–	–	–	–	B077
–	–	–	–	–	–	–	≧690	
–	–	–	–	–	–	–	–	
–	–	–	–	–	–	–	≦740	
–	–	–	–	–	–	–	–	
–	–	–	–	–	–	–	–	
–	–	–	A	–	205 (21)	–	≧450 (≧46)	B078
–	–	–	A	–	225 (23)	–	≧520 (≧53)	
–	–	–	–	–	205	–	≧450	
–	–	–	–	–	–	–	–	
–	–	–	–	–	–	–	–	
–	–	–	–	–	–	–	–	
–	–	–	–	–	–	–	–	

Type of Steel 鋼種	Country 国	Standard No. 規格番号及び年号	Designation 種類記号	Chemical Composition 化学成分（%）							
				C	Si	Mn	P	S	Cu	Cr	Ni
430	JP日	JIS G4304-91	SUS430	≦0.12	≦0.75	≦1.00	≦0.040	≦0.030	–	16.00～18.00	≦0.60
	US米										
		ASTM A240-94a	TYP.430 UNS S43000	≦0.12	≦1.00	≦1.00	≦0.040	≦0.030	–	16.00～18.00	≦0.75
	UK英	–	–	–	–	–	–	–	–	–	–
	GR独	–	–	–	–	–	–	–	–	–	–
	FR仏	–	–	–	–	–	–	–	–	–	–
	URソ	ГОСТ 5632-72	12X17	≦0.12	≦0.8	≦0.8	≦0.035	≦0.025	–	16.0～18.0	–
430Ti 430Al 430Nb	JP日	JIS G4304-91	SUS430LX	≦0.030	≦0.075	≦1.00	≦0.040	≦0.030	–	16.00～19.00	≦0.60
	US米	–	–	–	–	–	–	–	–	–	–
	UK英	–	–	–	–	–	–	–	–	–	–
	GR独	DIN 17441-85	X6CrTi7	≦0.8	–	–	–	–	–	16.0～18.0	–
			X6CrNb17	≦0.8	–	–	–	–	–	16.0～	–
	FR仏	–	–	–	–	–	–	–	–	–	–
	URソ	ГОСТ 5632-72	10X18CЮ	≦0.15	1.0～1.5	≦0.08	≦0.035	≦0.025	–	17.0～20.0	–
			08X17T	≦0.08	≦0.08	≦0.08	≦0.035	≦0.025	–	16.0～18.0	–
431	JP日	–	–	–	–	–	–	–	–	–	–
	US米	ASTM A240-94a	TYP.431 UNS S43100	≦0.20	≦1.00	≦1.00	≦0.040	≦0.030	–	15.00～17.00	1.25～2.50
	UK英	–	–	–	–	–	–	–	–	–	–
	GR独	–	–	–	–	–	–	–	–	–	–
	FR仏	–	–	–	–	–	–	–	–	–	–
	URソ	ГОСТ 5632-72	14X17H2	0.11～0.17	≦0.08	≦0.08	≦0.030	≦0.025	–	–	1.5～2.5
			20X17H2	0.17～0.25	≦0.08	≦0.08	≦0.035	≦0.025	–	1.60～1.80	1.5～2.5
434	JP日	JIS G4312-91	SUS434	≦0.12	≦1.00	≦1.00	≦0.040	≦0.030	–	16.00～18.00	–

1N/mm² = 1.01972 ×10⁻¹ kgf/mm²

Mo	V	Others	Mechanical Properties 機械的性質 Heat Treatment 熱処理	Thickness (t)mm 板厚	Min. Yield Strength N/mm² or MPa(kgf/mm²) 耐力 0.2% Proof Stress	1% Proof Stress	Tensile Strength N/mm² or MPa (kgf/mm²)	Index No. 索引番号
–	–	–	A	–	205 (21)	–	≧450 (≧46)	B079
–	–	–	–	–	205	–	≧450	
–	–	–	–	–	–	–	–	
–	–	–	–	–	–	–	–	
–	–	–	–	–	–	–	–	
–	–	–	–	–	–	–	–	
–	–	Ti or Nb 0.1～1.00	–	–	175 (18)	–	≧360 (≧37)	B080
–	–	–	–	–	–	–	–	
–	–	–	–	–	–	–	–	
–	–	Ti 7C ～1.20	A	–	270	–	450 ～600	
–	–	Nb –	–	–	270	–	450 ～600	
–	–	–	–	–	–	–	–	
–	–	Al 0.7～1.2	–	–	–	–	–	
–	–	Al 5C～0.8	–	–	–	–	–	
–	–	–	–	–	–	–	–	B081
–	–	–	–	–	245	–	≧415	
–	–	–	–	–	–	–	–	
–	–	–	–	–	–	–	–	
–	–	–	–	–	–	–	–	
–	–	–	–	–	–	–	–	
–	–	–	–	–	–	–	–	
0.75～1.25	–	–	–	–	205	–	≧450	B082

Type of Steel 鋼種	Country 国	Standard No. 規格番号及び年号	Designation 種類記号	Chemical Composition 化学成分（%） C	Si	Mn	P	S	Cu	Cr	Ni
434	US米										
	UK英	−	−	−	−	−	−	−	−	−	−
	GR独	DIN 17441-85	X6CrMo 171	≦0.08	−	−	−	−	−	16.00 ～ 18.00	−
	FR仏	−	−	−	−	−	−	−	−	−	−
	URソ	−	−	−	−	−	−	−	−	−	−
436	JP日	JIS G4304-91	SUS436L	≦0.025	≦1.00	≦1.00	≦0.040	≦0.030	−	16.00 ～ 18.00	≦0.60
	US米										
	UK英	−	−	−	−	−	−	−	−	−	−
	GR独	−	−	−	−	−	−	−	−	−	−
	FR仏	−	−	−	−	−	−	−	−	−	−
	URソ	−	−	−	−	−	−	−	−	−	−
442	JP日	−	−	−	−	−	−	−	−	−	−
	US米										
		ASTM A176-94	TYP.442 UNS S44200	≦0.20	≦1.00	≦1.00	≦0.040	≦0.040	−	18.00 ～ 23.00	≦0.60
	UK英	−	−	−	−	−	−	−	−	−	−
	GR独	−	−	−	−	−	−	−	−	−	−
	FR仏	−	−	−	−	−	−	−	−	−	−
	URソ	−	−	−	−	−	−	−	−	−	−
444	JP日	JIS G4304-91	SUS444	≦0.025	≦1.00	≦1.00	≦0.040	≦0.030	−	17.00 ～ 20.00	−
	US米										
		ASTM A240-94a	UNS S44400	≦0.025	≦1.00	≦1.00	≦0.040	≦0.030	−	17.50 ～ 19.5	≦1.00
	UK英	−	−	−	−	−	−	−	−	−	−
	GR独	−	−	−	−	−	−	−	−	−	−
	FR仏	−	−	−	−	−	−	−	−	−	−
	URソ	−	−	−	−	−	−	−	−	−	−

$1 N/mm^2 = 1.01972 \times 10^{-1} kgf/mm^2$

			Mechanical Properties 機械的性質					Index No. 索引番号
Mo	V	Others	Heat Treatment 熱処理	Thickness (t)mm 板厚	Min. Yield Strength N/mm^2 or MPa(kgf/mm^2) 耐力 0.2% Proof Stress	Min. Yield Strength N/mm^2 or MPa(kgf/mm^2) 耐力 1% Proof Stress	Tensile Strength N/mm^2 or MPa (kgf/mm^2)	
								B082
–	–	–	–	–	–	–	–	
0.9 ~1.3	–	–	A	–	275	240	450 ~650	
–	–	–	–	–	–	–	–	
–	–	–	–	–	–	–	–	
0.75~1.25	–	Ti+Nb+Zr 8(C+N)~ 0.80	–	–	205 (21)	–	≧450 (≧46)	B083
–	–	–	–	–	–	–	–	
–	–	–	–	–	–	–	–	
–	–	–	–	–	–	–	–	
–	–	–	–	–	–	–	–	
–	–	–	–	–	–	–	–	B084
–	–	–	–	–	275	–	≧515	
–	–	–	–	–	–	–	–	
–	–	–	–	–	–	–	–	
–	–	–	–	–	–	–	–	
–	–	–	–	–	–	–	–	
1.75~2.50	N ≦0.025	Ti+Nb+Zr 8(C+N)~ 0.80	–	–	245 (25)	–	≧410 (≧42)	B085
1.75~2.00	N ≦0.035	Cb+Ti 0.20+4(C +N)~0.8	–	–	275	–	≧415	
–	–	–	–	–	–	–	–	
–	–	–	–	–	–	–	–	
–	–	–	–	–	–	–	–	
–	–	–	–	–	–	–	–	

Type of Steel 鋼 種	Country 国	Standard No. 規格番号及び年号	Desig-nation 種類記号	Chemical Composition 化 学 成 分 （ % ）							
				C	Si	Mn	P	S	Cu	Cr	Ni
4 4 6	J P 日	J I S G4312-91	SUH446	≦0.20	≦1.00	≦1.50	≦0.040	≦0.030	–	23.00 ～ 27.00	–
		J I S G4304-87	SUSXM27	≦0.010	≦0.40	≦0.40	≦0.030	≦0.020	≦0.20	25.00 ～ 27.00	–
	U S 米										
		A S T M A176-94	TYP.446 UNS S44600	≦0.20	≦1.00	≦1.50	≦0.040	≦0.030	–	23.00 ～ 27.00	–
			X M27	≦0.010	≦0.40	≦0.40	≦0.020	≦0.020	≦0.20	25.00 ～ 27.50	≦0.75
		A S T M A240-94a	TYP.XM27 UNS S44627	≦0.010	≦0.40	≦0.40	≦0.020	≦0.020	≦0.20	25.00 ～ 27.00	≦0.50
			TYP.XM33 UNS S44626	≦0.06	≦0.75	≦0.75	≦0.040	≦0.020	≦0.20	25.00 ～ 27.00	≦0.50
	U K 独	–	–	–	–	–	–	–	–	–	–
	G R 独	–	–	–	–	–	–	–	–	–	–
	F R 仏	–	–	–	–	–	–	–	–	–	–
	U R ソ	ГОСТ 5632-72	15X25T	≦0.15	≦1.0	≦0.8	≦0.035	≦0.025	–	24.0～ 27.0	–

1N/mm² = 1.01972 ×10^{-1} kgf/mm²

Mo	V	Others	Mechanical Properties 機械的性質 Heat Treatment 熱処理	Thickness (t)mm 板厚	Min. Yield Strength N/mm² or MPa(kgf/mm²) 耐力 0.2% Proof Stress	1% Proof Stress	Tensile Strength N/mm² or MPa (kgf/mm²)	Index No. 索引番号
–	–	N ≦0.25	A	–	275 (28)	–	≧510 (≧52)	B086
17.5～25.0	N ≦0.025	Ti+Nb+Zr 8(C+N)～0.80	–	–	245 (28)	–	≧410 (≧42)	
–	–	N ≦0.25	–	–	275	–	≧515	
–	–	N ≦0.25	–	–	275	–	≧450	
0.75～1.50	Ni+Cu ≦0.50	Cb 0.05～0.20 N≦0.015	–	–	275	–	≧450	
0.75～1.50	Ti 0.20～1.00and ≧ 7(C+N)	N ≦0.04	–	–	310	–	≧470	
–	–	–	–	–	–	–	–	
–	–	–	–	–	–	–	–	
–	–	–	–	–	–	–	–	
–	–	Ti 5C ～0.90	–	–	–	–	–	

Type of Steel 鋼種	Country 国	Standard No. 規格番号及び年号	Designation 種類記号	Chemical Composition 化学成分（%）							
				C	Si	Mn	P	S	Cu	Cr	Ni
447	JP日	G4304-91	SUS447J1	≦0.010	≦0.40	≦0.40	≦0.030	≦0.030		28.50～32.00	–
	US米										
		ASTM A240-94a	UNS S44700	≦0.010	≦0.20	≦0.30	≦0.025	≦0.020	≦0.15	28.00～30.00	≦0.15
			UNS S44735	≦0.30	≦1.00	≦1.00	≦0.040	≦0.030	–	28.00～30.00	≦1.00
	UK独	–	–	–	–	–	–	–	–	–	–
	GR独	–	–	–	–	–	–	–	–	–	–
	FR仏	–	–	–	–	–	–	–	–	–	–
	URソ	–	–	–	–	–	–	–	–	–	–
448	JP日	–	–	–	–	–	–	–	–	–	–
	US米										
		ASTM A240-94a	S44800	≦0.010	≦0.20	≦0.30	≦0.025	≦0.020	≦0.15	28.00～30.00	2.0～2.5
	UK独	–	–	–	–	–	–	–	–	–	–
	GR独	–	–	–	–	–	–	–	–	–	–
	FR仏	–	–	–	–	–	–	–	–	–	–
	URソ	–	–	–	–	–	–	–	–	–	–
Others	JP日	JIS G4312-91	SUH21	≦0.10	≦1.50	≦1.00	≦0.040	≦0.030	–	17.00～21.00	≦0.60
		JIS G4304-87	SUS440A	0.60～0.75	≦1.00	≦1.00	≦0.040	≦0.030	–	16.00～18.00	–

1N/mm² = 1.01972 × 10^{-1} kgf/mm²

Mo	V	Others	Mechanical Properties 機械的性質 Heat Treatment 熱処理	Thickness (t)mm 板厚	Min. Yield Strength N/mm² or MPa(kgf/mm²) 耐力 0.2% Proof Stress	1% Proof Stress	Tensile Strength N/mm² or MPa (kgf/mm²)	Index No. 索引番号
1.50~2.50	–	N ≦0.015	–	–	–	–	–	B087
3.5~4.2	Cu+N ≦0.025	N ≦0.020	–	–	415	–	≧550	
3.60-4.20	Ti+Cb≦ 0.20-1.00 +6(C+N)	N ≦0.045	–	–	415	–	≧550	
–	–	–	–	–	–	–	–	
–	–	–	–	–	–	–	–	
–	–	–	–	–	–	–	–	
–	–	–	–	–	–	–	–	
–	–	–	–	–	–	–	–	B088
3.5~4.2	–	N ≦0.020	–	–	415	–	≧550	
–	–	–	–	–	–	–	–	
–	–	–	–	–	–	–	–	
–	–	–	–	–	–	–	–	
–	–	–	–	–	–	–	–	
–	–	Al 2.00~4.00	A	–	245	–	≧440	B089
–	–	–	–	–	245	–	≧590	

INDEX OF CHAPTER II

STEEL PLATES FOR BOILER AND PRESSURE VESSELS

第2部　ボイラ及び圧力容器用鋼板

索　　引

"Index Xo." stands for the
"Index No." in the last column
of each page.

「索引番号」は本文各頁の右端の
「索引番号」の欄の数字をあらわす。

Standard No. 規格番号	Grade グレード	Index No. 索引番号
〔JIS〕		
JIS-G3103	SB410(SB42)	B003
	SB450(SB46)	B004
	SB450M(SB46M)	B011
	SB480(SB49)	B006
	SB480M(SB49M)	B011
JIS-G3115	SPV235(SPV24)	B003
	SPV315(SPV32)	B006
	SPV355(SPV36)	B007
	SPV450(SPV46)	B008, B028
	SPV490(SPV50)	B008, B028
JIS-G3116	SG255(SG26)	B003
	SG295(SG30)	B004
	SG325(SG33)	B006
	SG365(SG37)	B008
JIS-G3118	SGV410(SGV42)	B003
	SGV450(SGV46)	B004
	SGV480(SGV49)	B006
JIS-G3119	SBV1A	B012
	SBV1B	B012
	SBV2	B014
	SBV3	B015
JIS-G3120	SQV1A	B012
	SQB1B	B012
	SQV2A	B014
	SQV2B	B014
	SQV3A	B015
	SQV3B	B015
JIS-G3124	SEV245(SEV25)	B009
	SEV295(SEV30)	B009
	SEV345(SEV35)	B009
JIS-G3126	SLA235A(SLA24A)	B029
	SLA235B(SLA24B)	B029
	SLA325A(SLA33A)	B029
	SLA325B(SLA33B)	B029
	SLA360(SLA37)	B029
JIS-G3127	SL2N255(SL2N26)	B032
	SL3N255(SL3N26)	B033
	SL3N275(SL3N28)	B033
	SL3N440(SL3N45)	B033
	SL5N590(SL5N60)	B034
	SL9N520(SL9N53)	B035
	SL9N590(SL9N60)	B035
JIS-G4109	SCMV1-1	B016
	SCMV1-2	B016
	SCMV2-1	B017
	SCMV2-2	B017
	SCMV3-1	B018
	SCMV3-2	B018
	SCMV4-1	B019
	SCMV4-2	B019
	SCMV5-1	B020
	SCMV5-2	B020
	SCMV6-1	B021
	SCMV6-2	B021
JIS-G4304	SUS201	B036
	SUS202	B037
	SUS302	B044
	SUS302B	B045
	SUS304	B046
	SUS304L	B047
	SUS304N1	B048
	SUS304N2	B049
	SUS305	B050
	SUS309S	B053
	SUS310	B054
	SUS310S	B055
	SUS316	B056
	SUS316JI	B061
	SUS316JIL	B061
	SUS316L	B057
	SUS316LN	B058
	SUS316N	B058
	SUS317	B062
	SUS317L	B063
	SUS321	B064
	SUS329J1	B065
	SUS319J2	B065
	SUS347	B067
	SUS403	B071
	SUS405	B072
	SUS409	B073
	SUS410	B074
	SUS410S	B075
	SUS429	B078
	SUS430	B079
	SUS430LX	B080
	SUS434	B082
	SUS436L	B083
	SUS440A	B092

Standard No. 規 格 番 号	Grade グレード	Index No. 索引番号
	SUS444	B085
	SUS631	B070
	SUSXM15JI	B069
	SUSXM27	B086
JIS-G4312	SUH21	B089
	SUH309	B052
	SUH310	B054
	SUH330	B066
	SUH409	B073
	SUH446	B086
〔ASTM〕		
ASTM-A167	301	B043
	302	B044
	302B	B045
	304	B046
	304L	B047
	305	B050
	308	B051
	309	B052
	309S	B053
	310	B054
	310S	B055
	316	B056
	316L	B057
	317	B062
	317L	B063
	321	B064
	347	B067
	348	B068
	381	B069
	XM15	B069
ASTM-A176	403	B071
	405	B072
	409	B073
	410	B074
	410S	B075
	429	B078
	430	B079
	442	B084
	444	B085
	447	B087
	448	B088
ASTM-A202M	Gr. A	B026
	Gr. B	B026
ASTM-A203M	Gr. A	B032
	Gr. B	B032
ASTM-A204M	Gr. A	B011
	Gr. B	B011
	Gr. C	B011
ASTM-A225M	Gr. C	B027
	Gr. D	B027
ASTM-A240	209	B038
	214	B039
	302	B044
	304	B046
	304L	B047
	304N	B048
	305	B050
	309S	B053
	310S	B055
	316	B056
	316H	B056
	316L	B057
	316N	B058
	317	B062
	317L	B063
	321	B064
	347	B067
	348	B068
	381	B069
	405	B072
	409	B073
	410	B074
	429	B078
	430	B079
	444	B085
	446	B086
	447	B087
	448	B088
	XM15	B069
	XM17	B040
	XM18	B040
	XM29	B041
ASTM-A285M	Gr. A	B001
	Gr. B	B001
	Gr. C	B002
ASTM-A299M	—	B007
ASTM-A302M	Gr. A	B012

Standard No. 規格番号	Grade グレード	Index No. 索引番号
	Gr. B	B012
	Gr. C	B014
	Gr. D	B015
ASTM-A387M	Gr. 2	B016
	Gr. 5	B021
	Gr. 7	B022
	Gr. 9	B023
	Gr. 11	B018
	Gr. 21	B020
	Gr. 22	B019
	Gr. 91	B023
ASTM-A442M	Gr. 55(Gr. 380)	B002
	Gr. 60(Gr. 415)	B003
ASTM-A455M	−	B007
ASTM-A515M	Gr. 55(Gr. 380)	B002
	Gr. 60(Gr. 415)	B003
	Gr. 65(Gr. 450)	B004
	Gr. 70(Gr. 415)	B006
ASTM-A516M	Gr. 55(Gr. 380)	B002
	Gr. 60(Gr. 415)	B003
	Gr. 65(Gr. 450)	B004
	Gr. 70(Gr. 415)	B006
ASTM-A517M	Gr. A	B028
	Gr. B	B028
	Gr. C	B028
	Gr. D	B028
	Gr. E	B028
	Gr. F	B028
	Gr. H	B028
	Gr. J	B028
	Gr. M	B028
	Gr. P	B028
	Gr. Q	B028
	Gr. S	B028
	Gr. T	B028
ASTM-A533M	TypeA	B012
	TypeB	B014
	TypeC	B015
	TypeD	B013
ASTM-A537M	Cl. 1	B006, B028
	Cl. 2	B028

Standard No. 規格番号	Grade グレード	Index No. 索引番号
ASTM-A543M	TypeB	B024
	TypeC	B024
ASTM-A562M	−	B002
ASTM-A612M	−	B029
ASTM-A645M	−	B034
ASTM-A662M	Gr. A	B029
	Gr. B	B029
	Gr. C	B029
ASTM-A724M	Gr. A	B028
	Gr. B	B028
	Gr. C	B028
ASTM-A734M	TypeA	B028
	TypeB	B028
ASTM-A735M	Cl. 1	B029
	Cl. 2	B029
	Cl. 3	B029
	Cl. 4	B029
ASTM-A736M	Gr. A	B029
	Gr. C	B029
ASTM-A737M	Gr. B	B006
	Gr. C	B008
ASTM-A738	Gr. A	B029
	Gr. B	B029
	Gr. C	B029
〔AISI〕		
MN-06-222	201	B036
	202	B037
	301	B043
	302	B044
	302B	B045
	304	B046
	304L	B047
	304N	B048
	305	B050
	308	B051
	309	B052
	309S	B053
	310	B054
	310S	B055

Standard No. 規格番号	Grade グレード	Index No. 索引番号	Standard No. 規格番号	Grade グレード	Index No. 索引番号
	316	B056		620Gr.20	B017
	316L	B057		620Gr.31	B017
	316N	B058		622Gr.31	B019
	316Ti	B059		622Gr.45	B019
	317	B062		625	B021
	317L	B063		Gr.271	B025
	321	B064		Gr.281	B024
	329	B065		Gr.282	B024
	330	B066			
	347	B067	BS-1501-Part 3	304S12	B047
	348	B068		304S15	B046
	403	B071		304S49	B049
	405	B072		304S62	B048
	410	B074		304S65	B048
	414	B076		316S16	B057
	420	B077		316S12	B056
	429	B078		316S37	B057
	430	B079		316S62	B058
	431	B081		316S66	B058
	434	B082		316S82	B057
	436	B083		316S49	B060
	442	B084		320S17	B059
	446	B086		321S12	B064
				321S49	B064
〔BS〕				321S87	B064
BS-1501-Part 1	141Gr.360	B001		347S17	B067
	151Gr.360	B001		347S49	B067
	151Gr.400	B003		347S67	B067
	151Gr.430	B004		403S17	B071
	154Gr.360	B001		405S17	B072
	154Gr.400	B003		460S52	B070
	154Gr.443	B004		NA15	B070
	161Gr.360	B001			
	161Gr.400	B003	〔DIN〕		
	161Gr.430	B004	DIN-17102	ESTE255	B029
	164Gr.360	B001		ESTE285	B029
	164Gr.400	B003		ESTE315	B029
	222Gr.400	B003		ESTE355	B029
	223Gr.460	B005		ESTE380	B029
	223Gr.490	B006		ESTE420	B029
	224Gr.430	B004		ESTE460	B029
	224Gr.460	B006		ESTE500	B029
	224Gr.490	B006		TSTE255	B029
	225Gr.490	B006		TSTE285	B029
				TSTE315	B029
BS-1501-Part 2	240	B011		TSTE355	B029
	261	B011		TSTE380	B029
	503	B033		TSTE420	B029
	509	B035		TSTE460	B029
	510	B035		TSTE500	B029

Standard No. 規格番号	Grade グレード	Index No. 索引番号
DIN-17155	13CrMo44	B017
	15Mo3	B010
	17Mn4	B005
	H I	B001
	H II	B003
	19Mn6	B007
DIN-17280	10Ni14	B033
	12Ni19	B034
	X8Ni9	B035
DIN-17440	X2CrNi1911	B047
	X2CrNiMo17132	B057
	X2CrNiN1810	B048
	X5CrNi1810	B046
	X5CrNi1911	B050
	X5CrNiMo17122	B056
	X6Cr13	B075
	X6CrAl13	B072
	X6CrMo171	B082
	X6CrNiMoNb17122	B060
	X6CrNiNb1810	B067
	X6CrNiMoTi17122	B059
	X6CrNiTi1810	B064
	X10Cr13	B074
DIN-17441	X6CrMo171	B082
	X6CrNb17	B080
	X6CrTi17	B080
〔NF〕		
NF-A35-573	Z2CN18-10	B047
	Z2CND17-12	B057
	Z2CND17-13	B057
	Z2CND19-15	B063
	Z6CND17-11	B056
	Z6CND17-12	B059
	Z6CNDNb17-12	B060
	Z6CNNb18-10	B067
	Z12CN17-07	B043
NF-A36-205	337AP	B002
	A37CP	B002
	A37FP	B002
	A42AP	B003
	A42CP	B003
	A42FP	B003
	A48AP	B005
	A48APR	B005
	A48CP	B005

Standard No. 規格番号	Grade グレード	Index No. 索引番号
	A48CPR	B005
	A48FP	B005
	A48FPR	B005
	A52AP	B007
	A52APR	B007
	A52CP	B007
	A52CPR	B007
	A52FP	B007
	A52FPR	B007
NF-A36-206	10CD9.10	B019
	15CD205	B015
	15CD405	B017
	15D3	B010
	15MDV4.05	B012
	18MD4.05	B012
	Z10CD505	B021
NF-A36-207	A510	B007
	A530	B007
	A550	B008
	A590	B008
NF-A36-208	0.5Ni285	B030
	0.5Ni355	B030
	1.5Ni285	B031
	1.5Ni355	B031
	3.5Ni285	B033
	3.5Ni355	B033
	9Ni490	B035
	9Ni585	B035
NF-A35-209	Z2CN18-10	B047
	Z2CN18-10AZ	B048
	Z3CMN18-08-07AZ	B042
	Z5CN18-09	B046
	Z5CN18-09AZ	B048
	Z6CNNb18-10	B067
	Z10CN18-09	B043
〔ГОСТ〕		
ГОСТ-5520	12K	B001
	15K	B001
	16K	B003
	18K	B004
	20K	B003
ГОСТ-5632	03X17H14M2	B057
	03X18H11	B050
	03X18H12	B050

Standard No. 規格番号	Grade グレード	Index No. 索引番号	Standard No. 規格番号	Grade グレード	Index No. 索引番号
	04X18H10	B047			
	06X18H11	B050			
	07X16H6	B043			
	07X18H9	B044			
	17X21Г7AH	B042			
	08X13	B071			
	08X16H13M2Б	B060			
	18X17H13M2T	B059			
	18X17H15M3T	B059			
	08X17T	B080			
	08X18H9	B044			
	08X18H10	B048			
	08X18H12Б	B067			
	08X18N10T	B064			
	08X18N12T	B064			
	09X16H15M3Б	B060			
	09X17H0	B070			
	09X17H01	B070			
	10X17H13M2T	B059			
	10X18C0	B080			
	10X23H18	B055			
	12X3	B071			
	12X17	B079			
	12X18	B064			
	12X18N10T	B064			
	12X18N12T	B064			
	14X17H2	B081			
	15X14AГ15	B039			
	15X17AГ14	B039			
	15X25T	B088			
	20X17H2	B081			
	20X22H13	B052			

CHAPTER III

STEEL PIPES AND TUBES

第 3 部

鋼　管

Abbreviations

略 語 表

1. Method of Manufacture 製造方法

2. Type 鋼 種

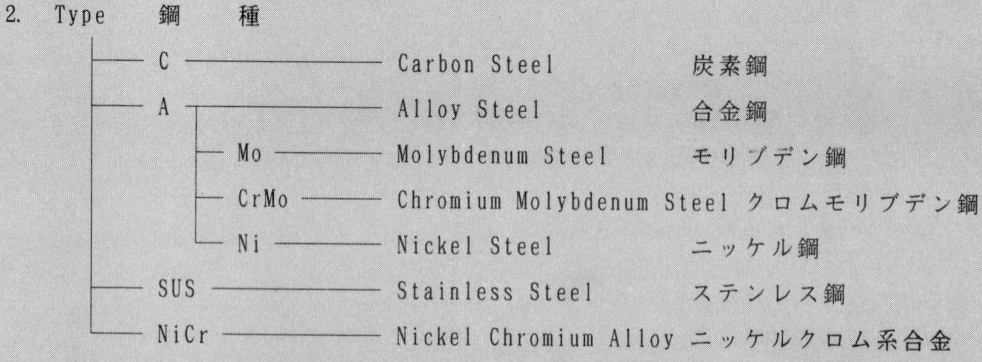

3. Designation 種類番号

Gr.: Grade グレード

Cl.: Class クラス

Typ.: Type タイプ

CHAPTER III CONTENTS　第3部　目次

1. STANDARDS LIST 収録各国規格リスト

1-1 JIS (Japan 日 本)

C8305-92 Rigid Steel Conduits
鋼製電線管

G3429-88 Seamless Steel Tubes for High Pressure Gas Cylinder
高圧ガス容器用継目無鋼管

G3439-88 Seamless Steel Oil Well Casing, Tubing and Drill Pipe
油井用継目無鋼管

G3441-88 Alloy Steel Tubes for Machine Purposes
機械構造用合金鋼鋼管

G3442-88 Galvanized Steel Pipes for Water Service
水道用亜鉛めっき鋼管

G3444-94 Carbon Steel Tubes for General Structural Purposes
一般構造用炭素鋼鋼管

G3445-88 Carbon Steel Tubes for Machine Structural Purposes
機械構造用炭素鋼鋼管

G3446-94 Stainless Steel Tubes for Machine and Structrural Purposes
機械構造用ステンレス鋼鋼管

G3447-88 Stainless Steel Sanitary Tubing
ステンレス鋼サニタリー管

G3448-88 Light Gauge Stainless Steel Pipes for Ordinary Piping
一般配管用薄肉ステンレス鋼鋼管

G3452-88 Carbon Steel Pipes for Ordinary Piping
一般配管用炭素鋼鋼管

G3454-88 Carbon Steel Pipes for Pressure Service
圧力配管用炭素鋼鋼管

G3455-88 Carbon Steel Pipes for High Pressure Service
高圧配管用炭素鋼鋼管

G3456-88 Carbon Steel Pipes for High Temprature Service
高温配管用炭素鋼鋼管

G3457-88 Arc Welded Carbon Steel Pipes
配管用アーク溶接炭素鋼鋼管

G3458-88 Alloy Steel Pipes
配管用合金鋼鋼管

G3459-88 Stainless Steel Pipes
配管用ステンレス鋼鋼管

G3460-88 Steel Pipes for Low Temprature Service
低温配管用鋼管

G3461-88 Carbon Steel Boiler and Heat Exchanger Tubes
ボイラ・熱交換器用炭素鋼鋼管

G3462-88 Alloy Steel Boiler and Heat Exchager Tubes
ボイラ・熱交換器用合金鋼鋼管

G3463-88 Stainless Steel Boiler and Heat Exchanger Tubes
ボイラ・熱交換器用ステンレス鋼鋼管

G3464-88 Steel Heat Exchanger Tubes for Low Temprature Service
低温熱交換器用鋼管

G3465-88 Seamless Steel Tubes for Drilling
試すい用継目無鋼鋼管

G3466-88 Carbon Steel Square Pipes for General Structural Purposes
一般構造用角型鋼管

G3467-88 Steel Tubes for Fire Heater
加熱炉用鋼管

G3468-88 Arc Welded Large Diameter Stainless Steel Pipes
配管用アーク溶接大径ステンレス鋼鋼管

G3469-92 Polyethylene Coated Steel Pipes
ポリエチレン被覆鋼管

G3472-88 Electric Resistance Welded Carbon Steel Tubes for Automobile Structural Purposes
自動車構造用電気抵抗溶接炭素鋼鋼管

G3473-88 Carbon Steel Tubes for Cylinder Barrels
シリンダーチューブ用炭素鋼鋼管

G3474-88 High Tensile Strength Steel Tubes for Tower Structural Purposes
鉄塔用高張力鋼鋼管

G4903-88 Seamless Nickel-Chromium-Iron Alloy Pipes
配管用継目無ニッケルクロム鉄合金管

G4904-91 Seamless Nickel-Chromium-Iron Alloy Heat Exchanger Tubes
熱交換器用継目無ニッケルクロム鉄合金管

1-2 ASTM（U.S.A. アメリカ）

A53-94 Pipe, Steel, Black and Hot-Dipped, Zinc-Coated, Welded and Seamless
継目無及び溶接鋼管，黒管及び亜鉛メッキ管

A105M-95 Forgings, Carbon Steel, for Piping Components
配管部品用炭素鋼鍛造品

A106-94a Seamless Carbon Steel Pipe for High Temperature Service
高温配管用継目無炭素鋼鋼管

A134- 93 Pipe, Steel, Electric-Fusion(Arc)-Welded
配管用アーク溶接鋼管

A135-93 Electric-Resistance-Welded Steel Pipe
配管用電気抵抗溶接鋼管

A139-93a Electric-Fusion(Arc)-Welded Steel Pipe
配管用アーク溶接鋼管

A161-94 Seamless Low-Carbon and Carbon-Molybdenum Steel Still Tubes for Refinery Service
精油所用継目無低炭素Mo加熱炉用鋼管

A178M-90a Electric-Resistance-Welded Carbon Steel Boiler Tubes
電気抵抗溶接ボイラ用炭素鋼鋼管

A179M-90a Seamless Low-Carbon and Carbon-Molybdenum Steel Still Tubes for Refinery Service
熱交換器及び凝縮器用継目無冷間引抜低炭素鋼鋼管

A181M-95 Forgings, Carbon Steel, for General Purposes Piping
汎用配管用炭素鋼鍛造品

A192M-91 Seamless Carbon Steel Boiler Tubes for High-Pressure Service
高圧用継目無炭素鋼ボイラー用鋼管

A199M-92 Seamless Cold-Drawn Intermediate Alloy-Steel Heat-Exchanger and Condenser Tubes
熱交換器及び凝縮器用継目無冷間引抜合金鋼鋼管

A200-94 Seamless Intermediate Alloy-Steel Still Tubes for Refinery Service
精油所用継目無合金鋼加熱炉用鋼管

A209M-91 Seamless Carbon-Molybdenum Alloy-Steel Boiler and Super-Heater Tubes
ボイラ及び過熱器用継目無Mo合金鋼鋼管

A210M-91 Seamless Medium-Carbon Steel Boiler and Super Heater Tubes
ボイラ及び過熱器用継目無中炭素鋼鋼管

A211-85 Spiral-Welded Steel or Iron Pipe
スパイラル溶接鋼管

A213M-94b Seamless Ferritic and Austenitic Alloy-Steel Boiler, Superheater, and Heat-Exchanger Tubes
ボイラ・過熱器及び熱交換器用継目無フェライト及びオーステナイト系合金鋼鋼管

A214M-90a Electric-Resistance-Welded Carbon Steel Heat-Exchanger and Condenser Tubes
熱交換及び凝縮器用電気抵抗溶接炭素鋼鋼管

A226M-90a Electric-Resistance-Welded Carbon Steel Boiler and Superheater Tubes for High-Pressure Service
高圧ボイラ及び過熱器用電気抵抗溶接炭素鋼鋼管

A249M-94a Welded Austenitic Steel Boiler, Superheater, Heat-Exchanger and Condenser Tubes
ボイラ・過熱器・熱交換器及び凝縮器用溶接オーステナイト系ステンレス鋼鋼管

A250M-91 Electric-Resistance-Welded Carbon-Molybdenum Alloy-Steel Boiler and Superheater Tubes
ボイラ及び過熱器用電気抵抗溶接C-Mo合金鋼鋼管

A252-93 Welded and Seamless Steel Pipe Piles
溶接及び継目無鋼管パイル

A266M-95 Forgings, Carbon Steel for Pressure Vessel Components
圧力容器用鍛造炭素鋼

A268-94 Seamless and Welded Ferritic Stainless Steel Tubing for General Service
一般用継目無及び溶接フェライト系ステンレス鋼鋼管

A269-94a Seamless and Welded Austenitic Stainless Steel Tubing for General Service
一般継目無及び溶接オーステナイト系ステンレス鋼鋼管

A270-90a Seamless and Welded Austenitic Stainless Steel Sanitary Tubing
衛生配管用継目無及び溶接オーステナイト系ステンレス鋼鋼管

A271-94 Seamless Austenitic Chromium-Nickel Still Steel Tubes for Refinery Service
精油所用継目無オーステナイト系Cr-Ni 加熱炉用鋼管

A312M-94b Seamless and Welded Austenitic Stainless Steel Pipe
配管用継目無及び溶接オーステナイト系ステンレス鋼鋼管

A333M-94 Seamless and Welded Steel Pipe for Low-Temperature Service
低温配管用継目無及び溶接鋼管

A334M-91 Seamless and Welded Carbon and Alloy-Steel Tubes for Low-Temperature Service
低温用継目無及び溶接炭素鋼及び合金鋼鋼管

A335M-94 Seamless Ferritic Alloy-Steel Pipe for High-Temperature Service
高温配管用継目無合金鋼鋼管

A336-95 Steel Forgings, Alloy, for Pressure and High-Temperature Parts
高圧高温用合金鋼鍛造品

A358M-94a Electric-Fusion-Welded Austenitic Chromium-Nickel Alloy Steel Pipe for High-Temperature Service
高温用オーステナイトCr-Ni 合金鋼アーク溶接鋼鋼管

A369M-92 Carbon and Ferritic Alloy Steel Forged and Bored Pipe for High-Temperature Service
高温配管用炭素鋼及び合金鍛造内削鋼管

A376M-93 Seamless Austenitic Steel Pipe for High-Temperature Central-Station Service
高温セントラルステーション用継目無オーステナイト鋼鋼管

A381-93 Metal-Arc-Welded Steel Pipe for Use with High-Pressure Transmission Systems
高圧トランスミッション用アーク溶接鋼管

A405-91 Seamless Ferritic Alloy-Steel Pipe Specially Heat Treated for High-Temperature Service
高温配管用特殊熱処理継目無フェライト系合金鋼鋼管

A409M-92 Welded Large Diameter Austenitic Steel Pipe for Corrosive or High-Temperature Service
耐食・高温用大径オーステナイト鋼溶接鋼管

A422-85 Butt Welded in Still Tubes for Refinery Service
精油所加熱炉用鋼管の突合溶接

A423M-91 Seamless and Electric-Welded Low-Alloy Steel Tubes
継目無及び電気溶接低合金鋼鋼管

A430M-91 Austenitic Steel Forged and Bored Pipe for High-Temperature Service
高温用オーステナイト鍛造内削鋼管

A452-88 Centrifugally Cast Iron Carbon Steel Pipe for High-Temperature Service
高温用鋳造炭素鋼鋼管

A500-93 Cold-Formed Welded and Seamless Carbon Steel Structural Tubing in Rounds and shapes
冷間成型溶接及び継目無炭素構造用円形及び角型鋼管

A501-93 Hot-Formed Welded and Seamless Carbon Steel Structural Tubing
熱間成型溶接及び継目無構造用炭素鋼鋼管

A511-90 Seamless Stainless Steel Machinery Tubing
機械構造用継目無ステンレス鋼鋼管

A512-94 Cold-Drawn Butt Weld Carbon Steel Mechanical Tubing
機械構造用冷間引抜鍛接鋼管

A513-94 Electric-Resistance-Welded Carbon and Alloy Steel Mechanical Tubing
電気抵抗溶接炭素鋼及び合金鋼機械構造用鋼管

A519-94 Seamless Carbon and alloy Steel Mechanical Tubing
継目無炭素鋼及び合金鋼機械構造用鋼管

A523-93 Plain End Seamless and Electric-Resistance-Welded Steel Pipe for High-Pressure Pipe Type Cable Circuits
高圧パイプ型ケーブル電線管用継目無及び電気抵抗溶接鋼管

A524-93 Seamless Carbon Steel Pipes for Atmospheric and Lower Temperatures
常低温用継目無炭素鋼鋼管

A539-90a Electric Resistance-Welded Coiled Steel Tubing for Gas and Fuel Oil Lines
ガス及び燃料ライン用コイル巻電気抵抗溶接鋼管

A554-94 Welded Stainless Steel Mechanical Tubing
機械構造用ステンレス鋼鋼管

A556M-90a Seamless Cold-Drawn Carbon Steel Feedwater Heater Tubes
給水加熱器用冷間引抜炭素鋼鋼管

A557M-90a Electric-Resistance-Welded Carbon Steel Feedwater Heat Tubes
給水加熱器用電気抵抗溶接鋼管

A587-93 Electric-Welded Low-Carbon Steel Pipe for the Chemical Industry
化学工業用電気溶接低炭素鋼鋼管

A589-94 Seamless and Welded Carbon Steel Water-Well Pipe
継目無及び溶接水井戸用炭素鋼鋼管

A595-93 Steel Tubes, Low-Carbon, Tapered for Structural Use
構造用低炭素鋼テーパ鋼管

A618-93 Hot-Formed Welded and Seamless High-Strength Low-Alloy Structural Tubing
低合金構造用熱間成型溶接及び継目無高張力鋼管

A632-90 Seamless and Welded Austenitic Stainless Steel Tubing(Small-Diameter) for General Service
一般用継目無及び溶接オーステナイトステンレス鋼小径鋼管

A660-91a Centrifugally Cast Carbon Steel Pipe for High Temperature Service
高温用鋳造炭素鋼鋼管

A671-94 Electric-Fusion-Welded Steel Pipe for Atmospheric and Lower Temperatures
常低温用配管用アーク溶接鋼管

A672-94 Electric-Fusion-Welded Steel Pipe for High-Pressure Service at Moderate Temperatures
中温度高圧配管用アーク溶接鋼管

A688M-91 Welded Austenitic Stainless Steel Feedwater Heater Tube
溶接オーステナイトステンレス鋼給水加熱鋼管

A691-93 Carbon and Alloy Steel Pipe, Electric-Fusion-Welded for High Pressure Service
高圧配管用アーク溶接炭素鋼及び合金鋼鋼管

A692-91 Seamless Medium-Strength Carbon Molybdenum Alloy-Steel Boiler and Superheater Tubes
継目無モリブデン鋼ボイラ及び過熱器用鋼管

A714-93 High-Strength Low-Alloy Welded and Seamless Steel Pipe
溶接及び継目無高張力低合金鋼鋼管

A727M-95 Forgings, Carbon Steel, for Piping Components with Inherent Notch Toughness
配管部品用炭素鋼靱性鍛造品

A731M-91 Seamless and Welded Ferritic Stainless Steel Pipe
継目無及び溶接フェライトステンレス鋼鋼管

A771-88 Austenitic-Stainless Steel Tubing for Breeder Reactor Core Components
増殖炉炉心用オーステナイト系ステンレス鋼鋼管

A778-90a Welded, Unannealed Austenitic Stainless Steel Tubular Products
熱処理なしオーステナイト系ステンレス鋼溶接鋼管

A789-94 Seamless and Welded Ferritic/Austenitic Stainless Steel Pipe for General Service
一般用継目無溶接フェライト・オーステナイト系ステンレス鋼鋼管

A790M-94 Seamless and Welded Ferritic/Austenitic Stainless Steel Pipe
継目無し溶接フェライト・オーステナイト系ステンレス鋼鋼管

A791-94 Welded Unannealed Ferritic Stainless Steel Tubing
溶接非焼なましフェライト系ステンレス鋼鋼管

A803-94 Welded Ferritic Stainless Steel Feed Water Heater Tubes
溶接フェライト系ステンレス鋼給水加熱器用鋼管

A813M-91 Single or Double-Welded Austenitic Stainless Steel Pipe
１回または２回溶接オーステナイト系ステンレス鋼鋼管

A814M-91 Cold-Worked Welded Austenitic Stainless Steel Pipe
冷間加工溶接オーステナイト系ステンレス鋼鋼管

A822-90 Seamless Cold-Drawn Carbon Steel Tubing for Hydraulic System Service
水系用継目無冷間引抜炭素鋼鋼管

A826-88 Austenitic and Ferritic Stainless Steel Duct Tubes for Breeder Reactor Core Compenents
増殖炉炉心用オーステナイト系及びフェライト系ステンレス鋼導管

A847-93 Cold-Formed Welded and Seamless High Strength, Low Alloy Structural Tubing
大気耐蝕性冷間加工溶接継目無高強度低合金構造用鋼鋼管

A851-90 High-Frequency Induction Welded, Unannealed, Austenitic Steel Condenser Tubes
高周波誘電溶接非焼なましオーステナイト系ステンレス鋼凝縮器用鋼管

A872-91 Centrifugally Cast Ferritic/Austenitic Stainless Steel Pipe for Corrosive Environments
腐蝕性環境用鋳造フェライト・オーステナイト系ステンレス鋼鋼管

B167-94a Nickel-Chromium-Iron Alloys Seamless Pipe and Tubes
ニッケル・クロム・鉄合金継目無鋼鋼管

B407-93 Nickel-Cromium Alloy Seamless Pipe and Tubes
ニッケル・クロム合金継目無鋼鋼管

B423-90 Nickel-Iron-Chromium-Molybdenum-Copper Alloy Seamless Pipe and Tube
ニッケル・鉄・クロム・モリブデン・銅合金継目無鋼鋼管

1-3 API（U.S.A. アメリカ）

Spec 5CT-89 Casing and Tubing
ケーシング及び鋼管

Spec 5D-88 Drill Pipe
掘削用鋼管

Spec 5L-90 Line Pipe
配管用鋼管

1-4 BS（U.K. イギリス）

534-90 Steel Pipes, Fittings and Specials for Water and Gas and Sewage
水・ガス・下水道用鋼管及び継手類

778-66 Steel Pipes and Joints for Hydraulic Purposes
水圧用鋼管及び継手

879-85 Steel pipes and Joints for Hydraulic Purposes
水圧用鋼管及び継手

1139-Part1-82 Tubes for Use in Scaffolding
足場用鋼管

1387-85 Steel Tubes and Tublars Suitable for Screwing to BS21 Pipe Threads
BS21のねじ切り用鋼管

1717-83 Steel Tubes for Cycle and Motor Cycle Purposes
自転車及び原動機付自転車用鋼管

90 94 95 93 a 91a 91

1864-88 Stainless Steel Milk Pipes and Fittings
ミルク用ステンレスパイプ及び継手

3059-78,87 Part 2-90 Steel Boiler and Super Heater Tubes
ボイラ及び過熱器用鋼管

3601-87 Carbon Steel Pipes for Pressure Purposes
圧力用炭素鋼鋼管

3602-87 Steel Pipes and Tubes for Pressure Purposes
圧力用鋼管

3603-91 Steel Pipes and Tubes for Pressure Purposes
圧力用鋼管

3604-Part1-90 Part2-91 Steel Pipes and Tubes for Pressure Purposes
圧力用鋼管

3605-Part1-91 Part2-92 Seamless and Welded Austenitic Stainless Steel Pipes and Tubes for Pressure purposes
圧力用オーステナイト系ステンレス継目無及び溶接鋼鋼管

3606-92 Steel Tubes for Heat Exchanger
熱鋼管器用鋼管

4127-94 Light Gauge Stainless Steel Tubes
薄肉ステンレス鋼鋼管

4825-91 Pipes and Fittings for the Food Industry
食品工業用鋼管及び継手

5242-Part1-87 Tubes for Fluid Power Cylinder Barrels
シリンダー用鋼管

6323-82 Seamless and Welded Steel Tubes for Automobile, Mechanical and General Engineering Purposes
自動車，機械構造及び一般エンジニアリング用継目無溶接鋼管

6363-83 Welded Cold Formed Steel Structural Hollow Sections
溶接冷間成形構造用中空形鋼

1-5 DIN（West Germany 西ドイツ）

1615-84 Welded Circular Unalloyed Steel Tubes not Subject to Special Requirement
特別規定のなしの非合金溶接鋼管

1626-84 Welded Circular Tubes of Non-Alloy Steels with Special Quality Requirements
特別品質要求付溶接非合金鋼鋼管

1628-84 Welded Circular Tubes of Non-Alloy Steels with Very High Quality Requirements
高品質要求付溶接非合金鋼鋼管

1629-84 Seamless Circular Tubes of Non-Alloy Steels with Special Quality Requirements
特別品質要求継目無非合金鋼鋼管

1630-84 Seamless Circular Tubes of Non-Alloy Steels with Very High Quality Requirements
高品質要求継目無非合金鋼鋼管

2391-81 Seamless Precision Steel Tubes
継目無精密鋼管

2393-81 Welded Precision Steel Tubes
溶接精密鋼管

2394-81 Welded Precision Steel Tubes
溶接精密鋼管

2440-78 Steel Tubes, Medium-Weight Suitable for Screwing
ねじ切用中肉鋼管

2441-78 Steel Tubes, Heavy-Weight Suitable for Screwing
ねじ切用厚肉鋼管

2462-81 Seamless Tubes of Stainless Steel
継目無ステンレス鋼鋼管

2463-81 Welded Tubes of Austenitic Stainless Steel
溶接オーステナイトステンレス鋼鋼管

17120-84 Welded Circular Steel Tubes for Structural Steel Work
一般構造用溶接鋼管

17121-84 Seamless Circular Steel Tubes for Structural Steel Work
一般構造用継目無鋼管

17123-86 Welded Circular Steel Tubes for Structural Steel Work
構造用溶接細粒鋼鋼管

17124-86 Seamless Circular Fine Grain Steel Tubes for Structural Steel Work
構造用継目無細粒鋼鋼管

17172-78 Steel Pipes for Pipe Lines for the Transport of Combustible Fluids and Gases
可燃性流体輸送パイプライン用鋼管

17173-85 Seamless Circular Tubes Made from Steels with Low Temperature Toughness
低温用継目無鋼管

17174-85 Welded Circular Tubes Made from Steel with Low Temperature Toughness
低温用溶接鋼管

17175-79 Seamless Steel Tubes for Elevated Temperatures
高温用継目無鋼管

17177-79 Electrically Resistance or Induction Welded Steel Tubes for Elevated Temperatures
高温用電気溶接鋼管

17455-85 General Purpose Welded Circular Stainless Steel Tubes
一般用ステンレス鋼溶接鋼管

17456-85 General Purpose Seamless Circular Stainless Steel Tubes
一般用ステンレス鋼継目無鋼管

17457-85 Welded Circular Austenitic Stainless Steel Tubes Subject to Special Requirements
特別規定付ステンレス鋼溶接鋼管

17458-85 Seamless Circular Austenitic Stainless Steel Tubes Subject to Special Requirement
特別規定付オーステナイト系ステンレス鋼継目無鋼鋼管

1-6 NF（France フランス）

A49-111-78 Steel Tubes-Plain End Seamless Tubes of Commercial Quality for General Purposes at Mean Pressure
常圧一般用市販品質プレンエンド継目無鋼管

A49-112-87 Steel Tubes-Plain End Seamless Hot Rolled Tubes with Special Delivery Conditions
特殊出荷条件付プレエンド熱間圧延継目無鋼管

A49-115-78 Steel Tubes-Hot-Finished Seamless Tubes Suitable for Threading
ねじ切用熱間仕上継目無鋼管

A49-117-85 Steel Tubes-Seamless Plain End Tubes for Pipe Lines and General Use Ferritic and Austenitic Stainless Steels
パイプライン及び一般プレエンド継目無フェライト系及びオーステナイト系ステンレス鋼鋼管

A49-141-78 Steel Tubes-Welded Plain End Tubes of Commercial Quality for General Purposes at Medium Pressure
常圧一般用市販品質プレエンド溶接鋼管

A49-142-87 Steel Tubes-Longitudinally Pressure Welded and Hot-finished Plain End Tubes
プレエンド縦加圧加熱仕上鋼管

A49-145-78 Steel Tubes-Hot-Finished Welded Tubes Suitable for Threading
ねじ切用熱間仕上溶接鋼管

A49-146-75 Steel Tubes-Unthreaded Plain Ended Welded Tubes for Liquid
流体ねじ無プレエンド溶接鋼管

A49-147-80 Steel Tubes-Plain End Longitudinally Welded Tubes for General Purposes, Austenitic Stainless Steels
パイプライン及び一般プレエンド縦溶接オーステナイト系ステレ

A49-148-80 Steel Tubes-Plain End Longitudinally Welded Tubes for Pipe Lines and General Use-Ferritic Stainless Steels
パイプライン及び一般用プレエンド縦溶接フェライト系ステンレス鋼鋼管

A49-150-85 Steel Tubes-Welded Tubes to Be Coated or Protected for Water Piping systems
水管用被覆または保護管原管用溶接鋼管

A49-207-81 Steel Tubes-Seamless or Longitudinally Welded Stainless Steel Tubes for Exchangers i Aircraft Construction
航空機熱交換用ステンレス鋼鋼管

A49-210-89 Steel Tubes-Seamless Cold-Drawn Tubes for Fluids Piping
流体輸送用冷間引抜継目無鋼管

A49-211-86 Steel Tubes-Seamless Plain End Carbon Steel Tubes for the Transport of Fluids at Elevated Temperature
高温流体輸送用プレンエンド継目無鋼管

A49-212-83 Steel Tubes-Seamless Unalloyed Steel Tubes for the at Medium Temperatures
中温用継目無非合金鋼鋼管

A49-213-90 Steel Tubes-Seamless Unalloyed and Ferritic Alloy Steel Tubes for Use at High Temperature
高温継目無非合金鋼及びフェライト系合金鋼鋼管

A49-214-78 Steel Tubes-Seamless Austenitic Steel Tubes for Use at High Temperatures
高温用継目無オーステナイト系ステンレス鋼鋼管

A49-215-81 Steel Tubes-Seamless Tubes for Ferritic Non-Alloy and Alloy Steel Heat Exchangers
熱交換器用継目無非合金鋼及びフェライト系合金鋼鋼管

A49-217-87 Steel Tubes-Seamless Tubes for Heat Exchanger
熱交換器用継目無鋼管

A49-218-79 Steel Tubes-Seamless Pipes for Furnaces-Austenitic Stainless Steels
加熱炉用継目無オーステナイト系鋼鋼管

A49-230-85 Steel Tubes-Plain End Seamless Tubes for Pressure Vessels and Piping Systems at Low Temperatures
低温圧力容器及び配管用プレンエンド継目無鋼管

A49-240-83 Steel Tubes-Longitudinally Buttwelding Plain End for Pressure Vessels and Pipe Systems Used at Low Temperatures
低温圧力容器及配管系用プレンエンド縦鍛接鋼管

A49-241-86 Steel Tubes-Longitudinally Pressure Welded Tubes in Non Alloyed Steel Grades for Fluid Piping up to 425℃
425 ℃以下配管用非合金縦加圧溶接鋼管

A49-242-85 Steel Tubes-Longitudinally Welded Tubes, D≦168.3mm, in Non-Alloy Steels, Used at Averagely Elevated Temperatures
外径168.3 mm以下の中温用縦溶接非合金鋼鋼管

A49-243-85 Steel Tubes-Longitudinally Welded Tubes, D≦168.3mm, in Non-Alloyed and Ferritic Alloyed Steels, Used at Elevated Temperatures
外径168.3 mm以下の高温用縦溶接非合金鋼及びフェライト系合金鋼鋼管

A49-245-86 Steel Tubes-Longitudinally Welded Tubes for Ferritic Non-Alloy and Alloy Steel Heat Exchangers
熱交換器用縦溶接フェライト系非合金鋼及び合金鋼鋼管

A49-249-80 Steel Tubes-Longitudinally Welded Plain End Tubes Used for Food Industry
食品工業用プレンエンド縦溶接オーステナイト系ステンレス鋼鋼管

A49-250-80 Steel Tubes-Welded Plain End Tubes of Commercial Quality with or without Special Delivery Conditions- D≦168.3mm
外径168.3 mm以上特殊出荷条件付またはなしの市販品質プレンエンド溶接鋼管

A49-253-82 Steel Tubes-Longitudinally Fusion Welded Non Alloy and Ferritic Alloy Steel Tubes for Use at Elevated Temperatures
高温用縦溶融溶接非合金鋼及びフェライト系合金鋼鋼管

A49-310-94 Steel Tubes-Seamless Precision Tubes for Mechanical Application
機械用継目無精密鋼管

A49-311-74 Steel Tubes-Seamless Tubes for Mechanical Application
機械用継目無鋼管

A49-312-93 Steel Tubes-Seamless Tubes for Mechanical Application-Supplementary Series
機械用継目無鋼管（補完用）

A49-317-80 Steel tubes-Seamless Plain End Tubes for Engineering Use, Austenitic Stainless Steels
機械用プレンエンド継目無オーステナイト系ステンレス鋼鋼管

A49-321-78 Steel Tubes-Jacks for Hydraulic Transmissions-Cold-Rolled or Cold-Drawn Seamless Tubes, Type "Standard"
水圧伝達装置用ジャッキ，冷間圧延または冷間引抜継目無鋼管（標準型）

A49-322-78 Steel Tubes-Jacks for Hydraulic Transmissions-Cold-Rolled or Cold-Drawn Seamless Tubes, Type "Suitable for Grinding in"
水圧伝達装置用ジャッキ，冷間圧延または冷間引抜継目無鋼管（すりあわせ型）

A49-323-78 Steel Tubes-Jacks for Hydraulic Transmissions-Cold-Rolled or Cold-Drawn Seamless Tubes, Type "Ready for Use"
水圧伝達装置用ジャッキ，冷間圧延または冷間引抜継目無鋼管（直接適用型）

A49-326-75 Steel Tubes-Jacks for Pneumatic Transmissions-Cold-Drawn Seamless Types, Type "Suitable for Grinding in"
空気伝達用ジャッキ，冷間引抜継目無鋼管（すりあわせ型）

A49-327-75 Steel Tubes-Jacks for Pneumatic Transmissions-Cold-Drawn Seamless Tubes, Type "Ready for Use"
空気伝達用ジャッキ，冷間引抜継目無鋼管（直接適用型）

A49-330-85 Steel Tubes-Seamless Cold-Drawn Tubes for Hydraulic and Pneumatic Power Systems
水圧及び空気圧用冷間引抜継目無鋼管

A49-341-75 Steel Tubes-Precision Welded Tubes for Mechanical Application
機械用精密溶接鋼管

A49-343-80 Steel Tubes-Longitudinally Welded Tubes D ≦168.3mm for Engineering Use
外径168.3 mm以下の機械用縦溶接鋼管

A49-400-82 Steel Tubes-Longitudinally Electric Resistance Welded Unalloyed Steel Tubes 17.2≦D ≦406.4mm for the Transport of Pressurized Fluids
外径17.2mm～406.4 mmの加圧流体輸送用電気抵抗溶接非合金鋼鋼管

A49-401-88 Steel Tubes-Longitudinally Fusion Welded Unalloyed Steel Tubes for Pipes and Pressure Vessels
圧力容器用縦溶接非合金鋼鋼管

A49-402-88 Steel Tubes-Spiral Fusion Welded Non-Alloy and Micro-Alloy Steel Tubes for Fluid Transporting Pipes and Pressure Vessels
液体輸送及び圧力容器用螺旋溶融溶接非合金及び低合金鋼鋼管

A49-501-86 Steel Tubes-Seamless or Welded Hot-Finished Structural Hollow Sections
継目無または溶接熱間仕上中空形鋼

A49-541-86 Steel Tubes-Cold-Finished Welded Structural Hollow Sections
構造用冷間仕上溶接中空形鋼

A49-643-87 Steel Tubes-Round, Square and Rectangular Steel Tubes for Ordinary Uses, Longitudinally Pressure Welded and Cold Formed from Flat Products
平板原料縦溶融溶接冷間成形丸・角・矩形一般用鋼管

A49-645-87 Steel Tubes-Round, Square and Rectangular Steel Tubes for Ordinary Uses, Longitudinally Pressure Welded and Cold Formed from Cold-Rolled Flat Pressure
冷間圧延平板原料冷間成形縦加圧溶接丸・角・矩形一般用鋼管

A49-647-79 Steel Tubes-Circular, Square, Rectangular or Oval in Ferritic or Austenitic Stainless Steels
構造用円形，方形，矩形または楕円形フェライト系オーステナイト系ステンレス鋼鋼管

1-7 ISO（International 国 際）

65-81 Carbon Steel Tubes Suitable for Screwing in Accordance with ISO 7/1
ねじ付炭素鋼鋼管

559-77 Welded or Seamless Steel Tubes for Water, Sewage and Gas
給排水・ガス用溶接又は継目無鋼管

2604/2-75 Steel Products for Pressure Purposes, Part 2 Wrought Seamless Tubes
圧力容器用継目無鋼管

2604/3-75 Steel Products for Pressure Purposes, Part 3 Electric Resistance and Induction-Welded Tubes
圧力容器用電気抵抗溶接鋼管

2604/5-78 Steel Products for Pressure Purposes, Part 5 Longitudinally Welded Austenitic Stainless Steel
圧力容器用縦溶接オーステナイト系ステンレス鋼鋼管

2937-74 Plain End Seamless Steel Tubes for Mechanical Application
機械用プレンエンド継目無鋼管

2938-74 Hollow Steel Bars for Machining
機械用中空鋼棒

3138-80 Oil and Natural Gas Industries-Steel Line Pipe
石油及び天然ガス用パイプライン

3304-85 Plain End Seamless Precision Steel Tubes
プレーンエンド継目無精密鋼管

3305-85 Plain End Welded Precision Steel Tubes
プレーンエンド溶接精密鋼管

3306-85 Plain End as Welded and Sized Precision Steel Tubes
プレーンエンド溶接定径精密鋼管

2. JIS Number and Corresponding Foreign Standards

JIS			ASTM			BS		
Standard Number 規格番号	Grade グレード	Type 鋼種	Standard Number 規格番号	Grade グレード	Type 鋼種	Standard Number 規格番号	Grade グレード	Type 鋼種
G3452	SGP	C	A53	Type F	C	–	–	–
G3454	STPG370 (STPG38)	C	A135 A587	GrA –	C C	3601 〃	ERW360 S360	C C
	STPG410 (STPG42)	C	A135 A524	GrB –	C C	3601 〃	ERW410 S410	C C
G3455	STS370 (STS38)	C						
	STS41[illegible] (STS4[illegible])	C						
	STS480 (STS49)	C						
G3456	STPT370 (STPT38)	C	A106	GrA	C	3602 〃 〃 〃	HFS360 CFS360 ERW360 CEW360	C C C C
	STPT410 (STPT42)	C	A106	GrB	C	3602 〃 〃 〃	HFS410 CFS410 ERW410 CEW410	C C C C
	STPT480 (STPT49)	C	A106	GrC	C	3602 〃 〃 〃	HFS460 CFS460 ERW460 CEW460	C C C C
G3457	STPY400 (STPY41)	C	A139	GrB	C	3601	SAW410	C
G3458	STPA12	Mo	A335	P1	Mo			
	STPA20	CrMo	A335	P2	CrMo			

JISと外国規格との対比

DIN			NF			ISO			Index Number
Standard Number 規格番号	Grade グレード	Type 鋼　種	Standard Number 規格番号	Grade グレード	Type 鋼　種	Standard Number 規格番号	Grade グレード	Type 鋼　種	索　引 番　号
1615	St33-2	C	A49-145	TS34-1	C	65	TW	C	C001
2441	St33-2	C	A49-146	TS34-a	C	559	TW0	C	
1626	St37.0	C	A49-112	TS37b	C	559	TS4	C	C002
1629	St37.0	C	A49-141	TS37a	C	2604/2	TS4	C	
17172	StE210.7	C	A49-142	TS37a	C	2604/3	TW4	C	
			〃	TS37b	C				
			A49-150	TSE235	C				
			A49-212	TS37c	C				
			A49-242	TS37c	C				
			A49-250	TSE24a	C				
			A49-400	TSE220	C				
1626	St44.0	C	A49-112	TS42b	C	559	TS9	C	
			A49-142	TS42b	C	2604/3	TW9	C	
			A49-212	TS42c	C	2604/3	TW9	C	
			A49-242	TS42c	C	3183	E24-1	C	
			A49-250	TSE26b	C	〃	E24-2	C	
			A49-400	TSE250	C				
1630	St37.4	C	A49-211	TU37b	C	2604/2	TS4	C	C003
			A49-410	TUE220b	C				
1630	St44.4	C	A49-210	TU42b	C	2604/2	TS9	C	
			A49-410	TUE250b	C				
1630	St52.4	C				2604/2	TS13	C	
17175	St35.8	C	A49-211	TU37b	C	2604/2	TS5	C	C004
17177	St37.8	C	A49-213	TU37c	C	2604/3	TW9H	C	
			A49-243	TS37c	C				
17175	St45.8	C	A49-211	TU42b	C	2604/2	TS9H	C	
17177	St42.8	C	A49-213	TU42c	C				
			A49-243	TS42c	C				
			A49-211	TU48b	C	2604/2	TS14	C	
			A49-213	TU48c	C				
									C005
									C006
			A49-213	TU15CD205	CrMo				

JIS			ASTM			BS		
Standard Number 規格番号	Grade グレード	Type 鋼種	Standard Number 規格番号	Grade グレード	Type 鋼種	Standard Number 規格番号	Grade グレード	Type 鋼種
G3458 (Continued)	STPA22	CrMo	A335	P12	CrMo	3604	HFS620-460	CrMo
						〃	CFS620-460	CrMo
						〃	ERW620-460	CrMo
						〃	CEW620-460	CrMo
						〃	HFS620-440	CrMo
						〃	CFS620-440	CrMo
						〃	ERW620-440	CrMo
						〃	CEW620-440	CrMo
	STPA23	CrMo	A335	P11	CrMo	3604	HFS621	CrMo
						〃	CFS621	CrMo
						〃	ERW621	CrMo
						〃	CEW621	CrMo
	STPA24	CrMo	A335	P22	CrMo	3604	HFS622	CrMo
						〃	CFS622	CrMo
	STPA25	CrMo	A335	P5	CrMo	3604	HFS625	CrMo
						〃	CFS625	CrMo
	STPA26	CrMo	A335	P9	CrMo	3604	HFS629-470	CrMo
						〃	CFS629-470	CrMo
G3460	STPL380 (STPL39)	C	A333	Gr1	C	3608	HFS410LT50	C
						〃	CFS410LT50	C
						〃	ERW410LT50	C
						〃	CEW410LT50	C
	STPL450 (STPL46)	Ni	A333	Gr3	Ni	3603	HFS503 LT100	Ni
						〃	CFS503 LT100	Ni
	STPL690	Ni	A333	Gr8	Ni	3605	HFS503 LT196	Ni
	(STPL70)					〃	CFS503 LT196	Ni
G3448 G3459	SUS		A312	TP304	SUS	3605	304S18	SUS
	304TP	SUS	A376	TP304	SUS	〃	304S25	SUS
	304TPD	SUS				4127	–	SUS
	SUS 304HTP	SUS	A312	TP304H	SUS	3605	304S59	SUS
			A376	TP304H	SUS			

DIN			NF			ISO			Index Number 索引番号
Standard Number 規格番号	Grade グレード	Type 鋼種	Standard Number 規格番号	Grade グレード	Type 鋼種	Standard Number 規格番号	Grade グレード	Type 鋼種	
17175	13CrMo44	CrMo							C006
			A49-213	TU10CD5.05	CrMo				
17175	10CrMo910	CrMo	A49-213	TU10CD9.10	CrMo				
			A49-21	TUZ12CD 5.05	CrMo	2604/2	TS37	CrMo	
			A49-213	TUZ10CD9	CrMo				
						2604/2 2604/3	TS6 TW6	C C	C007
			A49-230	TU10N14	Ni	2604/2	TS43	Ni	
			A49-230	TU26N9	Ni	2604/2	TS45	Ni	
2462 2463	X5CrNi189 X5CrNi189	SUS SUS	A49-117 A49-147 A49-230	TUZ6 CN18.09 TUZ CN18.09 TUZ6 CN18.09	SUS SUS SUS	2604/2	TS47	SUS	C008
						2604/2	TS48	SUS	

JIS			ASTM			BS		
Standard Number 規格番号	Grade グレード	Type 鋼　種	Standard Number 規格番号	Grade グレード	Type 鋼　種	Standard Number 規格番号	Grade グレード	Type 鋼　種
G3448	SUS		A312	TP304L	SUS	3605	304S14	SUS
G3459								
(Continued)	304LTP	SUS				〃	304S22	SUS
	SUS		A312					
	309TP	SUS						
	309STP	SUS						
	SUS		A312					
	310TP	SUS						
	310STP	SUS						
	SUS		A312	TP316	SUS	3605	316S18	SUS
	316TP	SUS	A376	TP316	SUS	〃		
	316TPD	SUS	A651	TP316	SUS			
	SUS		A312	TP316H	SUS			
	316HTP	SUS	A376	TP316H	SUS			
	SUS		A312	TP316L	SUS	3605	316S14	SUS
	316LTP	SUS				〃	316S22	SUS
	SUS		A312	TP317	SUS			
	317TP	SUS						
	SUS		A312	TP317L	SUS			
	317LTP	SUS						

DIN			NF			ISO			Index Number 索引番号
Standard Number 規格番号	Grade グレード	Type 鋼種	Standard Number 規格番号	Grade グレード	Type 鋼種	Standard Number 規格番号	Grade グレード	Type 鋼種	
2462 2463	X2CrNiMO 1810 X2CrNi189	SUS SUS	A49-117 A49-147 A49-230 A49-645 A49-647	TUZ2 CN18.10 TUZ2 CN18.10 TUZ2 CN18.10 TUZ2 CN18.10 TUZ2 CN18.10	SUS SUS SUS SUS SUS	2604/2	TS46	SUS	C008
			A49-117	TUZ12 CN24.12	SUS	2604/2	TS68	SUS	
2462 2463 〃	X5CrNiMo 1810 X5CrNiMo 1810 X5CrNiMo 1812 (17440)	SUS SUS SUS SUS	A49-117 A49-147 A49-230	TUZ6CND 17.11 TUZ6CND 17.11 TUZ6CND 17.11	SUS SUS SUS	2604/2 〃	TS60 TS61	SUS SUS	
						2604/2	TS63	SUS	
2462 〃 2463 〃	X2CrNiMo 1810 X2CrNiMo 1812 X2CrNiMo 1810 X2CrNiMo 1812	SUS SUS SUS SUS				2604/2 〃	TS57 TS58	SUS SUS	

JIS			ASTM			BS		
Standard Number 規格番号	Grade グレード	Type 鋼　種	Standard Number 規格番号	Grade グレード	Type 鋼　種	Standard Number 規格番号	Grade グレード	Type 鋼　種
G3448 G3459 (Continued)	SUS 321TP	SUS	A312 A376	TP321L TP321	SUS SUS	3605 3605	321S18 321S22	SUS SUS
	SUS 321HTP	SUS	A312 A376	TP321H TP321H	SUS SUS	3605	321S59	SUS
	SUS 329J1TP	SUS						
	SUS 329J2LTP	SUS						
	SUS 347TP	SUS	A312 A376	TP347 TP347	SUS SUS	3605 〃	347S18 347S18	SUS SUS
	SUS 347HTP	SUS	A312 A376	TP347H TP347H	SUS SUS	3605	347S59	SUS
	SUS 405TP	SUS						
G3468	SUS 304TPY	SUS	A409 A358	TP304 304	SUS SUS			
	SUS 304LTPY	SUS	A409 A358	TP304L 304L	SUS SUS			
	SUS 309STPY	SUS	A358	309	SUS			
	SUS 310STPY	SUS	A358	310	SUS			
	SUS 316TPY	SUS	A409 A358	TP316 316	SUS SUS			
	SUS 316LTPY	SUS	A409 A358	TP316L 316L	SUS SUS			
	SUS 317TPY	SUS	A409	TP317	SUS			
	SUS 317LTPY	SUS						
	SUS 321TPY	SUS	A409 A778 A358	TP321 TP321 321	SUS SUS SUS			

DIN			NF			ISO			Index Number 索引番号
Standard Number 規格番号	Grade グレード	Type 鋼種	Standard Number 規格番号	Grade グレード	Type 鋼種	Standard Number 規格番号	Grade グレード	Type 鋼種	
2462	X10CrNi Ti189	SUS	A49-117	TUZ6CNT 1810	SUS	2604/2	TS53	SUS	C008
2463	X10CrNi Ti189	SUS	A49-147	TSZ6CNT 18.10	SUS				
			A49-230	TSZ6CNT 18.10	SUS				
2462	X10CrNiMo Mo1810	SUS							
2462	X10CrNiNb 189	SUS				2604/2	TS50	SUS	
						2604/2	TS56	SUS	
						2604/5	TW47	SUS	C009
						2604/5	TW46	SUS	
						2604/5	TW60	SUS	
						〃	TW61	SUS	
						2604/5	TW57	SUS	

JIS			ASTM			BS		
Standard Number 規格番号	Grade グレード	Type 鋼　種	Standard Number 規格番号	Grade グレード	Type 鋼　種	Standard Number 規格番号	Grade グレード	Type 鋼　種
G3468 (Continued)	SUS 347TPY	SUS	A409	TP347	SUS			
			A778	TP347	SUS			
			A358	347	SUS			
	SUS 329J1TPY	SUS						
G3461	STB340 (STB35)	C	A161	LC	C	3059	HFS320	C
			A192	−	C	〃	CFS320	C
			A226	−	C	〃	ERW320	C
			A556	GrA2	C	〃	CEW320	C
			A557	GrA2	C	〃	S1 360	C
						〃	S2 360	C
						〃	ERW360	C
						〃	CEW360	C
						3606	ERW320	C
						〃	CEW320	C
						〃	CFS320	C
	STB410 (STB42)	C	A178	GrC	C	3059	S1 440	C
			A210	GrA1	C	〃	S2 440	C
			A556	GrB2	C	〃	ERW440	C
			A557	GrB2	C	〃	CEW440	C
						3602	HFS410	C
						〃	CFS410	C
						〃	ERW410	C
						〃	CEW410	C
						3606	ERW440	C
						〃	CEW440	C
						〃	CFS440	C
	STB510 (STB52)	C						
G4903	NCF600TP	A	B167	N06600	A			
	NCF800TP	A	B407	N08800	A			
	NCF825TP	A	B423	N08825	A			
G3462	STBA12	Mo	A209	T1	Mo	3606	ERW245	Mo
			A250	T1	Mo	〃	CEW245	Mo
						〃	CFS245	Mo
	STBA13	Mo	A209	T1-a	Mo			
			A250	T1-a	Mo			
			A692	−	Mo			
	STBA20	CrMo	A213	T2	CrMo			

DIN			N F			I S O			Index Number
Standard Number 規格番号	Grade グレード	Type 鋼　種	Standard Number 規格番号	Grade グレード	Type 鋼　種	Standard Number 規格番号	Grade グレード	Type 鋼　種	索　引 番　号
							TW50	SUS	C009
			A49-245	TS34e	C				C010
			〃	TS34c	C				
17175	St45.8	C	A49-213	TU42c	C	2604/2	TS9H	C	
17177	St42.8	C	A49-215	TU42c	C	〃	TW9H	C	
			A49-243	TS42c	C				
			A49-245	TS42c	C				
			〃	TS42c	C				
17175	19Mn5	C	A49-213	TU52C	C	2604/2	TS18	C	
			A49-248	TU52C	C				
									C011

JIS			ASTM			BS		
Standard Number 規格番号	Grade グレード	Type 鋼　種	Standard Number 規格番号	Grade グレード	Type 鋼　種	Standard Number 規格番号	Grade グレード	Type 鋼　種
G3462 (Continued)	STBA22	CrMo	A213	T12	CrMo	3059	S1 620	CrMo
						〃	S2 620	CrMo
						3059	ERW620	CrMo
						〃	CEW620	CrMo
						3604	HFS620-460	CrMo
						〃	CFS620-460	CrMo
						〃	ERW620-460	CrMo
						〃	CEW620-460	CrMo
						〃	HFS620-440	CrMo
						〃	CFS620-440	CrMo
						〃	ERW620-440	CrMo
						〃	CEW620-440	CrMo
						3606	ERW620	CrMo
						〃	CEW620	CrMo
						〃	CFW620	CrMo
	STBA23	CrMo	A199	T11	CrMo	3604	HFS621	CrMo
			A213	T11	CrMo	〃	CFS621	CrMo
						〃	ERW621	CrMo
						〃	CEW621	CrMo
						3606	ERW621	CrMo
						〃	CEW621	CrMo
						〃	CFS621	CrMo
	STBA24	CrMo	A199	T22	CrMo	3059	S1-622-440	CrMo
			A213	T22	CrMo	〃	S2-622-440	CrMo
						〃	S1-622-490	CrMo
						〃	S2-622-490	CrMo
	STBA25	CrMo	A199	T5	CrMo	3604	HFS625	CrMo
			A213	T5	CrMo	〃	CFS625	CrMo
						3606	CFS625	CrMo
	STBA26	CrMo	A199	T9	CrMo	3059	S1-629-470	CrMo
			A213	T9	CrMo	〃	S2-629-470	CrMo
						〃	S1-629-590	CrMo
						〃	S2-629-590	CrMo
						3604	HFS629-470	CrMo
G3463			A213	TP304	SUS	3605	304S18	SUS
	SUS		A249	TP304	SUS	〃	304S25	SUS
	304TB	SUS	A269	TP304	SUS	3606	LWHT304S22	SUS

DIN			N F			I S O			Index Number 索引番号
Standard Number 規格番号	Grade グレード	Type 鋼種	Standard Number 規格番号	Grade グレード	Type 鋼種	Standard Number 規格番号	Grade グレード	Type 鋼種	
17175	13CrMo44	CrMo							C011
			A49-213 A49-215	TU10CD TU10CD5.05	CrMo CrMo				
17175	10CrMo910	CrMo	A49-213 A49-215	Tu10CD9.10 Tu10CD9.10	CrMo CrMo				
						260 4/2	TS37	CrMo	
			A49-213 A49-215	TUZ10CD9 TUZ10CD9	CrMo CrMo				
2462 2463	X5CrNi189 X5CrNi189	SUS SUS	A49-230	TUZ6CN 18.09	SUS	2604/2	TS47	SUS	C012

JIS			ASTM			BS		
Standard Number 規格番号	Grade グレード	Type 鋼　種	Standard Number 規格番号	Grade グレード	Type 鋼　種	Standard Number 規格番号	Grade グレード	Type 鋼　種
G3463 (Continued)			A632	TP304	SUS	3606	LWCF304S22	SUS
			A688	TP304	SUS	〃	LWBC304S25	SUS
	SUS 304HTB	SUS	A213	TP304H	SUS	3059	CFS304S59	SUS
			A249	TP304H	SUS	3606	304S59	SUS
	SUS 304LTB	SUS	A213	TP304L	SUS	3605	304S14	SUS
			A249	TP304L	SUS	〃	304S22	SUS
			A269	TP304L	SUS	3606	LWHT304S22	SUS
			A632	TP304L	SUS	〃	LWCF304S22	SUS
			A688	TP304L	SUS	〃	LWBC304S22	SUS
						〃	CFS304S22	SUS
	SUS 309TB	SUS						
	SUS 309STB	SUS	A213	TP309S	SUS			
	SUS 310TB	SUS	A632	TP310	SUS			
	SUS 310STB	SUS	A213	TP310S	SUS			
	SUS 316TB	SUS	A213	TP316	SUS			
			A249	TP316	SUS	3605	316S18	SUS
			A269	TP316	SUS			
			A632	TP316	SUS	3606	LWHT316S25	SUS
			A688	TP316	SUS	〃	LWCF316S25	SUS
						〃	LWBC316S25	SUS
						〃	CFS316S25	SUS
						〃	LWHT316S30	SUS
						〃	LWCF316S30	SUS
						〃	LWBC316S30	SUS
						〃	CFS316S30	SUS

DIN			NF			ISO			Index Number 索　引 番　号
Standard Number 規格番号	Grade グレード	Type 鋼　種	Standard Number 規格番号	Grade グレード	Type 鋼　種	Standard Number 規格番号	Grade グレード	Type 鋼　種	
									C012
						2604/2	TS48	SUS	
2462 2463	X2CrNi189 X2CrNi189	SUS SUS	A49-207 〃 A49-230	TSZ2CN 18.10 TSZ2CN 18.10 TSZ2CN 18.10	SUS SUS SUS	2604/2	TS46	SUS	
						2604/2	TS68	SUS	
2462 2463 〃 17455 17456 17457 17458	X5CrNiMo 1810 X5CrNiMo 1810 X5CrNiMo 1812 X5CrNiMo 17122 X2CrNiMo 17132 X5CrNiMo 17122 X2CrNiMo 17132	SUS SUS SUS	A49-230	TUZ6CN 18.09	SUS	2604/2	TS60	SUS	

JIS			ASTM			BS		
Standard Number 規格番号	Grade グレード	Type 鋼　種	Standard Number 規格番号	Grade グレード	Type 鋼　種	Standard Number 規格番号	Grade グレード	Type 鋼　種
G3468 (Continued)	SUS		A213	TP316H	SUS	3059	CFS316S59	SUS
	316HTB	SUS	A249	TP316H	SUS			
			A213	TP316L	SUS	3605	316S14	SUS
	SUS							
	316LTB	SUS	A249	TP316L	SUS	〃	316S22	SUS
			A269	TP316L	SUS	3606	LWHT316S24	SUS
			A632	TP316L	SUS	〃	LWCF316S24	SUS
			A688	TP316L	SUS	〃	LWBC316S24	SUS
						〃	CFS316S24	SUS
						〃	LWHT316S29	SUS
						〃	LWCF316S29	SUS
						〃	LWBC316S29	SUS
						〃	CFS316S29	SUS
	SUS		A249	TP317	SUS			
	317TB	SUS	A632	TP317	SUS			
	SUS		A249	TP317	SUS			
	317LTB	SUS						
			A213	TP321	SUS	3605	321S18	SUS
	321TB	SUS	A249	TP321	SUS	〃	321S22	SUS
			A269	TP321	SUS	3606	LWHT321S22	SUS
			A632	TP321	SUS	〃	LWCF321S22	SUS
						〃	LWBC321S22	SUS
						〃	CFS321S22	
	SUS		A213	TP321H	SUS	3059	CFS321S59	SUS
	321HTB	SUS	A249	TP321H	SUS	3605	321S59	SUS
			A213	TP347	SUS	3605	347S18	SUS
	347TB	SUS	A249	TP347	SUS	〃	347S17	SUS
			A269	TP347	SUS	3606	LWHT347S17	SUS
			A632	TP347	SUS	〃	LWCF347S17	SUS

DIN			NF			ISO			Index Number 索引番号
Standard Number 規格番号	Grade グレード	Type 鋼種	Standard Number 規格番号	Grade グレード	Type 鋼種	Standard Number 規格番号	Grade グレード	Type 鋼種	
17455	X5CrNiMo 17122					2604/2	TS63	SUS	C012
17456						〃	TS61	SUS	
2462	X2CrNiMo 1810	SUS				2604/2	TS57	SUS	
	X2CrNiMo 1812	SUS				〃	TS58	SUS	
2463	X2CrNiMo 1810	SUS							
	X2CrNiMo 1812	SUS							
17457	X5CrNi 17122								
17458	〃								
2462	X10CrNi Ti189	SUS	A49-230	TUZ6CNT 18.10	SUS	2604/2	TS53	SUS	
2463	X10CrNi Ti189	SUS							
17457									
17458	X6CrNiTi 1810								
2462	X10CrNi Nb189	SUS	A49-207	TSZ6CNNb 18.10	SUS	2604/2	TS50		
17457			〃	TUZ6CNNb 18.10	SUS				
17458	X6CrNiNb 1810								

JIS			ASTM			BS		
Standard Number 規格番号	Grade グレード	Type 鋼　種	Standard Number 規格番号	Grade グレード	Type 鋼　種	Standard Number 規格番号	Grade グレード	Type 鋼　種
G3463 (Continued)						3606 3606	LWBC347S17 CFS347S17	SUS SUS
	SUS 347HTB	SUS	A213 A249	TP347H TP347H	SUS SUS	3059 3605	CFS347S59 347S59	SUS SUS
	SUS XM15J1TB	SUS						
	SUS 329J1TB	SUS						
	SUS 329J2LTB	SUS						
	SUS 405TB	SUS	A268	TP405	SUS			
	SUS 409TB	SUS	A268	TP409				
	SUS 410TB	SUS	A268	TP410	SUS			
	SUS 410TiTB	SUS						
	SUS 430TB	SUS	A268	TP430	SUS			
	SUS 444TB	SUS	A213	18Cr2Mo	SUS			
	SUS XM8TB	SUS						
	SUS XM27TB	SUS	A731 〃	18Cr2Mo TP439	SUS SUS			
G3467	STF410 (STF42)	C						
	STFA12	Mo	A161	T1	Mo			
	STFA22	CrMo						
	STFA23	CrMo	A200	T11	CrMo			
	STFA24	CrMo	A200	T22	CrMo			
	STFA25	CrMo	A200	T5	CrMo			
	STFA26	CrMo	A200	T9	CrMo			
	SUS 304TF	SUS	A271	TP304	SUS			
	SUS 304HTF		A271	TP304H	SUS			

DIN			NF			ISO			Index Number 索引番号
Standard Number 規格番号	Grade グレード	Type 鋼種	Standard Number 規格番号	Grade グレード	Type 鋼種	Standard Number 規格番号	Grade グレード	Type 鋼種	
									C012
						2604/2	TS56	SUS	
2462	X10Cr13	SUS							
2462	X8Cr7	SUS							
						2604/2	TS9H	C	C013
						2604/2	TS37	CrMo	
						2604/2	TS47	SUS	
			A49-218	TUZ6CN 18.09		2604/2	TS48	SUS	

JIS			ASTM			BS		
Standard Number 規格番号	Grade グレード	Type 鋼　種	Standard Number 規格番号	Grade グレード	Type 鋼　種	Standard Number 規格番号	Grade グレード	Type 鋼　種
G3467 (Continued)	SUS 309STF	SUS			SUS			
	SUS 310STF	SUS			SUS			
	SUS 316TF	SUS	A271	TP316	SUS			
	SUS 316HTF	SUS	A271	TP316H	SUS			
	SUS 321TF	SUS	A271	TP321	SUS			
	SUS 321HTF	SUS	A271	TP321H	SUS			
	SUS 347TF	SUS	A271	TP347	SUS			
	SUS 347HTF	SUS	A271	TP347H	SUS			
	NCF800TF	NiCr						
	NCF800HTF	NiCr						
G4904	NCF600TB NCF800TB NCF825TB	A A A	B167 B407 B423	N06600 N08800 N08825	A A A			
G3464	STBL380 (STBL39)	C	A334	Gr1	C	3603 〃 〃 〃	HFS410LT50 CFS410LT50 ERW410LT50 CEW410LT50	C C C C
	STBL450 (STBL46)	Ni	A334	Gr3	Ni	3603 〃	HF503LT100 CF503LT100	Ni Ni
	STBL690 (STBL70)	Ni	A334	Gr8	Ni	3603 〃	HFS509LT196 CFS509LT196	Ni Ni
G3447	SUS 304TBS	SUS	A270	TP304	SUS	1864	Gr. 1	SUS
	SUS 304LTBS	SUS	A270	TP304L	SUS			
	SUS 316TBS	SUS	A270	TP316	SUS	1864 〃	Gr. 5 Gr. 6	SUS SUS
	SUS 316LTBS	SUS	A270	TP316L	SUS			
G3444	STK290 (STK30)	C	A252 A500	Gr1 GrA	C C			

DIN			NF			ISO			Index Number
Standard Number 規格番号	Grade グレード	Type 鋼　種	Standard Number 規格番号	Grade グレード	Type 鋼　種	Standard Number 規格番号	Grade グレード	Type 鋼　種	索　引 番　号
									C013
						2604/2	TS68	SUS	
			A49-218	TUZ6CND 17.11		2604/2	TS61	SUS	
						2604/2	TS63	SUS	
						2604/2	TS53	SUS	
			A49-218	TUZ6CNT 18.10	SUS				
			A49-218	TUZ6CNNb 18.10	SUS	2604/2	TS50	SUS	
									C014
17173 17174	F0Ni14 X8Ni9	Ni Ni				2604/2	TS6	C	
			A49-230 A49-215	TU10N14 TU10N14	Ni Ni	2604/2	TS43	Ni	
			A49-230 A49-215	TUZ6N9 TUZ6N9	Ni Ni	2604/2	TS45	Ni	
			A49-249	TSZ6CN 18.09	SUS				C015
			A49-249	TSZ2CN 18.10	SUS				
			A49-249	TSZ2CND 17.11	SUS				
			A49-249	TSZ2CND 17.12	SUS				
			A49-642 A49-643	TS30.0 TS30.0	C C				C016

JIS			ASTM			BS		
Standard Number 規格番号	Grade グレード	Type 鋼　種	Standard Number 規格番号	Grade グレード	Type 鋼　種	Standard Number 規格番号	Grade グレード	Type 鋼　種
G3444 (Continued)								
	STK400 (STK41)	C	A252 A500 A501	Gr2 GrB —	C C C	6232	SAW4	C
	STK500 (STK51)	C				6323	SAW5	C
	STK490 (STK50)	C						
	STK540 (STK55)	C						
G3445	STKM11A	C	A512 A513	MT1010 MT1010	C C	1717	ERWC1	C
	STKM12A	C	A512 A513	MT1015 MT1015	C C	1717 6323	ERWC2 HFS3	C C
	STKM12B	C	A512 A513 A519	MT1015 MT1015 MT1015	C C C			
	STKM12C	C				1717 〃 6323 〃	CEWC2 CFSC3 CFS3 CFS3A	C C C C
	STKM13A	C	A312 A513	MT1020 MT1020	C C	1717	ERWC3	C
	STKM13B	C	A513	MT1020	C			
	STKM13C	C				1717 〃 6323	CEWC3 CFSC4 CFS4	C C C
	STKM14A	C	A513	MT1020	C	6323	HFS4	C

DIN			NF			ISO			Index Number 索引番号
Standard Number 規格番号	Grade グレード	Type 鋼種	Standard Number 規格番号	Grade グレード	Type 鋼種	Standard Number 規格番号	Grade グレード	Type 鋼種	
			A49-643	TS30a	C				C016
			A49-644	TS30.0	C				
			A49-645	TS30.0	C				
			〃	TS30E	C				
			〃	TS30ES	C				
			A49-643	TS37a	C				
			A49-643	TS47a	C				
2391	St30Si	C				3304	R28	C	C017
	St30A1	C				3305	〃	C	
2393	St28	C				3306	〃	C	
	RSt28	C							
2394	St28	C							
	USt28	C							
	RSt28	C							
2393	St37-2					3304	R33	C	
	RSt37-2					3305	〃	〃	
2394	St37-2					3306	〃	〃	
	USt37-2								
	RSt-2								
			A49-322	TU37b	C				
			A49-327	TU37b	C				
2391	St45	C	A49-326	TU37b	C	2937	TS4	C	
2393	St44-2	C	A49-330	TU37b	C	3304	R37	〃	
2394	St44-2	C	A49-343	TU37b	C	3305	〃	〃	
						3306	〃	〃	
						2937	TS9	C	
						3304	R42	〃	

JIS			ASTM			BS		
Standard Number 規格番号	Grade グレード	Type 鋼種	Standard Number 規格番号	Grade グレード	Type 鋼種	Standard Number 規格番号	Grade グレード	Type 鋼種
G3445 (Continued)	STKM14B	C				6323	HFS5	C
	STKM14C	C						
	STKM15A	C	A513 A519	1030 1030	C C			
	STKM15C	C						
	STKM16A	C	A519	1040	C			
	STKM16C	C						
	STKM17A	C	A519	1050	C	6323	HFS8	C
	STKM17C	C				6323	CFS8	C
	STKM18A	C	A519	1518	C	1717	ERWC5	C
	STKM18B	C						
	STKM18C	C						
	STKM19A	C	A519	1524	C			
	STKM19C	C						
	STKM20A	C						
G3441	SCr420TK	Cr						
	SCM415TK	CrMo						
	SCM418TK	CrMo				6323	CFS10	CrMo
	SCM420TK	CrMo	A519	5120	CrMo			
	SCM430TK	CrMo	A519	4130	CrMo			
	SCM435TK	CrMo	A519	4135	CrMo			
	SCM440TK	CrMo	A519	4140	CrMo	6323	CFS10	CrMo
G3446	SUS 304TKA	SUS	A511 A554	MT304 MT304	SUS SUS	6323 〃 〃 〃	LW13 LWCF13 LW15 LWCF15	SUS SUS SUS SUS
	SUS 304TKC	SUS						
	SUS 316TKA	SUS	A511 A554	MT316 MT316	SUS SUS	6323 〃	LW17 LWCF17	SUS SUS

DIN			NF			ISO			Index Number 索引番号
Standard Number 規格番号	Grade グレード	Type 鋼種	Standard Number 規格番号	Grade グレード	Type 鋼種	Standard Number 規格番号	Grade グレード	Type 鋼種	
						3305 3306	R42 〃	C 〃	C017
			A49-311 A49-312	TUXC35 TUXC35	C C				
			A49-310 A49-311 A49-312 A49-321 A49-323 A49-326 A49-330 A49-341 〃 A49-343	TU52b TU52b TU52b TU52b TU52b TU52b TU52b TS42a TS47a TS18M5	C C C C C C C C C C				
2391 2393 2394	ST52 ST52-3 ST52-3	C C C				2937 2938 3304 3305 3306	TS18 Gr.1 R50 R50 R50	C C C C C	
									C018
			A49-647	TSZ8C17	SUS				C019

JIS			ASTM			BS		
Standard Number 規格番号	Grade グレード	Type 鋼 種	Standard Number 規格番号	Grade グレード	Type 鋼 種	Standard Number 規格番号	Grade グレード	Type 鋼 種
G3446 (Continued)	SUS 316TKC	SUS						
	SUS 321TKA	SUS	A511 A554	MT321 MT321	SUS SUS	6323 〃	LW18 LWCF18	SUS SUS
	SUS 347TKA	SUS	A511 A554	MT347 MT347	SUS SUS			
	SUS 430TKA	SUS	A511 A554	MT430 MT430	SUS SUS			
	SUS 430TKC	SUS						
	SUS 410TKA	SUS	A511	MT410	SUS			
	SUS 410TKC	SUS						
	SUS 420J1TKA	SUS						
	SUS 420J2TKA	SUS						
	STKR41 STKR50	C C	A500	GrB	C	6363 〃 〃	34/26 43/36 50/45	C C C
G3472	STAM 290GA (30GA)	C				6323 〃	ERW1 CEW1	C C
	STAM 290GB (30GB)	C						
	STAM 340G (35G)	C				6323 〃 〃 〃	HFW2 HFW3 CEW2 CEW3	C C C C
	STAM 390G (40G)	C				6323 〃	HFW4 HFW4	C C
	STAM 440G (45G)	C				6323 〃	ERW4 CEW4	C C
	STAM 470G	C				6323	ERW5	C

DIN			NF			ISO			Index Number
Standard Number 規格番号	Grade グレード	Type 鋼　種	Standard Number 規格番号	Grade グレード	Type 鋼　種	Standard Number 規格番号	Grade グレード	Type 鋼　種	索　引 番　号
									C019
			A49-652	TS37a	C				C020
2393 〃 〃 2394 〃 〃	St28 USt28 RSt28 St28 USt28 RSt28					3305 3305	R28 R28	C C	C021
						3305	R33	C	
2393 2394	St44-2 St44-2					3305	R37	C	

JIS			ASTM			BS		
Standard Number 規格番号	Grade グレード	Type 鋼 種	Standard Number 規格番号	Grade グレード	Type 鋼 種	Standard Number 規格番号	Grade グレード	Type 鋼 種
G3472	(48G)							
(Continued)	STAM 500G (51G)	C				6323 〃 〃	HFW5 ERW5 CEW5	C C C
	STAM 440H (45H)	C				6323 〃	ERW4 CEW4	C C
	STAM 470H (48H)	C				6323	ERW5	C
	STAM 500H (51H)	C						
	STAM 540H (55H)	C						
G3473	STC370 (STC38)	C				5242/1 5242/3	HP1 HP1	C C
	STC440 (STC45)	C				5242/1 5242/3	HP4 HP4	C C
	STC510A (STC52A)	C				5242/1 5242/3	HP2 HP2	C C
	STC510B (STC52B)	C						
	STC540 (STC55)	C						
	STC590A (STC60A)	C				5242/1 5242/3	HP5 HP5	C C
	STC590B (STC60B)	C						

DIN			NF			ISO			Index Number 索引番号
Standard Number 規格番号	Grade グレード	Type 鋼種	Standard Number 規格番号	Grade グレード	Type 鋼種	Standard Number 規格番号	Grade グレード	Type 鋼種	
									C021
2391	St35					2937	TS4	C	C022

3. JIS Steel Pipes for Piping　配管用鋼管

Standard No. Year Designation 規格番号 年号 規格名称	Grade グレード	Mfg. Process 製造方法	Chemical Composition C	 Si	 Mn	 P
G3442－88 Galvanized Steel Pipe for Water Service 水道用亜鉛めっき鋼管	SGPW	－	Original pipes are JIS G3452 SGP.			
G3447－88 Stainless Steel Sanitary Tubing ステンレス鋼サニタリー管	SUS304TBS	S, A, E	≦0.08	≦1.00	≦2.00	≦0.040
	SUS304LTBS	S, A, E	≦0.030	≦1.00	≦2.00	≦0.040
	SUS316TBS	S, A, E	≦0.08	≦1.00	≦2.00	≦0.040
	SUS316LTBS	S, A, E	≦0.030	≦1.00	≦2.00	≦0.040
G3448－88 Light Gauge Stainless Steel Pipes for Ordinary Piping 一般配管用薄肉ステンレス鋼鋼管	SUS304TPD	A, E	≦0.08	≦1.00	≦2.00	≦0.045
	SUS316TPD	A, E	≦0.08	≦1.00	≦2.00	≦0.045
G3452－88 Carbon Steel Pipes for Ordinary Piping 一般配管用炭素鋼鋼管	SGP	E, B	－	－	－	≦0.040
G3454－88 Carbon Steel Pipes for Pressure Service 圧力配管用炭素鋼鋼管	STPG370 (STPG38)	S, E	≦0.25	≦0.35	0.30～0.90	≦0.040
	STPG410 (STPG42)	S, E	≦0.30	≦0.35	0.30～1.00	≦0.040
G3455－88 Carbon Steel Pipes for High Pressure Service 高圧配管用炭素鋼鋼管	STS370 (STS38)	S	≦0.25	0.10～0.35	0.30～1.10	≦0.035
	STS410 (STS42)	S	≦0.30	0.10～0.35	0.30～1.40	≦0.035
	STS480 (STS49)	S	≦0.33	0.10～0.35	0.30～1.50	≦0.035
G3456－88 Carbon Steel Pipes for High Temperature service 高温配管用炭素鋼鋼管	STPT370 (STPT38)	S, E	≦0.25	0.10～0.35	0.30～0.90	≦0.035
	STPT410 (STPT42)	S, E	≦0.30	0.10～0.35	0.30～1.00	≦0.035
	STPT480 (STPT49)	S	≦0.33	0.10～0.35	0.30～1.00	≦0.035
G3457－88 Arc Welded Carbon Steel Pipes 配管用アーク溶接炭素鋼鋼管	STPY400 (STPY41)	AM	≦0.25	－	－	≦0.040
G3458－88 Alloy Steel Pipes 配管用合金鋼鋼管	STPA12	S	0.01～0.20	0.01～0.50	0.30～0.80	≦0.035

1 N/mm² = 1.01972 × 10^{-1} kgf/mm²

化学成分（%）					Tensile Test N/mm²(kgf/mm²) 引張試験		Remarks 備考	Index No. 索引番号
S	Ni	Cr	Mo	Others	Min. Yield Point 最小降伏点	Tensile Strength 引張強さ		
–	–	–	–	–	–	–	–	C100
≦0.030	8.00～11.00	18.00～20.00	–	–	–	≧520 (≧53)	–	C101
≦0.030	9.00～13.00	18.00～21.00	–	–	–	≧480 (≧49)	–	
≦0.030	10.00～14.00	16.00～18.00	2.00～3.00	–	–	≧520 (≧53)	–	
≦0.030	12.00～16.00	16.00～18.00	2.00～3.00	–	–	≧480 (≧49)	–	
≦0.030	8.00～10.50	18.00～20.00	–	–	–	≧520 (≧53)	–	C102
≦0.030	10.00～14.00	16.00～18.00	2.00～3.00	–	–	≧520 (≧53)	–	
≦0.040	–	–	–	–	–	≧290 (≧30)	–	C103
≦0.040	–	–	–	–	215(22)	≧370 (≧38)	–	C104
≦0.040	–	–	–	–	245(25)	≧410 (≧42)	–	
≦0.035	–	–	–	–	215(22)	≧370 (≧38)	–	C105
≦0.035	–	–	–	–	245(25)	≧410 (≧42)	–	
≦0.035	–	–	–	–	275(28)	≧480 (≧49)	–	
≦0.035	–	–	–	–	215(22)	≧370 (≧38)	–	C106
≦0.035	–	–	–	–	245(25)	≧410 (≧42)	–	
≦0.035	–	–	–	–	275(28)	≧480 (≧49)	–	
≦0.040	–	–	–	–	225(23)	≧400 (≧41)	–	C107
≦0.035	–	–	0.45～0.65	–	205(21)	≧380 (≧39)	–	C108

Standard No. Year Designation 規格番号 年号 規格名称	Grade グレード	Mfg. Process 製造方法	Chemical Composition			
			C	Si	Mn	P
G3458−88 (Continued)	STPA20	S	0.01〜0.20	0.01〜0.50	0.30〜0.60	≦0.035
	STPA22	S	≦0.15	≦0.50	0.30〜0.60	≦0.035
	STPA23	S	≦0.15	0.50〜1.00	0.30〜0.60	≦0.030
	STPA24	S	≦0.15	≦0.50	0.30〜0.60	≦0.030
	STPA25	S	≦0.15	≦0.50	0.30〜0.60	≦0.030
	STPA26	S	≦0.15	0.25〜1.00	0.30〜0.60	≦0.030
G3459−88 Stainless Steel Pipes 配管用ステンレス鋼鋼管	SUS304TP	AM, S, E	≦0.08	≦1.00	≦2.00	≦0.040
	SUS304HTP	AM, S, E	0.04〜0.10	≦0.75	≦2.00	≦0.040
	SUS304LTP	AM, S, E	≦0.030	≦1.00	≦2.00	≦0.040
	SUS309TP	AM, S, E	≦0.15	≦1.00	≦2.00	≦0.040
	SUS309STP	AM, S, E	≦0.08	≦1.00	≦2.00	≦0.040
	SUS310TP	AM, S, E	≦0.15	≦1.50	≦2.00	≦0.040
	SUS310STP	AM, S, E	≦0.08	≦1.50	≦2.00	≦0.040
	SUS316TP	AM, S, E	≦0.08	≦1.00	≦2.00	≦0.040
	SUS316HTP	AM, S, E	0.04〜0.10	≦0.75	≦2.00	≦0.030
	SUS316LTP	AM, S, E	≦0.030	≦1.00	≦2.00	≦0.040
	SUS317TP	AM, S, E	≦0.08	≦1.00	≦2.00	≦0.040
	SUS317LTP	AM, S, E	≦0.030	≦1.00	≦2.00	≦0.040
	SUS321TP	AM, S, E	≦0.08	≦1.00	≦2.00	≦0.040
	SUS321HTP	AM, S, E	0.04〜1.00	≦0.75	≦2.00	≦0.030

1 N/mm² = 1.01972 × 10⁻¹ kgf/mm²

化学成分（%）					Tensile Test N/mm²(kgf/mm²) 引張試験		Remarks 備考	Index No. 索引番号
S	Ni	Cr	Mo	Others	Min. Yield Point 最小降伏点	Tensile Strength 引張強さ		
≦0.035	−	0.50〜0.80	0.40〜0.65	−	205(21)	≧410 (≧42)	−	C108
≦0.035	−	0.80〜1.25	0.40〜0.65	−	205(21)	≧410 (≧42)	−	
≦0.030	−	1.00〜1.50	0.45〜0.65	−	205(21)	≧410 (≧42)	−	
≦0.030	−	1.90〜2.60	0.87〜1.13	−	205(21)	≧410 (≧42)	−	
≦0.030	−	4.00〜6.00	0.45〜0.65	−	205(21)	≧410 (≧42)	−	
≦0.030	−	8.00〜10.00	0.90〜1.10	−	205(21)	≧410 (≧42)	−	
≦0.030	8.00〜11.00	18.00〜20.00	−	−	205(21)	≧520 (≧53)	−	C109
≦0.030	8.00〜11.00	18.00〜20.00	−	−	205(21)	≧520 (≧53)	−	
≦0.030	9.00〜13.00	18.00〜20.00	−	−	175(18)	≧480 (≧49)	−	
≦0.030	12.00〜15.00	22.00〜24.00	−	−	205(21)	≧520 (≧53)	−	
≦0.030	12.00〜15.00	22.00〜24.00	−	−	205(21)	≧520 (≧53)	−	
≦0.030	19.00〜22.00	24.00〜26.00	−	−	205(21)	≧520 (≧53)	−	
≦0.030	19.00〜22.00	24.00〜26.00	−	−	205(21)	≧520 (≧53)	−	
≦0.030	10.00〜14.00	16.00〜18.00	2.00〜3.00	−	205(21)	≧520 (≧53)	−	
≦0.030	11.00〜14.00	16.00〜18.00	2.00〜3.00	−	205(21)	≧520 (≧53)	−	
≦0.030	12.00〜16.00	16.00〜18.00	2.00〜3.00	−	175(18)	≧480 (≧49)	−	
≦0.030	11.00〜15.00	18.00〜20.00	3.00〜4.00	−	205(21)	≧520 (≧53)	−	
≦0.030	11.00〜15.00	18.00〜20.00	3.00〜4.00	−	175(18)	≧480 (≧49)	−	
≦0.030	9.00〜13.00	17.00〜19.00	−	Ti ≦ 5 × C	205(21)	≧520 (≧53)	−	
≦0.030	9.00〜13.00	17.00〜20.00	−	Ti 4×C〜0.60	205(21)	≧520 (≧53)	−	

Standard No. Year Designation 規格番号 年号 規格名称	Grade グレード	Mfg. Process 製造方法	Chemical Composition			
			C	Si	Mn	P
G3459－88 (Continued)	SUS347TP	AM, S, E	≦0.08	≦1.00	≦2.00	≦0.040
	SUS347HTP	AM, S, E	0.04～0.10	≦1.00	≦2.00	≦0.030
	SUS329J1TP	AM, S, E	≦0.08	≦1.00	≦1.50	≦0.040
	SUS329J2LTP	AM, S, E	≦0.030	≦1.00	≦1.50	≦0.040
	SUS405TP	AM, S, E	≦0.08	≦1.00	≦1.00	≦0.040
G3460－88 Steel Pipes for Low Temperature Service 低温配管用鋼管	STPL380 (STPL39)	S, E	≦0.25	≦0.35	≦1.35	≦0.035
	STPL450 (STPL46)	S	≦0.18	0.10～0.35	0.30～0.60	≦0.030
	STPL690 (STPL70)	S	≦0.13	0.10～0.35	≦0.90	≦0.030
G3468－88 Arc Welded Large Diameter Stainless Steel Pipes 配管用アーク溶接大径ステンレス鋼鋼管	SUS304TPY	AM	≦0.08	≦1.00	≦2.00	≦0.045
	SUS304LTPY	AM	≦0.030	≦1.00	≦2.00	≦0.045
	SUS309STPY	AM	≦0.08	≦1.00	≦2.00	≦0.045
	SUS310STPY	AM	≦0.08	≦1.50	≦2.00	≦0.045
	SUS316STPY	AM	≦0.08	≦1.00	≦2.00	≦0.045
	SUS316LTPY	AM	≦0.030	≦1.00	≦2.00	≦0.045
	SUS317TPY	AM	≦0.08	≦1.00	≦2.00	≦0.045
	SUS317LTPY	AM	≦0.030	≦1.00	≦2.00	≦0.045
	SUS321TPY	AM	≦0.08	≦1.00	≦2.00	≦0.045
	SUS347TPY	AM	≦0.08	≦1.00	≦2.00	≦0.045
	SUS329J1TPY	AM	≦0.08	≦1.00	≦1.50	≦0.045

1 N/mm² = 1.01972 × 10⁻¹ kgf/mm²

化 学 成 分 (%)					Tensile Test N/mm²(kgf/mm²) 引 張 試 験		Remarks 備 考	Index No. 索引番号
S	Ni	Cr	Mo	Others	Min. Yield Point 最小降伏点	Tensile Strength 引張強さ		
≦0.030	9.00～13.00	17.00～19.00	–	Nb≧10×C	205(21)	≧520 (≧53)	–	C109
≦0.030	9.00～13.00	17.00～20.00	–	Nb 8×C～1.00	205(21)	≧520 (≧53)	–	
≦0.030	3.00～6.00	23.00～28.00	1.00～3.00	–	390(40)	≧590 (≧60)	–	
≦0.030	4.50～7.50	21.00～26.00	2.50～4.00	N 0.08～0.30	450(46)	≧620 (≧63)	–	
≦0.030	–	11.50～14.50	–	Al 0.10～0.30	205(21)	≧410 (≧42)	–	
≦0.030	–	–	–	–	205(21)	≧380 (≧39)	Impact Tast J 2V, -45° ≧21	C110
≦0.030	3.20～3.80	–	–	–	245(25)	≧450 (≧46)	-100℃≧21	
≦0.030	8.50～9.50	–	–	–	520(53)	≧690 (≧76)	-100℃≧21	
≦0.030	8.00～10.50	18.00～20.00	–	–	205(21)	≧520 (≧53)	–	C111
≦0.030	8.00～13.00	18.00～20.00	–	–	175(18)	≧480 (≧49)	–	
≦0.030	12.00～15.00	22.00～24.00	–	–	205(21)	≧520 (≧53)	–	
≦0.030	19.00～22.00	24.00～26.00	–	–	205(21)	≧520 (≧53)	–	
≦0.030	10.00～14.00	16.00～18.00	2.00～3.00	–	205(21)	≧520 (≧53)	–	
≦0.030	12.00～15.00	16.00～18.00	2.00～3.00	–	175(18)	≧480 (≧49)	–	
≦0.030	11.00～15.00	18.00～20.00	3.00～4.00	–	205(21)	≧520 (≧53)	–	
≦0.030	11.00～15.00	18.00～20.00	3.00～4.00	–	205(21)	≧480 (≧49)	–	
≦0.030	9.00～13.00	17.00～19.00	–	Ti≧5×C	205(21)	≧520 (≧53)	–	
≦0.030	9.00～13.00	17.00～19.00	–	Nb≧10×C	205(21)	≧520 (≧53)	–	
≦0.030	3.00～6.00	23.00～28.00	1.00～3.00	–	390(40)	≧520 (≧53)	–	

Standard No. Year Designation 規格番号 年号 規格名称	Grade グレード	Mfg. Process 製造方法	Chemical Composition			
			C	Si	Mn	P
G3469－92 Polyethylene Coated Steel Pipes ポリエチレン被覆鋼管	P1H	－	Original pipes are JIS G3452－SGP, G3454-STPG, and G3457-STPY.			
	P2S	－				
	P1F	－				
G4903－88 Seamless Nickel-Chromium-Iron Alloy Pipes 配管用継目無ニッケルクロム鉄合金	NCF600TP	S	≦0.15	≦0.50	≦1.00	≦0.030
	NCF800TP	S	≦0.10	≦1.00	≦1.50	≦0.030
	NCF800HTP	S	0.05～0.10	≦1.00	≦1.50	≦0.030
	NCF825TP	S	≦0.05	≦0.50	≦1.00	≦0.030

1 N/mm² = 1.01972 × 10⁻¹ kgf/mm²

化 学 成 分 (%)					Tensile Test N/mm²(kgf/mm²) 引 張 試 験		Remarks 備 考		Index No. 索引番号
S	Ni	Cr	Mo	Others	Min. Yield Point 最小降伏点	Tensile Strength 引張強さ			
–	–	–	–	–	–	–		–	C112
–	–	–	–	–	–	–		–	
–	–	–	–	–	–	–		–	
≦0.015	≧72.00	6.00~ 10.00	Fe Balance	Cu≦0.50	205(21)	≧550(56)	H+N	t≦127mm	C113
					175(18)	≧520(53)	*	127mm≦t	
					245(25)	≧550(56)	C+N	t≦127mm	
					205(21)	≧550(56)	*	127mm<t	
≦0.015	30.00~ 35.00	19.00~ 23.00	Fe Balance	Cu≦0.75	175(18)	≧450(46)	H+N *		
			AL0.15~0.60	Ti 0.15 ~0.60	205(21)	≧520(53)	C+N *		
≦0.015	30.00~ 35.00	19.00~ 23.00	Fe Balance AL0.15~0.60	Cu≦0.75 Ti 0.15 ~0.60	175(18)	≧450(46)	H+S * or C+S *		
≦0.015	38.00~ 46.00	19.50~ 23.50	Fe Balance	Cu 1.50 ~3.00	175(18)	≧520(53)	H+N *		
			AL≦0.20	Ti 0.60 ~1.20	235(24)	≧580(59)	C+N *		

* S — Solubilized Heat Treatment
H+N — Normalized after Heat-finishing
C+N — Normalized after Cold-finishing

4. JIS Steel Tubes for Heat Transfer　熱伝達用鋼管

Standard No. Year Designation 規格番号 年号 規格名称	Grade グレード	Mfg. Process 製造方法	Chemical Composition			
			C	Si	Mn	P
G3461－88 Carbon Steel Boiler abd Heat Exchanger Tubes ボイラ，熱交換器用炭素鋼鋼管	STB35	S, E	≦0.18	≦0.35	0.30〜0.60	≦0.035
	STB42	S, E	≦0.32	≦0.35	0.30〜0.80	≦0.035
	STB52	S, E	≦0.25	≦0.35	1.00〜1.50	≦0.035
G3462－88 Alloy Steel Boiler and Heat Exchanger Tubes ボイラ，熱交換器用合金鋼鋼管	STBA12	S, E	0.10〜0.20	0.10〜0.50	0.30〜0.80	≦0.035
	STBA13	S, E	0.15〜0.25	0.10〜0.50	0.30〜0.80	≦0.035
	STBA20	S, E	0.10〜0.20	0.10〜0.50	0.30〜0.60	≦0.035
	STBA22	S, E	≦0.15	≦0.50	0.30〜0.60	≦0.035
	STBA23	S	≦0.15	0.50〜1.00	0.30〜0.60	≦0.030
	STBA24	S	≦0.15	≦0.50	0.30〜0.60	≦0.030
	STBA25	S	≦0.15	≦0.50	0.30〜0.60	≦0.030
	STBA26	S	≦0.15	0.25〜1.00	0.30〜0.60	≦0.030
G3463－88 Stainless Steel Boiler and Heat Exchanger Tubes ボイラ，熱交換器用ステンレス鋼鋼管	SUS304TB	S, A, E	≦0.08	≦1.00	≦2.00	≦0.040
	SUS304HTB	S, A, E	0.04〜0.10	≦0.75	≦2.00	≦0.040
	SUS304LTB	S, A, E	≦0.030	≦1.00	≦2.00	≦0.040
	SUS309TB	S, A, E	≦0.15	≦1.00	≦2.00	≦0.040
	SUS309STB	S, A, E	≦0.08	≦1.00	≦2.00	≦0.040
	SUS310TB	S, A, E	≦0.15	≦1.50	≦2.00	≦0.040
	SUS310STB	S, A, E	≦0.08	≦1.50	≦2.00	≦0.040
	SUS316TB	S, A, E	≦0.08	≦1.00	≦2.00	≦0.040
	SUS316HTB	S, A, E	0.04〜0.10	≦0.75	≦2.00	≦0.030

1 N/mm² = 1.01972 × 10⁻¹ kgf/mm²

化　学　成　分　(%)					Tensile Test N/mm²(kgf/mm²) 引張試験		Remarks 備考	Index No. 索引番号
S	Ni	Cr	Mo	Others	Min. Yield Point 最小降伏点	Tensile Strength 引張強さ		
≦0.035	−	−	−	−	175(18)	≧340 (≧35)	−	C114
≦0.035	−	−	−	−	255(26)	≧410 (≧42)	−	
≦0.035	−	−	−	−	295(30)	≧510 (≧52)	−	
≦0.035	−	−	0.45〜0.65	−	205(21)	≧380 (≧39)	−	C115
≦0.035	−	−	0.45〜0.65	−	205(21)	≧410 (≧42)	−	
≦0.035	−	0.50〜0.80	0.40〜0.65	−	205(21)	≧410 (≧42)	−	
≦0.035	−	0.80〜1.25	0.45〜0.65	−	205(21)	≧410 (≧42)	−	
≦0.030	−	1.00〜1.50	0.45〜0.65	−	205(21)	≧410 (≧42)	−	
≦0.030	−	1.90〜2.60	0.87〜1.13	−	205(21)	≧410 (≧42)	−	
≦0.030	−	4.00〜6.00	0.45〜0.65	−	205(21)	≧410 (≧42)	−	
≦0.030	−	8.00〜10.00	0.90〜1.10	−	205(21)	≧410 (≧42)	−	
≦0.030	8.00〜11.00	18.00〜20.00	−	−	205(21)	≧410 (≧53)	−	C116
≦0.030	8.00〜11.00	18.00〜20.00	−	−	205(21)	≧410 (≧53)	−	
≦0.030	9.00〜13.00	18.00〜20.00	−	−	175(18)	≧480 (≧49)	−	
≦0.030	12.00〜15.00	22.00〜24.00	−	−	205(21)	≧520 (≧53)	−	
≦0.030	12.00〜15.00	22.00〜24.00	−	−	205(21)	≧520 (≧53)	−	
≦0.030	19.00〜22.00	24.00〜26.00	−	−	205(21)	≧520 (≧53)	−	
≦0.030	19.00〜22.00	24.00〜26.00	−	−	205(21)	≧520 (≧53)	−	
≦0.030	10.00〜14.00	16.00〜18.00	2.00〜3.00	−	205(21)	≧520 (≧53)	−	
≦0.030	11.00〜14.00	16.00〜18.00	2.00〜3.00	−	205(21)	≧520 (≧53)	−	

Standard No. Year Designation 規格番号 年号 規格名称	Grade グレード	Mfg. Process 製造方法	Chemical Composition			
			C	Si	Mn	P
G3463−88 (Continued)	SUS316LTB	S, A, E	≦0.030	≦1.00	≦2.00	≦0.040
	SUS317TB	S, A, E	≦0.08	≦1.00	≦2.00	≦0.040
	SUS317LTB	S, A, E	≦0.030	≦1.00	≦2.00	≦0.040
	SUS321TB	S, A, E	≦0.08	≦1.00	≦2.00	≦0.040
	SUS321HTB	S, A, E	0.04〜0.10	≦0.75	≦2.00	≦0.030
	SUS347TB	S, A, E	≦0.08	≦1.00	≦2.00	≦0.030
	SUS347HTB	S, A, E	0.04〜0.10	≦1.00	≦2.00	≦0.030
	SUSXM15J1TB	S, A, E	≦0.08	3.00〜5.00	≦2.00	≦0.045
	SUS329J1TB	S, A, E	≦0.08	≦1.00	≦1.50	≦0.040
	SUS329J2JTB	S, A, E	≦0.030	≦1.00	≦1.50	≦0.040
	SUS405TB	S, A, E	≦0.080	≦1.00	≦1.00	≦0.040
	SUS409TB	S, A, E	≦0.080	≦1.00	≦1.00	≦0.040
	SUS410TB	S, A, E	≦0.15	≦1.00	≦1.00	≦0.040
	SUS410TiTB	S, A, E	≦0.08	≦1.00	≦1.00	≦0.040
	SUS430TB	S, A, E	≦0.12	≦0.75	≦1.50	≦0.040
	SUS444TB	S, A, E	≦0.025	≦1.00	≦1.00	≦0.040
	SUSXM8TB	S, A, E	≦0.08	≦1.00	≦1.00	≦0.040
	SUSXM27TB	S, A, E	≦0.010	≦0.40	≦0.40	≦0.030
G3464−88 Steel Heat Exchanger Tubes for Low Temperature Service 低温熱交換器用鋼管	STBL39	S, E	≦0.25	≦0.35	≦1.35	≦0.035
	STBL46	S	≦0.18	0.10〜0.35	0.30〜0.60	≦0.030

1 N/mm² = 1.01972 × 10⁻¹ kgf/mm²

化学成分 (%)					Tensile Test N/mm² (kgf/mm²) 引張試験		Remarks 備考	Index No. 索引番号
S	Ni	Cr	Mo	Others	Min. Yield Point 最小降伏点	Tensile Strength 引張強さ		
≦0.030	12.00～16.00	16.00～18.00	2.00～3.00	–	205(21)	≧520 (≧53)	–	C116
≦0.030	11.00～15.00	18.00～20.00	3.00～4.00	–	205(21)	≧520 (≧53)	–	
≦0.030	11.00～15.00	18.00～20.00	3.00～4.00	–	175(18)	≧480 (≧49)	–	
≦0.030	9.00～13.00	17.00～19.00	–	Ti≧5×C	205(21)	≧520 (≧53)	–	
≦0.030	9.00～13.00	17.00～20.00	–	Ti:4×C～0.60	205(21)	≧520 (≧53)	–	
≦0.030	9.00～13.00	17.00～20.00	–	Nb≧10×C	205(21)	≧520 (≧53)	–	
≦0.030	9.00～13.00	17.00～20.00	–	Nb:8×C～1.00	205(21)	≧520 (≧53)	–	
≦0.030	11.50～15.00	15.00～20.00	–	–	205(21)	≧520 (≧53)	–	
≦0.030	3.00～6.00	23.00～28.00	1.00～3.00	–	390(40)	≧590 (≧60)	–	
≦0.030	4.50～7.50	21.00～26.00	2.50～4.00	N 0.08～0.30	450(46)	≧620 (≧63)	–	
≦0.030	–	11.50～14.50	–	Al 0.10～0.30	205(21)	≧410 (≧42)	–	
≦0.030	–	10.50～11.75	–	Ti 6×C%～0.75	205(21)	≧410 (≧42)	–	
≦0.030	–	11.50～12.50	–	–	205(21)	≧410 (≧42)	–	
≦0.030	–	11.50～12.50	–	Ti 6×C%～0.75	205(21)	≧410 (≧42)	–	
≦0.030	–	16.00～18.00	–	–	205(21)	≧410 (≧42)	–	
≦0.030	–	17.00～20.00	1.75～2.50	N≦0.025	245(25)	≧410 (≧42)	–	
≦0.030	–	17.00～20.00	–	Ti 12×C%～1.10	205(21)	≧410 (≧42)	–	
≦0.020	–	25.00～27.50	0.75～1.80	N≦0.015	245(25)	≧410 (≧42)	–	
≦0.035	–	–	–	–	201(21)	≧380 (≧39)	-45℃21J (2.1kg-m)	C117
≦0.030	3.20～3.80	–	–	–	245(25)	≧450 (≧46)	-100℃18J (1.8kg-m)	

Standard No. Year Designation 規格番号 年号 規格名称	Grade グレード	Mfg. Process 製造方法	Chemical Composition			
			C	Si	Mn	P
G3464−88 (Continued)	STBL70	S	≦0.13	0.10〜0.35	≦0.90	≦0.030
G3467−88 Steel Tubes for Fired Heater 加熱炉用鋼管	STF42	S	≦0.30	0.10〜0.35	0.30〜1.00	≦0.035
	STFA12	S	0.10〜0.20	0.10〜0.50	0.30〜0.80	≦0.035
	STFA22	S	≦0.15	≦0.50	0.30〜0.60	≦0.035
	STFA23	S	≦0.15	0.50〜1.00	0.30〜0.60	≦0.030
	STFA24	S	≦0.15	≦0.50	0.30〜0.60	≦0.030
	STFA25	S	≦0.15	≦0.50	0.30〜0.60	≦0.030
	STFA26	S	≦0.15	0.25〜1.00	0.30〜0.60	≦0.030
	SUS304TF	S	≦0.08	≦1.00	≦2.00	≦0.040
	SUS304HTF	S	0.04〜0.10	≦0.75	≦2.00	≦0.040
	SUS309STF	S	≦0.15	≦1.00	≦2.00	≦0.040
	SUS310STF	S	≦0.15	≦1.50	≦2.00	≦0.040
	SUS316TF	S	≦0.08	≦1.00	≦2.00	≦0.040
	SUS316HTF	S	0.04〜0.10	≦0.75	≦2.00	≦0.030
	SUS321TF	S	≦0.08	≦1.00	≦2.00	≦0.040
	SUS321HTF	S	0.04〜0.10	≦0.75	≦2.00	≦0.030
	SUS347TF	S	≦0.08	≦1.00	≦2.00	≦0.040
	SUS347HTF	S	0.04〜0.10	≦0.75	≦2.00	≦0.030
	NCF800TF	S	≦0.10	≦1.00	≦1.50	≦0.030

1 N/mm² = 1.01972 × 10⁻¹ kgf/mm²

化学成分 (%)					Tensile Test N/mm² (kgf/mm²) 引張試験		Remarks 備考	Index No. 索引番号
S	Ni	Cr	Mo	Others	Min. Yield Point 最小降伏点	Tensile Strength 引張強さ		
≦0.030	8.50〜9.50	−	−	−	520 (53)	≧690 (≧70)	-196℃14J (1.4kg-m)	C117
≦0.035	−	−	−	−	245 (25)	≧410 (≧42)	−	C118
≦0.035	−	−	0.45〜0.65	−	205 (21)	≧380 (≧39)	−	
≦0.035	−	0.80〜1.25	0.45〜0.65	−	205 (21)	≧410 (≧42)	−	
≦0.030	−	1.00〜1.50	0.45〜0.65	−	205 (21)	≧410 (≧42)	−	
≦0.030	−	1.90〜2.60	0.87〜1.13	−	205 (21)	≧410 (≧42)	−	
≦0.030	−	4.00〜6.00	0.45〜0.65	−	205 (21)	≧410 (≧42)	−	
≦0.030	−	8.00〜10.00	0.90〜1.10	−	205 (21)	≧410 (≧42)	−	
≦0.030	8.00〜11.00	18.00〜20.00	−	−	205 (21)	≧520 (≧53)	−	
≦0.030	8.00〜11.00	18.00〜20.00	−	−	205 (21)	≧520 (≧53)	−	
≦0.030	12.00〜15.00	22.00〜24.00	−	−	205 (21)	≧520 (≧53)	−	
≦0.030	19.00〜22.00	24.00〜26.00	−	−	205 (21)	≧520 (≧53)	−	
≦0.030	10.00〜14.00	16.00〜18.00	2.00〜3.00	−	205 (21)	≧520 (≧53)	−	
≦0.030	11.00〜14.00	16.00〜18.00	2.00〜3.00	−	205 (21)	≧520 (≧53)	−	
≦0.030	9.00〜13.00	17.00〜19.00	−	Ti≧5×C%	205 (21)	≧520 (≧53)	−	
≦0.030	9.00〜13.00	17.00〜20.00	−	Ti:4×C〜0.60	205 (21)	≧520 (≧53)	−	
≦0.030	9.00〜13.00	17.00〜19.00	−	Nb≧10×C	205 (21)	≧520 (≧53)	−	
≦0.030	9.00〜13.00	17.00〜20.00	−	Nb:8×C〜1.00	205 (21)	≧520 (≧53)	−	
≦0.015	30.00〜35.00	19.00〜23.00	−	Cu≦0.75 Al:0.15〜0.60 Ti:0.15〜0.60	205 (21)	≧520 (≧53)	Cold-finished	
					175 (18)	≧520 (≧51)	Hot-finished	

Standard No. Year Designation 規格番号 年号 規格名称	Grade グレード	Mfg. Process 製造方法	Chemical Composition			
			C	Si	Mn	P
G3467－88 (Continued)	NCF800HTF	S	0.05～0.10	≦1.00	≦1.50	≦0.030
G4904－88 Seamless Nickel-Chromium-Iron Alloy Heat Exchanger Tubes 熱交換器用継目無ニッケルクロム鉄合金	NCF600TB	S	≦0.15	≦0.50	≦1.00	≦0.030
	NCF800TB	S	≦0.10	≦1.00	≦1.50	≦0.030
	NCF800HTB	S	0.05～0.10	≦1.00	≦1.50	≦0.030
	NCF825TB	S	≦0.05	≦0.50	≦1.00	≦0.030

1 N/mm² = 1.01972 × 10^{-1} kgf/mm²

化　学　成　分（%）					Tensile Test N/mm²(kgf/mm²) 引張試験		Remarks 備考	Index No. 索引番号
S	Ni	Cr	Mo	Others	Min. Yield Point 最小降伏点	Tensile Strength 引張強さ		
≦0.015	30.00～35.00	19.00～23.00	–	–	175 (18)	≧450 (≧46)	–	C118
≦0.015	≦72.00	14.00～17.00	Fe 6.00～10.00	Cu≦0.50	245(25)	≧550 (≧56)	N	C119
≦0.015	30.00～35.00	19.00～23.00	Fe Balance	Cu≦0.75 Al 0.15 ～0.60	205(21)	≧520 (≧53)	N	
≦0.015	30.00～35.00	19.00～23.00	Fe Bakance Al 0.15~0.60	Cu≦0.75 Ti 0.15 ～0.60	175(18)	≧450 (≧46)	S	
≦0.015	38.00～46.00	19.50～23.50	2.50～3.50 Fe Balance	Cu≦0.75 Al≦0.20 Ti 0.60 ～1.20	235(24)	≧580 (≧59)	N	

5. JIS Steel Pipes for Structural Purpose 構造用鋼管

Standard No. Year Designation 規格番号 年号 規格名称	Grade グレード	Mfg. Process 製造方法	Chemical Composition			
			C	Si	Mn	P
G3441-88 Alloy Steel for Machine Purposes 機械構造用合金鋼鋼管	SCr420TK	S, E	0.18～0.23	0.15～0.35	0.60～0.85	≦0.030
	SCM415TK	S, E	0.13～0.18	0.13～0.18	0.60～0.85	≦0.030
	SCM418TK	S, E	0.16～0.21	0.16～0.21	0.60～0.85	≦0.030
	SCM420TK	S, E	0.18～0.23	0.18～0.23	0.60～0.85	≦0.030
	SCM430TK	S, E	0.28～0.33	0.28～0.33	0.60～0.85	≦0.030
	SCM435TK	S, E	0.33～0.38	0.33～0.38	0.60～0.85	≦0.030
	SCM440TK	S, E	0.38～0.43	0.38～0.43	0.60～0.85	≦0.030
G3444-94 Carbon Steel Tubes for General Structural Purposes 一般構造用炭素鋼鋼管	STK290 (STK30)	S, E, B, A	–	–	–	≦0.050
	STK400 (STK41)	S, E, B, A	≦0.25	–	–	≦0.040
	STK500 (STK51)	S, E, B, A	≦0.24	≦0.35	0.30～1.00	≦0.040
	STK490 (STK50)	S, E, B, A	≦0.18	≦0.55	≦1.50	≦0.040
	STK540 (STK55)	S, E, B, A	≦0.23	≦0.55	≦1.50	≦0.040
G3445-94 Carbon Steel Tubes for Machine Structural Purposes 機械構造用炭素鋼鋼管	STKM11A	S, E, B	≦0.12	≦0.35	0.25～0.60	≦0.040
	STKM12A					
	STKM12B	S, E, B	≦0.20	≦0.35	0.25～0.60	≦0.040
	STKM12C					
	STKM13A					
	STKM13B	S, E, B	≦0.25	≦0.35	0.30～0.90	≦0.040
	STKM13C					
	STKM14A					
	STKM14B	S, E	≦0.30	≦0.35	0.30～1.00	≦0.040
	STKM14C					
	STKM15A	S, E	0.25～0.35	≦0.35	0.30～1.00	≦0.040
	STKM15C					

1 N/mm² = 1.01972×10^{-1} kgf/mm²

化　学　成　分　(%)					Tensile Test N/mm² (kgf/mm²) 引張試験		Remarks 備考	Index No. 索引番号
S	Ni	Cr	Mo	Others	Min. Yield Point 最小降伏点	Tensile Strength 引張強さ		
≦0.030	Ni (as Impurity) 不純物としてのNi ≦0.25	0.90～1.20	－	Cu (as Impurity) 不純物としてのCu ≦0.30	－	－	－	C120
≦0.030		0.90～1.20	0.15～0.30		－	－	－	
≦0.030		0.90～1.20	0.15～0.30		－	－	－	
≦0.030		0.90～1.20	0.15～0.30		－	－	－	
≦0.030		0.90～1.20	0.15～0.30		－	－	－	
≦0.030		0.90～1.20	0.15～0.30		－	－	－	
≦0.030		0.90～1.20	0.15～0.30		－	－	－	
≦0.050	－	－	－	－	－	≧290(20)	－	C121
≦0.040	－	－	－	－	235(24)	≧400 (≧41)	－	
≦0.040	－	－	－	－	355(36)	≧500 (≧49)	－	
≦0.040	－	－	－	－	315(32)	≧490 (≧50)	－	
≦0.040	－	－	－	－	390(40)	≧540 (≧53)	－	
≦0.040	－	－	－	－	－	≧290(30)	－	C122
≦0.040	－	－	－	－	175(18)	≧340 (≧35)	－	
					275(28)	≧390 (≧40)	－	
					355(36)	≧470 (≧48)	－	
≦0.040	－	－	－	－	215(22)	≧370 (≧38)	－	
					305(31)	≧440 (≧45)	－	
					380(39)	≧510 (≧52)	－	
≦0.040	－	－	－	－	245(25)	≧410 (≧42)	－	
					355(36)	≧500 (≧51)	－	
					410(42)	≧550 (≧56)	－	
≦0.040	－	－	－	－	275(28)	≧470 (≧48)	－	
					430(44)	≧580 (≧59)	－	

Standard No. Year Designation 規格番号 年号 規格名称	Grade グレード	Mfg. Process 製造方法	Chemical Composition C	 Si	 Mn	 P
G3445－88 (Continued)	STKM16A STKM16C	S, E	0. 35～0. 45	≦0. 40	0. 40～1. 00	≦0. 040
	STKM17A STKM17C	S, E	0. 45～0. 55	≦0. 40	0. 40～1. 00	≦0. 040
	STKM18A STKM18B STKM18C	S, E	≦0. 18	≦0. 55	≦1. 50	≦0. 040
	STKM19A STKM19C	S, E	≦0. 25	≦0. 55	≦1. 50	≦0. 040
	STKM20A	S, E	≦0. 25	≦0. 55	≦1. 60	≦0. 040
G3446－94 Stainless Steel Tubes for Machine and Structural Purpose 機械構造用ステンレス鋼管					Austenitic	
	SUS304TKA SUS304TKC	S, E, A E, A	≦0. 08	≦1. 00	≦2. 00	≦0. 040
	SUS316TKA SUS316TKC	S, E, A E, A	≦0. 08	≦1. 00	≦2. 00	≦0. 040
	SUS321TKA	S, E, A	≦0. 08	≦1. 00	≦2. 00	≦0. 040
	SUS347TKA	S, E, A	≦0. 08	≦1. 00	≦2. 00	≦0. 040
					Ferritic	
	SUS430TKA SUS430TKC	S, E, A E, A	≦0. 12	≦0. 75	≦1. 00	≦0. 040
					Martensitic	
	SUS410TKA SUS410TKC	S, E, A E, A	≦0. 15	≦1. 00	≦1. 00	≦0. 040
	SUS420J1TKA	S, E, A	0. 16～0. 25	≦1. 00	≦1. 00	≦0. 040
	SUS420J2TKA	S, E, A	0. 26～0. 40	≦1. 00	≦1. 00	≦0. 040

1 N/mm² = 1.01972 × 10^{-1} kgf/mm²

化　学　成　分　(%)					Tensile Test N/mm²(kgf/mm²) 引張試験		Remarks 備考	Index No. 索引番号
S	Ni	Cr	Mo	Others	Min. Yield Point 最小降伏点	Tensile Strength 引張強さ		
≦0.040	−	−	−	−	325(33)	≧510 (≧52)	−	C122
					460(47)	≧620 (≧63)	−	
≦0.040	−	−	−	−	345(35)	≧510 (≧52)	−	
					460(47)	≧620 (≧63)	−	
≦0.040	−	−	−	−	275(28)	≧440 (≧45)	−	
					315(32)	≧490 (≧50)	−	
					380(39)	≧510 (≧52)	−	
≦0.040	−	−	−	−	310(32)	≧490 (≧50)	−	
					410(42)	≧550 (≧56)	−	
≦0.040	−	−	−	Nb or V ≦0.15	390(40)	≧540 (≧55)	−	
オーステナイト系								C123
≦0.030	8.00〜11.00	18.00〜20.00	−	−	205(21)	≧520 (≧53)	−	
≦0.030	10.00〜14.00	16.00〜18.00	2.00〜3.00	−	205(21)	≧520 (≧53)	−	
≦0.030	9.00〜13.00	17.00〜19.00	−	Ti≧5×C%	205(21)	≧520 (≧53)	−	
≦0.030	9.00〜13.00	17.00〜19.00	−	Nb≧10×C%	205(21)	≧520 (≧53)	−	
フェライト系								
≦0.030	−	16.00〜18.00	−	−	245(25)	≧410 (≧42)	Ferrites and Martensites may contain Ni 0.6% or less. フェライト系及びマルテンサイト系は0.6%以下のNiを含有しても差支えない。	
マルテンサイト系								
≦0.030	−	11.50〜13.00	−	−	205(21)	≧410 (≧42)		
≦0.030	−	12.00〜14.00	−	−	225(22)	≧470 (≧46)		
≦0.030	−	12.00〜14.00	−	−	205(21)	≧540 (≧53)		

Standard No. Year Designation 規格番号 年号 規格名称	Grade グレード	Mfg. Process 製造方法	Chemical Composition			
			C	Si	Mn	P
G3466−88 Carbon Steel Square Pipes for General Structural Purposes 一般構造用角形鋼管	STKR400 (STKR41)	S, E, B, A	≦0.24	−	−	≦0.040
	STKR490 (STKR50)	S, E, B, A	≦0.18	≦0.55	≦1.50	≦0.040
G3472−88 Electric Resistance Welded Carbon Steel Tubes for Automobile Structural Purposes 自動車構造用電気抵抗溶接炭素鋼鋼管	STAM290GA (STAM30GA)	E	≦0.12	≦0.35	≦0.60	≦0.035
	STAM290GB (STAM30GB)	E	≦0.12	≦0.35	≦0.60	≦0.035
	STAM340G (STAM35G)	E	≦0.20	≦0.35	≦0.60	≦0.035
	STAM390G (STAM40G)	E	≦0.25	≦0.35	0.30〜0.90	≦0.035
	STAM440G (STAM45G)	E	≦0.25	≦0.35	0.30〜0.90	≦0.035
	STAM470G (STAM48G)	E	≦0.25	≦0.35	0.30〜0.90	≦0.035
	STAM500G (STAM51G)	E	≦0.30	≦0.35	0.30〜1.00	≦0.035
	STAM440H (STAM45H)	E	≦0.25	≦0.35	0.30〜0.90	≦0.035
	STAM470H (STAM48H)	E	≦0.25	≦0.35	0.30〜0.90	≦0.035
	STAM500H (STAM51H)	E	≦0.30	≦0.35	0.30〜1.00	≦0.035
	STAM540H (STAM55H)	E	≦0.30	≦0.35	0.30〜1.00	≦0.035
G3473−88 Carbon Steel Tubes for Cylinder Barrels シリンダーチューブ用炭素鋼鋼管	STC370 (STC38)	S	≦0.25	≦0.35	0.30〜0.90	≦0.040
	STC440 (STC45)	E	≦0.25	≦0.35	0.30〜0.90	≦0.040
	STC510A (STC52A)	S, E	≦0.25	≦0.35	0.30〜0.90	≦0.040
	STC510B (STC52B)	S, E	≦0.18	≦0.55	≦1.50	≦0.040
	STC540 (STC55)	S	≦0.25	≦0.55	≦1.60	≦0.040
	STC590A (STC60A)	S	≦0.25	≦0.35	0.30〜0.90	≦0.040

1 N/mm² = 1.01972 × 10^{-1} kgf/mm²

化 学 成 分 (%)					Tensile Test N/mm²(kgf/mm²) 引張試験		Remarks 備考	Index No. 索引番号
S	Ni	Cr	Mo	Others	Min. Yield Point 最小降伏点	Tensile Strength 引張強さ		
≦0.040	–	–	–	–	245(25)	≧410 (≧42)	–	C124
≦0.040	–	–	–	–	325(33)	≧490 (≧50)	–	
≦0.035	–	–	–	–	175(18)	≧290 (≧30)	A and B have different elongation value.	C125
≦0.035	–	–	–	–	175(18)	≧290 (≧30)		
≦0.035	–	–	–	–	195(24)	≧340 (≧35)	–	
≦0.035	–	–	–	Nb and/or V ≦0.15 (by agreement) 協議	235(20)	≧390 (≧40)	–	
≦0.035	–	–	–	–	305(31)	≧440 (≧45)	–	
≦0.035	–	–	–	–	325(33)	≧470 (≧48)	–	
≦0.035	–	–	–	–	355(36)	≧500 (≧51)	–	
≦0.035	–	–	–	–	355(36)	≧440 (≧45)	–	
≦0.035	–	–	–	–	410(42)	≧470 (≧48)	–	
≦0.035	–	–	–	–	430(44)	≧500 (≧51)	–	
≦0.035	–	–	–	–	480(49)	≧540 (≧55)	–	
≦0.040	–	–	–	–	215(22)	≧370 (≧38)	–	C126
≦0.040	–	–	–	–	305(31)	≧440 (≧45)	–	
≦0.040	–	–	–	–	380(39)	≧510 (≧52)	–	
≦0.040	–	–	–	–	380(39)	≧510 (≧52)	–	
≦0.040	–	–	–	Nb and/or V ≦0.15	390(40)	≧540 (≧55)	–	
≦0.040	–	–	–	–	490(50)	≧590 (≧60)	–	

Standard No. Year Designation 規格番号 年号 規格名称	Grade グレード	Mfg. Process 製造方法	Chemical Composition			
			C	Si	Mn	P
G3473−88 (Continued)	STC590B (STC60B)	S	≦0.25	≦0.55	≦1.50	≦0.040

1 N/㎜² = 1.01972 × 10^{-1} kgf/㎜²

化　学　成　分 (%)					Tensile Test N/㎜²(kgf/㎜²) 引　張　試　験		Remarks 備　　考	Index No. 索引番号
S	Ni	Cr	Mo	Others	Min. Yield Point 最小降伏点	Tensile Strength 引張強さ		
≦0.040	—	—	—	—	490(50)	≧590 (≧60)	—	C126

6. JIS Other Steel Pipes その他の鋼管

Standard No. Year Designation 規格番号 年号 規格名称	Grade グレード		Mfg. Process 製造方法	Chemical Composition C	Si	Mn	P
G3429－88 Seamless Steel Tubes for High Pressure Gas Cylinder 高圧ガス容器用継目無鋼管	STH11		S	≦0.50	0.10～0.35	≦1.80	≦0.035
	STH12		S	0.30～0.41	0.10～0.35	1.35～1.70	≦0.030
	STH21		S	0.25～0.35	0.15～0.35	0.40～0.90	≦0.030
	STH22		S	0.33～0.38	0.15～0.35	0.40～0.90	≦0.030
	STH31		S	0.35～0.40	0.10～0.50	1.20～1.90	≦0.030
G3439－88 Seamless Steel Oil Well Casing Tubing and Drill Pipe 油井用継目無鋼管	Casing and Tubing	Type1	S	－	－	－	－
		Type2	S	－	－	－	－
		Type3	S	－	－	－	－
		Type4	S	－	－	－	－
	Drill Pipe	Type1	S	－	－	－	－
		Type2	S	－	－	－	－
		Type3	S	－	－	－	－
G3465－88 Seamless Steel Tubes for Drilling 試すい用継目無鋼管	STM-C540 (STM-C55)		S	－	－	－	≦0.040
	STM-C640 (STM-C65)		S	－	－	－	≦0.040
	STM-R590 (STM-R60)		S	－	－	－	≦0.040
	STM-R690 (STM-R70)		S	－	－	－	≦0.040
	STM-R790 (STM-R80)		S	－	－	－	≦0.040
	STM-R840 (STM-R85)		S	－	－	－	≦0.040
C8305－92 Rigid Steel Conduit 鋼製電線管	－		E	≦0.10	≦0.035	≦0.50	≦0.040

1 N/mm² = 1.01972 × 10⁻¹ kgf/mm²

化学成分 (%)					Tensile Test N/mm² (kgf/mm²) 引張試験		Remarks 備考	Index No. 索引番号
S	Ni	Cr	Mo	Others	Min. Yield Point 最小降伏点	Tensile Strength 引張強さ		
≦0.035	–	–	–	–	–	–	–	C127
≦0.030	–	–	–	–	–	–	–	
≦0.030	≦0.25	0.80〜1.20	0.15〜0.30	–	–	–	–	
≦0.030	≦0.25	0.80〜1.20	0.15〜0.30	–	–	–	–	
≦0.030	0.50〜1.00	0.30〜0.60	0.15〜0.25	–	–	–	–	
–	–	–	–	–	175	≧270	–	C128
–	–	–	–	–	275	≧410	–	
–	–	–	–	–	380	≧520	–	
–	–	–	–	–	550	≧690	–	
–	–	–	–	–	315	≧520	–	
–	–	–	–	–	380	≧660	–	
–	–	–	–	–	520	≧690	–	
≦0.040	–	–	–	–	–	≧540 (≧55)	–	C129
≦0.040	–	–	–	–	–	≧640 (≧65)	–	
≦0.040	–	–	–	–	375(38)	≧590 (≧60)	–	
≦0.040	–	–	–	–	440(45)	≧690 (≧70)	–	
≦0.040	–	–	–	–	520(53)	≧790 (≧80)	–	
≦0.040	–	–	–	–	590(60)	≧830 (≧84)	–	
≦0.040	(JIS SPHT or SPCC)				–	–	–	C130

7. ASTM Steel Pipes for Piping　配管用鋼管

Standard No. Year Designation 規格番号 年号 規格名称	Grade グレード	Mfg. Process 製造方法	Chemical Composition C	Si	Mn	P
A53-93a Pipe Steel, Black and Hot-Dipped; Zine-Coated Welded and Seamless 継目無及溶接鋼管，黒管及亜鉛メッキ管	Type S Gr A	S	≦0.25	−	≦0.95	≦0.05
	〃 Gr B	S	≦0.30	−	≦1.20	≦0.05
	Type E Gr A	E	≦0.25	−	≦0.95	≦0.05
	〃 Gr B	E	≦0.30	−	≦1.20	≦0.05
	Type F	B	≦0.30	−	≦1.20	≦0.05
A105M-95 Forgings, Carbon Steel, for Piping Components 配管部品用炭素鋼鍛造品	−	F	≦0.35	0.10～0.35	0.60～1.05	≦0.040
A106- 94 Seamless Carbon Steel Pipe for High-Temperature Service 高温配管用継目無炭素鋼鋼管	Gr A	S	≦0.25	≧0.10	0.27～0.98	≦0.035
	Gr B	S	≦0.30	≧0.10	0.29～1.06	≦0.035
	Gr C	S	≦0.35	≧0.10	0.29～1.06	≦0.035
A134- 93 Pipe, Steel, Electric-Fusion(Arc)-Welded 配管用アーク溶接鋼管	A36 Structural Steel A283 Low and Intermrdiate Tensile Strength Carbon Steel Plates, A285 Pressure Vessel Plates. Carbon Steel Low and mediate Tensile A570 Hot-Rolled Carbon Steel Sheet and Strip, Structural Quality Other materials which satisfy ASTM standards are used. その他のASTM規格に適合する材料を使用して製造する。					
A135-93 Electric-Resistance-Welded Steel Pipe 配管用電気抵抗溶接鋼鋼管	Gr A	E	≦0.25	−	≦0.95	≦0.035
	Gr B	E	≦0.30	−	≦1.20	≦0.035
A139-93a Electric-Fusion(Are)Welded Steel Pipe 配管用アーク溶接鋼管	Gr A	A	−	−	≦1.00	≦0.035
	Gr B	A	−	−	≦1.00	≦0.035
	Gr C	A	−	−	≦1.20	≦0.035
	Gr D	A	−	−	≦1.30	≦0.035
	Gr E	A	−	−	≦1.40	≦0.035
A181M-95 Forgings, Carbon Steel, for General Purpose Piping 汎用配管用炭素鋼鍛造品	Cl.60	FB	≦0.35	0.10～0.35	≦1.10	≦0.05
	Cl.70	FB	≦0.35	0.10～0.35	≦1.10	≦0.05
A211-85 Spiral-Welded Steel or Iron Pipe スパイラル溶接鋼管 Note: ASTM A211 was discontinued in 1994.	A570-30	A	≦0.25	−	≦0.90	≦0.04
	A570-33	A	≦0.25	−	≦0.90	≦0.04
	A570-36	A	≦0.25	−	≦0.90	≦0.04
	A570-40	A	≦0.25	−	≦1.35	≦0.04
	A570-45	A	≦0.25	−	≦1.35	≦0.04
	A570-50	A	≦0.25	−	≦1.35	≦0.04
	A570-55	A	≦0.25	−	≦1.35	≦0.05

$1 MPa = 1.01972 \times 10^{-1} kgf/mm^2$

化　学　成　分　(%)					Tensile Test MPa or N/mm² 引張試験		Remarks 備考	Index No.
S	Ni	Cr	Mo	Others	Min. Yield Point 最小降伏点	Tensile Strength 引張強さ	(Similar to JIS)	索引番号
≦0.045	≦0.40	≦0.40	≦0.15	Cu≦0.40 V ≦0.08	205	≧330	(STPG370)	C201
≦0.045	≦0.40	≦0.40	≦0.15	Cu≦0.40 V ≦0.08	240	≧415	(STPG410)	
≦0.045	≦0.40	≦0.40	≦0.15	Cu≦0.40 V ≦0.08	205	≧330	(STPG370)	
≦0.045	≦0.40	≦0.40	≦0.15	Cu≦0.40 V ≦0.08	240	≧415	(STPG410)	
≦0.045	≦0.40	≦0.40	≦0.15	Cu≦0.40 V ≦0.08	205	≧330	(SGP)	
≦0.050	≦0.40	≦0.30	≦0.12	Cu≦0.40 V ≦0.03 Cb≦0.02	250	≧485	–	C202
≦0.035	≦0.40	≦0.40	≦0.15	Cu≦0.40 V ≦0.08	205	≧330	(STPT370)	C203
≦0.035	≦0.40	≦0.40	≦0.15	Cu≦0.40 V ≦0.08	240	≧415	(STPT410)	
≦0.035	≦0.40	≦0.40	≦0.15	Cu≦0.40 V ≦0.08	275	≧485	(STPT480)	
Shapes, and Bars Strength		–	–	–	–	–	–	C204
≦0.035	–	–	–	–	207	≧331	(STPG370)	C205
≦0.035	–	–	–	–	241	≧414	(STPG410)	
≦0.035	–	–	–	–	207	≧330	–	C206
≦0.035	–	–	–	–	240	≧415	(STPY400)	
≦0.035	–	–	–	–	290	≧415	–	
≦0.035	–	–	–	–	315	≧415	–	
≦0.035	–	–	–	–	360	≧455	–	
≦0.05	–	–	–	–	205	≧415	–	C208
≦0.05	–	–	–	–	250	≧485	–	
≦0.05	–	–	–	(Cu≧0.20)	205	≧340	–	C209
≦0.05	–	–	–	(Cu≧0.20)	230	≧360	–	
≦0.05	–	–	–	(Cu≧0.20)	250	≧360	–	
≦0.05	–	–	–	(Cu≧0.20)	275	≧380	–	
≦0.05	–	–	–	(Cu≧0.20)	310	≧415	–	
≦0.05	–	–	–	(Cu≧0.20)	345	≧450	(STPY410)	
≦0.05	–	–	–	(Cu≧0.20)	380	≧480	–	

Standard No. Year Designation 規格番号 年号 規格名称	Grade / UNS No. グレード	Mfg. Process 製造方法	Chemical Composition			
			C	Si	Mn	P
A270－90 Seamless and Welded Austenitic Stainless Steel Sanitary Tubing 衛生配管用継目無および溶接オーステナイト系ステンレス鋼鋼管	TP304 UNS S30400	S, W	≦0.08	≦0.75	≦2.00	≦0.040
	TP304L UNS S30403	S, W	≦0.035	≦0.75	≦2.00	≦0.040
	TP316 UNS S31600	S, W	≦0.08	≦0.75	≦2.00	≦0.040
	TP316L UNS S31603	S, W	≦0.035	≦0.75	≦2.00	≦0.040
A312M－94b Seamless and Welded Austenitic Stainless Steel Pipe 配管用継目無しおよび溶接オーステナイト系ステンレス鋼鋼管	TP304 S30400	S, E, AT	≦0.08	≦0.75	≦2.00	≦0.040
	TP304H S30409	S, E, AT	0.04～0.10	≦0.75	≦2.00	≦0.040
	TP304L S30403	S, E, AT	≦0.035	≦0.75	≦2.00	≦0.040
	TP304N S30451	S, E, AT	≦0.08	≦0.75	≦2.00	≦0.040
	TP304LN S30453	S, E, AT	≦0.035	≦0.75	≦2.00	≦0.040
	TP309Cb S30940	S, E, AT	≦0.08	≦0.75	≦2.00	≦0.045
	TP309H S30909	S, E, AT	0.04～0.10	≦0.75	≦2.00	≦0.040
	TP309HCb S30941	S, E, AT	0.04～0.10	≦0.75	≦2.00	≦0.045
	TP309S S30908	S, E, AT	≦0.08	≦0.75	≦2.00	≦0.045
	TP310Cb S31040	S, E, AT	≦0.08	≦0.75	≦2.00	≦0.045
	TP310H S31009	S, E, AT	0.04～0.10	≦0.75	≦2.00	≦0.040
	TP310HCb S31041	S, E, AT	0.04～0.10	≦0.75	≦2.00	≦0.045
	TP310S S31008	S, E, AT	≦0.08	≦0.75	≦2.00	≦0.045
	S31272	S, E, AT	0.08～0.12	0.3～0.7	1.5～2.00	≦0.030
	TP316 S31600	S, E, AT	≦0.08	≦0.75	≦2.00	≦0.040
	TP316H S31609	S, E, AT	0.04～0.10	≦0.75	≦2.00	≦0.040
	TP316L S31603	S, E, AT	≦0.035	≦0.75	≦2.00	≦0.040
	TP316N S31651	S, E, AT	≦0.08	≦0.75	≦2.00	≦0.040
	TP316LN S31653	S, E, AT	≦0.035	≦0.75	≦2.00	≦0.040
	TP317 S31700	S, E, AT	≦0.08	≦0.75	≦2.00	≦0.040
	TP317L S31703	S, E, AT	≦0.035	≦0.75	≦2.00	≦0.040
	TP321 S32100	S, E, AT	≦0.08	≦0.75	≦2.00	≦0.040
	TP321H S32109	S, E, AT	0.04～0.10	≦0.75	≦2.00	≦0.040

1 MP a = 1.01972 × 10^{-1} kgf/mm²

化　学　成　分　(%)					Tensile Test MPa or N/mm² 引　張　試　験		Remarks 備　考	Index No.
S	Ni	Cr	Mo	Others	Min. Yield Point 最小降伏点	Tensile Strength 引張強さ	(Similar to JIS)	索引番号
≦0.030	8.00〜 11.00	8.00〜 20.00	−	−	−	−	(SUS304TBS)	C210
≦0.030	8.00〜 13.00	8.00〜 20.00	−	−	−	−	(SUS304LTBS)	
≦0.030	10.00〜 14.00	16.00〜 18.00	−	−	−	−	(SUS316TBS)	
≦0.030	10.00〜 15.00	16.00〜 18.00	−	−	−	−	(SUS316LTBS)	
≦0.030	8.00~11.0	18.0~20.0	−	−	205	≧515	(SUS304TP)	C211
≦0.030	8.00~11.0	18.0~20.0	−	−	205	≧515	(SUS304HTP)	
≦0.030	8.00~13.0	18.0~20.0	−	−	170	≧485	(SUS304LTP)	
≦0.030	8.00~11.0	18.0~20.0	−	N 0.10〜0.16	240	≧550	−	
≦0.030	8.00~11.0	18.0~20.0	−	N 0.10〜0.16	205	≧515	−	
≦0.030	12.0~16.0	22.0~24.0	≦0.75	Cb 10C〜1.10	205	≧515	−	
≦0.030	12.0~15.0	22.0~24.0	−	−	205	≧515	−	
≦0.030	12.0~16.0	22.0~24.0	≦0.75	Cb 10C〜1.10	205	≧515	−	
≦0.030	12.0~15.0	22.0~24.0	≦0.75	−	205	≧515	(SUS309STP)	
≦0.030	19.0~22.0	24.0~26.0	≦0.75	Cb 10C〜1.10	205	≧515	−	
≦0.030	19.0~22.0	24.0~26.0	−	−	205	≧515	−	
≦0.030	19.0~22.0	24.0~26.0	≦0.75	Cb 10C〜1.10	205	≧515	−	
≦0.030	19.0~22.0	24.0~26.0	≦0.75	−	205	≧515	(SUS310STP)	
≦0.015	14.0~16.0	14.0~16.0	1.00〜1.40	Ti〜0.30〜0.60 B〜0.004〜0.008	200	≧450	−	
≦0.030	11.0~14.0	16.0~18.0	2.00〜3.00	−	205	≧515	(SUS316TP)	
≦0.030	11.0~14.0	16.0~18.0	2.00〜3.00	−	205	≧515	(SUS316HTP)	
≦0.030	10.0~15.0	16.0~18.0	2.00〜3.00	−	170	≧485	(SUS316LTP)	
≦0.030	11.0~14.0	16.0~18.0	2.00〜3.00	N 0.10〜1.16	240	≧550	−	
≦0.030	11.0~14.0	16.0~18.0	2.00〜3.00	N 0.10〜1.16	205	≧515	−	
≦0.030	11.0~14.0	18.0~20.0	3.00〜4.00	−	205	≧515	(SUS317TP)	
≦0.030	10.0~15.0	18.0~20.0	3.00〜4.00	−	205	≧515	(SUS317LTP)	
≦0.030	9.00~13.0	17.0~22.0	−	Ti 5C〜0.70	205	≧515	(SUS321TP)	
≦0.030	9.00~13.0	17.0~22.0	−	Ti 5C〜0.70	205	≧515	(SUS321HTP)	

Standard No. 規格番号	Year 年号	Designation 規格名称	Grade UNS. No. グレード	Mfg. Process 製造方法	Chemical Composition			
					C	Si	Mn	P
A312M－94b (Continued)			TP347 S34700	S, E, AT	≦0.08	≦0.75	≦2.00	≦0.040
			TP347H S34709	S, E, AT	0.04～0.10	≦0.75	≦2.00	≦0.040
			TP348 S34800	S, E, AT	≦0.08	≦0.75	≦2.00	≦0.040
			TP348H S34809	S, E, AT	0.04～0.10	≦0.75	≦2.00	≦0.040
			TPXM－10 S21900	S, E, AT	≦0.08	≦1.00	8.00~10.00	≦0.040
			TPXM－11 S21904	S, E, AT	≦0.04	≦1.00	8.00~10.00	≦0.040
			TPXM－15 S38100	S, E, AT	≦0.08	1.50～2.50	≦2.00	≦0.030
			TPXM－19 S20910	S, E, AT	≦0.060	≦1.00	4.00～6.00	≦0.040
			TPXM－29 S24000	S, E, AT	≦0.08	≦1.00	11.5~14.5	≦0.060
			S31254	S, E, AT	≦0.020	≦0.80	≦1.00	≦0.030
			S30615	S, E, AT	0.16～0.24	3.2～4.0	≦2.00	≦0.03
			S30815	S, E, AT	0.05～0.10	1.40～2.00	≦0.80	≦0.040
			S31050	S, E, AT	≦0.25	≦0.4	≦2.00	≦0.020
			S30600	S, E, AT	≦0.018	3.7～4.3	≦2.00	≦0.02
			S31725	S, E, AT	≦0.03	≦0.75	≦2.00	≦0.040
			S31726	S, E, AT	≦0.03	≦0.75	≦2.00	≦0.040
			S32615	S, E, AT	≦ 0.07	4.8～6.0	≦ 2.00	≦0.045
			S33228	S, E, AT	0.04～0.08	≦ 0.30	≦ 1.0	≦ 0.020
			S24565	S, E, AT	≦ 0.03	≦ 1.00	5.0～7.0	≦0.030
			S30415	S, E, AT	0.04～0.06	1.00～2.00	≦ 0.80	≦ 0.045
			S32654	S, E, AT	≦ 0.020	≦ 0.50	2.00～4.00	≦0.030

1 MPa = 1.01972 × 10^{-1} kgf/mm²

化学成分 (%)					Tensile Test MPa or N/mm² 引張試験		Remarks 備考	Index No.
S	Ni	Cr	Mo	Others	Min. Yield Point 最小降伏点	Tensile Strength 引張強さ	(Similar to JIS)	索引番号
≦0.030	9.00~13.0	17.0~20.0	–	Nb+Ta 10C～1.00	205	≧515	(SUS347TP)	C211
≦0.030	9.00~13.0	17.0~20.0	–	Nb+Ta 10C～1.00	205	≧515	(SUS347HTP)	
≦0.030	9.00~13.0	17.0~20.0	Cb+Ta 10C～1.00	Ta ≦0.10	205	≧515	–	
≦0.030	9.00~13.0	17.0~20.0	Cb+Ta 10C～1.00	Ta ≦0.10	205	≧515	–	
≦0.030	5.50～7.50	19.0~21.5	–	Ta 0.15～0.40	345	≧620	–	
≦0.030	5.50～7.50	19.0~21.5	–	Ta 0.15～0.40	345	≧620	–	
≦0.030	17.5~18.5	17.0~19.0	–	–	205	≧515	–	
≦0.030	11.5~13.5	20.5~23.5	1.50～3.00	Cb+Ta 0.10～0.30 N 0.20～0.40 V 0.10～0.30	380	≧690	–	
≦0.030	2.25～3.75	17.0~19.0	–	N 0.20～0.40	380	≧690	–	
≦0.010	17.5~18.5	19.5~20.5	6.00～6.50	N 0.18～0.22 Cu 0.50～1.00	300	≧650	–	
≦0.03	13.5～16.0	17.0～19.5	–	Al 0.8～1.5	275	≧620	–	
≦0.030	10.0~12.0	20.0~22.0	–	N 0.14～0.20 Ce 0.30～0.08	310	≧600	–	
≦0.015	20.5~26.0	24.0~26.0	1.6～2.6	N 0.09～0.15	t≦6.25mm 270 6.25mm<t 255	t≦6.25mm ≧580 6.25<t ≧540	–	
≦0.02	14.0~15.5	17.0~18.5	≦0.20	～ Cu ≦0.50	240	≧540	–	
≦0.030	13.5~17.5	18.0~20.0	4.0～5.0	N ≦0.10 V ≦0.75	205	≧515	–	
≦0.030	13.5~17.5	17.0~20.0	4.0～5.0	N ≦0.10 V ≦0.75	240	≧550	–	
≦ 0.030	19.0～22.0	16.5～19.5	0.3～1.5	Cu 1.5～2.5	220	≧ 550	–	
≦ 0.015	31.0～33.0	26.0～28.0	Al ≦ 0.025	Nb + Ta 0.6～1.0 Ce 0.05～0.10	185	≧ 500	–	
≦ 0.010	16.0～18.0	23.0～25.0	–	Nb + Ta ≦ 0.1 N 0.4～0.6	415	≧ 795	–	
≦ 0.030	9.00～10.0	18.0～19.0	Ce 0.03～0.08	N 0.12～0.18	290	≧ 600	–	
≦ 0.005	21.0～23.0	24.0～25.0	7.00～8.00	N 0.45～0.55	430	≧ 750	–	

Standard No. Year Designation 規格番号 年号 規格名称	Grade / UNS No. グレード	Mfg. Process 製造方法	Chemical Composition			
			C	Si	Mn	P
A333M－94 Seamless and Welded Steel Pipe for Low-Temperature Service 低温配管用継目無および溶接鋼管	Gr. 1	S, E, AT	≦0.30	－	0.40～1.06	≦0.025
	Gr. 3	S, E, AT	≦0.19	0.18～0.37	0.30～0.64	≦0.025
	Gr. 4	S	≦0.12	0.08～0.37	0.50～1.05	≦0.025
	Gr. 6	S, E, AT	≦0.30	≧0.10	0.29～1.06	≦0.025
	Gr. 7	S, E, AT	≦0.19	0.13～0.32	≦0.90	≦0.025
	Gr. 8	S, E, AT	≦0.13	0.13～0.32	≦0.90	≦0.025
	Gr. 9	S, E, AT	≦0.20	－	0.40～1.06	≦0.025
	Gr. 10	S, E, AT	≦0.20	0.10～0.35	1.15～1.50	≦0.035
	Gr. 11	S, E, AT	≦ 0.10	≦ 0.35	≦ 0.60	≦ 0.025
A335M－94 Seamless Ferritic Alloy-Steel Pipe for High-Temperature Service 高温配管用継目無合金鋼鋼管	P1 / K11522	S	0.10～0.20	0.10～0.50	0.30～0.80	≦0.025
	P2 / K11547	S	0.10～0.20	0.10～0.30	0.30～0.61	≦0.025
	P5 /K41545	S	≦0.15	≦0.50	0.30～0.60	≦0.025
	P5b /K51545	S	≦0.15	1.00～2.00	0.30～0.60	≦0.025
	P5c /K41245	S	≦0.12	≦0.50	0.30～0.60	≦0.025
	P9 /S50400	S	≦0.15	0.25～1.00	0.30～0.60	≦0.025
	P11/K11597	S	0.05～0.15	0.50～1.00	0.30～0.60	≦0.025
	P12/K11562	S	0.05～0.15	≦0.50	0.30～0.61	≦0.025
	P15 /K11578	S	0.05～0.15	1.15～1.65	0.30～0.60	≦0.025
	P21 /K31545	S	0.05～0.15	≦0.50	0.30～0.60	≦0.025
	P22 /K21590	S	0.05～0.15	≦0.50	0.30～0.60	≦0.025
	P91 /K91560	S	0.08～0.12	0.20～0.50	0.30～0.60	≦0.020
A358M－94a Electric-Fusion-Welded Austenitic Chromium-Nickel Alloy Steel Pipe for High-Temperature Service 高温オーステナイト系クロムニッケル合金鋼アーク溶接鋼管 〔Ref. ASTM-A240M(Plate)〕	304 /S30400	A	－	－	－	－
	304L/S30403	A	－	－	－	－
	304N/S30451	A	－	－	－	－
	304LN/S30453	A	－	－	－	－
	304H/S30409	A	－	－	－	－
	309Cb/S30940	A	－	－	－	－
	309S/S30908	A	－	－	－	－
	310Cb/S31040	A	－	－	－	－
	310S /S31008	A	－	－	－	－
	316/S31600	A	－	－	－	－
	316L/S31603	A	－	－	－	－
	316N/S31651	A	－	－	－	－
	316LN/S31653	A	－	－	－	－

化　学　成　分　(%)					Tensile Test MPa or N/mm² 引張試験		Remarks 備　考 (Similar to JIS)	Index No. 索引番号
S	Ni	Cr	Mo	Others	Min. Yield Point 最小降伏点	Tensile Strength 引張強さ		
≦0.025	–	–	(STPL 380)	–	205	≧380	Impact Test (J) 2mmV	C212
≦0.025	3.18～3.82	–	(STPL 450)	–	240	≧450	18	
≦0.025	0.47～0.98	0.44～1.01	–	Cu : 0.40～0.75 Al : 0.40～0.30	240	≧415	18	
≦0.025	–	–	–	–	240	≧415	18	
≦0.025	2.03～2.57	–	–	–	240	≧415	18	
≦0.025	8.40～9.60	–	(STPL 690)	–	240	≧450	18	
≦0.025	1.60～2.24	–	–	Cu : 0.40～1.25	515	≧690	18	
≦0.025	≦0.25	≦0.15	–	–	450	≧550	18	
≦0.025	35.0～37.0	≦0.50	≦0.50	Co≦0.50	240	≧450	18	
≦0.025	–	–	0.44～0.65		205	≧380	(STPA12)	C213
≦0.025	–	0.50～0.81	0.44～0.65		205	≧380	(STPA20)	
≦0.025	–	4.00～6.00	0.45～0.65		205	≧415	(STPA25)	
≦0.025	–	4.00～6.00	0.45～0.65		205	≧415	–	
≦0.025	–	4.00～6.00	0.45～0.65		205	≧415	–	
≦0.025	–	8.00～10.00	0.90～1.00		205	≧415	(STPA26)	
≦0.025	–	1.00～1.50	0.44～0.65		205	≧415	(STPA23)	
≦0.025	–	0.80～1.25	0.44～0.65		220	≧415	(STPA22)	
≦0.025	–		0.44～0.65		205	≧415	–	
≦0.025	–	2.65～3.35	0.80～1.06		205	≧415	–	
≦0.025	–	1.90～3.35	0.80～1.06		205	≧415	(STPA24)	
≦0.010	≦0.40	8.00～9.50	0.85～1.05	N 0.030～0.070 Al≦0.04 Cb 0.06～0.10 V 0.18～0.25	415	≧585	–	
–	–	–	–	–	–	–	(SUS304TPY)	C214
–	–	–	–	–	–	–	(SUS304LTPY)	
–	–	–	–	–	–	–	–	
–	–	–	–	–	–	–	–	
–	–	–	–	–	–	–	–	
–	–	–	–	–	–	–	–	
–	–	–	–	–	–	–	(SUS309STPY)	
–	–	–	–	–	–	–	–	
–	–	–	–	–	–	–	–	
–	–	–	–	–	–	–	(SUS316STPY)	
–	–	–	–	–	–	–	(SUS316SLTPY)	
–	–	–	–	–	–	–	–	
–	–	–	–	–	–	–	–	

Standard No. Year Designation 規格番号 年号 規格名称	Grade/UNS No. グレード	Mfg. Process 製造方法	Chemical Composition			
			C	Si	Mn	P
A358M－94a(Continued)	316H/S31609	A	－	－	－	－
	321/S32100	A	－	－	－	－
	347/S34700	A	－	－	－	－
	348/S34800	A	－	－	－	－
	XM-19/S22100	A	－	－	－	－
	XM-29/S28300	A	－	－	－	－
	S31254	A	－	－		－
	S30815	A	－	－	－	－
	S31725	A	－	－	－	－
	S31726	A	－	－	－	－
	S30600	A	－	－	－	－
	S24565	A	－	－	－	－
	S30415	A	－	－	－	－
	S32654	A	－	－	－	－
A369M－92 Carbon and Ferritic Alloy Steel Forged and Bored Pipe for High-Temperature Service 高温配管用炭素鋼及び合金鋼鍛造内削鋼管	FPA	FB	≦0.25	≧0.10	0.27～0.98	≦0.035
	FPB	FB	≦0.30	≧0.10	0.29～1.06	≦0.035
	FP1	FB	0.10～0.20	0.10～0.50	0.30～0.80	≦0.025
	FP2	FB	0.10～0.20	0.10～0.30	0.30～0.61	≦0.025
	FP5	FB	≦0.15	≧0.50	0.30～0.60	≦0.025
	FP9	FB	≦0.15	0.50～1.00	0.80～0.60	≦0.030
	FP11	FB	≦0.15	0.50～1.00	0.80～0.60	≦0.025
	FP12	FB	≦0.15	≦0.50	0.80～0.61	≦0.025
	FP21	FB	≦0.15	≦0.50	0.80～0.60	≦0.025
	FP22	FB	≦0.15	≦0.50	0.80～0.60	≦0.025
	FP91	FB	0.08～0.12	0.20～0.50	0.30～0.60	≦0.025
A376M－93 Seamless Austenitic Steel Pipe for High-Temperature Central-Station Service 高温セントラルステーション用継目無オーステナイト鋼鋼管	TP304	S	≦0.08	≦0.75	≦2.00	≦0.040
	TP304H	S	0.04～0.10	≦0.75	≦2.00	≦0.040
	TP304N	S	≦0.08	≦0.75	≦2.00	≦0.040
	TP304LN	S	≦0.035	≦0.75	≦2.00	≦0.040
	TP316	S	≦0.08	≦0.75	≦2.00	≦0.040
	TP316H	S	0.04～0.10	≦0.75	≦2.00	≦0.040
	TP316N	S	≦0.08	≦0.75	≦2.00	≦0.040
	TP316LN	S	≦0.035	≦0.75	≦2.00	≦0.040
	TP321	S	≦0.08	≦0.75	≦2.00	≦0.040
	TP321H	S	0.04～0.10	≦0.75	≦2.00	≦0.040
	TP347	S	≦0.08	≦0.75	≦2.00	≦0.040
	TP347H	S	0.04～0.10	≦0.75	≦2.00	≦0.040
	TP348	S	≦0.08	≦0.75	≦2.00	≦0.040
	16-8-2H	S	0.05～0.10	≦0.75	≦2.00	≦0.040
	S31725	S	≦0.03	≦0.75	≦2.00	≦0.040
	S31726	S	≦0.03	≦0.75	≦2.00	≦0.040

1 MPa = 1.01972 × 10^{-1} kgf/mm²

化 学 成 分 (%)					Tensile Test MPa or N/mm² 引張試験		Remarks 備考	Index No.
S	Ni	Cr	Mo	Others	Min. Yield Point 最小降伏点	Tensile Strength 引張強さ	(Similar to JIS)	索引番号
–	–	–	–	–	–	–	–	
–	–	–	–	–	–	–	–	
–	–	–	–	–	–	–	(SUS347TPY)	
–	–	–	–	–	–	–	–	
–	–	–	–	–	–	–	–	
–	–	–	–	–	–	–	–	
–	–	–	–	–	–	–	–	
–	–	–	–	–	–	–	–	
–	–	–	–	–	–	–	–	
–	–	–	–	–	–	–	–	
–	–	–	–	–	–	–	–	
–	–	–	–	–	–	–	–	
–	–	–	–	–	–	–	–	
–	–	–	–	–	–	–	–	
≦0.035	–	–	–	–	210	≧330	–	C215
≦0.035	–	–	–	–	240	≧410	–	
≦0.025	–	–	0.44～0.65	–	210	≧380	–	
≦0.025	–	0.50～0.81	0.44～0.65	–	210	≧380	–	
≦0.025	–	4.00～6.00	0.45～0.65	–	210	≧410	–	
≦0.030	–	8.00～10.0	0.90～1.10	–	210	≧410	–	
≦0.025	–	1.00～1.50	0.44～0.65	–	210	≧410	–	
≦0.025	–	0.80～1.25	0.44～0.65	–	220	≧415	–	
≦0.025	–	2.65～3.35	0.80～1.06	–	210	≧410	–	
≦0.025	–	1.90～2.60	0.87～1.13	–	210	≧410	–	
≦0.025	≦0.40	8.00～9.50	V 0.18~0.25 Cb0.06~0.10	N 0.03～0.07 Al ≦0.04	415	≧585	–	
≦0.030	8.00~11.00	18.0~20.0	–	–	205	≧515	(SUS304TP)	C216
≦0.030	8.00~11.0	18.0~20.0	–	–	205	≧515	(SUS304HTP)	
≦0.030	8.00~11.0	18.0~20.0	–	N : 0.10～0.16	240	≧550	–	
≦0.030	8.00~11.00	18.0~20.00	–	N : 0.10～0.16	205	≧515	–	
≦0.030	11.0~14.0	16.0~18.0	2.00～3.00	–	205	≧515	(SUS316TP)	
≦0.030	11.0~14.0	16.0~18.0	2.00～3.00	–	205	≧550	(SUS316HTP)	
≦0.030	11.0~14.0	16.0~18.0	2.00～3.00	N : 0.10～0.16	240	≧515	–	
≦0.030	11.0~14.00	16.0~18.00	2.00～3.00	N : 0.10～0.16	205	≧550	–	
≦0.030	9.00～13.0	17.0~20.0	–	Ti : 5×C～0.16	205	≧515	(SUS321TP)	
≦0.030	9.00～13.0	17.0~20.0	–	Ti : 4×C～0.16	205	≧515	(SUS321HTP)	
≦0.030	9.00～13.0	17.0~20.0	–	Nb+Ta : 10×C~1.00	205	≧515	(SUS347TP)	
≦0.030	9.00～13.0	17.0~20.0	–	Nb+Ta : 8×C~1.00	205	≧515	(SUS347HTP)	
≦0.030	9.00～13.0	17.0~20.0	Ta ≦0.10	Nb+Ta : 10×C~1.00	205	≧515	–	
≦0.030	7.50～9.50	14.5~16.5	–	–	205	≧515	–	
≦0.030	13.5~17.5	18.0~20.0	Cu ≦0.75	N ≦0.10	205	≧515	(SUS317TP)	
≦0.030	13.5~17.5	18.0~20.0	Cu ≦0.75	N 0.10～0.20	240	≧550	(SUS317TP)	

Standard No. Year Designation 規格番号 年号 規格名称	Grade/ UNS No. グレード	Mfg. Process 製造方法	Chemical Composition			
			C	Si	Mn	P
A381−93 Metal-Arc Welded Steel Pipe for Use with High-Pressure Transmission Systems 高圧トランスミッション用アーク溶接鋼管						
	Class Y35	A	−	−	−	−
	〃 Y42	A	−	−	−	−
	〃 Y46	A	−	−	−	−
	〃 Y48	A	≦0.26	−	≦1.40	≦0.025
	〃 Y50	A	−	−	−	−
	〃 Y52	A	−	−	−	−
	〃 Y56	A	−	−	−	−
	〃 Y60	A	−	−	−	−
	〃 Y65	A	−	−	−	−
A405−91 Seamless Ferritic Alloy-Steel Pipe Specially Heat Treated for High-Temperature Service 高温配管用特殊熱処理継目無フェライト系合金鋼鋼管	P24	S	≦0.15	0.10〜0.35	0.30〜0.60	≦0.025
A409M−92 Welded Large Diameter Austenitic Steel Pipe for Corrosive or High-Temperature Service 耐食・高温用大径オーステナイト鋼溶接鋼管	TP304/S30400	A	≦0.08	≦0.75	≦2.00	≦0.045
	TP304L/S30403	A	≦0.035	≦0.75	≦2.00	≦0.045
	TP309Cb/ S30940	A	≦0.08	≦0.75	≦2.00	≦0.045
	TP309S/S30908	A	≦0.08	≦0.75	≦2.00	≦0.045
	TP310Cb/ S31040	A	≦0.08	≦0.75	≦2.00	≦0.045
	TP310S/S31008	A	≦0.08	≦0.75	≦2.00	≦0.045
	TP316/S31600	A	≦0.08	≦0.75	≦2.00	≦0.040
	TP316L/S31603	A	≦0.035	≦0.75	≦2.00	≦0.040
	TP317/S31703	A	≦0.08	≦0.75	≦2.00	≦0.040
	TP321/S32100	A	≦0.08	≦0.75	≦2.00	≦0.040
	TP347/S34700	A	≦0.08	≦0.75	≦2.00	≦0.040
	TP348/S34800	A	≦0.08	≦0.75	≦2.00	≦0.040
	S31254	A	≦0.020	≦0.80	≦1.00	≦0.030
	S30815	A	0.05〜0.10	14.0〜2.00	≦0.80	≦0.040
	S31725	A	≦0.03	≦0.75	≦2.00	≦0.045
	S31726	A	≦0.03	≦0.75	≦2.00	≦0.045
	S24565	A	≦0.03	≦1.00	5.0〜7.0	≦0.030

1 MPa = 1.01972 × 10^{-1} kgf/mm²

化学成分（%）					Tensile Test MPa or N/mm² 引張試験		Remarks 備考	Index No.
S	Ni	Cr	Mo	Others	Min. Yield Point 最小降伏点	Tensile Strength 引張強さ	(Similar to JIS)	索引番号
								C217
–	–	–	–	–	240	≧415	–	
–	–	–	–	–	290	≧415	–	
–	–	–	–	–	316	≧435	–	
≦0.025	–	–	–	–	330	≧430	–	
–	–	–	–	–	345	≧440	–	
–	–	–	–	–	360	≧455	–	
–	–	–	–	–	385	≧490	–	
–	–	–	–	–	415	≧515	–	
–	–	–	–	–	450	≧535	–	
≦0.025	–	0.80～1.25	0.87～1.13	V : 0.15～0.25	205 345	≧415 ≧350	– (Heat Treated)	C218
≦0.030	8.00～11.0	18.0～20.0	–	–	205	≧515	(SUS304TPY)	C219
≦0.030	8.00～13.0	18.0～20.0	–	–	170	≧485	(SUS304LTPY)	
≦0.030	12.0～16.0	22.0～24.0	≦0.75	Cu≦0.75 Cb+Ta10C～1.10	205	≧515	–	
≦0.030	12.0～15.0	22.0～24.0	≦0.75	Cu≦0.75	205	≧515	–	
≦0.030	19.0～22.0	24.0～26.0	≦0.75	Cu≦0.75 Cb+Ta10C～1.10	205	≧515	–	
≦0.030	19.0～22.0	24.0～26.0	≦0.75	Cu≦0.75	205	≧515	–	
≦0.030	11.0～14.0	16.0～18.0	2.00～3.00	–	205	≧515	(SUS316TPY)	
≦0.030	10.0～15.0	16.0～18.0	2.00～3.00	–	170	≧485	(SUS316LTPY)	
≦0.030	11.0～14.0	18.0～20.0	3.00～4.00	–	205	≧515	(SUS317TPY)	
≦0.030	9.0～13.0	17.0～20.0	–	Ti:5×C～0.7	205	≧515	(SUS321TPY)	
≦0.030	9.0～13.0	17.0～20.0	–	Nb+Ta:10×C～1.00	205	≧515	(SUS347TPY)	
≦0.030	9.0～13.0	17.0～20.0	Ta≦0.10,	Nb+Ta:10×C～1.00	205	≧515	–	
≦0.010	17.50～18.50	19.50～20.50	6.00～6.50	Cu : 0.50～1.00 N : 0.180～0.220	300	≧650	–	
≦0.030	10.0～12.0	20.0～22.0	–	Ce 0.03～0.08 N 0.14～0.20	310	≧600	–	
≦0.030	13.0～17.0	18.0～20.0	4.0～5.0	Cu≦0.75 N ≦0.10	205	≧515	–	
≦0.030	13.5～17.5	17.0～20.0	4.0～5.0	Cu≦0.75 N 0.10～0.20	245	≧550	–	
≦0.010	16.0～18.0	23.0～25.0	4.0～5.0	Cb≦0.10 N 0.4～0.6	415	795	–	

Standard No. Year Designation 規格番号 年号 規格名称	Grade グレード	Mfg. Process 製造方法	Chemical Composition			
			C	Si	Mn	P
A430M－91 Austenitic Steel Forged and Bored Pipe for High-Temperature Service 高温用オーステナイト鍛造内削鋼管	FP304	FB	0.04～0.10	≦0.75	≦2.00	≦0.040
	FP304H	FB	0.04～0.10	≦0.75	≦2.00	≦0.040
	FP304N	FB	0.04～0.10	≦0.75	≦2.00	≦0.040
	FP316	FB	0.04～0.10	≦0.75	≦2.00	≦0.040
	FP316H	FB	0.04～0.10	≦0.75	≦2.00	≦0.040
	FP316N	FB	0.04～0.10	≦0.75	≦2.00	≦0.040
	FP321	FB	0.04～0.10	≦0.75	≦2.00	≦0.040
	FP321H	FB	0.04～0.10	≦0.75	≦2.00	≦0.040
	FP347	FB	0.04～0.10	≦0.75	≦2.00	≦0.040
	FP347H	FB	0.04～0.10	≦0.75	≦2.00	≦0.040
	FP16-8-2H	FB	0.05～0.10	≦0.75	≦2.00	≦0.040
A524－93 Seamless Carbon Steel Pipe for Atmospheric and Lower Temperatures 常低温用継目無炭素鋼鋼管	－	S	≦0.21	0.10～0.40	0.90～1.35	≦0.035
A539－90a Electric Resistance-Welded Coiled Steel Tubing for Gas and Fuel Oil Lines ガス及び燃料ライン用コイル巻溶接鋼管	－	E	≦0.15	－	≦0.63	≦0.050
A587－93 Electric-Welded Low-Carbon Steel Pipe for the Chemical Industry 化学工業用電気溶接低炭素鋼鋼管	－	E	≦0.15	－	0.27～0.63	≦0.035
A671－94 Electric-Fusion-Welded Steel Pipe for Atmospheric and Lower Temperatures 常低温配管用アーク溶接鋼管	CA 55	A	Plain carbon			
	CB 60	A	Plain carbon, killed			
	CB 65	A	Plain carbon, killed			
	CB 70	A	Plain carbon, killed			
	CC 60	A	Plain carbon, killed, fine grain			
	CC 65	A	Plain carbon, killed, fine grain			
	CC 70	A	Plain carbon, killed, fine grain			
	CD 70	A	manganese-silicon, normalized			
	CD 80	A	manganese-silicon, quenched and tempered			
	CE 55	A	Plain carbon			
	CE 60	A	Plain carbon			
	CF 65	A	nickel steel			
	CF 70	A	nickel steel			
	CF 66	A	nickel steel			
	CF 71	A	nickel steel			
	CJ101	A	alloy steel, quenched and tempered			
	CJ102	A	alloy steel, quenched and tempered			

1 MPa $=1.01972\times10^{-1}$ kgf/mm²

化学成分 (%)					Tensile Test MPa or N/mm² 引張試験		Remarks 備考	Index No.
S	Ni	Cr	Mo	Others	Min. Yield Point 最小降伏点	Tensile Strength 引張強さ	(Similar to JIS)	索引番号
≦0.030	8.00~11.0	18.0~20.0	−	−	205	≧485	−	C220
≦0.030	8.00~11.0	18.0~20.0	−	−	205	≧485	−	
≦0.030	8.00~11.0	18.0~20.0	−	N:0.10~0.16	205	≧485	−	
≦0.030	11.00~14.0	16.0~18.0	2.00~3.00	−	205	≧485	−	
≦0.030	11.00~14.0	16.0~18.0	2.00~3.00	−	205	≧485	−	
≦0.030	11.00~14.0	16.0~18.0	2.00~3.00	N:0.10~0.16	205	≧485	−	
≦0.030	9.00~13.0	17.0~20.0	−	Ti:4×C~0.60	205	≧485	−	
≦0.030	9.00~13.0	17.0~20.0	−	Ti:4×C~0.60	205	≧485	−	
≦0.030	9.00~13.0	17.0~20.0	−	Cb:8×C~1.00	205	≧485	−	
≦0.030	9.00~13.0	17.0~20.0	−	Cb:8×C~1.00	205	≧485	−	
≦0.030	7.50~9.50	14.5~16.5	1.5~2.0	−	205	≧485	−	
≦0.035	−	−	−	−	≧95.2mm: 240 < 95.2m: 205	≧95.2mm: 414~586 < 95.2mm: 380~550	(STPG410)	C221
≦0.060	−	−	−	−	241	≧310	−	C222
≦0.035	−	−	−	Al 0.02~0.100	207	≧331	(STPG370)	C223
A285	Gr.C	−	−	−	−	−	−	C224
A515	Gr.60	−	−	−	−	−	−	
A515	Gr.65	−	−	−	−	−	−	
A515	Gr.70	−	−	−	−	−	−	
A516	Gr.60	−	−	−	−	−	−	
A516	Gr.65	−	−	−	−	−	−	
A516	Gr.70	−	−	−	−	−	−	
A537	Gr.1	−	−	−	−	−	−	
A537	Gr.2	−	−	−	−	−	−	
A442	Gr.55	−	−	−	−	−	−	
A442	Gr.60	−	−	−	−	−	−	
A203	Gr.A	−	−	−	−	−	−	
A203	Gr.B	−	−	−	−	−	−	
A203	Gr.D	−	−	−	−	−	−	
A203	Gr.E	−	−	−	−	−	−	
A517	Gr.A	−	−	−	−	−	−	
A517	Gr.B							

Standard No. Year Designation 規格番号 年号 規格名称	Grade グレード	Mfg. Process 製造方法	Chemical Composition C	 Si	 Mn	 P
A671－94 (Continued)	CJ103	A	alloy steel, quenched and tempered			
	CJ104	A	alloy steel, quenched and tempered			
	CJ105	A	alloy steel, quenched and tempered			
	CJ106	A	alloy steel, quenched and tempered			
	CJ107	A	alloy steel, quenched and tempered			
	CJ108	A	alloy steel, quenched and tempered			
	CJ109	A	alloy steel, quenched and tempered			
	CJ110	A	alloy steel, quenched and tempered			
	CJ111	A	alloy steel, quenched and tempered			
	CJ112	A	alloy steel, quenched and tempered			
	CJ113	A	alloy steel, quenched and tempered			
	CK 75	A	carbon-manganese-silicon			
	CP 65	A	alloy stell, age-hardening, normalized and precipitation heat treated			
	CP 75	A	alloy stell, age-hardening, quenched and precipitation heat treated			
A672－94 Electric-Fusion-Welded Steel Pipe for High-Pressure Service at Moderate Temperatures 中温度高圧配管用アーク溶接鋼管	A 45	A	Plain carbon			
	A 50	A	Plain carbon			
	A 55	A	Plain carbon			
	B 55	A	Plain carbon, killed			
	B 60	A	Plain carbon, killed			
	B 65	A	Plain carbon, killed			
	B 70	A	Plain carbon, killed			
	C 55	A	Plain carbon, killed, fine grain			
	C 60	A	Plain carbon, killed, fine grain			
	C 65	A	Plain carbon, killed, fine grain			
	C 70	A	Plain carbon, killed, fine grain			
	D 70	A	manganese-Silicon-normalized			
	D 80	A	manganese-Silicon-Q & T			
	E 55	A	Plain carbon			
	E 60	A	Plain carbon			
	H 75	A	manganese-molybdenum-normalized			
	H 80	A	manganese-molybdenum-normalized			
	J 80	A	manganese-molybdenum-Q & T			
	J 90	A	manganese-molybdenum-Q & T			
	J 100	A	manganese-molybdenum-Q & T			
	K 75	A	chromium-manganese-silicon			
	K 85	A	chromium-manganese-silicon			
	L 65	A	molybdenum			
	L 70	A	molybdenum			

1 MPa = 1.01972 × 10^{-1} kgf/mm²

化 学 成 分 (%)					Tensile Test MPa or N/mm² 引張試験		Remarks 備考	Index No.
S	Ni	Cr	Mo	Others	Min. Yield Point 最小降伏点	Tensile Strength 引張強さ	(Similar to JIS)	索引番号
A517	Gr. C	–	–	–	–	–	–	C224
A517	Gr. D	–	–	–	–	–	–	
A517	Gr. E	–	–	–	–	–	–	
A517	Gr. F	–	–	–	–	–	–	
A517	Gr. G	–	–	–	–	–	–	
A517	Gr. H	–	–	–	–	–	–	
A517	Gr. J	–	–	–	–	–	–	
A517	Gr. K	–	–	–	–	–	–	
A517	Gr. L	–	–	–	–	–	–	
A517	Gr. M	–	–	–	–	–	–	
A517	Gr. P	–	–	–	–	–	–	
A299	–	–	–	–	–	–	–	
A736	Gr. 2	–	–	–	–	–	–	
A736	Gr. 3	–	–	–	–	–	–	
A285	Gr. A	–	–	–	–	–	–	C225
A285	Gr. B	–	–	–	–	–	–	
A285	Gr. C	–	–	–	–	–	–	
A515	Gr. 55	–	–	–	–	–	–	
A515	Gr. 60	–	–	–	–	–	–	
A515	Gr. 65	–	–	–	–	–	–	
A515	Gr. 70	–	–	–	–	–	–	
A516	Gr. 55	–	–	–	–	–	–	
A516	Gr. 60	–	–	–	–	–	–	
A516	Gr. 65	–	–	–	–	–	–	
A516	Gr. 70	–	–	–	–	–	–	
A537	Gr. 1	–	–	–	–	–	–	
A537	Gr. 2	–	–	–	–	–	–	
A442	Gr. 55	–	–	–	–	–	–	
A442	Gr. 60	–	–	–	–	–	–	
A302	Gr. A	–	–	–	–	–	–	
A302	Gr. B, C or D	–	–	–	–	–	–	
A533	Gr. C1-1	–	–	–	–	–	–	
A533	Gr. C1-2	–	–	–	–	–	–	
A533	Gr. C1-3	–	–	–	–	–	–	
A202	Gr. A	–	–	–	–	–	–	
A202	Gr. B	–	–	–	–	–	–	
A204	Gr. A	–	–	–	–	–	–	
A204	Gr. B	–	–	–	–	–	–	

Standard No. Year Designation 規格番号 年号 規格名称	Grade グレード	Mfg. Process 製造方法	Chemical Composition			
			C	Si	Mn	P
A672－94 (Continued)	L 75	A	molybdenum			
	N 75	A	manganese-silicon			
A691－93 Carbon and Alloy Steel Pipe, Electric-Fusion-Welded for High Pressure Service at High Temperatures 高温高圧配管用アーク溶接炭素鋼及び合金鋼鋼管	CM－65	A	carbon-molybdenum steel			
	CM－70	A	carbon-molybdenum steel			
	CM－75	A	carbon-molybdenum steel			
	CMSH－70	A	carbon-manganese-silicon steel, normalized			
	CMSH－75	A	carbon-manganese-silicon steel			
	CMSH－80	A	carbon-manganese-silicon steel, Q & T			
	1/2CR	A	1/2% chromium, 1/2% molybdenum steel			
	1 CR	A	1% chromium, 1/2% molybdenum steel			
	1 1/4CR	A	1-1/4% chromium, 1/2% molybdenum steel			
	2 1/4CR	A	2-1/4% chromium, 1% molybdenum steel			
	3 CR	A	3% chromium, 1% molybdenum steel			
	5 CR	A	5% chromium, 1/2% molybdenum steel			
	9 CR	A	9% chromium, 1% molybdenum steel			
A714－93 High-Strengh Low-Alloy Welded and Seamless Steel Pipe 溶接及び継目無高張力低合金鋼鋼管	Gr. I	S, E, B	≦0.22	－	≦1.25	－
	Gr. II	S, E, B	≦0.22	≦0.30	0.85～1.25	≦0.04
	Gr. III	S, E, B	≦0.23	≦0.30	≦1.35	≦0.04
	Gr. IV	S, E, B	≦0.10	≦0.60	－	0.03～0.08
	Gr. V	S, E, B	≦0.16	－	0.40～1.01	≦0.035
	Gr. VI	S, E	≦0.15	－	0.50～1.00	≦0.035
	Gr. VII	S, E	≦0.12	0.25～0.75	0.20～0.50	0.07～0.15
	Gr. VIII	S, E	≦0.19	0.30～0.65	0.80～1.25	≦0.045
A727M－95 Forgings, Carbon Steel, for Piping Components with Inherent Notch Toughness 配管部品用炭素鋼靱性鍛造品	－	F	≦0.25	0.15～0.30	0.90～1.35	≦0.035
A731M－91 Seamless and Welded Ferritic Stainless Steel pipe 継目無及び溶接フェライト径ステンレス鋼鋼管	18Gr2Mo	S, E, AT	≦0.025	≦1.00	≦1.00	≦0.040
	TPXM33	S, E, AT	≦0.06	≦0.75	≦0.75	≦0.04

1 MPa = 1.01972×10^{-1} kgf/mm²

化 学 成 分 (%)					Tensile Test MPa or N/mm² 引 張 試 験		Remarks 備 考	Index No.
S	Ni	Cr	Mo	Others	Min. Yield Point 最小降伏点	Tensile Strength 引張強さ	(Similar to JIS)	索引番号
A204	Gr. C	–	–	–	–	–	–	C 225
A299	–	–	–	–	–	–	–	
A204	Gr. A	–	–	–	–	–	–	C 226
A204	Gr. B	–	–	–	–	–	–	
A204	Gr. C	–	–	–	–	–	–	
A537	Gr. 1	–	–	–	–	–	–	
A299	–	–	–	–	–	–	–	
A537	Gr. 2	–	–	–	–	–	–	
A387	Gr. 2	–	–	–	–	–	–	
A387	Gr. 12	–	–	–	–	–	–	
A387	Gr. 11	–	–	–	–	–	–	
A387	Gr. 22	–	–	–	–	–	–	
A387	Gr. 21	–	–	–	–	–	–	
A387	Gr. 5	–	–	–	–	–	–	
A387	Gr. 9	–	–	–	–	–	–	
≦0.05	–	–	–	Cu : ≧0.20	245	≧485	–	C 227
≦0.05	–	–	–	Cu : ≧0.20 V : ≧0.02	345	≧485	–	
≦0.05	–	–	–	Cu : ≧0.20 V : ≧0.02	345	≧450	–	
≦0.05	0.20〜0.50	0.80〜1.20	–	Cu : 0.25〜0.45	250	≧400	–	
≦0.04	≧1.65	–	–	Cu : ≧0.80	Type F 275 Type, ES 315	≧380 ≧450	–	
≦0.045	0.40〜1.10	≦0.30	0.10〜0.20	Cu : 0.30〜1.00	315	≧450	–	
≦0.05	≦0.65	0.3〜1.25	–	Cu : 0.25〜0.55	315	≧450	–	
≦0.05	≦0.40	0.40〜0.65	–	Cu : 0.25〜0.40 V : 0.02〜0.10	345	≧485	–	
≦0.025	–	–	–	–	250	415〜585	–	C 228
≦0.030	≦1.00	17.5〜19.5	1.75〜2.50	N ≦0.035 Ti+Nb 0.20+4(C+N) 〜0.80	275	≧415	–	C 229
≦0.02	≦0.50	25.0〜27.0	0.75〜1.50	Cu≦0.20 N ≦0.040 Ti 7×(C+N) 〜1.00	275	≧450	–	

Standard No. Year Designation 規格番号 年号 規格名称	Grade / UNS No. グレード	Mfg. Process 製造方法	Chemical Composition			
			C	Si	Mn	P
A731M－91 (Continued)	TPXM27	S, E, AT	≦0.01	≦0.40	≦0.40	≦0.02
	TP439	S, E, AT	≦0.07	≦1.00	≦1.00	≦0.030
	29-4	S, E, AT	≦0.010	≦0.20	≦0.30	≦0.025
	29-4-2	S, E, AT	≦0.010	≦0.20	≦0.30	≦0.020
	26-3-3	S, E, AT	≦0.030	≦1.00	≦1.00	≦0.040
	S41500	S, E, AT	≦0.05	≦0.06	0.50～1.0	≦0.030
A778－90a Welded, Unannealed Austenitic Stainless Steel Tublar Products 熱処理なしオーステナイト系ステンレス鋼溶接鋼管	TP304L/S30403	A	≦0.030	≦1.00	≦2.00	≦0.045
	TP316L/S31603	A	≦0.030	≦1.00	≦2.00	≦0.045
	TP317L/S31703	A	≦0.030	≦1.00	≦2.00	≦0.045
	TP321/S32100	A	≦0.08	≦1.00	≦2.00	≦0.045
	TP347/S34700	A	≦0.08	≦1.00	≦2.00	≦0.045
A822－90 Seamless Cold-Drawn Carbon Steel Tubing for Hydraulic Sytem Service 水系用継目無冷間引抜炭素鋼鋼管	－	E, S	≦0.18	－	0.27～0.63	≦0.048

1 MPa = 1.01972 × 10^{-1} kgf/mm²

化学成分 (%)					Tensile Test MPa or N/mm² 引張試験		Remarks 備考	Index No.
S	Ni	Cr	Mo	Others	Min. Yield Point 最小降伏点	Tensile Strength 引張強さ	(Similar to JIS)	索引番号
≦0.02	≦0.50	2.50～2.75		Cu≦0.20 N ≦0.015 Nb 0.05 ～0.20	275	≧450	(SUSXM27TB)	C229
≦0.030	≦1.00	17.00～19.00	–	Al≦0.15 N ≦0.04 Ti 0.20+4(C+N)～1.10	205	≧415	(SUSXM8TB)	
≦0.020	≦0.015	28.0～30.0	3.5～4.2	Cu≦0.15 N ≦0.020	415	≧515		
≦0.025	2.0～2.5	28.0～30.0	3.5～4.2	Cu≦0.15 N ≦0.020	415	≧550		
≦0.030	1.0～3.50	25.0～28.0	3.0～3.50	N ≦0.040 Ti+Cb 0.20～100 and C(C+N)min.	450	≧580	–	
≦0.030	3.5～5.5	11.5～14.0	0.5～1.0	–	620	≧795		
≦0.030	8.00~13.00	18.00~20.00	–	N ≦0.10	170	≧485	(SUS304LTPY)	C231
≦0.030	10.00~15.00	16.00~20.00	2.00～3.00	N ≦0.10	170	≧485	(SUS316LTPY)	
≦0.030	10.00~14.00	18.00~20.00	3.00～4.00	N ≦0.10	205	≧515	–	
≦0.030	9.00~12.00	17.00~19.00	–	Ti:5×C～0.70	205	≧515	(SUS321TPY)	
≦0.030	9.00~13.00	17.00～19.00	–	Nb+Ta: 10×C～1.10	205	≧515	(SUS347TPY)	
≦0.058					170	≧310	–	C232

8. ASTM Steel Tubes for Heat Transfer　熱伝達用鋼管

StandardNo. 規格番号 / Year 年号 / Designation 規格名称	Grade グレード	Mfg. Process 製造方法	Chemical Composition C	Si	Mn	P
A161－94 Seamless Low-Carbon and Cardon-Molybdenum Steel Still Tubes for Refinery Service 精油所用継目無低炭素Mo加熱炉用鋼管	Low-Carbon	S	0.10～0.20	≦0.25	0.30～0.80	≦0.035
	Gr.T1	S	0.10～0.20	0.10～0.50	0.30～0.80	≦0.025
A178M－90a Electric-Resistance-Welded Carbon Steel Boiler Tubes 電気抵抗溶接ボイラー用炭素鋼鋼管	Gr.A	E	0.06～0.18	－	0.27～0.63	≦0.035
	Gr.C	E	≦0.35	－	≦0.80	≦0.035
	Gr.D	E	≦0.27	≦0.10	1.00～1.50	≦0.030
A179M－90a Seamless Cold-Drawn Low-Carbon Steel Heat-Exchanger and Condenser Tubes 熱交換器および凝縮器用継目無冷間引抜低炭素鋼鋼管	－	S	0.06～0.18	－	0.27～0.63	≦0.035
A192M－91 Seamless Carbon Steel Boiler Tubes for High-Pressure Service 高圧用継目無炭素鋼ボイラー用鋼管	－	S	0.06～0.18	≦0.25	0.27～0.63	≦0.035
A199M－92 Seamless Cold-Drawn Intermediate Alloy Steel Heat-Exchanger and Condenser Tubes 熱交換器および凝縮器用継目無冷間引抜合金鋼鋼管	T4	S	0.05～0.15	0.50～1.00	0.30～0.60	≦0.025
	T5	S	≦0.15	≦0.50	0.30～0.60	≦0.025
	T7	S	≦0.15	0.50～1.00	0.30～0.60	≦0.025
	T9	S	≦0.15	0.25～1.00	0.30～0.60	≦0.025
	T11	S	0.05～0.15	0.50～1.00	0.30～0.60	≦0.025
	T21	S	0.05～0.15	≦0.50	0.30～0.60	≦0.025
	T22	S	0.05～0.15	≦0.50	0.30～0.60	≦0.025
	T91	S	0.08～0.12	0.20～0.50	0.30～0.60	≦0.020
A200－94 Seamless Intermediate Alloy-Steel Still Tubes for Refinery Service 精油所用継目無合金鋼加熱炉用鋼管	T4	S	≦0.15	0.50～1.00	0.30～0.60	≦0.025
	T5	S	≦0.15	≦0.50	0.30～0.60	≦0.025
	T7	S	≦0.15	0.50～1.00	0.30～0.60	≦0.025
	T9	S	≦0.15	0.25～1.00	0.30～0.60	≦0.025
	T91	S	0.08～0.12	0.20～0.50	0.30～0.60	≦0.020
	T11	S	≦0.15	0.50～1.00	0.30～0.60	≦0.025
	T21	S	≦0.15	≦0.50	0.30～0.60	≦0.025

1 M P a = 1.01972 × 10^{-1} kgf/mm²

化学成分 (%)					Tensile Test MPa or N/mm² 引張試験		Remarks 備考	Index No.
S	Ni	Cr	Mo	Others	Min. Yield Point 最小降伏点	Tensile Strength 引張強さ	(Similar to JIS)	索引番号
≦0.035	–	–	–	–	179	≧324	(STB340)	C233
≦0.025	–	–	0.44～0.65	–	207	≧379	(STFA12)	
≦0.035	–	–	–	–	–	–	–	C234
≦0.035	–	–	–	–	255	≧415	(STB410)	
≦0.015	0.10～0.15	–	–	–	275	≧485	–	
≦0.035	–	–	–	–	–	–	–	C235
≦0.035	–	–	–	–	–	–	(STB340)	C236
≦0.025	–	2.15～2.85	0.44～0.65	–	170	≧415	–	C237
≦0.025	–	4.00～6.00	0.45～0.65	–	170	≧415	(STBA25)	
≦0.025	–	6.00～8.00	0.45～0.65	–	170	≧415	–	
≦0.025	–	8.00～10.00	0.90～1.10	–	170	≧415	(STBA26)	
≦0.025	–	1.00～1.50	0.44～0.65	–	170	≧415	(STBA23)	
≦0.025	–	2.65～3.35	0.80～1.06	–	170	≧415	–	
≦0.025	–	1.90～2.60	0.87～1.13	–	170	≧415	(STBA24)	
≦0.010	≦0.40	8.00～9.50	0.85～1.05	V 0.18～0.25 Cb 0.06～0.10 N 0.03～0.07 AL ≦0.04	170	≧415	–	
≦0.025	–	2.15～2.85	0.44～0.65	–	172	≧414	–	C238
≦0.025	–	4.00～6.00	0.45～0.65	–	172	≧414	(STFA25)	
≦0.025	–	6.00～8.00	0.45～0.65	–	172	≧414	–	
≦0.025	–	8.00～10.00	0.90～1.10	–	172	≧414	(STFA26)	
≦0.010	≦0.40	8.00～9.00	0.85～1.05	V 0.18～0.25 Cb 0.06～0.10 N 0.03～0.07 AL ≦0.04	414	≧585	–	
≦0.025	–	1.00～1.50	0.44～0.65	–	172	≧414	(STFA23)	
≦0.025	–	2.65～3.35	0.80～1.06	–	172	≧414	–	

Standard No. 規格番号	Year 年号	Designation 規格名称	Grade/ UNS No. グレード	Mfg. Process 製造方法	Chemical Composition C	Si	Mn	P
A200	94	(Continued)	T22	S	≦0.15	≦0.50	0.30〜0.60	≦0.030
A209M	91	Seamless Carbon-Molybdenum Alloy-Steel Boiler and Superheater Tubes ボイラーおよび加熱器用継目Mo合金鋼鋼管	T1	S	0.10〜0.20	0.10〜0.50	0.30〜0.80	≦0.025
			T1a	S	0.15〜0.25	0.10〜0.50	0.30〜0.80	≦0.025
			T1b	S	≦0.14	0.10〜0.50	0.30〜0.80	≦0.025
A210M	91	Seamless Medium-Carbon Steel Boiler and Superheater Tubes ボイラーおよび過熱器用継目無中炭素鋼鋼管	Gr.A-1	S	≦0.27	≦0.10	≦0.93	≦0.035
			Gr.C	S	≦0.35	≦0.10	0.29〜1.06	≦0.035
A213M	94b	Seamless Ferritic and Austenitic Alloy Steel Boiler, Superheater, and Heat-Exchanger Tubes ボイラー，過熱器および熱交換器用継目フェライト系およびオーステナイト系合金鋼鋼管	T2	S	0.10〜0.20	0.10〜0.30	0.30〜0.61	≦0.025
			T5	S	≦0.15	≦0.50	0.30〜0.60	≦0.025
			T5b	S	≦0.15	1.00〜2.00	0.30〜0.60	≦0.025
			T5c	S	≦0.15	≦0.50	0.30〜0.60	≦0.025
			T7	S	≦0.15	0.50〜1.00	0.30〜0.60	≦0.025
			T9	S	≦0.15	0.25〜1.00	0.30〜0.60	≦0.025
			T11	S	≦0.15	0.50〜1.00	0.30〜0.60	≦0.025
			T12	S	≦0.15	≦0.50	0.30〜0.61	≦0.025
			T17	S	0.15〜0.25	0.15〜0.35	0.30〜0.61	≦0.025
			T21	S	≦0.15	≦0.50	0.30〜0.60	≦0.025
			T22	S	≦0.15	≦0.50	0.30〜0.60	≦0.025
			T91	S	0.08〜0.12	0.20〜0.50	0.30〜0.60	≦0.020
			18Cr2Mo	S	≦0.025	≦1.00	≦1.00	≦0.040
			TP201 /S20100	S	≦0.15	≦1.00	5.50〜7.50	≦0.060
			TP202 /S20200	S	≦0.15	≦1.00	7.50〜10.0	≦0.060
			TP304 /S30400	S	≦0.08	≦0.75	≦2.00	≦0.040
			TP304H /S30409	S	0.04〜0.10	≦0.75	≦2.00	≦0.040
			TP304N /S30451	S	≦0.08	≦0.75	≦2.00	≦0.040
			TP304L /S30403	S	≦0.035	≦0.75	≦2.00	≦0.040
			TP304LN/S30453	S	≦0.035	≦0.75	≦2.00	≦0.040
			TP309Cb/S30940	S	≦0.08	≦0.75	≦2.00	≦0.045
			TP309H /S30909	S	0.04〜0.10	≦0.75	≦2.00	≦0.045

1 MPa = 1.01972×10⁻¹kgf/mm²

化　学　成　分　(%)					Tensile Test MPa or N/mm² 引張試験		Remarks 備　考	Index No.
S	Ni	Cr	Mo	Others	Min. Yield Point 最小降伏点	Tensile Strength 引張強さ	(Similar to JIS)	索引番号
≦0.030	−	1.90〜2.60	0.87〜1.13	−	172	≧414	(STFA24)	C238
≦0.025	−	−	0.44〜0.65	−	205	≧380	(STBA12)	C239
≦0.025	−	−	0.44〜0.65	−	220	≧415	(STBA13)	
≦0.025	−	−	0.44〜0.65	−	195	≧365	−	
≦0.035	−	−	−	−	255	≧415	(STB410)	C240
≦0.035	−	−	−	−	275	≧485	−	
≦0.025	−	0.50〜0.81	0.44〜0.65	−	205	≧415	(STBA20)	C241
≦0.025	−	4.00〜6.00	0.44〜0.65	−	205	≧415	(STBA25)	
≦0.025	−	4.00〜6.00	0.44〜0.65	−	205	≧415	−	
≦0.025	−	4.00〜6.00	0.44〜0.65	Ti4×C 〜0.70	205	≧415	−	
≦0.025	−	6.00〜8.00	0.44〜0.65	−	205	≧415	−	
≦0.025	−	8.00〜10.00	0.90〜1.10	−	205	≧415	(STBA26)	
≦0.025	−	1.00〜1.50	0.44〜0.65	−	205	≧415	(STBA23)	
≦0.025	−	0.80〜1.25	0.44〜0.65	−	205	≧415	(STBA22)	
≦0.025	−	0.80〜1.25		V ≧0.15	205	≧415	−	
≦0.025	−	2.65〜3.35	0.80〜1.06	−	205	≧415	−	
≦0.025	−	1.90〜2.60	0.87〜1.13	−	205	≧415	(STBA24)	
≦0.010	≦0.40	8.00〜9.50	0.85〜1.05	V 0.18 〜0.25 Cb 0.06 〜0.1 N 0.030〜0.070 AL ≦0.04	415	≧580		
≦0.030	−	17.5〜19.5	1.75〜2.50	N ≦0.035 Ni+Cu≦1.00 Ti+Cb 0.20+4(C+N) 〜0.80	275	≧415	(SUS444TB)	
≦0.030	3.50〜5.50	16.0〜18.0	−	N ≧0.25	260	≧655	−	
≦0.030	4.00〜6.00	17.0〜19.0	−	N ≧0.25	310	≧620	−	
≦0.030	8.00〜11.0	18.0〜20.0	−	−	205	≧515	(SUS304TB)	
≦0.030	8.00〜11.0	18.0〜20.0	−	−	205	≧515	(SUS304HTB)	
≦0.030	8.00〜11.0	18.0〜20.0	−	N:0.10〜0.16	205	≧515	−	
≦0.030	8.00〜13.0	18.0〜20.0	−	−	205	≧515	(SUS304LTB)	
≦0.030	8.00〜11.0	18.0〜20.0	−	N:0.10〜0.16	205	≧515	−	
≦0.030	12.00〜16.00	22.00〜24.00	≦0.75	Cb+Ta 10C〜1.10	205	≧415	−	
≦0.030	12.00〜15.00	22.00〜24.00	≦0.75	−	205	≧515	−	

StandardNo. Year Designation 規格番号 年号 規格名称	Grade / UNS No. グレード	Mfg. Process 製造方法	Chemical Composition			
			C	Si	Mn	P
A213M−94b(continued)	TP309HCb/S30941	S	0.04～0.10	≦0.75	≦2.00	≦0.045
	TP309S/S30908	S	≦0.08	≦0.75	≦2.00	≦0.040
	TP310Cb/S31040	S	≦0.08	≦0.75	≦2.00	≦0.045
	TP310H/S31009	S	0.04～0.10	≦0.75	≦2.00	≦0.045
	TP310HCb/S31041	S	0.04～0.10	≦0.75	≦2.00	≦0.045
	TP310HCbN/S31042	S	0.04～0.10	≦0.75	≦2.00	≦0.030
	TP310S/S31008	S	≦0.08	≦0.75	≦2.00	≦0.045
	S31272	S	0.08～0.12	0.3～0.7	1.5～2.0	≦0.030
	TP316/S31600	S	≦0.08	≦0.75	≦2.00	≦0.040
	TP316H/S31609	S	0.04～0.10	≦0.75	≦2.00	≦0.040
	TP316L/S31603	S	≦0.035	≦0.75	≦2.00	≦0.040
	TP316N/S31651	S	≦0.08	≦0.75	≦2.00	≦0.040
	TP316LN/S31653	S	≦0.035	≦0.75	≦2.00	≦0.040
	TP317/S31700	S	≦0.06	≦0.75	≦2.00	≦0.040
	TP317L/S31703	S	≦0.035	≦0.75	≦2.00	≦0.040
	TP321/S32100	S	≦0.08	≦0.75	≦2.00	≦0.040
	TP321H/S32109	S	0.04～0.10	≦0.75	≦2.00	≦0.040
	TP347/S34700	S	≦0.08	≦0.75	≦2.00	≦0.040
	TP347H/S34709	S	0.04～0.10	≦0.75	≦2.00	≦0.040
	TP347HFG	S	0.06～0.10	≦0.75	≦2.00	≦0.040
	TP348/S34800	S	≦0.08	≦0.75	≦2.00	≦0.040
	TP348H/S34809	S	0.04～0.10	≦0.75	≦2.00	≦0.040
	XM-15/S38100	S	≦0.08	1.50～2.50	≦2.00	≦0.030
	S30615	S	0.16～0.24	3.2～4.0	≦2.00	≦0.030
	S30815	S	0.05～0.10	1.40～2.00	≦0.80	≦0.040
	S31050	S	≦0.25	≦0.4	≦2.00	≦0.020
	S21500	S	0.06～0.15	0.2 ～1.0	5.5～7.0	≦0.040
	S31725	S	≦0.03	≦0.75	≦2.00	≦0.040
	S31726	S	≦0.03	≦0.75	≦2.00	≦0.040
	S32615	S	≦0.07	4.8 ～6.0	≦2.00	≦0.045
	S33228	S	0.04～0.08	≦0.30	≦1.0	≦0.020
	XM-19/S20910	S	≦0.06	≦1.00	4.00～6.00	≦0.04
A214M Electric-Resistance-Welded Carbon Steel Heat-Exchanger and Condenser Tubes 熱交換器及び凝縮器用電気抵溶溶接炭素鋼鋼管	−	E	≦0.18	−	0.27～0.68	≦0.050

1MPa＝1.01972×10^{-1}kgf/mm²

化　学　成　分　(%)					Tensile Test MPa or N/mm² 引張試験		Remarks 備考	Index No.
S	Ni	Cr	Mo	Others	Min. Yield Point 最小降伏点	Tensile Strength 引張強さ	(Similar to JIS)	索引番号
≦0.030	12.00～16.00	22.00～24.00	≦0.75	Cb＋Ta 10C～1.10	205	≧515	–	C241
≦0.030	12.00～15.00	24.00～26.00	≦0.75	–	205	≧515	(SUS309STB)	
≦0.030	19.00～22.0		–	Cb＋Ta 10C～1.10	205	≧515	–	
≦0.030	19.00～22.00	24.00～26.00	≦0.75	–	205	≧515	–	
≦0.030	19.00～22.00	24.00～26.00	–	Cb＋Ta 10C～1.10	205	≧515	–	
≤0.030	17.00～23.00	24.00～26.00	–	Cb＋Ta 0.20～0.60 N 0.15～0.35	295	≥655	–	
≦0.030	19.00～22.00	24.00～26.00	≦0.75	–	205	≧515	(SUS310STB)	
≤0.015	14.00～16.0	14.00～16.00	1.0～1.4	Ti 0.3~0.6: B 0.004~0.008	200	≥450		
≦0.030	11.00～14.0	16.0～18.0	2.00～3.00	–	205	≧515	(SUS316TB)	
≦0.030	11.0～14.0	16.0～18.0	2.00～3.00	–	205	≧515	(SUS316HTB)	
≦0.030	10.0～15.0	16.0～18.0	2.00～3.00	–	205	≧515	(SUS316LTB)	
≦0.030	11.0～14.0	16.0～18.0	2.00～3.00	N:0.10～0.16	205	≧515	–	
≦0.030	11.0～14.0	16.0～18.0	2.00～3.00	N:0.10～0.16	205	≧515	–	
≤0.030	11.0～14.0	18.0～20.0	3.00～4.00	–	205	≥515	–	
≤0.030	11.0～15.0	18.0～20.0	3.00～4.00	–	205	≥515	–	
≦0.030	9.00～13.00	17.0～20.0	–	Ti:5×C～0.60	205	≧515	(SUS321TB)	
≦0.030	9.00～13.00	17.0～20.0	–	Ti:5×C～0.60	205	≧515	(SUS321HTB)	
≦0.030	9.00～13.00	17.0～20.0	–	Nb＋Ta:10×C～1.00	205	≧515	(SUS347TB)	
≦0.030	9.00～13.00	17.0～20.0	–	Nb＋Ta:8×C～1.0	205	≧515	(SUS347HTB)	
≤0.030	9.00～13.00	17.00～20.00	–	Cb+Ta: 8×C～1.0	205	≧550	–	
≦0.030	9.00～13.00	17.0～20.0	–	Nb＋Ta:10×C~1.00, Ta≦0.10	205	≧515	–	
≦0.030	9.00～13.00	17.0～20.0	–	Nb＋Ta:8×C~1.00 Ta≦0.10	205	≧515	–	
≦0.030	17.50～18.00	17.00～19.00	–	–	205	≧515	–	
≤0.030	13.5～16.0	17.0～19.5	–	Al 0.8～1.5	275	≥620	–	
≦0.030	10.0～12.0	20.0～22.0	–	N 0.14 ～0.20 Ce 0.03 ～0.08	310	≧600	–	
≦0.020	20.5～23.5	24.0～26.0	1.6 ～2.6	Ta 0.09 ～0.15	t ≦6.25mm 270 6.25mm＜ t 255	t ≦6.25mm – 6.25mm＜t ≧540	–	
≦0.030	9.0～11.0	14.0～16.0	0.8 ～1.2	Cb 0.75 ～1.25 V 0.15 ～0.40 B 0.003～0.007	230	≧540	–	
≦0.030	13.0～17.0	18.0～20.0	4.0 ～5.0	N ≦0.10 Cu≦0.75	205	≧515	–	
≦0.030	13.5～17.5	17.0～20.0	4.0 ～5.0	N 0.10～0.20 Cu≦0.75	240	≧550	–	
≦0.030	19.0～22.0	16.5～19.5	0.3 ～1.5	Cu 1.5～2.5	220	≧550	–	
≦0.015	31.0～33.0	26.0～28.0	–	Al ≥ 0.025 Cb+Ta: 0.6～1.0 Ce 0.05～0.10	185	≧500	–	
≦0.03	11.5～13.5	20.5～23.5	1.50～3.00	V 0.10～0.30 Cb+Ta: 0.10～0.30 N 0.20～0.40	380	≧690	–	
≦0.060	–	–	–	–	–	–	–	C242

StandardNo. 規格番号	Year 年号	Designation 規格名称	Grade/ UNS No. グレード	Mfg. Process 製造方法	Chemical Composition			
					C	Si	Mn	P
A226M－90a Electric-Resistance-Welded Carbon Steel Boiler and Superheater Tubes for High-Pressure Service 高圧ボイラーおよび過熱器用電気抵抗溶接炭素鋼鋼管			−	E	0.06〜0.18	≦0.25	0.27〜0.63	≦0.035
A249M－94a Welded Austenitic Steel Boiler, Superheater, Heat-Exchanger and Condenser Tubes ボイラー過熱器，熱交換器および凝縮器用溶接オーステナイト系ステンレス鋼鋼管			TP201/S20100	E, AT	≦0.15	≦1.00	5.50〜7.50	≦0.060
			TP202/S20200	E, AT	≦0.15	≦1.00	7.50〜10.00	≦0.060
			TP304/S30400	E, AT	≦0.08	≦0.75	≦2.00	≦0.040
			TP304H/S30409	E, AT	0.04〜0.10	≦0.75	≦2.00	≦0.040
			TP304L/S30403	E, AT	≦0.035	≦0.75	≦2.00	≦0.040
			TP304N/S30451	E, AT	≦0.08	≦0.75	≦2.00	≦0.040
			TP304LN/S30453	E, AT	≦0.035	≦0.75	≦2.00	≦0.040
			TP305/S30500	E, AT	≦0.12	≦1.00	≦2.00	≦0.040
			TP309Cb/S30940	E, AT	≦0.08	≦0.75	≦2.00	≦0.045
			TP309H/S30909	E, AT	0.04〜0.10	≦0.75	≦2.00	≦0.040
			TP309HCb/ S30941	E, AT	0.04〜0.10	≦0.75	≦2.00	≦0.045
			TP309S/S30908	E, AT	≦0.08	≦0.75	≦2.00	≦0.045
			TP310Cb/S31040	E, AT	≦0.08	≦0.75	≦2.00	≦0.040
			TP310H/S31009	E, AT	0.04〜0.10	≦0.75	≦2.00	≦0.040
			TP310HCb/ S31049	E, AT	0.04〜0.10	≦0.75	≦2.00	≦0.045
			TP310S/S31008	E, AT	≦0.08	≦0.75	≦2.00	≦0.045
			TP316/S31600	E, AT	≦0.08	≦0.75	≦2.00	≦0.040
			TP316H/S31609	E, AT	0.04〜0.10	≦0.75	≦2.00	≦0.040
			TP316L/S31603	E, AT	≦0.035	≦0.75	≦2.00	≦0.040
			TP316N/S31651	E, AT	≦0.08	≦0.75	≦2.00	≦0.040
			TP316LN/S31653	E, AT	≦0.035	≦0.75	≦2.00	≦0.040
			TP317/S31700	E, AT	≦0.08	≦0.75	≦2.00	≦0.040
			TP317L/S31703	E, AT	≦0.035	≦0.75	≦2.00	≦0.040
			TP321/S32100	E, AT	≦0.08	≦0.75	≦2.00	≦0.040
			TP321H/S32109	E, AT	0.04〜0.10	≦0.75	≦2.00	≦0.040
			TP347/S34700	E, AT	≦0.08	≦0.75	≦2.00	≦0.040
			TP347H/S34709	E, AT	0.04〜0.10	≦0.75	≦2.00	≦0.040

$1\,MPa = 1.01972 \times 10^{-1}\,kgf/mm^2$

化　学　成　分　(%)					Tensile Test MPa or N/mm^2 引張試験		Remarks 備考 (Similar to JIS)	Index No. 索引番号
S	Ni	Cr	Mo	Others	Min. Yield Point 最小降伏点	Tensile Strength 引張強さ		
≦0.035	−	−	−	−	−	−	(STB340)	C243
≦0.030	3.50〜5.50	16.0〜18.0	−	N ≧0.25	260	≧655	−	C244
≦0.030	4.00〜6.00	17.0〜19.0	−	N ≧0.25	260	≧620	−	
≦0.030	8.00〜11.00	18.0〜20.0	−	−	205	≧515	(SUS304TB)	
≦0.030	8.00〜11.00	18.0〜20.0	−	−	205	≧515	(SUS304HTB)	
≦0.030	8.00〜13.00	18.0〜20.0	−	−	170	≧485	(SUS304LTB)	
≦0.030	8.00〜11.00	18.0〜20.0	−	N:0.10〜0.16	240	≧550	−	
≦0.030	8.00〜13.00	18.0〜20.0	−	N:0.10〜0.16	205	≧515	−	
≦0.030	10.00〜13.00	17.0〜19.0	−	−	205	≧515	−	
≦0.030	12.0〜16.0	22.0〜24.0	−	Ti+Cb 10C 〜1.00	205	≧515	−	
≦0.030	12.0〜15.0	22.0〜24.0	−	N:0.10〜0.16	205	≧515	−	
≦0.030	12.0〜16.0	22.0〜24.0	≦0.75	Ti+Cb 10C 〜1.00	205	≧515	−	
≦0.030	12.0〜15.0	22.0〜24.0	≦0.75	−	205	≧515	(SUS309STB)	
≦0.030	19.0〜22.0	24.0〜26.0	≦0.75	Ti+Cb 10C 〜1.00	205	≧515	−	
≦0.030	19.0〜22.0	24.0〜26.0	−	N:0.10〜0.16	205	≧515	−	
≦0.030	19.0〜22.0	24.0〜26.0	≦0.75	Ti+Cb 10C 〜1.00	205	≧515	−	
≦0.030	19.0〜22.0	24.0〜26.0	≦0.75	−	205	≧515	(SUS310STB)	
≦0.030	11.0〜14.0	16.0〜18.0	2.00〜3.00	−	205	≧515	(SUS316TB)	
≦0.030	11.0〜14.0	16.0〜18.0	2.00〜3.00	−	205	≧515	(SUS316HTB)	
≦0.030	10.0〜15.0	16.0〜18.0	2.00〜3.00	−	205	≧485	(SUS316LTB)	
≦0.030	11.0〜14.0	16.0〜18.0	−	N:0.10〜0.16	205	≧550	−	
≦0.030	10.0〜15.0	16.0〜18.0	2.00〜3.00	N:0.10〜0.16	205	≧515	−	
≦0.030	11.0〜14.0	18.0〜20.0	3.00〜4.00	−	205	≧515	(SUS317TB)	
≦0.030	11.0〜15.0	18.0〜20.0	3.00〜4.00	−	205	≧515	(SUS317LTB)	
≦0.030	9.0〜13.0	17.0〜20.0	−	Ti:5×C 〜0.70	205	≧515	(SUS321TB)	
≦0.030	9.0〜13.0	17.0〜20.0	−	Ti:5×C 〜0.60	205	≧515	(SUS321HTB)	
≦0.030	9.0〜13.0	17.0〜20.0	−	Nb+Ta: 10×C 〜1.00	205	≧515	(SUS347TB)	
≦0.030	9.0〜13.0	17.0〜20.0	−	Nb+Ta: 8×C 〜1.00	205	≧515	(SUS347HTB)	

Standard No. 規格番号	Year 年号	Designation 規格名称	Grade/ UNS No. グレード	Mfg. Process 製造方法	Chemical Composition C	Si	Mn	P
A249M－94a (Continued)			TP348/S34800	E, AT	≦0.08	≦0.75	≦2.00	≦0.040
			TP348H/S34809	E, AT	0.04～0.10	≦0.75	≦2.00	≦0.040
			XM-15/S38100	E, AT	≦0.08	1.50～2.50	≦2.00	≦0.030
			XM-19/S20910	E, AT	≦0.06	≦1.00	4.00～6.00	≦0.040
			XM-29/S24000	E, AT	≦0.08	≦1.00	11.5～14.5	≦0.040
			S30615	E, AT	0.16～0.24	3.2～4.0	≦2.00	≦0.030
			S31050	E, AT	≦0.025	≦0.4	≦2.00	≦0.020
			S31254	E, AT	≦0.020	≦0.80	≦1.00	≦0.030
			S30815	E, AT	0.05～0.10	1.40～2.00	≦0.80	≦0.040
			S31725	E, AT	≦0.03	≦0.75	≦2.00	≦0.045
			S31726	E, AT	≦0.03	≦0.75	≦2.00	≦0.045
			S24565	E, AT	≦0.03	≦1.00	5.0～7.0	≦0.030
			S33228	E, AT	0.04～0.08	≦0.30	≦1.0	≦0.020
			S30415	E, AT	0.04～0.06	1.00～2.00	≦0.80	≦0.045
			S32654	E, AT	≦0.020	≦0.50	2.00～4.00	≦0.030
A250－91 Electric-Resistance-Welded Carbon-Molybdenum Alloy-Steel Boiler and Superheater Tubes ボイラーおよび過熱器用電気抵抗溶接炭素Mo合金鋼鋼管			T1	E	0.10～0.20	0.10～0.50	0.30～0.80	≦0.025
			T1a	E	0.15～0.25	0.10～0.50	0.30～0.80	≦0.025
			T1b	E	≦0.14	0.10～0.50	0.30～0.80	≦0.025
			T2	E	0.10～0.20	0.10～0.30	0.30～0.61	≦0.025
			T11	E	0.05～0.15	0.50～1.00	0.30～0.60	≦0.025
			T12	E	0.05～1.15	≦0.50	0.30～0.61	≦0.030
			T22	E	≦0.15	≦0.50	0.30～0.60	≦0.025
A266－95 Forgings, Carbon Steel for Pressure Vessel Components 圧力容器用鍛造炭素鋼			1	F B	≦0.35	0.15～0.35	0.40～1.05	≦0.040
			2	F B	≦0.35	0.15～0.35	0.40～1.05	≦0.040
			3	F B	≦0.45	≦0.35	0.50～0.90	≦0.040
			4	F B	≦0.30	0.15～0.35	0.80～1.35	≦0.035
A268－94 Seamless and Welded Ferritic Stainless Steel Tubing for General Service 一般用継目無および溶接フェライト系ステンレス鋼鋼管			TP405/S40500	S, E, AT	≦0.08	≦0.75	≦1.00	≦0.040
			TP410/S41000	S, E, AT	≦0.15	≦0.75	≦1.00	≦0.040
			TP429/S42900	S, E, AT	≦0.12	≦0.75	≦1.00	≦0.040
			TP430/S43000	S, E, AT	≦0.12	≦0.75	≦1.00	≦0.040
			TP443/S44300	S, E, AT	≦0.20	≦0.75	≦1.00	≦0.040
			TP446-1/S44600	S, E, AT	≦0.20	≦0.75	≦0.50	≦0.040
			TP446-2/S44600	S, E, AT	≦0.12	≦0.75	≦0.50	≦0.040

1 MP a = 1.01972 × 10^{-1} kgf/mm²

化 学 成 分 (%)					Tensile Test MPa or N/mm² 引張試験		Remarks 備 考 (Similar to JIS)	Index No. 索引番号
S	Ni	Cr	Mo	Others	Min. Yield Point 最小降伏点	Tensile Strength 引張強さ		
≦0.030	9.00～13.00	17.00～20.0	–	N+Ta:10×C~1.00, Ta≦0.10	205	≧515	–	C244
≦0.030	9.00～13.00	17.00～20.0	–	b+Ta:8×C~1.00, Ta≦0.10	205	≧515	–	
≦0.030	17.50～18.50	17.00～19.0	–	–	205	≧515	–	
≦0.030	11.50～13.50	20.50～23.50	1.50～3.00	V :0.10 ～0.30 N :0.20 ～0.40 Nb+Ta : 0.10～0.30	205	≧515	–	
≦0.030	2.25～3.75	17.00～19.00	–	N :0.20 ～0.40	205	≧515	–	
≦0.030	13.50～16.00	17.00～19.50	–	Al:0.8~1.5	275	≧620	–	
≦0.015	20.5～23.5	24.0～26.0	1.6～2.6	N 0.09～0.15	t≦6.25mm 270 6.25mm< t 255	t≦6.25mm ≧580 6.25mm<t ≧540	–	
≦0.010	17.50～18.50	19.50～20.50	6.00～6.50	N :0.180～0.220 Cu: 0.50～1.00	300	≧650	–	
≦0.030	10.0～12.00	20.0～22.0	–	N : 0.14～0.20 Cu: 0.03～0.08	310	≧600	–	
≦0.030	13.00～17.00	18.0～20.0	4.0～5.0	N ≦0.10 Cu≦0.75	205	≧515	– –	
≦0.030	13.00～17.00	17.0～20.0	4.0～5.0	N : 0.10～0.20 Cu≦0.75	240	≧550	–	
≦0.010	16.0～18.0	23.0～25.0	4.0～5.0	Ti≦0.1 N 0.4~0.6	415	≧795	–	
≦0.015	31.0～33.0	26.0～28.0	–	Ti :0.6~1.0 Ce :0.05~0.10 Al≦0.025	185	≧500	–	
≦0.030	9.00～10.00	18.0～19.0	–	N :0.12~0.18 Cu:0.03~0.08	290	≧600	–	
≦0.005	21.0～23.0	24.0～25.0	7.00～8.00	N:0.45~0.55 Cu:0.30~0.60	430	≧750	–	
≦0.025	–	–	0.44～0.65	–	205	≧380	(STBA12)	C245
≦0.025	–	–	0.44～0.65	–	220	≧415	(STBA13)	
≦0.025	–	–	0.44～0.65	–	195	≧365	–	
≦0.020	–	0.50～0.81	0.44～0.65	–	205	≧415	–	
≦0.020	–	0.80～1.25	0.44～0.65	–	205	≧415	–	
≦0.020	–	0.80～1.25	0.44～0.65	–	220	≧415	–	
≦0.020	–	1.90～2.60	0.87～1.13	–	205	≧415	–	
≦0.040	–	–	–	–	205	415～585	–	C246
≦0.040	–	–	–	–	250	485～655	–	
≦0.040	–	–	–	–	260	515～690	–	
≦0.040	–	–	–	–	250	485～655	–	
≦0.030	≦0.50	11.5～13.5	–	Al 0.10 ～0.30	205	≧415	(SUS405TB)	C247
≦0.030	≦0.50	11.5～13.5	–	–	215	≧415	(SUS410TB)	
≦0.030	≦0.50	14.0～16.0	–	–	240	≧415	–	
≦0.030	≦0.50	16.0～18.0	–	–	240	≧415	(SUS430TB)	
≦0.030	≦0.50	18.0～23.0	–	Cu 0.90 ～1.25	275	≧485	–	
≦0.030	≦0.50	23.0～30.0	–	N 0.10 ～0.25	275	≧485	–	
≦0.030	≦0.50	23.0～30.0	–	N 0.15 ～0.25	275	≧450	–	

StandardNo. 規格番号	Year 年号	Designation 規格名称	Grade / UNS No. グレード	Mfg. Process 製造方法	Chemical Composition C	Si	Mn	P
A268	94	(Continued)	S40800	S, E, AT	≦0.08	≦1.00	≦1.00	≦0.045
			TP409/S40900	S, E, AT	≦0.08	≦1.00	≦1.00	≦0.045
			S41500	S, E, AT	≦0.05	≦0.60	0.50～1.0	≦0.03
			TP439/S43035	S, E, AT	≦0.07	≦1.00	≦1.00	≦0.040
			TP430Ti/S43036	S, E, AT	≦0.10	≦1.00	≦1.00	≦0.040
			TPXM-27/S44627	S, E, AT	≦0.01	≦0.40	≦0.40	≦0.02
			XM-33/S44626	S, E, AT	≦0.026	≦1.75	≦0.75	≦0.040
			18Cr-2Mo/S44400	S, E, AT	≦0.025	≦1.00	≦1.00	≦0.040
			29-4/S44700	S, E, AT	≦0.010	≦0.20	≦0.30	≦0.025
			29-4-2/S44800	S, E, AT	≦0.010	≦0.20	≦0.30	≦0.025
			26-3-3/S44660	S, E, AT	≦0.025	≦0.75	≦1.00	≦0.040
			25-4-4/S44635	S, E, AT	≦0.025	≦0.75	≦1.00	≦0.040
			S44735	S, E, AT	≦0.030	≦1.00	≦1.00	≦0.040
			S32803	S, E, AT	≦0.015	≦0.5	≦0.5	≦0.020
A269	94a	Seamles and Welded Austenitic Stainless Steel Tubing for General Service 一般用継目無および溶接オーステナイト系ステンレス鋼鋼管	TP304/S30400	S, W	≦0.08	≦0.75	≦2.00	≦0.040
			TP304L/S30403	S, W	≦0.035	≦0.75	≦2.00	≦0.040
			TP304LN/S30453	S, W	≦0.035	≦0.75	≦2.00	≦0.040
			TP316/S31600	S, W	≦0.08	≦0.75	≦2.00	≦0.040
			TP316L/S31603	S, W	≦0.035	≦0.75	≦2.00	≦0.040
			TP316LN/S31653	S, W	≦0.035	≦0.75	≦2.00	≦0.040
			TP317/S31700	S, W	≦0.08	≦0.75	≦2.00	≦0.040
			TP321/S32100	S, W	≦0.08	≦0.75	≦2.00	≦0.040
			TP347/S34700	S, W	≦0.08	≦0.75	≦2.00	≦0.040
			TP348/S34800	S, W	≦0.08	≦0.75	≦2.00	≦0.040
			XM-10/S21900	S, W	≦0.08	≦1.00	8.00～10.00	≦0.060
			XM-11/S21903	S, W	≦0.04	≦1.00	8.00～10.00	≦0.060

$1 MPa = 1.01972 \times 10^{-1} kgf/mm^2$

化学成分 (%)					Tensile Test MPa or N/mm^2 引張試験		Remarks 備考	Index No.
S	Ni	Cr	Mo	Others	Min. Yield Point 最小降伏点	Tensile Strength 引張強さ	(Similar to JIS)	索引番号
≦0.045	≦0.80	11.5~13.0	–	Ti12C ~1.10	205	≧380	–	C247
≦0.045	≦0.50	10.50~11.7	–	Ti6C~0.75	205	≧380	–	
≦0.03	0.5~1.0	11.5~14.0	–	–	620	≧795	–	
≦0.030	≦0.50	17.00~19.00	Al ≦0.15 N ≦0.04	Ti 0.20 + 4(C+N)~1.0	205	≧415	≦ –	
≦0.030	≦0.75	16.0~19.5	–	Ti:5×C~0.75	240	≧415	–	
≦0.020	≦0.50	25.0~27.5	0.75~1.50	Cu≦0.20 N ≦0.015 Nb 0.05~0.20	275	≧450	(SUSXM27TB)	
≦0.020	≦0.50	25.0~27.0 Cu≦0.20, N≦0.040	0.75~1.50	Ti:7×(C+N) min0.20 max1.00	310	≧470	–	
≦0.030	≦1.00	17.5~19.5	1.75~2.50 Ti+Nb:0.20	N ≦0.035 +4×(C+N)~0.80	275	≧415	–	
≦0.020	≦0.15	28.0~30.0	3.5~4.2	Cu≦0.15 N ≦0.20	415	≧550	–	
≦0.020	2.0~2.5	28.0~30.0	3.5~4.2	Cu≦0.15 N ≦0.20	415	≧550	–	
≦0.030	1.50~3.50	25.0~27.0	3.0~4.0	N ≦0.045 Ti+Nb=0.20~1.00 〃 ≧6×(C+N)	450	≧585	–	
≦0.030	3.5~4.5	24.5~26.0	3.5~4.5 Ti+Nb:0.20	N ≦0.035 +4×(C+N)~0.80	515	≧620	–	
≦0.040	≦1.00	28.00~30.00	3.60~4.20	N ≦0.045 Ti+Nb=0.20~1.00 〃 ≧6×(C+N)	415	≧515	–	
≦0.005	3.0~4.0	28.0~29.0	1.8~2.5	N≦0.020 Cb 0.15~0.50	500	≧600		
≦0.030	8.00~11.0	18.0~20.0	–	–	–	–	(SUS304TB)	C248
≦0.030	8.00~13.0	18.0~20.0	–	–	–	–	(SUS304LTB)	
≦0.030	8.00~13.0	18.0~20.0	–	N 0.10~0.16	–	–	–	
≦0.030	11.0~14.0	16.0~18.0	2.00~3.00	–	–	–	(SUS316TB)	
≦0.030	10.0~15.0	16.0~18.0	2.00~3.00	–	–	–	(SUS316LTB)	
≦0.030	10.0~15.0	16.0~18.0	2.00~3.00	N 0.10~0.16	–	–	–	
≦0.030	11.0~14.0	18.0~20.0	3.00~4.00	–	–	–	(SUS317TB)	
≦0.030	9.00~13.0	17.0~20.0	–	Ti:5×C ~0.70	–	–	(SUS321TB)	
≦0.030	9.00~13.0	17.0~20.0	–	Nb+Ta:10×C ~ 1.00	–	–	(SUS347TB)	
≦0.030	9.00~13.0	17.0~20.0	–	Nb+Ta:10×C~1.0 Ta≦0.1	–	–	–	
≦0.030	5.50~7.50	19.00~21.50	–	N:0.15~0.40	–	–	–	
≦0.030	5.50~7.50	19.00~21.50	–	N:0.15~0.40	–	–	–	

StandardNo. 規格番号	Year 年号	Designation 規格名称	Grade/ UNS No. グレード	Mfg. Process 製造方法	Chemical Composition C	Si	Mn	P
A269−94a (Continued)			XM−15/S38100	S, W	≦0.08	1.50〜2.50	≦2.00	≦0.030
			XM−19/S20910	S, W	≦0.06	≦1.00	4.00〜6.00	≦0.040
			XM−29/S24000	S, W	≦0.06	≦1.00	11.50〜14.50	≦0.060
			S31254	S, W	≦0.020	≦0.80	≦1.00	≦0.030
			S31725	S, W	≦0.03	≦0.75	≦2.00	≦0.040
			S31726	S, W	≦0.03	≦0.75	≦2.00	≦0.040
			S32654	S, W	≤0.020	≤0.5	2.00〜4.00	≤0.030
			S30600	S, W	≦0.18	3.7〜4.3	≦2.00	≦0.020
			S24565	S, W	≤0.03	≤1.00	5.0〜7.0	≤0.030
A271−94		Seamless Austenitic Chromium-Nickel Steel Still Tubes for Refinery Service 精油所用継目無オーテスナイト系ニッケル，クローム加熱用鋼管	TP304/S30400	S	≦0.08	≦0.75	≦2.00	≦0.040
			TP304H/S30409	S	0.04〜0.10	≦0.75	≦2.00	≦0.040
			TP316/S31600	S	≦0.08	≦0.75	≦2.00	≦0.040
			TP316H/S31609	S	0.04〜0.10	≦0.75	≦2.00	≦0.040
			TP321/S32100	S	≦0.08	≦0.75	≦2.00	≦0.040
			TP321H/S32109	S	0.04〜0.10	≦0.75	≦2.00	≦0.040
			TP347/S34700	S	≦0.08	≦0.75	≦2.00	≦0.040
			TP347H/S34709	S	0.04〜0.10	≦0.75	≦2.00	≦0.040
A334M−91		Seamless and Welded Carbon and Alloy-Steel Tubes for Low-Temperature Service 低温用継目および溶接炭素鋼および合金鋼鋼管	Gr. 1	S, E, AT	≦0.30	−	0.40〜1.06	≦0.05
			〃 3	S, E, AT	≦0.19	0.18〜0.37	0.31〜0.64	≦0.05
			〃 6	S, E, AT	≦0.30	≦0.10	0.29〜1.06	≦0.048
			〃 7	S, E, AT	≦0.19	0.13〜0.32	≦0.90	≦0.04
			〃 8	S, E, AT	≦0.18	0.13〜0.32	≦0.90	≦0.045
			〃 9	S, E, AT	≦0.20	−	0.40〜1.06	≦0.045
			〃 11	S, E, AT	≦0.1	≦0.35	≦0.60	≦0.025
A336M−94		Steel Forgings, Alloy, for Pressure and High-Temperature Parts 高圧高温用合金鋼鍛造品	Gr. F1	F	0.20〜0.30	0.20〜0.35	0.60〜0.80	≦0.040
			F11, C1.2	F	0.10〜0.20	0.50〜1.00	0.30〜0.80	≦0.040
			F11, C1.3	F	0.10〜0.20	0.50〜1.00	0.30〜0.80	≦0.040
			F11, C1.1	F	≦0.15	0.50〜1.00	0.30〜0.60	≦0.030
			F12	F	0.10〜0.20	0.10〜0.60	0.30〜0.80	≦0.040
			F5	F	≦0.15	≦0.50	0.31〜0.60	≦0.030
			F5A	F	≦0.25	≦0.50	≦0.60	≦0.040
			F9	F	≦0.15	0.50〜1.00	0.30〜0.60	≦0.030
			F6	F	≦0.12	≦1.00	≦1.00	≦0.040
			F21, C1.3	F	≦0.15	≦0.50	0.30〜0.60	≦0.030
			F21, C1.1	F	≦0.15	≦0.50	0.30〜0.60	≦0.030

$1MPa = 1.01972 \times 10^{-1} kgf/mm^2$

化　学　成　分　(%)					Tensile Test MPa or N/mm^2 引張試験		Remarks 備考	Index No.
S	Ni	Cr	Mo	Others	Min. Yield Point 最小降伏点	Tensile Strength 引張強さ	(Similar to JIS)	索引番号
≦0.030	17.50～18.5	17.0～19.0	–	–	–	–	–	C248
≦0.030	11.50～13.50	20.50～23.50	1.50～3.00	Nb+Ta:0.10～0.30 V: 0.10 ～0.30 N: 0.20 ～0.40	–	–	–	
≦0.030	2.20～3.75	17.0～19.0	–	N: 0.20 ～0.40	–	–	–	
≦0.010	17.50～18.50	19.50～20.50	6.00～6.50	N :0.180～0.220 Cu 0.50 ～1.00	–	–	–	
≦0.030	13.00～17.00	18.0～20.0	4.0～5.0	N ≦0.20 Cu≦0.75	–	–	–	
≦0.030	13.5～17.5	17.0～20.0	4.0～5.0	N 0.10～0.20 Cu≦0.75	–	–	–	
≦0.005	21.0～23.0	24.0～25.0	7.00～8.00	N 0.45～0.55 Cu 0.30～0.60	–	–	–	
≦0.020	14.0～15.5	17.0～18.5	≦0.20	Cu≦0.50	–	–	–	
≦0.010	16.0～18.0	23.0～25.0	4.0～5.0	N 0.4～0.6 Cb+Ta ≦0.1	–	–	–	
≦0.030	8.00～11.00	18.0～20.0	–	–	205	≧515	(SUS304TF)	C249
≦0.030	8.00～11.00	18.0～20.0	–	–	205	≧515	(SUS304HTF)	
≦0.030	11.0～14.00	16.0～18.0	–	–	205	≧515	(SUS316TF)	
≦0.030	11.0～14.00	16.0～18.0	–	–	205	≧515	(SUS316HTF)	
≦0.030	9.00～13.00	17.0～20.0	–	Ti:5×C～0.60	205	≧515	(SUS321TF)	
≦0.030	9.00～13.00	17.0～20.0	–	Ti:4×C～0.60	205	≧515	(SUS321HTF)	
≦0.030	9.00～13.00	17.0～20.0	–	Cb+Ta: 10×C～1.0	205	≧515	(SUS347TF)	
≦0.030	9.00～13.0	17.0～20.0	–	Cb+Ta: 8×C～1.0	205	≧515	(SUS347HTF)	
≦0.060	–	–	–	(STBL380)	205	≧380	Impact Test(J) 2V 18	C250
≦0.050	3.18～3.82	–	–	(STBL450)	240	≧450	〃 18	
≦0.058	–	–	–	–	240	≧415	〃 18	
≦0.050	2.03～2.57	–	–	–	240	≧450	〃 18	
≦0.045	8.40～9.60	–	–	(STBL690)	520	≧690	〃 18	
≦0.050	1.60～2.24	–	–	(Cu0.75 ～1.25)	315	≧435	〃 18	
≤0.025	35.0～37.0	≤0.50	≤0.50	Co≤0.50	240	≤450 -		
≦0.040	–	–	0.40～0.60	–	275	485～660	–	C251
≦0.040	–	1.00～1.50	0.45～0.65	–	275	485～660	–	
≦0.040	–	1.00～1.50	0.45～0.65	–	310	515～690	–	
≦0.030	–	1.00～1.50	0.44～0.65	–	205	415～585	–	
≦0.040	–	0.80～1.10	0.45～0.65	–	275	485～660	–	
≦0.030	≦0.50	4.0 ～6.0	0.45～0.65	–	250	415～585	–	
≦0.030	≦0.50	4.0 ～6.0	0.45～0.65	–	345	550～725	–	
≦0.030	–	8.00～10.00	0.90～1.10	–	380	585～760	–	
≦0.030	≦0.50	11.5～13.5	–	–	380	585～760	–	
≦0.030	–	2.65～3.25	0.80～1.06	–	310	515～690	–	
≦0.030	–	2.65～3.25	0.80～1.06	–	205	415～585	–	

Standard No. 規格番号 / Year 年号 / Designation 規格名称	Grade グレード	Mfg. Process 製造方法	Chemical Composition C	Si	Mn	P
A336M -94 (Continued)	F22 C1.3	F	≦0.15	≦0.50	0.30～0.60	≦0.030
	F22A C1.1	F	≦0.15	≦0.50	0.30～0.60	≦0.030
	FXM27 Cb	F	≦0.01	≦0.40	≦0.40	≦0.020
	F91	F	0.08～0.12	0.20～0.50	0.30～0.60	≦0.025
	F3V	F	0.10～0.15	≦0.10	0.30～0.60	≦0.020
	F22V	F	0.11～0.15	≦0.10	0.30～0.60	≦0.015
	F304	F	≦0.08	≦1.00	≦2.00	≦0.040
	F304H	F	0.04～0.10	≦1.00	≦2.00	≦0.045
	F304L	F	≦0.035	≦1.00	≦2.00	≦0.040
	F304N	F	≦0.08	≦0.75	≦2.00	≦0.030
	F304LN	F	≦0.030	≦1.00	≦2.00	≦0.040
	F309H	F	0.04～0.10	≦1.00	≦2.00	≦0.040
	F310	F	≦0.10	≦1.00	≦2.00	≦0.040
	F310H	F	0.04～0.10	≦1.50	≦2.00	≦0.030
	F316	F	≦0.08	≦1.00	≦2.00	≦0.040
	F316H	F	0.04～0.10	≦1.00	≦2.00	≦0.045
	F316L	F	≦0.035	≦1.00	≦2.00	≦0.040
	F316N	F	≦0.08	≦0.75	≦2.00	≦0.030
	F316LN	F	≦0.030	≦1.00	≦2.00	≦0.040
	F347	F	≦0.08	≦0.85	≦2.00	≦0.040
	F347H	F	0.04～0.10	≦1.00	≦2.00	≦0.040
	F321	F	≦0.08	≦0.85	≦2.50	≦0.035
	F321H	F	0.04～0.10	≦1.00	≦2.00	≦0.040
	F348	F	≦0.08	≦1.00	≦2.00	≦0.040
	F348H	F	0.04～0.10	≦1.00	≦2.00	≦0.040
	FXM-19	F	≦0.06	≦1.00	5.00～10.00	≦0.040
	FXM-11	F	≦0.04	≦1.00	5.00～10.00	≦0.060
	F46	F	≦0.018	3.70～4.30	≦2.00	≦0.020
A422 -64 (Reapproved 1985) Butt Welds in Still Tubes for Refinery Service 精油所用加熱炉用鋼管の突合せ溶接	－	－	A161 Seamless Low-Carbon and Carbon- A200 Seamless Intermediate Alloy-Steel A271 Seamless Austenitic Chromium-			

1 MPa = 1.01972 × 10⁻¹ kgf/mm²

化学成分 (%)					Tensile Test MPa or N/mm² 引張試験		Remarks 備考	Index No.
S	Ni	Cr	Mo	Others	Min. Yield Point 最小降伏点	Tensile Strength 引張強さ	(Similar to JIS)	索引番号
≦0.030	–	2.00～2.50	0.90～1.10	–	310	510～690	–	C251
≦0.030	–	2.00～2.50	0.90～1.10	–	205	415～585	–	
≦0.020	≦0.50	25.0～27.5	0.75～1.50	Cu≦0.20 N ≦0.015	240	415～585	–	
≦0.025	≦0.40	8.00～9.50	0.85～1.05	V 0.18～0.25 Cb 0.0 ～0.10 N 0.03～0.07 Al≦0.04	415	585～760	–	
≦0.020	–	2.65～3.25	0.90～1.10	V 0.20 ～0.30 B 0.001～0.003 Ti 0.015～0.035	415	585～760	–	
≦0.010	≦0.25	2.00～2.50	0.90～1.10	V 0.25～0.35 Cb ≦0.07 B ≦0.0020 Ti ≦0.030 Cu ≦0.20 Ca ≦0.015	415	585～760	–	
≦0.030	8.00～11.0	18.00～20.00	–	–	205	≧485	–	
≦0.030	8.00～12.0	19.00～20.00	–	–	205	≧485	–	
≦0.030	8.00～13.0	18.00～20.00	–	–	170	≧450	–	
≦0.030	8.00～11.0	18.00～20.00	–	N 0.10～0.16	240	≧550	–	
≦0.030	8.00～11.0	18.00～20.00	–	N 0.10～0.16	205	≧485	–	
≦0.030	12.00～15.0	22.00～24.00	–	–	205	≧485	–	
≦0.030	19.00～22.00	24.00～26.00	–	–	205	≧515	–	
≦0.030	19.00～22.00	24.00～26.00	–	–	205	≧485	–	
≦0.030	10.00～14.00	16.00～18.00	2.00～3.00	–	205	≧485	–	
≦0.030	10.00～14.00	16.00～18.00	2.00～3.00	–	205	≧485	–	
≦0.030	10.00～15.00	16.00～18.00	2.00～3.00	–	170	≧450	–	
≦0.030	11.00～14.00	16.00～18.00	2.00～3.00	N 0.10～0.16	240	≧550	–	
≦0.030	10.00～14.0	16.00～18.00	2.00～3.00	N 0.10～0.16	205	≧485	–	
≦0.030	9.00～12.00	17.00～19.00	–	Nb10×C～0.60	205	≧485	–	
≦0.030	9.00～13.00	17.00～20.00	–	Nb+Ta:8×C～1.00	205	≧485	–	
≦0.030	≧9.00	≧17.00	–	Ti:5×C～0.60	205	≧485	–	
≦0.030	9.00～12.00	≧17.00	–	Ti:4×C～0.60	205	≧485	–	
≦0.030	9.00～13.00	17.00～20.00	–	Nb+Ta:8×C～1.00 Ta≦0.10	205	≧485	–	
≦0.030	9.00～13.00	17.00～20.00	–	Nb+Ta:8×C～1.00 Ta≦0.10	170	≧450	–	
≦0.030	11.50～13.50	20.50～23.50	1.50～3.00	N 0.20 ～0.40 V 0.10 ～0.30 Na+Ta:0.10～0.30	380	≧690	–	
≦0.030	5.50～7.50	19.00～21.50	–	N 0.15～0.40	345	≧380	–	
≦0.020	14.00～15.50	17.00～18.50	≦0.20	–	220	540～690	–	
Molybdenum Steel Still Tube for Refinery Service Still Tube for Refinery Service Nickel Steel Still Tube for Refinery Service					–	–	–	C252

Standard No. 規格番号	Year 年号	Designation 規格名称	Grade / UNS No. グレード	Mfg. Process 製造方法	Chemical Composition C	Si	Mn	P
A423M-91		Seamless and Electric Welded Low-Alloy Steel Tubes 継目無および電気溶接低合金鋼管	Gr. 1	S, E	≦0.15	≧0.10	≦0.55	0.06～0.16
			Gr. 2	S, E	≦0.15	–	0.50～1.00	≦0.040
A556M-90a		Seamless Cold-Drawn Carbon Steel Feedwater Heater Tubes 給水加熱用冷間引抜炭素鋼管	Gr. A2	S	≦0.18	–	0.27～0.63	≦0.035
			Gr. B2	S	≦0.27	≧0.10	0.29～0.93	≦0.035
			Gr. C2	S	≦0.30	≧0.10	0.29～1.06	≦0.035
A557M-90a		Electric-Resistance-Welded Carbon Steel Feedwater Heater Tubes 給水加熱用電気抵抗溶接鋼管	Gr. A2	S	≦0.18	–	0.27～0.63	≦0.035
			Gr. B2	S	≦0.30	–	0.27～0.93	≦0.035
			Gr. C2	S	≦0.35	–	0.27～1.06	≦0.035
A632-90		Seamless and Welded Austenitic Stainless Steel Tubing(Small Diameter)for General Service 一般用継目無およびオーステナイト系ステンレス鋼管（小口径）	TP304	S, W	≦0.80	≦0.75	≦2.00	≦0.040
			TP304L	S, W	≦0.040	≦0.75	≦2.00	≦0.040
			TP310	S, W	≦0.15	≦0.75	≦2.00	≦0.040
			TP316	S, W	≦0.08	≦0.75	≦2.00	≦0.040
			TP316L	S, W	≦0.040	≦0.75	≦2.00	≦0.040
			TP317	S, W	≦0.08	≦0.75	≦2.00	≦0.040
			TP321	S, W	≦0.08	≦0.75	≦2.00	≦0.040
			TP347	S, W	≦0.08	≦0.75	≦2.00	≦0.040
			TP348	S, W	≦0.08	≦0.75	≦2.00	≦0.040
A688M-91		Welded Austenitic Stainless Steel Feedwater Heater Tube 溶接オーステナイトステンレス鋼給水加熱鋼管	TP304 /S30400	E, AT	≦0.80	≦0.75	≦2.00	≦0.040
			TP304L/S30403	E, AT	≦0.035	≦0.75	≦2.00	≦0.040
			TP304LN/S30453	E, AT	≦0.035	≦0.75	≦2.00	≦0.040
			TP316 /S31600	E, AT	≦0.08	≦0.75	≦2.00	≦0.040
			TP316L/S31603	E, AT	≦0.035	≦0.75	≦2.00	≦0.040
			TP316LN/S31653	E, AT	≦0.035	≦0.75	≦2.00	≦0.040
			TPXM-29/S28300	E, AT	≦0.060	≦1.00	11.50～14.50	≦0.060
			TP304N/S30451	E, AT	≦0.08	≦0.75	≦2.00	≦0.040
			TP316N/S31651	E, AT	≦0.08	≦0.75	≦2.00	≦0.040
A692-91		Seamless Medium-Strength Carbon-Molybdenum Alloy-Steel Boiler and Superheater Tubes 継目無中強度ﾓﾘﾌﾞﾃﾞﾝ鋼ﾎﾞｲﾗ及び過熱器用鋼管	–	S	0.17～0.26	0.18～0.37	0.46～0.94	≦0.045

$1\ MPa = 1.01972 \times 10^{-1}\ kgf/mm^2$

化　学　成　分　(%)					Tensile Test MPa or N/mm² 引張試験		Remarks 備考	Index No. 索引番号
S	Ni	Cr	Mo	Others	Min. Yield Point 最小降伏点	Tensile Strength 引張強さ	(Similar to JIS)	
≦0.060	0.20～0.70	0.24～1.31	－	Cu 0.20 ～0.60	255	≧415	－	C253
≦0.05	0.40～1.00	－	≧0.10	Cu 0.30 ～1.00	255	≧415	－	
≦0.035	－	－	－	－	180	≧320	(STB340)	C254
≦0.035	－	－	－	－	260	≧410	(STB410)	
≦0.035	－	－	－	－	280	≧480	－	
≦0.035	－	－	－	－	180	≧325	(STB340)	C255
≦0.035	－	－	－	－	255	≧415	(STB410)	
≦0.035	－	－	－	－	275	≧485	－	
≦0.030	8.0 ～11.0	18.0～20.0	－	－	205	≧515	(SUS304TB)	C256
≦0.030	8.0 ～13.0	18.0～20.0	－	－	170	≧485	(SUS304LTB)	
≦0.030	19.0～22.0	24.0～26.0	－	－	205	≧515	(SUS310TB)	
≦0.030	11.0～14.0	16.0～18.0	2.00～3.00	－	205	≧515	(SUS316TB)	
≦0.030	10.0～15.0	16.0～18.0	2.00～3.00	－	170	≧485	(SUS316LTB)	
≦0.030	11.0～14.0	18.0～20.0	3.00～4.00	－	205	≧515	(SUS317TB)	
≦0.030	9.0～13.0	17.0～20.0	－	Ti:5 ×C～0.60	205	≧515	(SUS321TB)	
≦0.030	9.0～13.0	17.0～20.0	－	Na+Ta:10×C～1.0	205	≧515	(SUS347TB)	
≦0.030	9.0～13.0	17.0～20.0	－	Na+Ta:10×C～1.0 Ta≦0.10	205	≧515	－	
≦0.030	8.0 ～11.0	18.00～20.00	－	－	205	≧515	(SUS304TB)	C257
≦0.030	8.0 ～13.0	18.00～20.00	－	－	175	≧485	(SUS304LTB)	
≦0.030	8.0 ～13.0	18.00～20.00	－	N:0.10～0.16	205	≧515	－	
≦0.030	11.00～14.00	16.00～18.00	2.00～3.00	－	205	≧515	(SUS316TB)	
≦0.030	10.00～15.00	16.00～18.00	2.00～3.00	－	175	≧485	(SUS316LTB)	
≦0.030	10.00～15.00	16.00～18.00	3.00～3.00	N:0.10～0.16	205	≧515	－	
≦0.030	2.25～3.75	17.00～19.00	－	N:0.20～0.40	380	≧690	－	
≦0.030	8.00～11.00	18.00～20.00	－	N:0.10～0.16	240	≧550	－	
≦0.030	11.00～14.00	16.00～18.00	2.00～3.00	N:0.10～0.16	240	≧550	－	
≦0.045	－	－	0.42～0.68	－	290	441～579	(STBA13)	C258

StandardNo. Year Designation 規格番号 年号 規格名称	Grade / UNS No. グレード	Mfg. Process 製造方法	Chemical Composition			
			C	Si	Mn	P
A803－94 Welded Ferritic Stainless Steel Feed Water Heater Tubes 溶接フェライト径ステンレス鋼給水加熱器用鋼管	TP409 /S40900	E, W	≦0.08	≦1.00	≦1.0	≦0.045
	TP439 /S43035	E, W	≦0.07	≦1.00	≦1.00	≦0.040
	TPXM-27/S44627	E, W	≦0.01	≦0.40	≦0.40	≦0.02
	TPXM-33/S44626	E, W	≦0.06	≦0.75	≦0.075	≦0.040
	25-4-4/S44635	E, W	≦0.025	≦0.75	≦1.00	≦0.040
	26-3-3/S44660	E, W	≦0.030	≦1.00	3.0～4.0	≦0.040
	29-4 / S44700	E, W	≦0.010	≦0.20	≦0.30	≦0.025
	29-4-2/S44800	E, W	≦0.010	≦0.20	≦0.30	≦0.025
	18-2 / S44400	E, W	≦0.025	≦1.00	≦1.00	≦0.040
	29-4C /S44735	E, W	≦0.030	≦1.00	≦1.00	≦0.040
A851－90 High-Frequency Induction Welded, Unannealed, Austenitic Steel Condenser Tubes 高周波誘電溶接非焼なましオーステナイト鋼凝縮器用鋼管	TP304 /S30400	E, W	≦0.08	≦0.75	≦2.00	≦0.040
	TP304L/S30403	E, W	≦0.08	≦0.75	≦2.00	≦0.040

1MPa＝1.01972×10⁻¹kgf/mm²

化学成分（%）					Tensile Test MPa or N/mm² 引張試験		Remarks 備考 (Similar to JIS)	Index No. 索引番号
S	Ni	Cr	Mo	Others	Min. Yield Point 最小降伏点	Tensile Strength 引張強さ		
≦0.045	≦0.50	10.50～11.75	−	Ti 6C ～0.75	205	≧380	−	C259
≦0.040	≦0.030	17.00～19.00	Ti 0.20+ 4(C+N)~1.10	Al≦0.15 N ≦0.14	205	≧380	−	
≦0.02	≦0.50	25.0～27.5	Mo0.75~1.50 Cb0.05~0.20	Cu≦0.2 N ≦0.015	275	≧450	−	
≦0.020	0.75～1.00	25.0～27.5	Mo0.75~1.50 Ti 0.26 ≦ 7(C+N)≦1.00	Cu≦0.20 N ≦0.040	310	≧470	−	
≦0.030	3.5～4.5	24.5～26.0	3.5～4.5	N ≦0.040 Ti+Cb= 0.2+4(C+N) 0.20～1.00	450	≧585	−	
≦0.030	3.5～4.5	24.5～26.0	3.5～4.5	N ≦0.040 Ti+Cb= 6+(C+N) 0.20～1.00	450	≧585	−	
≦0.025	≦0.15	28.0～30.0	3.5～4.2	Cu≦0.15 N ≦0.025	415	≧550	−	
≦0.020	2.0～2.5	28.0～30.0	3.5～4.2	≦0.20	415	≧550	−	
≦0.030	≦1.00	17.5～19.5	1.75～2.50	Cu≦0.15 N ≦0.020	415	≧550	(Ti+Cb) 0.20+4(C+N)~0.80	
≦0.030	≦1.00	28.0～30.0	3.60～4.20	N ≦0.045	415	≧515	0.20≦Ti+Cb =4(C+N)～1.00	
≦0.030	8.00～11.0	18.0～20.0	−	−	205	515	−	C260
≦0.030	8.00～11.0	18.0～20.0	−	−	170	485	−	

9. ASTM Steel Pipes for Structural Purposes　構造用鋼管

Standard No. Year Designation 規格番号 年号 規格名称	Grade グレード	Mfg. Process 製造方法	Chemical Composition C	 Si	 Mn	 P
A252－93 Welded and Seamless Steel Pipe Piles 溶接および継目無鋼管パイル	1	S.W	–	–	–	≦0.050
	2	S.W	–	–	–	≦0.050
	3	S.W	–	–	–	≦0.050
A500－93 Cold-Formed Welded and Seamless Carbon Steel Structural Tubing in Rounds and Shapes 冷間成型溶接及び継目無炭素構造用円形及角形鋼管	A	S.E.AT	≦0.26	–	–	≦0.035
	B	S.E.AT	≦0.26	–	–	≦0.035
	C	S.E.AT	≦0.26	–	≦1.35	≦0.035
	D	S.E.AT	≦0.26	–	–	≦0.035
A501－93 Hot-Formed Welded and Seamless Carbon Steel Structural Tubing 熱間成型溶接および継目無構造用炭素鋼鋼管	–	S.W	≦0.26	–	–	≦0.035
A511－90 Seamless Stainless Steel Mechanical Tubing 機械構造用継目無ステンレス鋼鋼管	MT302	S	0.08～0.20	≦1.00	≦2.00	≦0.040
	MT303Se	S	≦0.15	≦1.00	≦2.00	≦0.040
	MT304	S	≦0.08	≦1.00	≦2.00	≦0.040
	MT304L	S	≦0.035	≦1.00	≦2.00	≦0.040
	MT305	S	≦0.12	≦1.00	≦2.00	≦0.040
	MT309S	S	≦0.08	≦1.00	≦2.00	≦0.040
	MT310S	S	≦0.08	≦1.00	≦2.00	≦0.040
	MT316	S	≦0.08	≦1.00	≦2.00	≦0.040
	MT316L	S	≦0.035	≦1.00	≦2.00	≦0.040
	MT317	S	≦0.08	≦1.00	≦2.00	≦0.040
	MT321	S	≦0.08	≦1.00	≦2.00	≦0.040
	MT347	S	≦0.08	≦1.00	≦2.00	≦0.040
	MT403	S	≦0.15	≦0.50	≦1.00	≦0.040
	MT410	S	≦0.15	≦1.00	≦1.00	≦0.040
	MT414	S	≦0.15	≦1.00	≦1.00	≦0.040
	MT416Se	S	≦0.15	≦1.00	≦1.25	≦0.060
	MT431	S	≦0.20	≦1.00	≦1.00	≦0.040
	MT440A	S	0.60～0.75	≦1.00	≦1.00	≦0.040
	MT405	S	≦0.08	≦1.00	≦1.00	≦0.040
	MT429	S	≦0.12	≦1.00	≦1.00	≦0.040
	MT430	S	≦0.12	≦1.00	≦1.00	≦0.040
	MT443	S	≦0.20	≦1.00	≦1.00	≦0.040
	MT446-1	S	≦0.20	≦1.00	≦1.50	≦0.040
	MT446-2	S	≦0.12	≦1.00	≦1.50	≦0.040
	29-4	S	≦0.010	≦0.20	≦0.30	≦0.025

1 MPa = 1.01972 × 10^{-1} kgf/mm²

化学成分 (%)					Tensile Test MPa or N/mm² 引張試験		Remarks 備考 (Similar to JIS)	Index No. 索引番号
S	Ni	Cr	Mo	Others	Min. Yield Point 最小降伏点	Tensile Strength 引張強さ		
−	−	−	−	−	205	≧345	(STK290)	C261
−	−	−	−	−	240	≧414	(STK400)	
−	−	−	−	−	310	≧310	−	
≦0.035	−	−	−	(Cu≧0.20)	228	≧310	(STK290)	C262
≦0.035	−	−	−	(Cu≧0.20)	290	≧400	(STK400) (STKR400)	
≦0.035	−	−	−	(Cu≧0.20)	317	≧417	−	
≦0.035	−	−	−	(Cu≧0.20)	−	−	−	
≦0.035	−	−	−	(Cu≧0.20)	250	≧400	STK400	C263
≦0.030	8.0〜10.0	17.0〜19.0	−	−	207	≧517	−	C264
≦0.030	8.0〜11.0	17.0〜19.0	−	−	207	≧517	−	
≦0.030	8.0〜11.0	18.0〜20.0	−	−	207	≧517	(SUS304TK)	
≦0.030	8.0〜13.0	18.0〜20.0	−	−	207	≧517	−	
≦0.030	10.0〜13.0	17.0〜19.0	−	−	207	≧517	−	
≦0.030	12.0〜15.0	22.0〜24.0	−	−	207	≧517	Austenitic	
					207	≧517		
≦0.030	19.0〜22.0	24.0〜26.0	−	−	207	≧517	−	
≦0.030	11.0〜14.0	16.0〜18.0	2.0〜3.0	−	207	≧517	(SUS316TK)	
≦0.030	10.0〜15.0	16.0〜18.0	2.0〜3.0	−	207	≧517	−	
≦0.030	11.0〜14.0	18.0〜20.0	3.0〜4.0	−	207	≧517	−	
≦0.030	9.0〜13.0	17.0〜20.0	−	Ti:5×C〜0.60	207	≧517	(SUS321TK)	
≦0.030	9.0〜13.0	17.0〜20.0	−	Nb+Ta:10×C〜1.00	207	≧517	(SUS347TK)	
≦0.030	≦0.50	11.5〜13.0	≦0.60	−	207	≧414	−	
≦0.030	≦0.50	11.5〜13.5	−	−	207	≧414	(SUS410TK)	
≦0.030	1.25〜2.50	11.5〜13.5	−	−	448	≧680	−	
≦0.060	≦0.50	12.0〜14.0	−	Se:0.12〜0.20	241	≧414	−	
≦0.030	1.25〜2.50	15.0〜17.0	−	−	621	≧724	Martensitic	
≦0.030	−	16.0〜18.0	≦0.75	−	379	≧655	−	
≦0.030	≦0.50	11.5〜14.5	−	Al:0.10〜0.30	207	≧414	−	
≦0.030	≦0.50	14.0〜16.0	−	−	241	≧414	−	
≦0.030	≦0.50	16.0〜18.0	−	−	241	≧414	(SUS430TK)	
≦0.030	≦0.50	18.0〜23.0	−	−	276	≧483	−	
≦0.030	≦0.50	23.0〜30.0	−	−	276	≧483	Ferritic	
≦0.030	≦0.50	23.0〜30.0	−	−	276	≧483	−	
≦0.020	≦0.15	28.0〜30.0	3.5〜4.2	Cu≦0.15 N ≦0.020	379	≧483	−	

Standard No. Year Designation 規格番号 年号 規格名称	Grade グレード	Mfg. Process 製造方法	Chemical Composition C	 Si	 Mn	 P
A511－90 (Continued)	29-4-2	S	≦0.010	≦0.20	≦0.30	≦0.025
A512－94 Cold-Drawn Buttweld Carbon Steel Mechanical Tubing 機械構造用冷間引抜鍛接鋼管	MT 1010	B	0.05～0.15	－	0.30～0.60	≦0.040
	MT 1015	B	0.10～0.20	－	0.30～0.60	≦0.040
	MTX1015	B	0.10～0.20	－	0.60～0.90	≦0.040
	MT 1020	B	0.15～0.25	－	0.30～0.60	≦0.040
	MTX1020	B	0.15～0.25	－	0.70～1.00	≦0.040
	1008	B	≦0.10	－	0.25～0.50	≦0.040
	1010	B	0.08～0.13	－	0.30～0.60	≦0.040
	1012	B	0.10～0.15	－	0.30～0.60	≦0.040
	1015	B	0.12～0.18	－	0.30～0.60	≦0.040
	1016	B	0.12～0.18	－	0.60～0.90	≦0.040
	1018	B	0.14～0.20	－	0.60～0.90	≦0.040
	1019	B	0.14～0.20	－	0.70～0.90	≦0.040
	1020	B	0.17～0.23	－	0.30～0.60	≦0.040
	1021	B	0.17～0.23	－	0.60～0.90	≦0.040
	1025	B	0.22～0.28	－	0.30～0.60	≦0.040
	1026	B	0.21～0.28	－	0.60～0.90	≦0.040
	1030	B	0.27～0.34	－	0.60～0.90	≦0.040
	1035	B	0.31～0.38	－	0.60～0.90	≦0.040
	1110	B	0.08～0.15	－	0.30～0.60	≦0.040
	1115	B	0.13～0.20	－	0.60～0.90	≦0.040
	1117	B	0.14～0.20	－	1.00～1.30	≦0.040
A513－94 Electric-Resistance-Welded Carbon and Alloy Steel Mechanical Tubing 電気抵抗溶接炭素鋼および合金鋼機械構造用鋼管	MT 1010	E	0.05～0.15	－	0.30～0.60	≦0.035
	MT 1015	E	0.10～0.20	－	0.30～0.60	≦0.035
	MTX1015	E	0.10～0.20	－	0.60～0.90	≦0.035
	MT 1020	E	0.15～0.25	－	0.30～0.60	≦0.035
	MTX1020	E	0.15～0.25	－	0.70～1.00	≦0.035
	1008	E	≦0.10	－	0.25～0.50	≦0.035
	1010	E	0.08～0.13	－	0.30～0.60	≦0.035
	1012	E	0.10～0.15	－	0.30～0.60	≦0.035

$1 MPa = 1.01972 \times 10^{-1} kgf/mm^2$

化学成分 (%)					Tensile Test MPa or N/mm^2 引張試験		Remarks 備考 (Similar to JIS)	Index No. 索引番号
S	Ni	Cr	Mo	Others	Min. Yield Point 最小降伏点	Tensile Strength 引張強さ		
≦0.020	2.0～2.5	28.0～30.0	3.5～4.2	Cu≦0.020 N ≦0.020	379	≧483		C264
≦0.045	–	–	–	(STKM11A)	400	434-689	Stress Relief Annealed Tube	C265
≦0.045	–	–	–	(STKM12A)	407	450-689	Stress Relief Annealed Tube	
≦0.045	–	–	–	(STKM13A)	–	–	–	
≦0.045	–	–	–	–	–	–	–	
≦0.045	–	–	–	–	–	–	–	
≦0.045	–	–	–	–	–	–	–	
≦0.045	–	–	–	–	–	–	–	
≦0.045	–	–	–	–	–	–	–	
≦0.045	–	–	–	–	–	–	–	
≦0.045	–	–	–	–	421	462-689	Stress Relief Annealed Tube	
≦0.045	–	–	–	–	427	469-689	Stress Relief Annealed Tube	
≦0.045	–	–	–	–	–	–	–	
≦0.045	–	–	–	–	–	–	–	
≦0.045	–	–	–	–	–	–	–	
≦0.045	–	–	–	–	462	496-896	Stress Relief Annealed Tube	
≦0.045	–	–	–	–	–	–	–	
≦0.045	–	–	–	–	483	552-896	Stress Relief Annealed Tube	
≦0.045	–	–	–	–	–	–	–	
≦0.130	–	–	–	–	400	434-689	Stress Relief Annealed Tube	
≦0.130	S:min 0.08	–	–	–	427	469-689	Stress Relief Annealed Tube	
≦0.130	–	–	–	–	–	–	–	
≦0.035	–	–	–	–	–	–	(STKM11)	C266
≦0.035	–	–	–	–	–	–	(STKM12)	
≦0.035	–	–	–	–	–	–	–	
≦0.035	–	–	–	–	–	–	(STKM13)	
≦0.035	–	–	–	–	(As-welded Tubing)		–	
≦0.035	–	–	–	–	207	≧290	–	
≦0.035	–	–	–	–	221	≧310	–	
≦0.035	–	–	–	–	–	–	–	

Standard No. Year Designation 規格番号 年号 規格名称	Grade グレード	Mfg. Process 製造方法	Chemical Composition			
			C	Si	Mn	P
A513−94 (Continued)	MTX1015	E	0.12〜0.18	−	0.30〜0.60	≦0.035
	1016	E	0.12〜0.19	−	0.60〜0.90	≦0.035
	1017	E	0.14〜0.21	−	0.30〜0.60	≦0.035
	1018	E	0.14〜0.21	−	0.60〜0.90	≦0.035
	1019	E	0.14〜0.21	−	0.70〜1.00	≦0.035
	1020	E	0.17〜0.23	−	0.30〜0.60	≦0.035
	1021	E	0.17〜0.24	−	0.60〜0.90	≦0.035
	1022	E	0.17〜0.24	−	0.70〜1.00	≦0.035
	1023	E	0.19〜0.26	−	0.30〜0.60	≦0 035
	1024	E	0.18〜0.26	−	1.30〜1.65	≦0.035
	1025	E	0.21〜0.28	−	0.30〜0.60	≦0.035
	1026	E	0.21〜0.28	−	0.60〜0.90	≦0.035
	1027	E	0.21〜0.29	−	1.20〜1.55	≦0.035
	1030	E	0.27〜0.35	−	0.60〜0.90	≦0.035
	1033	E	0.29〜0.37	−	0.70〜1.00	≦0.035
	1035	E	0.31〜0.39	−	0.60〜0.90	≦0.035
	1040	E	0.36〜0.44	−	0.47〜0.55	≦0.040
	1050	E	0.55〜0.60	−	0.60〜0.90	≦0.040
	1060	E	0.55〜0.66	−	0.60〜0.90	≦0.040
	1340	E	0.38〜0.43	0.15〜0.35	1.60〜1.90	≦0.035
	1524	E	0.18〜0.24	−	1.35〜1.65	≦0.040
	4118	E	0.18〜0.23	0.15〜0.35	0.70〜0.90	≦0.035
	4130	E	0.28〜0.33	0.15〜0.35	0.40〜0.60	≦0.035
	4140	E	0.38〜0.43	0.15〜0.35	0.75〜1.00	≦0.035
	5130	E	0.23〜0.33	0.15〜0.35	0.70〜0.90	≦0.035
	8620	E	0.18〜0.23	0.15〜0.35	0.70〜0.90	≦0.035
	8630	E	0.28〜0.33	0.15〜0.35	0.70〜0.90	≦0.035
A519−94 Seamless Carbon and Alloy Steel Mechanical Tubing 継目無炭素鋼および合金鋼機械構造用鋼管	MT 1010	S	0.05〜0.15	−	0.30〜0.60	≦0.040
	MT 1015	S	0.10〜0.20	−	0.30〜0.60	≦0.040
	MTX1015	S	0.10〜0.20	−	0.60〜0.90	≦0.040
	MT 1020	S	0.15〜0.25	−	0.30〜0.60	≦0.040
	MTX1020	S	0.15〜0.25	−	0.70〜1.00	≦0.040
	1008	S	≦0.10	−	0.30〜0.50	≦0.040
	1010	S	0.08〜0.13	−	0.30〜0.60	≦0.040
	1012	S	0.10〜0.15	−	0.30〜0.60	≦0.040
	1015	S	0.13〜0.18	−	0.30〜0.60	≦0.040
	1016	S	0.13〜0.18	−	0.60〜0.90	≦0.040
	1017	S	0.15〜0.20	−	0.30〜0.60	≦0.040
	1018	S	0.15〜0.20	−	0.60〜0.90	≦0.040
	1019	S	0.15〜0.20	−	0.70〜1.00	≦0.040

1 MPa = 1.01972 × 10⁻¹ kgf/mm²

化学成分 (%)					Tensile Test MPa or N/mm² 引張試験		Remarks 備考	Index No.
S	Ni	Cr	Mo	Others	Min. Yield Point 最小降伏点	Tensile Strength 引張強さ	(Similar to JIS)	索引番号
≦0.035	–	–	–	–	241	≧331	–	C266
≦0.035	–	–	–	–	–	–	–	
≦0.035	–	–	–	–	–	–	–	
≦0.035	–	–	–	–	–	–	–	
≦0.035	–	–	–	–	–	–	–	
≦0.035	–	–	–	–	276	≧359	–	
≦0.035	–	–	–	–	–	–	–	
≦0.035	–	–	–	–	–	–	–	
≦0.035	–	–	–	–	–	–	–	
≦0.035	–	–	–	–	–	–	–	
≦0.035	–	–	–	–	276	≧386	–	
≦0.035	–	–	–	–	310	≧427	(STKM14)	
≦0.035	–	–	–	–	–	–	–	
≦0.035	–	–	–	–	310	≧429	(STKM15)	
≦0.035	–	–	–	–	–	–	–	
≦0.035	–	–	–	–	345	≧455	–	
≦0.050	–	–	–	–	345	≧645	–	
≦0.050	–	–	–	–	–	–	–	
≦0.050	–	–	–	–	–	–	–	
≦0.040	–	–	–	–	379	≧496	–	
≦0.050	–	–	–	–	345	≧455	–	
≦0.040	–	0.40〜0.60	0.08〜0.15	–	–	–	–	
≦0.040	–	0.80〜1.10	0.15〜0.25	–	379	≧490	–	
≦0.040	–	–	–	–	485	≧621	–	
≦0.040	0.40〜0.70	0.80〜1.10	0.15〜0.25	–	–	–	–	
≦0.040	0.40〜0.75	0.40〜0.60	0.15〜0.25	–	–	–	–	
≦0.040	0.40〜0.70	0.40〜0.60	0.15〜0.25	–	–	–	–	
≦0.050	–	–	–	–	–	–	(STKM11)	C267
≦0.050	–	–	–	–	–	–	(STKM12)	
≦0.050	–	–	–	–	–	–	–	
≦0.050	–	–	–	–	–	–	(STKM13)	
≦0.050	–	–	–	–	–	–	–	
≦0.050	–	–	–	–	–	–	–	
≦0.050	–	–	–	–	–	–	–	
≦0.050	–	–	–	–	–	–	–	
≦0.050	–	–	–	–	–	–	–	
≦0.050	–	–	–	–	–	–	–	
≦0.050	–	–	–	–	–	–	–	
≦0.050	–	–	–	–	–	–	–	
≦0.050	–	–	–	–	–	–	–	

Standard No. Year Designation 規格番号 年号 規格名称	Grade グレード	Mfg. Process 製造方法	Chemical Composition			
			C	Si	Mn	P
A519−94 (Continued)	1020	S	0.18～0.23	−	0.30～0.60	≦0.040
	1021	S	0.18～0.23	−	0.60～0.90	≦0.040
	1022	S	0.18～0.23	−	0.70～1.00	≦0.040
	1025	S	0.22～0.28	−	0.30～0.60	≦0.040
	1026	S	0.22～0.28	−	0.60～0.90	≦0.040
	1030	S	0.28～0.34	−	0.60～0.90	≦0.040
	1035	S	0.32～0.38	−	0.60～0.90	≦0.040
	1040	S	0.37～0.44	−	0.60～0.90	≦0.040
	1045	S	0.43～0.50	−	0.60～0.90	≦0.040
	1050	S	0.48～0.55	−	0.60～0.90	≦0.040
	1518	S	0.15～0.21	−	1.10～1.40	≦0.040
	1524	S	0.19～0.25	−	1.35～1.65	≦0.040
	1541	S	0.36～0.44	−	1.35～1.65	≦0.040
	1330	S	0.28～0.33	0.15～0.35	1.60～1.90	≦0.040
	1335	S	0.33～0.38	0.15～0.35	1.60～1.90	≦0.040
	1340	S	0.38～0.43	0.15～0.35	1.60～1.90	≦0.040
	1345	S	0.43～0.48	0.15～0.35	1.60～1.90	≦0.040
	3140	S	0.38～0.43	0.15～0.35	0.70～0.90	≦0.040
	E3310	S	0.08～0.13	0.15～0.35	0.45～0.60	≦0.025
	4012	S	0.09～0.14	0.15～0.35	0.75～1.00	≦0.040
	4023	S	0.20～0.25	0.15～0.35	0.70～0.90	≦0.040
	4024	S	0.20～0.25	0.15～0.35	0.70～0.90	≦0.040
	4027	S	0.25～0.30	0.15～0.35	0.70～0.90	≦0.040
	4028	S	0.25～0.30	0.15～0.35	0.70～0.90	≦0.040
	4037	S	0.35～0.40	0.15～0.35	0.70～0.90	≦0.040
	4042	S	0.40～0.45	0.15～0.35	0.70～0.90	≦0.040
	4047	S	0.45～0.50	0.15～0.35	0.70～0.90	≦0.040
	4063	S	0.60～0.67	0.15～0.35	0.75～1.00	≦0.040
	4118	S	0.18～0.23	0.15～0.35	0.70～0.90	≦0.040
	4130	S	0.28～0.33	0.15～0.35	0.40～0.60	≦0.040
	4135	S	0.33～0.38	0.15～0.35	0.70～0.90	≦0.040
	4137	S	0.35～0.40	0.15～0.35	0.70～0.90	≦0.040
	4140	S	0.38～0.43	0.15～0.35	0.75～1.00	≦0.040
	4142	S	0.40～0.45	0.15～0.35	0.75～1.00	≦0.040
	4145	S	0.43～0.48	0.15～0.35	0.75～1.00	≦0.040
	4147	S	0.45～0.50	0.15～0.35	0.75～1.00	≦0.040
	4150	S	0.48～0.53	0.15～0.35	0.75～1.00	≦0.040
	4320	S	0.17～0.22	0.15～0.35	0.45～0.65	≦0.040
	4337	S	0.35～0.40	0.15～0.35	0.60～0.80	≦0.040
	E4337	S	0.35～0.40	0.15～0.35	0.65～0.85	≦0.025

1 MPa = 1.01972 × 10^{-1} kgf/mm²

化 学 成 分 (%)					Tensile Test MPa or N/mm² 引 張 試 験		Remarks 備 考	Index No.
S	Ni	Cr	Mo	Others	Min. Yield Point 最小降伏点	Tensile Strength 引張強さ	(Similar to JIS)	索引番号
≦0.050	–	–	–	–	–	–	–	C267
≦0.050	–	–	–	–	–	–	–	
≦0.050	–	–	–	–	–	–	–	
≦0.050	–	–	–	–	–	–	–	
≦0.050	–	–	–	–	–	–	(STKM14)	
≦0.050	–	–	–	–	–	–	(STKM15)	
≦0.050	–	–	–	–	–	–	–	
≦0.050	–	–	–	–	–	–	(STKM16)	
≦0.050	–	–	–	–	–	–	–	
≦0.050	–	–	–	–	–	–	(STKM17)	
≦0.050	–	–	–	–	–	–	(STKM18)	
≦0.050	–	–	–	–	–	–	(STKM19)	
≦0.050	–	–	–	–	–	–	–	
≦0.050	–	–	–	–	–	–	–	
≦0.040	–	–	–	–	–	–	–	
≦0.040	–	–	–	–	–	–	–	
≦0.040	–	–	–	–	–	–	–	
≦0.040	1.10～1.40	0.55～0.75	–	–	–	–	–	
≦0.025	3.25～3.75	1.40～1.75	–	–	–	–	–	
≦0.040	–	–	0.15～0.25	–	–	–	–	
0.040~0.035	–	–	0.20～0.30	–	–	–	–	
≦0.050	–	–	0.20～0.30	–	–	–	–	
≦0.040	–	–	0.20～0.30	–	–	–	–	
0.035~0.050	–	–	0.20～0.30	–	–	–	–	
≦0.040	–	–	0.20～0.30	–	–	–	–	
≦0.040	–	–	0.20～0.30	–	–	–	–	
≦0.040	–	–	0.20～0.30	–	–	–	–	
≦0.040	–	–	0.20～0.30	–	–	–	–	
≦0.040	–	0.40～0.60	0.08～0.15	–	–	–	–	
≦0.040	–	0.80～1.10	0.15～0.25	–	–	–	(SCM430TK)	
≦0.040	–	0.80～1.10	0.15～0.25	–	–	–	(SCM435TK)	
≦0.040	–	0.80～1.10	0.15～0.25	–	–	–	–	
≦0.040	–	0.80～1.10	0.15～0.25	–	–	–	(SCM440TK)	
≦0.040	–	0.80～1.10	0.15～0.25	–	–	–	–	
≦0.040	–	0.80～1.10	0.15～0.25	–	–	–	–	
≦0.040	–	0.80～1.10	0.15～0.25	–	–	–	–	
≦0.040	–	0.80～1.10	0.15～0.25	–	–	–	–	
≦0.040	1.65～2.00	0.40～0.60	0.20～0.30	–	–	–	–	
≦0.040	1.65～2.00	0.70～0.90	0.20～0.30	–	–	–	–	
≦0.025	1.65～2.00	0.70～0.90	0.20～0.30	–	–	–	–	

Standard No. Year Designation 規格番号 年号 規格名称	Grade グレード	Mfg. Process 製造方法	Chemical Composition			
			C	Si	Mn	P
A519−94 (Continued)	4340	S	0.38～0.43	0.15～0.35	0.60～0.80	≦0.040
	E4340	S	0.38～0.43	0.15～0.35	0.65～0.85	≦0.025
	4422	S	0.20～0.25	0.15～0.35	0.70～0.90	≦0.040
	4427	S	0.24～0.29	0.15～0.35	0.70～0.90	≦0.040
	4520	S	0.18～0.23	0.15～0.35	0.45～0.65	≦0.040
	4615	S	0.13～0.18	0.15～0.35	0.45～0.65	≦0.040
	4617	S	0.15～0.20	0.15～0.35	0.45～0.65	≦0.040
	4620	S	0.17～0.22	0.15～0.35	0.45～0.65	≦0.040
	4621	S	0.18～0.23	0.15～0.35	0.70～0.90	≦0.040
	4718	S	0.16～0.21	0.15～0.35	0.70～0.90	≦0.040
	4720	S	0.17～0.22	0.15～0.35	0.50～0.70	≦0.040
	4815	S	0.13～0.18	0.15～0.35	0.40～0.60	≦0.040
	4817	S	0.15～0.20	0.15～0.35	0.40～0.60	≦0.040
	4820	S	0.18～0.23	0.15～0.35	0.50～0.70	≦0.040
	5015	S	0.12～0.17	0.15～0.35	0.30～0.50	≦0.040
	5046	S	0.43～0.50	0.15～0.35	0.75～1.00	≦0.040
	5115	S	0.13～0.18	0.15～0.35	0.70～0.90	≦0.040
	5120	S	0.17～0.22	0.15～0.35	0.70～0.90	≦0.040
	5130	S	0.28～0.33	0.15～0.35	0.70～0.90	≦0.040
	5132	S	0.30～0.35	0.15～0.35	0.60～0.80	≦0.040
	5135	S	0.33～0.38	0.15～0.35	0.60～0.80	≦0.040
	5140	S	0.38～0.43	0.15～0.35	0.70～0.90	≦0.040
	5145	S	0.43～0.48	0.15～0.35	0.70～0.90	≦0.040
	5147	S	0.45～0.52	0.15～0.35	0.70～0.95	≦0.040
	5150	S	0.48～0.53	0.15～0.35	0.70～0.90	≦0.040
	5155	S	0.50～0.60	0.15～0.35	0.70～0.90	≦0.040
	5160	S	0.55～0.65	0.15～0.35	0.75～1.00	≦0.040
	E50100	S	0.95～1.10	0.15～0.35	0.25～0.45	≦0.025
	E51100	S	0.95～1.10	0.15～0.35	0.25～0.45	≦0.025
	E52100	S	0.95～1.10	0.15～0.35	0.25～0.45	≦0.025
	6118	S	0.16～0.21	0.15～0.35	0.50～0.70	≦0.040
	6120	S	0.17～0.22	0.15～0.35	0.70～0.90	≦0.040
	6150	S	0.48～0.53	0.15～0.35	0.70～0.90	≦0.040
	E7140	S	0.38～0.43	0.15～0.35	0.50～0.70	≦0.025
	8115	S	0.13～0.18	0.15～0.35	0.70～0.90	≦0.040
	8615	S	0.13～0.18	0.15～0.35	0.70～0.90	≦0.040
	8617	S	0.15～0.20	0.15～0.35	0.70～0.90	≦0.040
	8620	S	0.18～0.23	0.15～0.35	0.70～0.90	≦0.040
	8622	S	0.20～0.25	0.15～0.35	0.70～0.90	≦0.040

1 MPa = 1.01972×10⁻¹kgf/mm²

化　学　成　分　(%)					Tensile Test MPa or N/mm² 引張試験		Remarks 備　考	Index No.
S	Ni	Cr	Mo	Others	Min. Yield Point 最小降伏点	Tensile Strength 引張強さ	(Similar to JIS)	索引番号
≦0.040	1.65～2.00	0.70～0.90	0.20～0.30	－	－	－	－	C267
≦0.025	1.65～2.00	0.70～0.90	0.20～0.30	－	－	－	－	
≦0.040	－	－	0.25～0.45	－	－	－	－	
≦0.040	－	－	0.35～0.45	－	－	－	－	
≦0.040	－	－	0.45～0.60	－	－	－	－	
≦0.040	1.65～2.00	－	0.20～0.30	－	－	－	－	
≦0.040	1.65～2.00	－	0.20～0.30	－	－	－	－	
≦0.040	1.65～2.00	－	0.20～0.30	－	－	－	－	
≦0.040	1.65～2.00	－	0.20～0.30	－	－	－	－	
≦0.040	0.90～1.20	0.35～0.55	0.34～0.40	－	－	－	－	
≦0.040	0.90～1.20	0.35～0.55	0.15～0.25	－	－	－	－	
≦0.040	3.25～3.75	－	0.20～0.30	－	－	－	－	
≦0.040	3.25～3.75	－	0.20～0.30	－	－	－	－	
≦0.040	3.25～3.75	－	0.20～0.30	－	－	－	－	
≦0.040	－	0.30～0.50	－	－	－	－	－	
≦0.040	－	0.20～0.30	－	－	－	－	－	
≦0.040	－	0.70～0.90	－	－	－	－	－	
≦0.040	－	0.70～0.90	－	－	－	－	(SCr420TK)	
≦0.040	－	0.80～1.10	－	－	－	－	－	
≦0.040	－	0.75～1.00	－	－	－	－	－	
≦0.040	－	0.80～1.05	－	－	－	－	－	
≦0.040	－	0.70～0.90	－	－	－	－	－	
≦0.040	－	0.70～0.90	－	－	－	－	－	
≦0.040	－	0.85～1.15	－	－	－	－	－	
≦0.040	－	0.70～0.90	－	－	－	－	－	
≦0.040	－	0.70～0.90	－	－	－	－	－	
≦0.040	－	0.70～0.90	－	－	－	－	－	
≦0.025	－	0.40～0.60	－	－	－	－	－	
≦0.025	－	0.90～1.15	－	－	－	－	－	
≦0.025	－	1.30～1.60	－	－	－	－	－	
≦0.040	－	0.50～0.70	－	V:0.10～1.15	－	－	－	
≦0.040	－	0.70～0.90	－	V≧0.10	－	－	－	
≦0.040	－	0.80～1.10	－	V≧0.15	－	－	－	
≦0.025	－	1.40～1.80	－	V :0.30 ～0.40 Al:0.95 ～1.30	－	－	－	
≦0.040	0.20～0.40	0.30～0.50	0.08～0.15	－	－	－	－	
≦0.040	0.40～0.70	0.40～0.60	0.15～0.25	－	－	－	－	
≦0.040	0.40～0.70	0.40～0.60	0.15～0.25	－	－	－	－	
≦0.040	0.40～0.70	0.40～0.60	0.15～0.25	－	－	－	－	
≦0.040	0.40～0.70	0.40～0.60	0.15～0.25	－	－	－	－	

Standard No. 規格番号	Year 年号	Designation 規格名称	Grade グレード	Mfg. Process 製造方法	Chemical Composition			
					C	Si	Mn	P
A519－94 (Continued)			8625	S	0.23～0.28	0.15～0.35	0.70～0.90	≦0.040
			8627	S	0.25～0.30	0.15～0.35	0.70～0.90	≦0.040
			8630	S	0.28～0.33	0.15～0.35	0.70～0.90	≦0.040
			8637	S	0.35～0.40	0.15～0.35	0.75～1.00	≦0.040
			8640	S	0.38～0.43	0.15～0.35	0.75～1.00	≦0.040
			8642	S	0.40～0.45	0.15～0.35	0.75～1.00	≦0.040
			8645	S	0.43～0.48	0.15～0.35	0.75～1.00	≦0.040
			8650	S	0.48～0.53	0.15～0.35	0.75～1.00	≦0.040
			8655	S	0.50～0.60	0.15～0.35	0.75～1.00	≦0.040
			8660	S	0.55～0.65	0.15～0.35	0.75～1.00	≦0.040
			8720	S	0.18～0.23	0.15～0.35	0.70～0.90	≦0.040
			8735	S	0.33～0.38	0.15～0.35	0.75～1.00	≦0.040
			8740	S	0.38～0.43	0.15～0.35	0.75～1.00	≦0.040
			8742	S	0.40～0.45	0.15～0.35	0.75～1.00	≦0.040
			8822	S	0.20～0.25	0.15～0.35	0.75～1.00	≦0.040
			9255	S	0.50～0.60	1.80～2.20	0.70～0.95	≦0.040
			9260	S	0.55～0.65	1.80～2.20	0.70～1.00	≦0.040
			9262	S	0.55～0.65	1.80～2.20	0.75～1.00	≦0.040
			E9310	S	0.80～0.13	0.18～0.35	0.45～0.65	≦0.025
			9840	S	0.38～0.43	0.18～0.35	0.70～0.90	≦0.040
			9850	S	0.48～0.53	0.18～0.35	0.70～0.90	≦0.040
			50B40	S	0.38～0.42	0.15～0.35	0.75～1.00	≦0.040
			50B44	S	0.43～0.48	0.15～0.35	0.75～1.00	≦0.040
			50B46	S	0.43～0.50	0.15～0.35	0.75～1.00	≦0.040
			50B50	S	0.48～0.53	0.15～0.35	0.74～1.00	≦0.040
			50B60	S	0.55～0.65	0.15～0.35	0.75～1.00	≦0.040
			51B60	S	0.55～0.65	0.15～0.35	0.75～1.00	≦0.040
			81B45	S	0.43～0.48	0.15～0.35	0.75～1.00	≦0.040
			86B45	S	0.43～0.48	0.15～0.35	0.75～1.00	≦0.040
			94B15	S	0.13～0.18	0.15～0.35	0.75～1.00	≦0.040
			94B17	S	0.15～0.20	0.15～0.35	0.75～1.00	≦0.040
			94B30	S	0.28～0.33	0.15～0.35	0.75～1.00	≦0.040
			94B40	S	0.38～0.43	0.15～0.35	0.75～1.00	≦0.040
			1118	S	0.14～0.20	－	1.30～1.60	≦0.040
			11L18	S	0.14～0.20	－	1.30～1.60	≦0.040
			1132	S	0.27～0.34	－	1.35～1.65	≦0.040
			1137	S	0.32～0.39	－	1.35～1.65	≦0.040
			1141	S	0.37～0.45	－	1.35～1.65	≦0.040
			1144	S	0.40～0.48	－	1.35～1.65	≦0.040
			1213	S	≦0.13	－	0.70～1.10	0.07～0.12

1 M P a = 1. 01972 × 10^{-1} kgf/mm²

化 学 成 分 (%)					Tensile Test MPa or N/mm² 引 張 試 験		Remarks 備 考	Index No.
S	Ni	Cr	Mo	Others	Min. Yield Point 最小降伏点	Tensile Strength 引張強さ	(Similar to JIS)	索引番号
≦0. 040	0. 40~0. 70	0. 40~0. 60	0. 15~0. 25	−	−	−	−	C267
≦0. 040	0. 40~0. 70	0. 40~0. 60	0. 15~0. 25	−	−	−	−	
≦0. 040	0. 40~0. 70	0. 40~0. 60	0. 15~0. 25	−	−	−	−	
≦0. 040	0. 40~0. 70	0. 40~0. 60	0. 10~0. 25	−	−	−	−	
≦0. 040	0. 40~0. 70	0. 40~0. 60	0. 15~0. 25	−	−	−	−	
≦0. 040	0. 40~0. 70	0. 40~0. 60	0. 15~0. 25	−	−	−	−	
≦0. 040	0. 40~0. 70	0. 40~0. 60	0. 15~0. 25	−	−	−	−	
≦0. 040	0. 40~0. 70	0. 40~0. 60	0. 15~0. 25	−	−	−	−	
≦0. 040	0. 40~0. 70	0. 40~0. 60	0. 15~0. 25	−	−	−	−	
≦0. 040	0. 40~0. 70	0. 40~0. 60	0. 15~0. 25	−	−	−	−	
≦0. 040	0. 40~0. 70	0. 40~0. 60	0. 20~0. 30	−	−	−	−	
≦0. 040	0. 40~0. 70	0. 40~0. 60	0. 20~0. 30	−	−	−	−	
≦0. 040	0. 40~0. 70	0. 40~0. 60	0. 20~0. 30	−	−	−	−	
≦0. 040	0. 40~0. 70	0. 40~0. 60	0. 20~0. 30	−	−	−	−	
≦0. 040	0. 40~0. 70	0. 40~0. 60	0. 20~0. 40	−	−	−	−	
≦0. 040	−	−	−	−	−	−	−	
≦0. 040	−	−	−	−	−	−	−	
≦0. 040	−	0. 25~0. 40	−	−	−	−	−	
≦0. 025	3. 00~3. 50	1. 00~1. 40	0. 08~0. 15	−	−	−	−	
≦0. 040	0. 85~1. 15	0. 70~0. 90	0. 20~0. 30	−	−	−	−	
≦0. 040	0. 85~1. 15	0. 70~0. 90	0. 20~0. 30	−	−	−	−	
≦0. 040	−	0. 40~0. 60	−	−	−	−	−	
≦0. 040	−	0. 40~0. 60	−	−	−	−	−	
≦0. 040	−	0. 20~0. 30	−	−	−	−	−	
≦0. 040	−	0. 40~0. 60	−	−	−	−	−	
≦0. 040	−	0. 40~0. 60	−	−	−	−	−	
≦0. 040	−	0. 70~0. 90	−	−	−	−	−	
≦0. 040	0. 20~0. 40	0. 35~0. 55	0. 08~0. 15	−	−	−	−	
≦0. 040	0. 40~0. 70	0. 40~0. 60	0. 15~0. 25	−	−	−	−	
≦0. 040	0. 30~0. 60	0. 30~0. 50	0. 08~0. 15	−	−	−	−	
≦0. 040	0. 30~0. 60	0. 30~0. 50	0. 08~0. 15	−	−	−	−	
≦0. 040	0. 30~0. 60	0. 30~0. 50	0. 08~0. 15	−	−	−	−	
≦0. 040	0. 30~0. 60	0. 30~0. 50	0. 08~0. 15	−	−	−	−	
0. 08~0. 13	−	−	−	−	−	−	−	
0. 08~0. 13	−	−	−	Pb:0. 15~0. 35	−	−	−	
0. 08~0. 18	−	−	−	−	−	−	−	
0. 08~0. 13	−	−	−	−	−	−	−	
0. 08~0. 13	−	−	−	−	−	−	−	
0. 24~0. 33	−	−	−	−	−	−	−	
0. 24~0. 33	−	−	−	−	−	−	−	

Standard No. Year Designation 規格番号 年号 規格名称	Grade グレード	Mfg. Process 製造方法	Chemical Composition			
			C	Si	Mn	P
A519−94 (Continued)	12L14	S	≦0.15	−	0.85〜1.15	0.04〜0.09
	1215	S	≦0.09	−	0.75〜1.05	0.04〜0.09
A554−94 Welded Stainless Steel Mechanical Tubing 機械構造用溶接ステンレス鋼鋼管	MT301	E, AT	≦0.15	≦1.00	≦2.00	≦0.040
	MT302	E, AT	≦0.15	≦1.00	≦2.00	≦0.040
	MT304	E, AT	≦0.08	≦1.00	≦2.00	≦0.040
	MT304L	E, AT	≦0.035	≦1.00	≦2.00	≦0.040
	MT305	E, AT	≦0.12	≦1.00	≦2.00	≦0.040
	MT309S	E, AT	≦0.08	≦1.00	≦2.00	≦0.040
	MT309S-Cb	E, AT	≦0.08	≦1.00	≦2.00	≦0.040
	MT310S	E, AT	≦0.08	≦1.00	≦2.00	≦0.040
	MT316	E, AT	≦0.08	≦1.00	≦2.00	≦0.040
	MT316L	E, AT	≦0.035	≦1.00	≦2.00	≦0.040
	MT317	E, AT	≦0.08	≦1.00	≦2.00	≦0.040
	MT321	E, AT	≦0.08	≦1.00	≦2.00	≦0.040
	MT330	E, AT	≦0.15	≦1.00	≦2.00	≦0.040
	MT347	E, AT	≦0.08	≦1.00	≦2.00	≦0.040
	MT429	E, AT	≦0.12	≦1.00	≦1.00	≦0.040
	MT430	E, AT	≦0.12	≦1.00	≦1.00	≦0.040
	MT430Ti	E, AT	≦0.10	≦1.00	≦1.00	≦0.040
A595−93 Steel Tubes, Low-Carbon, Tapered for Structural Use 構造用低炭素鋼鋼テーパ鋼管	Gr. A	W	0.15〜0.25	≦0.04	0.30〜0.90	≦0.035
	Gr. B	W	0.15〜0.25	≦0.04	0.40〜1.35	≦0.035
	Gr. C	W	≦0.12	0.25〜0.75	0.20〜0.15	0.07〜0.15
A618−93 Hot-Formed Welded and Seamless High-Strength Low-Alloy Structural Tubing 低合金構造用熱間成型溶接および継目無高張力鋼管	Gr. Ia	S, E, B	≦0.15	−	≦1.00	≦0.15
	Gr. Ib	S, E, B	≦0.20	−	≦13.5	≦0.025
	Gr. Ⅱ	S, E, B	≦0.22	≦0.30	0.85〜1.25	≦0.025
	Gr. Ⅲ	S, E, B	≦0.23	≦0.30	≦1.35	≦0.025
A847−93 Cold-Formed Welded and Seamless High Strength, Low Alloy Structural Tubing with Improved Atmospheric Corrosion Resistance 大気耐蝕性冷間加工溶接継目無高強度低合金構造用鋼鋼管	−	Wor S	≦0.40	−	≦1.35	≦0.015

$1 MPa = 1.01972 \times 10^{-1} kgf/mm^2$

化学成分（%）					Tensile Test MPa or N/mm^2 引張試験		Remarks 備考	Index No.
S	Ni	Cr	Mo	Others	Min. Yield Point 最小降伏点	Tensile Strength 引張強さ	(Similar to JIS)	索引番号
0.26～0.35	－	－	－	Pb:0.15～0.35	－	－	－	C267
0.26～0.35	－	－	－	－	－	－	－	
≦0.030	6.0～8.0	16.0～18.0	－	－	207	≧517	－	C268
≦0.030	8.0～10.0	17.0～19.0	－	－	207	≧517	－	
≦0.030	8.0～11.0	18.0～20.0	－	－	207	≧517	(SUS304TK)	
≦0.030	8.0～13.0	18.0～20.0	－	－	172	≧483	－	
≦0.030	10.0～13.0	17.0～19.0	－	－	207	≧517	－	
≦0.030	12.0～15.0	22.0～24.0	－	－	207	≧517	Austenitie	
≦0.030	12.0～15.0	22.0～24.0	－	Nb+Ta:10×C～1.00	207	≧517	－	
≦0.030	19.0～22.0	24.0～26.0	－	－	207	≧517	－	
≦0.030	11.0～14.0	16.0～18.0	2.0～3.0	－	207	≧517	(SUS316TK)	
≦0.030	10.0～15.0	16.0～18.0	2.0～3.0	－	172	≧483	－	
≦0.030	11.0～14.0	18.0～20.0	2.0～3.0	－	207	≧517	－	
≦0.030	9.0～13.0	17.0～20.0	－	Ti:5×C～0.60	207	≧517	(SUS321TK)	
≦0.030	33.0～36.0	14.0～16.0	－	－	207	≧517	－	
≦0.030	9.0～13.0	17.0～20.0	－	Nb+Ta:10×C～1.00	207	≧517	(SUS347TK)	
≦0.030	≦0.50	14.0～16.0	－	－	241	≧414	－	
≦0.030	≦0.50	16.0～18.0	－	－	241	≧414	(SUS430TK)	
≦0.030	≦0.075	16.0～19.5	－	Ti:5×C～0.75	207	≧414	Ferritie	
≦0.035	－	－	－	－	380	≧450	－	C269
≦0.035	－	－	－	－	410	≧480	－	
≦0.025	≦0.65	0.30～1.25	－	Cu:0.25～0.55 Cr:0.30～1.25 Ni:≦0.65	410	≧480	－	
≦0.025	－	－	－	Cu≧0.20	t≦19.05	t≦19.05	－	
≦0.025	－	－	－	Cu≧0.20	345	≧485	－	
≦0.025	－	－	－	Cu≧0.20 V ≧0.02	t>19.05 315	t>19.05 ≧460	－	
≦0.025	－	－	－	V ≧0.02	345	≧450	－	
≦0.05	－	－	－	Cu≧0.20	345	≧483	－	C270

10. ASTM Other Steel Pipes　その他の鋼管

Standard No. Year Designation 規格番号 年号 規格名称	Grade グレード	Mfg. Process 製造方法	Chemical Composition C	Si	Mn	P
A452−88 Centrifugally Cast Austenitic Steel Cold-Wrought Pipe for High-Temperature Service 高温用鋳造低温鍛造オーステナイト鋼鋼管	Gr. TP304H	E	0.04〜0.10	≦0.75	≦2.00	≦0.040
	Gr. TP347H	E	0.04〜0.10	≦0.75	≦2.00	≦0.040
	Gr. TP316H	E	0.04〜0.10	≦0.75	≦2.00	≦0.040
A523−93 Plain End Seamless and Electric-Resistance-Welded Steel Pipe for High-Pressure Pipe-Type Cable Circuits 高圧パイプ型ケーブル電線管用継目無および電気抵抗溶接鋼管	Gr. A	S	≦0.22	−	≦0.90	≦0.040
		E	≦0.21			
	Gr. B	S	≦0.27	−	≦1.15	≦0.040
		E	≦0.26			
A589−93 Seamless and Welded Carbon Steel Water Well Pipe 継目無および溶接水井戸用炭素鋼管	Butt W	B	−	−	−	≦0.050
	Gr. A	S, E	−	−	−	≦0.050
	Gr. B	S, E	−	−	−	≦0.050
A771−88 Austenitic Stainless Steel Tubing for Breeder Reactor Core Components 増殖炉炉心用オーステナイト系ステンレス鋼鋼管	Type 316	S	0.040〜0.060	0.50〜0.75	1.50〜2.00	≦0.020
	S 38660	S	0.030〜0.050	0.50〜1.00	1.65〜2.35	≦0.040
A608−91a Centrifugally Cast Iron-Chromium-Nickel High-Alloy Tubing for Pressure Application at High Temperature						
	HC30	E	0.25〜0.35	0.50〜2.00	0.5〜1.0	≦0.04
	HD50	E	0.45〜0.55	0.50〜2.00	≦1.50	≦0.04

$1 MPa = 1.01972 \times 10^{-1} kgf/mm^2$

化　学　成　分　（%）					Tensile Test MPa or N/mm² 引張試験		Remarks 備考	Index No.
S	Ni	Cr	Mo	Others	Min. Yield Point 最小降伏点	Tensile Strength 引張強さ	(Similar to JIS)	索引番号
≦0.030	8.00～11.0	18.0～20.0	–	–	207	≧517	–	C271
≦0.030	9.00～13.0	17.0～20.0	–	Cb＋Ta8C～1.00	207	≧517	–	
≦0.030	11.0～14.0	16.0～18.0	2.00～3.00	–	207	≧517	–	
≦0.050	–	–	–	–	205	≧330	–	C272
≦0.050	–	–	–	–	240	≧415	–	
≦0.060	–	–	–	–	172	≧310	–	C273
≦0.060	–	–	–	–	207	≧331	–	
≦0.060	–	–	–	–	241	≧413	–	
≦0.010	13.0～14.0	17.0～18.0	2.00～3.00	Nb≦0.050 Ta≦0.020 N ≦0.010 Al≦0.050 As≦0.030 B ≦0.0010 Co≦0.050 Cu≦0.04 V ≦0.02	Annealed 205~345(R·T) 1.05~205 (540℃) 20%Cold Worked: 550~760(R·T) 415~585 (540℃)	Annealed 515~690(R·T) 380~480 (540℃) 760~860(R·T) 515~690 (540℃)	–	C274
≦0.010	14.5～16.5	12.5～14.5	1.50～2.50	Ti0.10～0.40 Nb≦0.050 Ta≦0.020 N ≦0.005 Al≦0.050 As≦0.030 B ≦0.020 Co≦0.050 Cu≦0.04 V ≦0.05	20% Cold Worked 515~760(R·T) 415~585 (540℃)	20% Cold Worked 655~830(R·T) 480~690 (540℃)		

S	Ni	Cr	Mo	Others	Tensile Test(MPa or N/mm²) 引張試験 760℃	871℃	982℃	1093℃	Index No.
≦0.04	≦4.0	26～30	≦0.50	–	≧36.5	≧20.4	≧11.0	–	C275
≦0.04	4～ 7	26～30	≦0.50	–	≧51.4	≧17.7	≧6.27	–	

Standard No. 規格番号	Year 年号	Designation 規格名称	Grade グレード	Mfg. Process 製造方法	Chemical Composition C	Si	Mn	P
高温高圧用鋳鉄ニッケルクロム鋼合金 A608 – 91a (Continued)			HE35	E	0.30～0.40	0.50～2.00	≦1.50	≦0.04
			HF30	E	0.25～0.30	0.50～2.00	≦1.50	≦0.04
			HH30	E	0.25～0.35	0.50～2.00	≦1.50	≦0.04
			HH33	E	0.28～0.38	0.50～2.00	≦1.50	≦0.04
			HI35	E	0.30～0.40	0.50～2.00	≦1.50	≦0.04
			HK30	E	0.25～0.30	0.50～2.00	≦1.50	≦0.04
			HK40	E	0.35～0.45	0.50～2.00	≦1.50	≦0.04
			HL30	E	0.25～0.35	0.50～2.00	≦1.50	≦0.04
			HL40	E	0.35～0.45	0.50～2.00	≦1.50	≦0.04
			HN40	E	0.35～0.45	0.50～2.00	≦1.50	≦0.04
			HT50	E	0.40～0.60	0.50～2.00	≦1.50	≦0.04
			HU50	E	0.40～0.60	0.50～2.00	≦1.50	≦0.04
			HW50	E	0.40～0.60	0.50～2.00	≦1.50	≦0.04
			HX50	E	0.40～0.60	0.50～2.00	≦1.50	≦0.04
A660 – 91a Centrifugally Cast Carbon Steel Pipe for High-Temperature Service 高温用鋳造炭素鋼鋼管								
			WCA	E	≦0.25	≦0.60	≦0.70	≦0.035
			WCB	E	≦0.30	≦0.60	≦1.00	≦0.035
			WCC	E	≦0.25	≦0.60	≦1.20	≦0.035
A789 – 94 Seamless and Welded Ferritic/Austenitic Stainless Steel Pipe for General Service 汎用継目無溶接フェライトオーステナイト系ステンレス鋼鋼管			S31803	E, S, W	≦0.030	≦1.0	≦2.0	≦0.030
			S31500	E, S, W	≦0.030	1.40～2.0	1.20～2.00	≦0.030
			S32500	E, S, W	≦0.040	≦1.5	≦1.0	≦0.040
			S31200	E, S, W	≦0.030	≦2.0	≦1.0	≦0.045
			S31260	E, S, W	≦0.030	≦1.00	≦0.75	≦0.030
			S32304	E, S, W	≦0.030	≦2.50	≦1.0	≦0.040
			S32740	E, S, W	≦0.030	≦0.80	≦1.0	≦0.030
			S32750	E, S, W	≦0.030	≦1.2	≦0.8	≦0.035
			S32760	E, S, W	≦0.05	≦1.00	≦1.00	≦0.030
			S32900	E, S, W	≦0.08	≦1.00	≦0.75	≦0.040
			S32950	E, S, W	≦0.03	≦2.00	≦0.60	≦0.030
A790M – 94 Seamless and Welded Ferritic/Austenitic Stainless Steel Pipe 継目無し溶接ﾌｪﾗｲﾄ/ｵｰｽﾃﾅｲﾄ系ｽﾃﾝﾚｽ鋼鋼管			S31803	E, S, W	≦0.030	≦1.0	≦2.0	≦0.030
			S31500	E, S, W	≦0.030	1.40～2.10	1.200~2.00	≦0.030

1 M P a = 1.01972 × 10^{-1} kgf/mm²

化　学　成　分　（%）					Tensile Test MPa or N/mm² 引　張　試　験				Index No. 索引番号
S	Ni	Cr	Mo	Others	760℃	871℃	982℃	1093℃	
≦0.04	8～11	26～30	≦0.50	–	–	–	–	–	C 275
≦0.04	9～12	19～23	≦0.50	–	≧179	≧99.9	–	–	
≦0.04	11～14	24～28	≦0.50	–	–	≧52.7	≧27.2	–	
≦0.04	12～14	24～26	≦0.50	–	–	≧138	≧56.5	≧27.6	
≦0.04	14～18	26～30	≦0.50	–	–	≧138	≧56.5	–	
≦0.04	19～22	23～47	≦0.50	–	≧179	≧96.5	≧51.7	≧24.8	
≦0.04	19～22	23～27	≦0.50	–	≧200	≧114	≧60.7	≧28.9	
≦0.04	18～22	28～32	≦0.50	–	–	–	–	–	
≦0.04	18～22	28～32	≦0.50	–	–	–	–	–	
≦0.04	23～27	19～23	≦0.50	–	–	–	–	–	
≦0.04	33～37	15～19	≦0.50	–	–	–	–	–	
≦0.04	37～41	17～21	≦0.50	–	–	–	–	–	
≦0.04	58～62	10～14	≦0.50	–	–	–	–	–	
≦0.04	64～68	15～19	≦0.50	–	–	–	–	–	
					Tensile Test (MPa or N/mm²)		Remarks 備　考		C 276
≦0.035	–	–	–	–	207	≧414	for High Temperature Service 高温用		
≦0.035	–	–	–	–	248	≧483			
≦0.035	–	–	–	–	276	≧483			
≦0.020	4.50～6.50	21.0～23.0	2.50～3.50	N 0.08～0.20	450	≧620	for High Temperature Service 高温用		C 277
≦0.030	4.25～5.25	18.0～19.0	2.50～3.50	N 0.05～0.1	440	≧630			
≦0.040	4.50～6.20	24.0～27.0	2.90～3.90	N 0.10～0.25 Cu 1.5 ～2.5	550	≧760			
≦0.030	5.50～6.50	24.0～26.0	1.20～2.00	0.14～0.20	450	≧690			
≦0.030	5.50～7.50	24.0～26.0	2.50～3.50	N 0.10～0.30 Cu 0.20～0.80 W 0.10～0.50	450	≧690			
≦0.040	3.0～5.5	21.5～24.5	0.05～0.06	N 0.24～0.32 Cu≦0.5	550	≧800			
≦0.020	6.0～8.0	24.0～26.0	2.50～3.50	N 0.24～0.32 Cu 0.20～0.80 W 1.50～2.50	550	≧800			
≦0.020	6.0～8.0	24.0～26.0	3.0～5.0	N 0.24～0.32 Cu≦0.5	550	≧880			
≦0.010	6.00～8.00	24.00～26.00	3.00～4.00	N 0.20～0.30 Cu 0.50～1.00 W 0.50～1.00	550	750～895			
≦0.030	2.50～5.00	23.00～28.00	1.00～2.00	–	485	≧620			
≦0.010	3.50～5.20	26.00～29.00	1.00～2.50	N 0.15～0.35	480	≧690			
≦0.020	4.50～6.50	21.0～23.0	2.50～3.50	N 0.08～0.20	450	≧620	for High Temperature Service 高温用		C 278
≦0.030	4.25～5.25	18.0～19.0	2.50～3.50	N 0.05～0.1	440	≧630			

Standard No. Year Designation 規格番号 年号 規格名称	Grade/ UNS - No. グレード	Mfg. Process 製造方法	Chemical Composition			
			C	Si	Mn	P
A790－94 (Continued)	S32500	E, S, W	≦0.040	≦1.5	≦1.0	≦0.040
	S31200	E, S, W	≦0.030	≦2.0	≦1.0	≦0.045
	S31260	E, S, W	≦0.030	≦1.0	≦0.75	≦0.030
	S32304	E, S, W	≦0.030	≦2.50	≦1.0	≦0.040
	S32740	E, S, W	≦0.030	≦0.80	≦1.0	≦0.030
	S32750	E, S, W	≦0.030	≦1.2	≦0.8	≦0.035
	S32760	E, S, W	≦0.05	≦1.00	≦1.00	≦0.030
	S32900	E, S, W	≦0.08	≦1.00	≦0.75	≦0.040
	S32950	E, S, W	≦0.03	≦2.00	≦0.60	≦0.030
A791－94 Welded Unannealed Ferritic Stainless Steel Tubing 溶接非焼なましフェライト系ステンレス鋼鋼管	TP409/ S40900	E, W	≦0.08	≦1.00	≦1.0	≦0.045
	TP439/ S40335	E, W	≦0.07	≦1.00	≦1.00	≦0.040
	TPXM-27/ S44627	E, W	≦0.01	≦0.40	≦0.40	≦0.02
	TPXM-33/ S44626	E, W	≦0.06	≦0.75	≦0.75	≦0.040
	25-4-4/ S44635	E, W	≦0.025	≦0.75	≦1.00	≦0.040
	26-3-3/ S44660	E, W	≦0.030	≦1.00	≦1.00	≦0.040
	29-4/ S44700	E, W	≦0.010	≦0.20	≦0.30	≦0.025
	29-4-2/ S44800	E, W	≦0.010	≦0.20	≦0.30	≦0.025
	18-2/ S44400	E, W	≦0.030	≦1.00	≦1.00	≦0.040
	29-4C/ S44735	E, W	≦1.00	≦1.00	≦1.00	≦0.040

1 MPa = 1.01972 × 10⁻¹ kgf/mm²

化　学　成　分　(%)					Tensile Test MPa or N/mm² 引張試験		Remarks 備考 (Similar to JIS)	Index No. 索引番号
S	Ni	Cr	Mo	Others	Min. Yield Point 最小降伏点	Tensile Strength 引張強さ		
≦0.040	4.50～6.20	24.0～27.0	2.90～3.90	N 0.10～0.25 Cu 1.5 ～2.5	550	≧760		C278
≦0.030	5.50～6.50	24.0～26.0	1.20～2.00	0.14～0.20	450	≧690		
≦0.030	5.50～6.50	24.0～26.0	2.50～3.50	N 0.10～0.30 Cu 0.20～0.30 W 0.10～0.50	450	≧690		
≦0.040	3.0～5.5	21.5～24.5	0.05～0.06	N 0.24～0.32 Cu ≦0.5	400	≧600		
≦0.020	6.0～8.0	24.0～26.0	2.50～3.50	N 0.24～0.32 Cu 0.20～0.80 W 1.50～2.50	550	≧800		
≦0.020	6.0～8.0	24.0～26.0	3.0～5.0	N 0.24～0.32 Cu ≦0.5	550	≧800		
≦0.010	6.00～8.00	24.00～26.00	3.00～4.00	N 0.20～0.30 Cu 0.50～1.00 W 0.50～1.00	550	750～895		
≦0.030	2.50～5.00	23.00～28.00	1.00～2.00	－	485	≧620		
≦0.010	3.50～5.20	26.00～29.00	1.00～2.50	N 0.15～0.35	480	≧690		
≦0.045	≦0.50	10.50～11.75	－	Ti 6C～0.75	205	≧380	for High Temperature Service 高温用	C279
≦0.030	≦0.50	17.00～19.00	N ≦0.04 Al ≦0.15	Ti 0.20＋ 4(C＋N)～1.10	205	≧415		
≦0.02	≦0.50	25.0～27.0	0.75～1.50	Cu ≦0.2 N ≦0.015	275	≧450		
≦0.020	≦0.50	25.0～27.0	Mo 0.75～1.50 Ti 0.20≦7(C+N) ≦1.00	Cu ≦0.20 N ≦0.040	310	≧415		
≦0.030	3.5～4.5	24.5～26.0	3.5～4.5	N ≦0.035 Ti＋Cb 0.2＋4(C+N)～0.80	515	≧620		
≦0.030	1.0～3.50	25.0～28.0	3.0～4.0	N ≦0.040 0.20≦Ti＋Cb ＝6(C＋N)≦1.00	450	≧585		
≦0.15	28.0～30.0	3.5～4.2	－	Cu ≦0.15 N ≦0.020	415	≧550		
≦0.20	2.0～2.5	28.0～30.0	3.5～4.2	Cu ≦0.15 N ≦0.020	415	≧550		
≦1.00	≦1.00	28.0～30.0	3.60～4.20	N ≦0.045 0.20≦(Ti＋Cb) ＝6(C＋N)≦1.00	240	≧415		
≦1.00	≦1.00	17.5～19.5	1.75～2.50	N ≦0.035 0.20≦(Ti＋Cb) ＝6(C＋N)≦1.00	415	≧515		

Standard No. Year Designation 規格番号 年号 規格名称	Grade/ UNS No. グレード	Mfg. Process 製造方法	Chemical Composition			
			C	Si	Mn	P
A813M−91 Single or Double-Welded Austenitic Stainless Steel Pipe 1回または2回溶接オーステナイト系ステンレス鋼鋼管	TP304/S30400	E, W	≦0.08	≦0.75	≦2.00	≦0.045
	TP304H/S30409	E, W	0.04〜0.10	≦0.75	≦2.00	≦0.045
	TP304L/S30403	E, W	≦0.035	≦0.75	≦2.00	≦0.045
	TP304N/S30451	E, W	≦0.08	≦0.75	≦2.00	≦0.045
	TP304LN/S30453	E, W	≦0.035	≦0.75	≦2.00	≦0.045
	TP309Cb/S30940	E, W	≦0.08	≦0.75	≦2.00	≦0.045
	TP309S/S30908	E, W	≦0.08	≦0.75	≦2.00	≦0.045
	TP310Cb/S31040	E, W	≦0.08	≦0.75	≦2.00	≦0.045
	TP310S/S31008	E, W	≦0.08	≦0.75	≦2.00	≦0.045
	TP316/S31600	E, W	≦0.08	≦0.75	≦2.00	≦0.045
	TP316H/S31609	E, W	0.04〜0.10	≦0.75	≦2.00	≦0.045
	TP316L/S31603	E, W	≦0.035	≦0.75	≦2.00	≦0.045
	TP316N/S31651	E, W	≦0.035	≦0.75	≦2.00	≦0.045
	TP316LN/S31653	E, W	≦0.035	≦0.75	≦2.00	≦0.045
	TP317/S31700	E, W	≦0.08	≦0.75	≦2.00	≦0.045
	TP317L/S31703	E, W	≦0.035	≦0.75	≦2.00	≦0.045
	TP321/S32100	E, W	≦0.08	≦0.75	≦2.00	≦0.045
	TP321H/S32109	E, W	0.04〜0.10	≦0.75	≦2.00	≦0.045
	TP347/S34700	E, W	≦0.08	≦0.75	≦2.00	≦0.045
	TP347H/S34709	E, W	0.04〜0.10	≦0.75	≦2.00	≦0.045
	TP348/S34800	E, W	≦0.08	≦0.75	≦2.00	≦0.045
	TP348H/S34809	E, W	0.04〜0.10	≦0.75	≦2.00	≦0.045
	TPXM10/S21900	E, W	≦0.08	≦1.00	8.00~10.00	≦0.040
	TPXM11/S21903	E, W	≦0.04	≦1.00	8.00~10.00	≦0.040

1 MPa = 1.01972 × 10⁻¹ kgf/mm²

化　学　成　分　(%)					Tensile Test MPa or N/mm² 引張試験		Remarks 備考	Index No.
S	Ni	Cr	Mo	Others	Min. Yield Point 最小降伏点	Tensile Strength 引張強さ	(Similar to JIS)	索引番号
≦0.030	8.00～16.0	18.0～20.0	−	−	170	≧485	for High Temperature and Corrosion Resistance Service 高温耐蝕用	C280
≦0.030	8.00～16.0	18.0～20.0	−	−	170	≧485		
≦0.030	8.00～13.0	18.0～20.0	−	−	205	≧515		
≦0.030	8.00～11.0	18.0～20.0	−	−	205	≧515		
≦0.030	8.00～11.0	18.0～20.0	−	Cb＋Ta 10C～1.10	205	≧515		
≦0.030	12.00~16.00	22.00~24.00	≦0.75	Cb＋Ta 10C～1.10 Cu≦0.75	205	≧515		
≦0.030	12.00~15.00	22.00~24.00	≦0.75	−	205	≧515		
≦0.030	19.00~22.00	24.00~26.00	≦0.75	Cb＋Ta 10C～1.10 Cu≦0.75	205	≧515		
≦0.030	19.00~22.00	24.00~26.00	≦0.75	−	205	≧515		
≦0.030	10.00~14.0	16.0～18.0	2.00～3.00	−	205	≧515		
≦0.030	10.0～14.0	16.0～18.0	2.00～3.00	−	205	≧515		
≦0.030	10.0～14.0	16.0～18.0	2.00～3.00	−	205	≧515		
≦0.030	10.0～14.0	16.0～18.0	2.00～3.00	N 0.10～0.16	205	≧515		
≦0.030	10.0～14.0	16.0～18.0	2.00～3.00	N 0.10～0.16	205	≧515		
≦0.030	11.00~14.0	18.0～20.0	3.00～4.00	−	205	≧515		
≦0.030	11.00~15.0	18.0～20.0	3.00～4.00	−	205	≧515		
≦0.030	9.00～13.0	17.0～20.0	−	Ti 5C～0.70	205	≧515		
≦0.030	9.00～13.0	17.0～20.0	−	Ti 4C～0.60	205	≧515		
≦0.035	9.00～13.0	17.0～20.0	−	Cb＋Ta 10C～1.00	205	≧515		
≦0.035	9.00～13.0	17.0～20.0	−	Cb＋Ta 8C～1.0	205	≧515		
≦0.035	9.00～13.0	17.0～20.0	−	Cb＋Ta 10C～1.00	205	≧515		
≦0.035	9.00～13.0	17.0～20.0	−	Ta≦0.10 Cb＋Ta 8C～1.00	205	≧515		
≦0.030	5.50～7.50	19.00~21.50	−	N 0.15～0.40	345	≧620		
≦0.030	5.50～7.50	19.00~21.50	−	N 0.15～0.40	345	≧620		

Standard No. Year Designation 規格番号 年号 規格名称	Grade UNS NO. グレード	Mfg. Process 製造方法	Chemical Composition			
			C	Si	Mn	P
A813M－91 (Continued)	TPXM15/S38100	E, W	≦0.08	1.50～2.00	≦2.00	≦0.040
	TPXM19/S20910	E, W	≦0.060	≦1.00	11.50~14.50	≦0.040
	TPXM29/S24000	E, W	≦0.080	≦1.00	11.50~14.50	≦0.060
	S31254	E, W	≦0.020	≦0.80	≦1.00	≦0.030
	S30815	E, W	≦0.10	≦0.80	1.40～2.00	≦0.040
A814－91 Cold-Worked Welded Austenitic Stainless Steel Pipe 冷間加工溶接オーステナイト系ステンレス鋼鋼管	TP304 /S30400	E, W	≦0.08	≦0.75	≦2.00	≦0.045
	TP304H/S30409	E, W	0.04～0.10	≦0.75	≦2.00	≦0.045
	TP304L/S30403	E, W	≦0.035	≦0.75	≦2.00	≦0.045
	TP304N/S30451	E, W	≦0.08	≦0.75	≦2.00	≦0.045
	TP304LN/S30453	E, W	≦0.035	≦0.75	≦2.00	≦0.045
	TP309Cb/S30940	E, W	≦0.08	≦0.75	≦2.00	≦0.045
	TP309S/S30908	E, W	≦0.08	≦0.75	≦2.00	≦0.045
	TP310Cb/S31040	E, W	≦0.08	≦0.75	≦2.00	≦0.045
	TP310S/S31008	E, W	≦0.08	≦0.75	≦2.00	≦0.045
	TP316 /S31600	E, W	≦0.08	≦0.75	≦2.00	≦0.045
	TP316H/S31609	E, W	0.04～0.10	≦0.75	≦2.00	≦0.045
	TP316L /S31603	E, W	≦0.035	≦0.75	≦2.00	≦0.045
	TP316N/S31651	E, W	≦0.035	≦0.75	≦2.00	≦0.045
	TP316LN/S31653	E, W	≦0.035	≦0.75	≦2.00	≦0.045
	TP317 /S31700	E, W	≦0.08	≦0.75	≦2.00	≦0.045
	TP317L/S31703	E, W	≦0.035	≦0.75	≦2.00	≦0.045
	TP321 /S32100	E, W	≦0.08	≦0.75	≦2.00	≦0.045
	TP321H/S32109	E, W	0.04～0.10	≦0.75	≦2.00	≦0.045

1 MPa = 1.01972 × 10^{-1} kgf/mm²

化学成分（%）					Tensile Test MPa or N/mm² 引張試験		Remarks 備考 (Similar to JIS)	Index No. 索引番号
S	Ni	Cr	Mo	Others	Min. Yield Point 最小降伏点	Tensile Strength 引張強さ		
≦0.030	17.50~18.50	17.0～19.0	–	–	205	≧515		C280
≦0.030	11.50~13.50	20.50~23.50	1.50～3.00	N 0.20～0.40 V 0.10～0.30 Cb＋Ta 0.10～0.30	380	≧690		
≦0.030	2.25～3.75	17.0～19.0	–	N 0.20～0.40 V 0.10～0.30 Cb＋Ta 0.10～0.30	380	≧690		
≦0.010	17.50~18.50	19.50~20.50	6.00～6.50	N 0.180～0.220 Cu 0.50～1.00	300	≧650		
≦0.030	10.0～12.0	20.0～22.0	–	N 0.14～0.20 Cu 0.06～0.08	310	≧600		
≦0.030	8.00～16.0	18.0～20.0	–	–	170	≧485	for High Temperature and Corrosion Resistance Service 高温耐蝕用	C281
≦0.030	8.00～16.0	18.0～20.0	–	–	170	≧485		
≦0.030	8.00～13.0	18.0～20.0	–	–	205	≧515		
≦0.030	8.00～11.0	18.0～20.0	–	–	205	≧515		
≦0.030	8.00～11.0	18.0～20.0	–	Cb＋Ta 10C～1.10	205	≧515		
≦0.030	12.00~16.00	22.00~24.00	≦0.75	Cb＋Ta 10C～1.10 Cu≦0.75	205	≧515		
≦0.030	12.00~15.00	22.00~24.00	≦0.75	–	205	≧515		
≦0.030	19.00~22.00	24.00~26.00	≦0.75	Cb＋Ta 10C～1.10 Cu≦0.75	205	≧515		
≦0.030	19.00~22.00	24.00~26.00	≦0.75	–	205	≧515		
≦0.030	10.00~14.0	16.0～18.0	2.00～3.00	–	205	≧515		
≦0.030	10.00~14.0	16.0～18.0	2.00～3.00	–	205	≧515		
≦0.030	10.00~14.0	16.0～18.0	2.00～3.00	–	205	≧515		
≦0.030	10.00~14.0	16.0～18.0	2.00～3.00	N 0.10～0.16	205	≧515		
≦0.030	10.00~14.0	16.0～18.0	2.00～3.00	N 0.10～0.16	205	≧515		
≦0.030	11.00~14.0	18.0～20.0	3.00～4.00	–	205	≧515		
≦0.030	11.00~15.0	18.0～20.0	3.00～4.00	–	205	≧515		
≦0.030	9.00～13.0	17.0～20.0	–	Ti 5C～0.70	205	≧515		
≦0.030	9.00～13.0	17.0～20.0	–	Ti 4C～0.60	205	≧515		

Standard No. Year Designation 規格番号 年号 規格名称	Grade / UNS NO. グレード	Mfg. Process 製造方法	Chemical Composition			
			C	Si	Mn	P
A814－91 (Continued)	TP347/S34700	E, W	≦0.08	≦0.75	≦2.00	≦0.045
	TP347H/S34709	E, W	0.04～0.10	≦0.75	≦2.00	≦0.045
	TP348 /S34800	E, W	≦0.08	≦0.75	≦2.00	≦0.045
	TP348H/S34809	E, W	0.04～0.10	≦0.75	≦2.00	≦0.045
	TPXM10/S21900	E, W	≦0.08	≦1.00	8.00~10.00	≦0.040
	TPXM11/S21903	E, W	≦0.04	≦1.00	8.00~10.00	≦0.040
	TPXM15/ S38100	E, W	≦0.08	1.50～2.00	≦2.00	≦0.040
	TPXM19/S20910	E, W	≦0.060	≦1.00	11.50~14.50	≦0.040
	TPXM29/S24000	E, W	≦0.080	≦1.00	11.50~14.50	≦0.060
	S31254	E, W	≦0.020	≦0.80	≦1.00	≦0.030
	S30815	E, W	≦0.10	≦0.80	1.40～2.00	≦0.040
A826－88 Austenitic and Ferritic Stainless Steel Duct Tubes for Breeder Reactor Core Components 増殖炉炉心用オーステナイト系及びフェライト系ステンレス鋼導管	TP316	S	0.040~0.060	0.50～0.75	1.00～2.00	≦0.040
	S38660	S	0.030~0.050	0.50～1.00	1.65～2.35	≦0.040

1 MPa = 1.01972 × 10^{-1} kgf/mm²

化　学　成　分　(%)					Tensile Test MPa or N/mm² 引　張　試　験		Remarks 備　　考 (Similar to JIS)	Index No. 索引番号
S	Ni	Cr	Mo	Others	Min. Yield Point 最小降伏点	Tensile Strength 引張強さ		
≦0.035	9.00〜13.0	17.0〜20.0	–	Cb+Ta 10C〜1.00	205	≧515		C281
≦0.035	9.00〜13.0	17.0〜20.0	–	Cb+Ta 8C〜1.0	205	≧515		
≦0.035	9.00〜13.0	17.0〜20.0	–	Cb+Ta 10C〜1.00	205	≧515		
≦0.035	9.00〜13.0	17.0〜20.0	–	Ta≦0.10 Cb+Ta 8C〜1.00	205	≧515		
≦0.030	5.50〜7.50	19.00〜21.50	–	N 0.15〜0.40	345	≧620		
≦0.030	5.50〜7.50	19.00〜21.50	–	N 0.15〜0.40	345	≧620		
≦0.030	17.50〜18.50	17.0〜19.0	–	–	205	≧515		
≦0.030	11.50〜13.50	20.50〜23.50	1.50〜3.00	N　0.20〜0.40 V　0.10〜0.30 Cb+Ta 0.10〜0.30	380	≧690		
≦0.030	2.25〜3.75	17.0〜19.0	–	N　0.20〜0.40 V　0.10〜0.30 Cb+Ta 0.10〜0.30	380	≧690		
≦0.010	17.50〜18.50	19.50〜20.50	6.00〜6.50	N　0.180〜0.220 Cu 0.50〜1.00	300	≧650		
≦0.030	10.0〜12.0	20.0〜22.0	–	N　0.14〜0.20 Cu 0.03〜0.08	310	≧600		
≦0.010	13.0〜14.0	17.0〜18.0	2.00〜3.00	Cb≦0.050 Ta≦0.20 N ≦0.010 Al≦0.050 As≦0.030 B ≦0.0020 Co≦0.050 Cu≦0.04 V ≦0.05	585-795	690〜895	–	C282
≦0.010	14.5〜16.5	12.5〜14.5	1.50〜2.50	Cb≦0.050　B ≦0.020 Ta≦0.20　Co≦0.050 N ≦0.005　Cu≦0.04 Al≦0.0050　V ≦0.05 As≦0.030		–	–	

Standard No. Year Designation 規格番号 年号 規格名称	Grade/ UNS No. グレード	Mfg. Process 製造方法	Chemical Composition			
			C	Si	Mn	P
A826－88 (Continued)	S42100	S	0.17～0.23	0.20～0.30	0.40～0.70	≦0.040
	T91	S	0.08～0.12	0.20～0.50	0.30～0.60	≦0.020
A872－91 Centrifugally Cast Ferrific/Austenitic Stainless Steel Pipe for Corrosive Environments 腐食性環境用鋳造フェライト・オーステナイト系ステンレス鋼鋼管	J93183	－	≦0.030	≦2.0	≦2.00	≦0.040
	J93550	－	≦0.030	≦2.0	≦2.00	≦0.040
B167 94a Nickel-Chromium-Iron Alloys UNS N06600, N06601, N0690, N06025, and N06045 Seamless Pipe and Tube ニッケル・クロム・鉄合金鋼継目無鋼管	N06600	S	≦0.15	≦0.5	≦1.0	－
	N06601	S	≦0.10	≦0.5	≦1.5	－
	N06690	S	≦0.05	≦0.5	≦0.5	－
	N06025	S	0.15～0.25	≦0.5	≦0.15	≦0.020
	N06045	S	0.05～0.12	2.50～3.0	≦1.0	≦0.020
B407－93 Nickel-Chromium Alloy Seamless Pipe and Tube ニッケル・クロム合金鋼継目無鋼管	N08800	S	≦0.10	≦0.015	≦1.5	－
	N08810	S	0.05～0.10	≦0.015	≦1.5	－
	N08811	S	0.06～0.10	≦0.015	≦1.5	－
	N08801	S	≦0.10	≦1.00	≦1.5	－
B423－90 Nickel-Iron-Chromium-Molybdenum-Copper Alloy Seamless Pipe and Tube ニッケル・鉄・クロム・モリブデン・銅合金鋼継目無鋼管	N08825	S	≦0.05	≦0.5	≦1.0	－
	N08821	S	≦0.25	≦0.5	≦1.0	－

1 MPa = 1.01972 × 10^{-1} kgf/mm²

化　学　成　分　(%)					Tensile Test MPa or N/mm² 引張試験		Remarks 備考 (Similar to JIS)	Index No. 索引番号
S	Ni	Cr	Mo	Others	Min. Yield Point 最小降伏点	Tensile Strength 引張強さ		
≦0.010	0.30~0.80	11.0~12.5	0.80~1.25	Cb ≦0.050 W 0.40~0.60 Al ≦0.050 V 0.25~0.35	–	–	–	C283
≦0.010	≦0.40	8.00~9.50	0.85~1.05	Cb 0.06~0.10 N 0.030~0.070 Al ≦0.040 V 0.18~0.25	–	–	–	
≦0.030	4.00~6.00	20.0~23.0	2.00~4.00	N 0.08~0.25 Cu ≦1.00 Co 0.50~1.50	450	≧620	–	C284
≦0.030	5.00~8.00	23.0~26.0	2.00~4.00	N 0.08~0.25 Cu ≦1.00 Co 0.50~1.50	450	≧620	–	

S	Ni	Cr	Mo / Others	Outside Diameter	Min. Yield Point Hot-Worked	Min. Yield Point Cold-Worked	Tensile Strength Hot-Worked	Tensile Strength Cold-Worked	Remarks	Index No.
≦0.015	≧72.0	14.0~17.0	Fe 6.0~10.0 Cu ≦0.5	≦127mm	205	240	≧550	≧515	(NCF600TP) (NCF600TB)	C285
				>127mm	170	205	≧550	≧550		
≦0.015	58.0~63.0	21.0~25.0	Al 1.0~1.7 Cu ≦0.1	All	205	205	≧550	≧550		
≦0.015	≧58.0	27.0~31.0	Fe 7.0~11.0 Cu ≦0.5	≦127mm	205	240	≧586	≧586	–	
				>127mm	170	205	≧515	≧515		
≦0.010	Remainder	24.0~26.0	Ti 0.1~0.2 Zr 0.01~0.10 Y 0.05~0.12	All	350	350	≧660	≧660		
≦0.010	≦45.0	26.0~29.0	Ce 0.03~0.09 N 0.05~0.12	All	240	240	≧620	≧620		

S	Ni	Cr	Mo	Others	Min. Yield Point		Tensile Strength		Remarks	Index No.
≦0.015	30.0~35.0	19.0~23.0	Fe ≧39.5	Al 0.15~0.60 Ti 0.15~0.60	170	205	≧450	≧520	(NCF800TP) (NCF800TB)	C286
≦0.015	30.0~35.0	19.0~23.0	Fe ≧39.5	Al 0.15~0.60 Ti 0.15~0.60	170		≧450		–	
≦0.015	30.0~35.0	19.0~23.0	Fe ≧39.5	Al 0.15~0.60 Ti 0.15~0.60	170		≧450		–	
≦0.015	30.0~34.0	19.0~22.0	–	Fe ≧39.5 Cu ≦0.5 Ti 0.75~1.50	170		≧450		–	
≦0.03	38.0~46.0	19.5~23.5	2.5~3.5	Fe ≧22.0 Cu 1.5~3.0 Al ≦0.2 Ti 0.6~1 2	172	242	≧517	≧586	(NCF825TP) (NCF825TB)	C287
≦0.03	39.0~46.0	20.0~22.0	5.0~6.5	Fe ≧22.0 Cu 1.5~3.0 Al ≦0.2 Ti 0.6~1.2	–	235	–	≧545	–	

11. API Steel Tubes 鋼管

Standard No. Year Designation 規格番号 年号 規格名称	Grade グレード	Mfg. Process 製造方法	Chemical Composition			
			C	Si	Mn	P
Spec 5CT −89 Casing and Tubing ケシーシング及び鋼管	H-40	S, E	−	−	−	≦0.030
	J-55	S, E	−	−	−	≦0.030
	K-55	S, E	−	−	−	≦0.030
	N-80	S, E	−	−	−	≦0.030
	C-75-1	S, E	≦0.50	≦0.45	≦1.90	≦0.030
	C-75-2	S, E	≦0.43	≦0.45	≦1.90	≦0.030
	C-75-3	S, E	≦0.38	−	0.75〜1.00	≦0.030
	C-75-9Cr	S	≦0.15	≦1.00	0.30〜0.60	≦0.020
	C-75-13Cr	S	0.15〜0.22	≦1.00	0.25〜1.00	≦0.020
	L-80-1	S, E	≦0.43	≦0.45	≦1.90	≦0.030
	L-80-9Cr	S	≦0.15	≦1.00	0.30〜0.60	≦0.020
	L-80-13Cr	S	0.15〜0.22	≦1.00	0.25〜1.00	≦0.020
	C-90-1	S	≦0.35	−	≦1.00	≦0.020
	C-90-2	S	≦0.50	−	≦1.90	≦0.030
	C-95	SpE	≦0.45	≦0.45	≦1.90	≦0.030
	T-95-1	S	≦0.35	−	≦1.20	≦0.020
	T-95-2	S	≦0.50	−	≦1.90	≦0.030
	P-105	S	−	−	−	≦0.030
	P-110	S	−	−	−	≦0.030
	Q-125-1	S, E	≦0.35	−	≦1.00	≦0.020
	Q-125-2	S, E	≦0.35	−	≦1.00	≦0.020
	Q-125-3	S, E	≦0.50	−	≦1.90	≦0.020
	Q-125-4	S, E	≦0.50	−	≦1.90	≦0.030
Spec 5D-88 Drill Pipe 掘削用鋼管	E-75	S	−	−	−	≦0.040
	X-95	S				
	G-105	S				
	S-135	S				
Spec 5L-90 Line Pipe 配管用鋼管	A25Cl I	S	≦0.21	−	0.30〜0.60	≦0.045
	A25Cl II		≦0.21	−	0.30〜0.60	0.045~0.80
	A		≦0.22	−	≦0.90	≦0.04
	B		≦0.27	−	≦1.15	≦0.04
	X42 (Non-expanded)		≦0.29	−	≦1.25	≦0.04
	X42 (Cold expanded)		≦0.29	−	≦1.25	≦0.04
	X46 (Non-expanded)		≦0.31	−	≦1.35	≦0.04
	X46 (Cold expanded)		≦0.29	−	≦1.25	≦0.04

1 MP a =1.01972×10^{-1}kgf/mm²

化　学　成　分　(%)					Tensile Test MPa(kgf/mm²) 引　張　試　験		Remarks 備　考 (Similar to JIS)	Index No. 索引番号
S	Ni	Cr	Mo	Others	Min. Yield Point 最小降伏点	Tensile Strength 引張強さ		
≦0.030	－	－	－	－	276～552	≧414	－	C288
≦0.030	－	－	－	－	379～552	≧517	－	
≦0.030	－	－	－	－	379～552	≧655	－	
≦0.030	－	－	－	－	552～758	≧689	－	
≦0.030	≦0.50	≦0.50	0.15～0.40	Cu≦0.50	517～620	≧655	－	
≦0.030	－	－	－	－	517～620	≧655	－	
≦0.030	－	0.80～1.10	0.15～0.25	－	517～620	≧655	－	
≦0.010	≦0.50	8.00～10.0	0.90～1.10	Cu≦0.25	517～620	≧655	－	
≦0.010	≦0.50	12.0～14.0	－	Cu≦0.25	517～620	≧655	－	
≦0.030	≦0.25	－	－	Cu≦0.35	552～655	≧655	－	
≦0.010	≦0.50	8.00~10.00	0.90～1.10	Cu≦0.25	552～655	≧655	－	
≦0.010	≦0.50	12.0～14.0	－	Cu≦0.25	552～655	≧655	－	
≦0.010	≦0.99	≦1.20	≦0.75	－	620～724	≧690	－	
≦0.010	≦0.99	No Limit	No limit	－	620～724	≧690	－	
≦0.030	－	－	－	－	655～758	≧724	－	
≦0.010	≦0.99	0.40～1.50	0.25～0.85	－	655～758	≧724	－	
≦0.010	≦0.99	－	－	－	655～758	≧724	－	
≦0.030	－	－	－	－	724～931	≧827	－	
≦0.030	－	－	－	－	758～965	≧862	－	
≦0.010	≦0.99	≦1.20	≦0.75	－	860～1035	≧930	－	
≦0.020	≦0.99	No Limit	No Limit	－	860～1035	≧930	－	
≦0.020	≦0.99	No Limit	No Limit	－	860～1035	≧930	－	
≦0.020	≦0.99	No Limit	No Limit	－	860～1035	≧930	－	
≦0.030	－	－	－	－	517～724	≧689	－	C289
					655～862	≧724	－	
					724～931	≧793	－	
					931～1138	≧1000	－	
≦0.06	－	－	－	－	172	≧310	－	C300
≦0.06	－	－	－	－	172	≧310	－	
≦0.05	－	－	－	－	207	≧331	(STPG370)	
≦0.05	－	－	－	－	241	≧413	(STPG410)	
≦0.05	－	－	－	－	289	≧413	－	
≦0.05	－	－	－	－	289	≧413	－	
≦0.05	－	－	－	－	317	≧434	－	
≦0.05	－	－	－	－	317	≧413	－	

Standard No. 規格番号	Year 年号	Designation 規格名称	Grade グレード	Mfg. Process 製造方法	Chemical Composition			
					C	Si	Mn	P
Spec 5L-90 (Continued)			X52 (Non-expanded)	S	≦0.31	—	≦1.35	≦0.04
			X52 (Cold expanded)		≦0.29	—	≦1.25	≦0.04
			X56		≦0.26	—	≦1.35	≦0.04
			X60		≦0.26	—	≦1.35	≦0.04
			X65		—	—	—	—
			X70		—	—	—	—
			X80		—	—	—	—
			A25Cl I	W	≦0.21	—	0.30～0.60	≦0.045
			A25Cl II		≦0.21	—	0.30～0.60	0.045~0.080
			A		≦0.21	—	≦0.90	≦0.04
			B		≦0.26	—	≦1.15	≦0.04
			X42		≦0.28	—	≦1.25	≦0.04
			X46 (Non-expanded)		≦0.30	—	≦1.35	≦0.04
			X46 (Cold expanded)		≦0.28	—	≦1.25	≦0.04
			X52 (Non-expanded)		≦0.30	—	≦1.35	≦0.04
			X52 (Cold expanded)		≦0.28	—	≦1.25	≦0.04
			X56		≦0.26	—	≦1.35	≦0.04
			X60		≦0.26	—	≦1.35	≦0.04
			X65		≦0.26	—	≦1.40	≦0.04
			X70		≦0.23	—	≦1.60	≦0.04
			X80		≦0.18	—	≦1.80	≦0.030

1 MPa = 1.01972 × 10^{-1} kgf/mm²

化　学　成　分　(%)					Tensile Test MPa(kgf/mm²) 引張試験		Remarks 備考	Index No.
S	Ni	Cr	Mo	Others	Min. Yield Point 最小降伏点	Tensile Strength 引張強さ	(Similar to JIS)	索引番号
≦0.05	−	−	−	−	358	≧455	−	C300
≦0.05	−	−	−	−	358	≧455	−	
≦0.05	−	−	Cb ≦0.005	V ≦0.005 Ti ≦0.005	386	≧489	−	
≦0.05	−	−	Cb ≦0.005	V ≦0.005 Ti ≦0.005	413	≧517	−	
−	−	−	−	−	448	≧530	−	
−	−	−	−	−	482	≧565	−	
−	−	−	−	−	551	620~827	−	
≦0.06	−	−	−	−	172	≧310	−	
≦0.06	−	−	−	−	172	≧310	−	
≦0.05	−	−	−	−	207	≧331	−	
≦0.05	−	−	−	−	241	≧413	−	
≦0.05	−	−	−	−	289	≧413	−	
≦0.05	−	−	−	−	289	≧413	−	
≦0.05	−	−	−	−	317	≧434	−	
≦0.05	−	−	−	−	358	≧455	−	
≦0.05	−	−	−	−	358	≧455	−	
≦0.05	−	−	−	−	386	≧489	−	
≦0.05	−	−	−	−	413	≧517	−	
≦0.05	−	−	−	−	448	≧530	−	
≦0.05	−	−	−	−	482	≧565	−	
≦0.018	−	−	−	−	551	≧620	−	

12. BS Steel Pipes for Piping 配管用鋼管

Standard No. Year Designation 規格番号 年号 規格名称	Grade グレード	Mfg. Process 製造方法	Chemical Composition C	 Si	 Mn	 P
534－90 Steel Pipes, Fittings and Specials for Water and Gas and Sewage 水・ガス・下水道用鋼管および継手類	－	S, E B, A	－	Original Pipes are defined by BS3601 （原管はBS. 3601 の規定による）		
778－66 Steel Pipes and Joints for Hydraulic Purposes 水圧用鋼管および継手	－	S, E	－	Original Pipes are defined by HFS22 （原管はBS. 3601 （旧版） のHFS22 及び		
1387－67 Steel Tubes and Tubulars Suitable for Screwing to BS 21 Pipe Threads BS21のねじ切り用鋼管	－	S, W	－	－	－	≦0.060
1864－66 Stainless Steel Milk Pipes and Fittings ミルク用ステンレスパイプ及び継手	Gr 1	S, W	≦0.06	0.2～1.0	0.5～2.0	≦0.040
	Gr 2	S, W	≦0.08	0.2～1.0	0.5～2.0	≦0.040
	Gr 3	S, W	≦0.08	0.2～1.0	0.5～2.0	≦0.040
	Gr 4	S, W	≦0.08	0.2～1.0	0.5～2.0	≦0.040
	Gr 5	S, W	≦0.07	0.2～1.0	0.5～2.0	≦0.040
	Gr 6	S, W	≦0.08	0.2～0.8	1.0～2.0	≦0.040
	Gr 7	S, W	≦0.08	≦1.50	≦2.0	≦0.040
	Gr 8	S, W	≦0.08	≦1.50	≦2.0	≦0.040
	Gr 9	S, W	≦0.08	≦1.50	≦2.0	≦0.040
3601－87 Steel Pipes and Tubes for Pressure Purposes: Carbon Steel with Specified Room Temperature Properties 圧力用鋼管：室温の性質を規定した炭素鋼	BW320	B	≦0.16	－	0.30～0.70	≦0.040
	ERW320	E	≦0.16	－	0.30～0.70	≦0.040
	ERW360	E	≦0.17	≦0.35	0.40～0.80	≦0.050
	ERW430	E	≦0.21	≦0.35	0.40～1.20	≦0.040
	S360	S	≦0.17	≦0.35	0.40～0.80	≦0.040
	S430	S	≦0.21	≦0.35	0.40～1.20	≦0.040
	SAW430	A	≦0.25	≦0.50	≦1.20	≦0.040

1 N/mm² or MPa = 1.01972×10^{-1} kgf/mm²

化　学　成　分　(%)					Tensile Test N/mm² 引張試験		Remarks 備考	Index No.
S	Ni	Cr	Mo	Others	Min. Yield Point 最小降伏点	Tensile Strength 引張強さ	(Similar to JIS)	索引番号
	–	–	–	–	–	–	–	C301
and 27, CDS22 and ERW 22 and 27 in BS3601 27, CDS22, ERW22及び27による)				–	–	–	–	C302
≦0.060	–	–	–	–	–	325~460	(SGF)	C303
≦0.035	8.0～11.0	18.0～20.0	–	–	–	–	(SUS304TBS)v	C304
≦0.5	9.0～11.0	18.0～20.0	–	–	–	–	Free Cutting Steel	
≦0.035	8.0～11.0	17.0～19.0	–	Ti:5×C～0.70	–	–	–	
≦0.030	10.0～13.0	17.0～19.0	–	Ti:5×C～0.70	–	–	–	
≦0.035	10.0～13.0	16.5～18.5	2.25～3.0	–	–	–	(SUS316TBS)	
≦0.030	11.0～14.0	16.5～18.5	2.5～3.0	–	–	–	(SUS316TBS)	
≦0.5	≧80	17.0～21.0	–	–	–	–	Free Cutting Steel	
≦0.5	≧80	17.0～20.0	2.0～3.0	–	–	–	Free Cutting Steel	
≦0.040	≧80	17.0～20.0	2.0～3.0	–	–	–	–	
≦0.040	–	–	–	–	195	320~460	–	C305
≦0.040	–	–	–	–	195	320~460	–	
≦0.050	–	–	–	–	235	360~500	(STPG370)	
≦0.040	–	–	–	–	195	430~570	(STPG410)	
≦0.040	–	–	–	t ≦16mm 16mm < t ≦ 40mm 40mm < t ≦ 65mm	235 225 215	360~550	(STPG370)	
≦0.040	–	–	–	t ≦16mm 16mm < t ≦ 40mm 40mm < t ≦ 65mm	275 265 255	430~570	(STPG410)	
≦0.040	–	–	–	t ≦16mm 16mm < t ≦ 40mm 40mm < t ≦ 65mm	275 265 255	430~570	(STPY400)	

Standard No. Year 規格番号 年号	Designation 規格名称	Grade グレード	Mfg. Process 製造方法	Chemical Composition C	Si	Mn	P
3602 Steel Pipes and Tubes for Pressure Purposes:Carbon and Carbon Manganese Steel With Specified Elevated Temperature Properties 圧力用鋼管・高温の性質を規定した炭素および炭素マンガン鋼鋼管	Part1−78 Seamless Electric Resistance Welded and Induction Welded Tubes 第１部 継目無電気抵抗溶接及びｲﾝﾀﾞｸｼｮﾝ溶接鋼管	HFS360	S	≦0.17	≦0.35	0.40〜0.80	≦0.045
		CFS360					
		CEW360	E				
		HFS430	S	≦0.21	≦0.35	0.40〜1.20	≦0.045
		CFS430					
		ERW430	E				
		CEW430					
		HFS460	S	≦0.22	≦0.35	0.80〜1.40	≦0.045
		CFS460					
		ERW460	E				
		CEW460					
		500Nb	S	≦0.23	≦0.35	0.80〜1.50	≦0.045
	Part2−87 Submegred Arc Welded Tubes 第２部ｻﾌﾞﾏｰｼﾞﾄﾞｱｰｸ溶接管	SAW410	AM	≦0.20	≦0.35	0.50〜1.30	≦0.040
		SAW460	AM	≦0.20	≦0.40	0.60〜1.50	≦0.040
3603−77 Steel Pipes and Tubes for Pressure Purposes: Carbon and Alloy Steel with Specified Low Temperature Properties 圧力用鋼管：低温の性質を規定した炭素鋼及び合金鋼		HFS410LT50	S	≦0.20	≦0.35	0.60〜1.20	≦0.045
		CFS410LT50					
		ERW410LT50	E				
		CEW410LT50					
		HFS503LT100	S	≦0.15	0.15〜0.35	0.30〜0.80	≦0.025
		CFS503LT100					
		HFS509LT196	S	≦0.10	0.15〜0.30	0.30〜0.80	≦0.025
		CFS509LT196					
3604−78 Steel Pipes and Tubes for Pressure Purposes:Ferritic Alloy Steel with Specified Elevated Temperatare Properties 圧力用鋼管：高温の性質を規定したフェライト系合金鋼		HFS620-460	S	0.10〜0.15	0.10〜0.35	0.40〜0.70	≦0.040
		CFS620-460					
		ERW620-460	E				
		CEW620-460					
		HFS620-440	S	0.10〜0.18	0.10〜0.35	0.40〜0.70	≦0.040
		CFS620-440					
		ERW620-440	E				
		CEW620-440					
		HFS621	S	≦0.15	0.50〜1.00	0.30〜0.60	≦0.040
		CFS621					
		ERW621	E				
		CEW621					

1 N/mm^2 or MPa = $1.01972 \times 10^{-1} kgf/mm^2$

化　学　成　分　(%)					Tensile Test N/mm^2 引張試験		Remarks 備考	Index No.
S	Ni	Cr	Mo	Others	Min. Yield Point 最小降伏点	Tensile Strength 引張強さ	(Similar to JIS)	索引番号
≦0.045	−	−	−	−	215	360~500	(STPT370)	C306
≦0.045	−	−	−	−	215	410~550	(STPT410) (STB 410) (STB 410)	
≦0.045	−	−	−	−	280	460~600	(STPT480)	
≦0.045	−	−	−	Nb 0.015～0.10	340	490~630	−	
≦0.040	−	−	−	−	245	410~550	−	
≦0.040	−	−	−	−	280	460~600	−	
≦0.045	−	−	−	Al≧0.015	235	410~530	2mmV, -50℃, 27J (STPL39, STBL39)	C307
≦0.020	3.25～3.75	−	−	−	245	440~590	2mmV, -50℃, 39J 〃 -100℃, 27J (STPL46, STBL46)	
≦0.020	8.50～9.50	−	−	−	510	690~845	2mmV, -100℃, 55J 〃 -150℃, 47J 〃 -196℃, 39J (STPL70, STBL70)	
≦0.040	−	0.70～1.10	0.45～0.65	Almet: ≦0.02	18 (N)	460~610	(STPA22) (STBA22)	C308
≦0.040	−	0.70～1.10	0.45～0.65	Almet: ≦0.02	295 (N+T)	440~590	(STPA22) (STBA22)	
≦0.040	−	1.00～1.50	0.45～0.65	Almet: ≦0.02	275 (N+T)	420~520	(STPA23) (STBA23)	

Standard No. 規格番号	Year 年号	Designation 規格名称	Grade グレード	Mfg. Process 製造方法	Chemical Composition C	Si	Mn	P
3604−78 (Continued)			HFS660 CFS660	S	0.10〜0.15	0.10〜0.35	0.40〜0.70	≦0.040
			HFS622 CFS622	S	0.08〜0.15	≦0.50	0.40〜0.70	≦0.040
			HFS625 CFS625	S	≦0.15	≦0.50	0.30〜0.60	≦0.030
			HFS629-470 CFS629-470	S	≦0.15	0.25〜1.00	0.30〜0.60	≦0.030
			HFS629-590 CFS629-590	S	≦0.15	0.25〜1.00	0.30〜0.60	≦0.030
			HFS762 CFS762	S	0.17〜0.23	≦0.50	≦1.00	≦0.030
3605−78 Seamless and Welded Austenitic Stainless Steel Pipes and Tubes for Pressure Purposes 圧力用オーステナイトステンレス継目無管及び溶接管		Seamless Pipes with Specified Room Temperature Preperties 室温の性質を規定した継目無管	304S14	S	≦0.03	0.20〜1.00	0.05〜2.00	≦0.040
			304S18	S	≦0.06	0.20〜1.00	0.05〜2.00	≦0.040
			316S14	S	≦0.03	0.20〜1.00	0.05〜2.00	≦0.040
			316S18	S	≦0.07	0.20〜1.00	0.05〜2.00	≦0.040
			321S18	S	≦0.08	0.20〜1.00	0.05〜2.00	≦0.040
			347S18	S	≦0.08	0.20〜1.00	0.05〜2.00	≦0.040
		Seamless Pipes with Specified Elevated Temperature Properties 高温の性質を規定した継目無管	304S14	S	≦0.03	0.20〜1.00	0.05〜2.00	≦0.040
			304S18	S	≦0.06	0.20〜1.00	0.05〜2.00	≦0.040
			304S59	S	0.04〜0.09	0.20〜1.00	0.05〜2.00	≦0.040
			316S14	S	≦0.03	0.20〜1.00	0.04〜2.00	≦0.040
			316S18	S	≦0.07	0.20〜1.00	0.05〜2.00	≦0.040
			316S59	S	0.04〜0.09	0.20〜1.00	0.05〜2.00	≦0.040
			321S18	S	≦0.08	0.20〜1.00	0.05〜2.00	≦0.040

1 N/mm² or MPa = 1.01972 × 10^{-1} kgf/mm²

化学成分 (%)					Tensile Test N/mm² 引張試験		Remarks 備考	Index No.
S	Ni	Cr	Mo	Others	Min. Yield Point 最小降伏点	Tensile Strength 引張強さ	(Similar to JIS)	索引番号
≦0.040	–	0.30～0.60	0.50～0.70	V:0.22～0.28 Almet:≦0.02	305 (N+T)	460~610	–	C 308
≦0.040	–	2.00～2.50	0.90～1.20	Al: ≦0.02	275 (N+T)	490~640	(STPA24) (STBA24)	
≦0.030	–	4.00～6.00	0.45～0.65	Al: ≦0.02	170 (A)	450~600	(STPA25) (STBA25)	
≦0.030	–	8.00~10.00	0.90～1.10	Al≦0.02	190 (A)	470~620	(STPA26) (STBA26)	
≦0.030	–	8.00~10.00	0.90～1.10	Al≦0.02	410 (N+T)	590~740	–	
≦0.030	0.30～0.80	10.00~12.50	0.90～1.20	V:0.25～0.35 W ≦0.70	470 (N+T)	720~875	–	
≦0.030	10.0～13.0	17.0～19.0	–	–	205	490~690	(SUS304LTP) (SUS304LTB)	C 309
≦0.030	9.0～12.0	17.0～19.0	–	–	235	490~690	(SUS304 TP) (SUS304 TB)	
≦0.030	12.0～15.0	16.0～18.5	2.0～3.0	–	215	490~690	(SUS316LTP) (SUS316LTB)	
≦0.030	11.0～14.0	16.0～18.5	2.0～3.0	–	245	510~710	(SUS316 TP) (SUS316 TB)	
≦0.030	10.0～13.0	17.0～19.0	–	Ti:5×C～0.60	235	510~710	(SUS321 TP) (SUS321 TB)	
≦0.030	10.0～13.0	17.0～19.0	Nb:10×C～20×C or 1.00		245	510~710	(SUS347 TP) (SUS347 TB)	
≦0.030	10.0～13.0	17.0～19.0	–	–	205	510~690	(SUS304LTP) (SUS304LTB)	
≦0.030	9.0～12.0	17.0～19.0	–	–	235	490~690	(SUS304 TP) (SUS304 TB)	
≦0.030	9.0～12.0	17.0～19.0	–	–	235	490~690	(SUS304HTP) (SUS304HTB)	
≦0.030	12.0～15.0	16.0～18.5	2.0～3.0	–	215	490~690	(SUS316LTP) (SUS316LTB)	
≦0.030	11.0～14.0	16.0～18.5	2.0～3.0	–	245	510~710	(SUS316 TP) (SUS316 TB)	
≦0.030	12.0～14.0	16.0～18.0	2.0～2.75	B:0.001～0.006	245	510~710	–	
≦0.030	10.0～13.0	17.0～19.0	–	Ti:5×C～0.60	235	510~710	(SUS321 TP) (SUS321 TB)	

Standard No. Year 規格番号 年号	Designation 規格名称	Grade グレード	Mfg. Process 製造方法	Chemical Composition C	 Si	 Mn	 P
3605−78 (Continued)		321S59	S	0.04〜0.09	0.20〜1.00	0.05〜2.00	≦0.040
		347S18	S	≦0.08	0.20〜1.00	0.05〜2.00	≦0.040
		347S59	S	0.04〜0.09	0.20〜1.00	0.05〜2.00	≦0.040
	Welded Pipes with Specified Room Temperature Properties 室温の性質を規定した溶接管	304S22	W	≦0.03	0.20〜1.00	0.05〜2.00	≦0.040
		304S25	W	≦0.06	0.20〜1.00	0.05〜2.00	≦0.040
		316S22	W	≦0.03	0.20〜1.00	0.05〜2.00	≦0.040
		316S26	W	≦0.07	0.20〜1.00	0.05〜2.00	≦0.040
		321S22	W	≦0.08	0.20〜1.00	0.05〜2.00	≦0.040
		347S17	W	≦0.08	0.20〜1.00	0.05〜2.00	≦0.040
	Welded Pipes with Specified Elevated Temperature Properties 高温の性質を規定した溶接管	304S22	W	≦0.03	0.20〜1.00	0.05〜2.00	≦0.040
		304S25	W	≦0.06	0.20〜1.00	0.05〜2.00	≦0.040
		316S22	W	≦0.03	0.20〜1.00	0.05〜2.00	≦0.040
		316S26	W	≦0.07	0.20〜1.00	0.05〜2.00	≦0.040
		321S22	W	≦0.08	0.20〜1.00	0.05〜2.00	≦0.040
		347S17	W	≦0.08	0.20〜1.00	0.05〜2.00	≦0.040

1 N/mm^2 or MPa = 1.01972×10^{-1}kgf/mm^2

化 学 成 分 (%)					Tensile Test N/mm^2 引 張 試 験		Remarks 備 考	Index No.
S	Ni	Cr	Mo	Others	Min. Yield Point 最小降伏点	Tensile Strength 引張強さ	(Similar to JIS)	索引番号
≦0.030	10.0~13.0	17.0~19.0	−	Ti:5×C~0.06	235	510~710	(1010 ℃S.H.T) (SUS321HTB)	C 309
					195	490~690	(1105 ℃S.H.T) (SUS321HTB)	
≦0.030	10.0~13.0	17.0~19.0	Nb:10×C~20×C or 1.00		245	510~710	(SUS347 TP) (SUS347 TB)	
≦0.030	11.0~14.0	17.0~19.0	Nb:10×C~20×C or 1.00		245	510~710	(SUS347HTP) (SUS347HTB)	
≦0.030	9.0~12.0	17.0~19.0	−	−	205	490~690	(SUS304LTP) (SUS304LTB)	
≦0.030	8.0~11.0	−	−	−	235	490~690	(SUS304 TP) (SUS304 TB)	
≦0.030	11.0~14.0	16.0~18.5	2.0~3.0	−	215	490~690	(SUS316LTP) (SUS316LTB)	
≦0.030	10.0~13.0	16.0~18.5	2.0~3.0	−	245	510~710	(SUS316 TP) (SUS316 TB)	
≦0.030	9.0~12.0	17.0~19.0	−	Ti:5×C~0.60	235	510~710	(SUS321 TP) (SUS321 TB)	
≦0.030	9.0~12.0	17.0~19.0	−	Nb:10×C~1.00	245	510~710	(SUS347 TP) (SUS347 TB)	
≦0.030	9.0~12.0	17.0~19.0	−	−	205	490~690	(SUS304LTP) (SUS304LTB)	
≦0.030	8.0~11.0	17.0~19.0	−	−	235	490~690	(SUS304 TP) (SUS304 TB)	
≦0.030	11.0~14.0	16.0~18.5	2.0~3.0	−	215	490~690	(SUS316LTP) (SUS316LTB)	
≦0.030	10.0~13.0	16.0~18.5	2.0~3.0	−	245	510~710	(SUS316 TP) (SUS316 TB)	
≦0.030	9.0~12.0	17.0~19.0	−	Ti:5×C~0.60	235	510~710	(SUS321 TP) (SUS321 TB)	
≦0.030	9.0~12.0	17.0~19.0	−	Nb:10×C~1.00	245	510~710	(SUS347 TP) (SUS347 TB)	

13. BS Steel Tubes for Heat Transfer 熱伝導用鋼管

Standard No. 規格番号	Year Designation 年号 規格名称	Grade グレード	Mfg. Process 製造方法	Chemical Composition C	Si	Mn	P
3059 Steel boiler and Superheater Tubes ボイラ及び過熱器用鋼管	Part1－87 Low Tensile Carbon Steel Tube Without Specified Elevated Temperature Properties	HFS320	S	≦0.16	－	0.30～0.70	≦0.050
		CFS320	E	≦0.16			
		ERW320	S	≦0.17	≦0.35	0.40～0.80	≦0.045
		CEW320	E				
	Part2－90 Carbon, Alloy and Austenitic Stainless Steel Tubes with Specified Elevated Temperature Properties	S1 360	S	≦0.17	0.10～0.35	0.40～0.80	≦0.035
		S2 360					
		ERW360	E				
		CEW360					
		S1 440	S	0.12～0.18	0.10～0.35	0.90～1.20	≦0.035
		S2 440					
		ERW440	E				
		CEW440					
		S1 243	S	0.12～0.20	0.10～0.35	0.40～0.80	≦0.035
		S2 243					
		ERW243	E				
		CEW243					
		S1 620-460	S	0.10～0.15	0.10～0.35	0.40～0.70	≦0.030
		S2 620-460					
		ERW620-460	E				
		CEW620-460					
		S1 622-490	S	0.08～0.15	≦0.50	0.40～0.70	≦0.030
		S2 622-490					
		S1 622-490	S				
		S2 622-490					
		S1 629-470	S	≦0.15	0.25～1.00	0.30～0.60	≦0.030
		S2 629-470					
		S1 629-590	S				
		S2 629-590					
		S1 762	S	0.17～0.23	≦0.50	≦1.00	≦0.030
		S2 762					
		CFS304S51	S	0.04～0.10	≦1.00	≦2.00	≦0.040
		CFS316S51		0.04～0.10	≦1.00	≦2.00	≦0.040
		CFS316S52	S	0.04～0.10	≦1.00	≦2.00	≦0.040
		CFS321S51 (1010)	S	0.04～0.10	≦1.00	≦2.00	≦0.040
		CFS321S51 (1105)					
		CFS347S51	S	0.04～0.10	≦1.00	≦2.00	≦0.040

1 N/mm² or MPa = 1.01972 × 10^{-1} kgf/mm²

化学成分 (%)					Tensile Test N/mm² 引張試験		Remarks 備考	Index No. 索引番号
S	Ni	Cr	Mo	Others	Min. Yield Point 最小降伏点	Tensile Strength 引張強さ	(Similar to JIS)	
≦0.050	–	–	–	–	195	325~480	(STB35)	C310
≦0.045	–	–	–	–	215	360~500	(STB35)	
≦0.035	–	–	–	–	235	360~500	(STB35)	
≦0.035	–	–	–	–	245	440~580	(STB42)	
≦0.035	–	–	0.25～0.35	Al : ≦0.012	275	480~630	–	
≦0.030	–	0.70～1.10	0.45～0.65	Al : ≦0.020	180	460~610	(STBA22)	
≦0.030	–	2.00～2.50	0.90～1.20	Al : ≦0.020	275	490~640	(STBA24)	
≦0.030	–	8.00~10.00	0.90～1.10	Al : ≦0.020	185	470~620	(STBA26)	
					400	590~740	–	
≦0.030	0.3～0.8	10.00~12.50	0.80～1.20	V : 0.25～0.35	470	720~870	–	
≦0.030	8.0～11.0	17.0～19.0	–	–	240	510~710	(SUS304HTB)	
≦0.030	10.5～13.5	16.5～18.5	2.00～2.75	–	240	510~710	(SUS316HTB)	
≦0.030	10.5～13.5	16.5～18.5	2.00～2.50	0.0015≦B≦0.006	240	510~710	–	
≦0.030	9.0～12.0	17.0～19.0	–	5C≦Ti≦0.8	235	510~710	(SUS321HTB)	
					190	490~690		
≦0.030	9.0～13.0	17.0～19.0	–	10C≦Nb≦1.2	240	510~710	(SUS347HTB)	

Standard No. Year Designation 規格番号 年号 規格名称	Grade グレード	Mfg. Process 製造方法	Chemical Composition			
			C	Si	Mn	P
3059 (Continued)	CFS215S15	S	0.06～0.15	0.20～1.00	5.50～7.00	≦0.040
3606－92 Steel Tubes for Heat Exchanger 熱交換器用鋼管	ERW320	E	≦0.16	–	0.30～0.70	≦0.040
	CEW320	E	(Rimmed≦0.19)			
	CFS320	S	≦0.16			
	ERW400	E	≦0.12	0.10～0.35	0.90～1.20	≦0.020
	CEW400	E				
	CFS400	S				
	ERW440	E	0.12～0.18	0.10～0.35	0.90～1.20	≦0.035
	CEW440	E				
	CFS440	S				
	ERW243	E	0.12～0.20	0.10～0.35	0.40～0.80	≦0.040
	CEW243	E				
	CFS243	S				
	CEW261	E	0.06～0.10	0.10～0.35	0.60～0.80	≦0.020
	CFS261	S				
	ERW620	E	0.10～0.15	0.10～0.35	0.40～0.70	≦0.040
	CEW620	E				
	CFS620	S				
	ERW621	E	0.10～0.15	0.50～1.00	0.30～0.60	≦0.040
	CEW621	E				
	CFS621	S				
	CFS622	S	0.08～0.15	≦0.50	0.40～0.70	≦0.040
	CFS625	S	≦0.15	≦0.50	0.30～0.60	≦0.030
	LWHT304S11	W	≦0.030	≦1.00	≦2.00	≦0.040
	LWCF304S11	W				
	LWBC304S11	W				
	CFS 304S11	S				
	LWHT304S31	W	≦0.07	≦1.00	≦2.00	≦0.040
	LWCF304S31	W				
	LWBC304S31	W				
	CFS 304S31	S				
	LWHT316S11	W	≦0.030	≦1.00	≦2.00	≦0.040
	LWCF316S11	W				
	LWBC316S11	W				
	CFS 316S11	S				

$1\ N/mm^2$ or $MPa = 1.01972 \times 10^{-1} kgf/mm^2$

化　学　成　分　(%)					Tensile Test N/mm^2 引張試験		Remarks 備考	Index No.
S	Ni	Cr	Mo	Others	Min. Yield Point 最小降伏点	Tensile Strength 引張強さ	(Similar to JIS)	索引番号
≦0.030	9.0～11.0	14.0～16.0	0.80～1.20	Nb:0.75～1.25 V :0.15～1.25 B :0.003～0.009	270	540~740	−	C310
≦0.040	−	−	−	−	195	320~460	(STB35) (N)	C311
≦0.020	≦0.30	≦0.20	≦0.10	Al: ≦0.04 Cu: ≦0.25 Sn: ≦0.025	230	400~520	−	
≦0.035	−	−	−	−	265	440~560	(STB42) (N)	
≦0.040	≦0.30	−	0.25～0.35	Cu: ≦0.25 Al: ≦0.012 Sn: ≦0.03	275	480~630	− (N, NT)	
≦0.020	≦0.30	≦0.20	0.40～0.60	B 0.002~0.006 Ti: ≦0.06 Cu ≦0.25 Al: ≦0.06 Sn ≦0.03	400	540~690	− (N)	
≦0.040	≦0.30	0.70～1.10	0.45～0.65	Cu: ≦0.25 Al: ≦0.020 Sn: ≦0.03	180	460~610	(STBA22) (N, NT)	
≦0.040	≦0.30	1.00～1.50	0.45～0.65	Cu: ≦0.25 Al: ≦0.020 Sn: ≦0.03	275	420~570	(STBA23) (N, NT)	
≦0.040	≦0.30	2.00～2.50	0.90～1.20	Al: ≦0.02 Cu ≦0.25 Sn ≦0.03	275	490~640	(STBA24) (NT)	
≦0.030	≦0.30	4.00～6.00	0.45～0.65	Al: ≦0.02 Cu ≦0.25 Sn ≦0.03	170	450~600	(STBA25) (A)	
≦0.030	9.0～12.0	17.0～19.0	−	−	205	490~690	(SUS304LTB) (Q)	
≦0.030	8.0～11.0	17.0～19.0	−	−	235	490~690	(SUS304 TB) (Q)	
≦0.030	11.0～14.0	16.5～18.5	2.00～2.50	−	215	490~690	(SUS316LTB) (Q)	

Standard No. Year Designation 規格番号 年号 規格名称	Grade グレード	Mfg. Process 製造方法	Chemical Composition			
			C	Si	Mn	P
3606－78 (Continued)	LWHT316S31	W	≦0.07	≦1.00	≦2.00	≦0.040
	LWCF316S31	W				
	LWBC316S31	W				
	CFS 316S31	S				
	LWHT316S13	W	≦0.03	≦1.00	≦2.00	≦0.040
	LWCF316S13	W				
	LWBC316S13	W				
	CFS 316S13	S				
	LWHT316S33	W	≦0.07	≦1.00	≦2.00	≦0.040
	LWCF316S33	W				
	LWBC316S33	W				
	CFS 316S33	S				
	LWHT321S31	W	≦0.08	≦1.00	≦2.00	≦0.040
	LWCF321S31	W				
	LWBC321S31	W				
	CFS 321S31	S				
	LWHT347S31	W	≦0.08	≦1.00	≦2.00	≦0.040
	LWCF347S31	W				
	LWBC347S31	W				
	CFS 347S31	S				
4127－94 Light Gauge Stainless Steel Tubes 薄肉ステンレス鋼チューブ	316S31	S, W	≦0.07	≦1.00	≦2.00	≦0.040
	304S15	S, W	≦0.06	≦1.00	≦2.00	≦0.040
4825－91 Pipes and Fittings for the Food Industry and other hygienic applications 食品工業用鋼管及び継手	－	S, W	Any of the following grades 304S11, 304S31, 316S11,			

1 N/mm² or MPa = 1.01972×10^{-1} kgf/mm²

化学成分（%）					Tensile Test MPa or N/mm² 引張試験		Remarks 備考	Index No. 索引番号
S	Ni	Cr	Mo	Others	Min. Yield Point 最小降伏点	Tensile Strength 引張強さ	(Similar to JIS)	
≦0.030	10.5～13.5	16.5～18.5	2.00～2.50	–	245	510~710	(SUS316 TB) (Q)	C311
≦0.030	11.5～14.5	16.5～18.5	2.50～3.00	–	215	490~690	(SUS316LTB) (Q)	
≦0.030	11.0～14.0	16.5～18.5	2.50～3.00	–	245	510~710	(SUS316 TB) (Q)	
≦0.030	9.0～12.0	17.0～19.0	–	Ti:5×C～0.8	235	510~710	(SUS321 TB) (Q)	
≦0.030	9.0～13.0	17.0～19.0	–	Nb 10×C～1.00	245	510~710	(SUS347 TB) (Q)	
≦0.030	10.5～13.5	16.5～18.5	2.00～2.50	–	245	510~710		C312
≦0.030	8.0～11.0	17.0～19.0	–	–	215	490~690	(SUS304TP)	
of material shall be selected by the purchaser : 316S13, 316S31, 316S33, 304S22, 304S25, 316S22, 316S26								C313

14. BS Steel Pipes for Structual Purposes 構造用鋼管

Standard No. Year Designation 規格番号 年号 規格名称	Grade グレード	Mfg. Process 製造方法	Chemical Composition			
			C	Si	Mn	P
879－85 Steel Pipes and Joints for Hyrauric Purposes 水圧用鋼管及び継手	HFS27	S	Pipes in BS 3601(old edition)are used. Other pipes may be used by agreement.			
	HFS35	S				
	ERW27	E				
	ERW26	EFW				
1139／1－82 Tubes for Use in Scaffolding 足場用鋼管	－	S, W	≦0.20	≦0.30	－	≦0.060
1717－83 Steel Tubes for Cycle and Motor Cycle Purposes 自転車及び原動機付自転車用鋼管 ① Cold finish (hard) ② Cold finish (soft) ③ Annealed ④ Normalized	CFSC3	S, W	≦0.20	≦0.35	≦0.9	≦0.05
	CFSC4	S, W	≦0.25	≦0.35	0.6～1.0	≦0.050
	CFSC6	S, W	≦0.29	≦0.35	≦1.5	≦0.050
	ERWC1	E	≦0.13	－	≦0.60	≦0.050
	ERWC2	E	≦0.16	－	≦0.70	≦0.050
	ERWC3	E	≦0.20	≦0.35	≦0.90	≦0.050
	ERWC5	E	≦0.15	≦0.35	≦1.20	≦0.040
	CRWC1	E	≦0.13	－	≦0.60	≦0.050
	CRWC2	E	≦0.16	－	≦0.70	≦0.050
	CRWC3	E	≦0.20	≦0.35	≦0.90	≦0.050
5242 Part1－87 Tubes for Fluid Power Cylinder Barrels シリンダー用鋼管	HP1	S, W	≦0.25*	－	－	≦0.05
* Welded Tubes	HP2	S, W	≦0.25*	－	－	≦0.05

1 N/mm² or MPa = 1.01972×10⁻¹ kgf/mm²

化学成分 (%)						Tensile Test N/mm² 引張試験		Remarks 備考	Index No.
S	Ni	Cr	Mo	Others		Min. Yield Point 最小降伏点	Tensile Strength 引張強さ	(Similar to JIS)	索引番号
–	–	–	–	–		–	–	–	C314
≦0.060	–	–	–	–		205	340~460	–	C315
≦0.050	–	–	–	–	①	360	≧450	(STKM12C)	C316
					④	205	≧360		
≦0.050	–	–	–	–	①	410	≧520	(STKM13C)	
					④	235	≧410		
≦0.050	–	–	0.15~0.25	–	①	620	≧720	–	
					–	–	–		
≦0.050	–	–	–	–	–	195	≧305	(STKM11A)	
					③	145	≧275		
					④	155	≧285		
≦0.050	–	–	–	–	–	255	≧340	(STKM12A)	
					③	155	≧305		
					④	195	≧325		
≦0.050	–	–	–	–	–	305	≧400	–	
					③	195	≧345		
					④	215	≧360		
≦0.040	–	–	–	–	–	350	≧430	(STKM18A)	
					④	265	≧360		
≦0.050	–	–	–	–	①	325	≧400	(STKM11A)	
					③	145	≧275		
					④	155	≧285		
≦0.050	–	–	–	–	①	350	≧420	(STKM12C)	
					③	155	≧305		
					④	195	≧325		
≦0.050	–	–	–	–	①	360	≧450	(STKM13C)	
					③	195	≧345		
					④	215	≧360		
≦0.05	–	–	Ec≦0.50	t ≦10mm		235	≧360	(STC370)	C317
				10 < t ≦20		225			
				20 < t ≦50		215			
≦0.05	–	–	Ec≦0.50	t ≦10mm		335	≧490	(STC510A)	
				10 < t ≦20		315			
				20 < t ≦50		285			

$$EC\ (\text{Carbon Equivalent}) = C + \frac{Mn}{b} + \frac{Ni+Cu}{15} + \frac{Cr+Mo+V}{5}$$

Standard No. Year Designation 規格番号 年号 規格名称	Grade グレード	Mfg. Process 製造方法	Chemical Composition			
			C	Si	Mn	P
5242 Part1－87 (Continued)	HP3	S, W	≦0.25＊	－	－	≦0.05
	HP4	S, W	≦0.25＊	－	－	≦0.05
	HP4	S, W	≦0.25＊	－	－	≦0.05
	HP6	S, W	≦0.25＊	－	－	≦0.05
	HP7	S,W		Chemical Composition and Manufacturer		
	HP8	S,W				
5242 Part3－78 Tubes for Fluid Power Cylinder Barrels (for Minining) シリンダー用鋼管（鉱山用） *Welded Tubes	HP1	－	≦0.25＊	－	－	≦0.05
	HP2	－	≦0.25＊	－	－	≦0.05
	HP3	－	≦0.25＊	－	－	≦0.05
	HP4	－	≦0.25＊	－	－	≦0.05
	HP5	－	≦0.25＊	－	－	≦0.05
	HP6	－	≦0.25＊	－	－	≦0.05
	HP7	－	－	－	－	≦0.05
	HP8	－	－	－	－	≦0.05
6323－82 Seamless and Welded Steel Tubes for Automobile, Mechanical and General Engineering Purposes 自転車，機械構造及び，一般エンジニアリング用継目無溶接鋼管 ① Cold finish (hard) ② Cold finish (soft) ③ Annealed ④ Normalized ⑤ Welded and Sized (fully softened) ⑥ Cold finish (fully softened) HFW : Hot finished welded HFS : Hot finish seamless CFS : Cold finish seamless ERW : Electric resistance weld induction weld CEW : Cold finished ERW SAW : Submarged arc weld	HFW2	AT	≦0.16	－	≦0.70	≦0.050
	HFW3	AT	≦0.20	≦0.35	≦0.90	≦0.050
	HFW4	AT	≦0.25	≦0.35	≦1.20	≦0.050
	HFW5	AT	≦0.23	≦0.35	≦1.50	≦0.050
	HFS3	S	≦0.20	≦0.35	≦0.90	≦0.050
	HFS4	S	≦0.25	≦0.35	≦1.20	≦0.050
	HFS5	S	≦0.23	≦0.35	≦1.50	≦0.050
	HFS8	S	0.45～0.55	≦0.35	0.50～0.90	≦0.050
	CFS3	S	≦0.20	≦0.35	≦0.90	≦0.050
	CFS3A	S	≦0.20	0.10～0.35	0.60～1.00	≦0.050
	CFS4	S	≦0.25	≦0.35	≦1.20	≦0.050

1 N/mm² or MPa = 1.01972 × 10^{-1} kgf/mm²

化　学　成　分　(%)					Tensile Test N/mm² 引張試験		Remarks 備考	Index No.
S	Ni	Cr	Mo	Others	Min. Yield Point 最小降伏点	Tensile Strength 引張強さ	(Similar to JIS)	索引番号
≦0.05	—	—	Ec≦0.55	t ≦10mm 10<t≦20 20<t≦50	460 450 420	≧550	—	C317
≦0.05	—	—	Ec≦0.50	—	380	≧450	(STC440)	
≦0.05	—	—	Ec≦0.50	—	440	≧550	—	
≦0.05	—	—	Ec≦0.55	—	540	≧640	(STC590A)	
as agreed upon by purchaser					740	≧770	—	
					740	≧850	—	
≦0.05			Ec<0.50	t ≦10mm 10<t≦20 20<t≦50	235 225 215	≧360	(STC370)	
≦0.05	—	—	Ec<0.50	t ≦10mm 10<t≦20 20<t≦50	335 315 285	≧490	(STC510A)	
≦0.05	—	—	Ec<0.55	t ≦10mm 10<t≦20 20<t≦50	460 450 420	≧550	—	
≦0.05	—	—	Ec<0.50	—	385	≧450	(STC440)	
≦0.05	—	—	Ec≦0.50	—	440	≧550	—	
≦0.05	—	—	Ec≦0.55	—	540	≧640	(STC590A)	
≦0.05	—	—	—	—	650	≧780	—	
≦0.05	—	—	—	—	750	≧845	—	
≦0.050	—	—	—	—	195	≧325	(STAM340G)	C318
≦0.050	—	—	—	—	215	≧360	(STAM340G)	
≦0.050	—	—	—	—	235	≧410	(STAM390G)	
≦0.050	—	—	—	—	340	≧490	(STAM500G)	
≦0.050	—	—	—	—	215	≧360	(STKM12A)	
≦0.050	—	—	—	—	235	≧410	(STKM14A)	
≦0.050	—	—	—	—	340	≧490	(STKM14A)	
≦0.050	—	—	—	—	340	≧540	(STKM17A)	
≦0.050	—	—	—	— ①	360	≧450	(STKM12C)	
				②	285	≧400		
				③	170	≧340		
				④	205	≧360		
≦0.050	—	—	—	— ①	360	≧450	(STKM12)	
				②	285	≧400		
				③	170	≧340		
				④	215	≧360		
≦0.050	—	—	—	— ①	410	≧520	(STKM13)	
				②	315	≧450		
				③	195	≧400		
				④	235	≧410		

$$Ec\ (\text{Carbon Equivalent}) = C + \frac{Mn}{b} + \frac{Ni+Cu}{15} + \frac{Cr+Mo+V}{5}$$

Standard No. Year Designation 規格番号 年号 規格名称	Grade グレード	Mfg. Process 製造方法	Chemical Composition C	 Si	 Mn	 P
LW : Longitudinal welded stainless LWCF : Cold finished longitudinal welded seamless	CFS5	S	≦0.23	≦0.50	≦1.50	≦0.050
	CFS6	S	0.30～0.40	≦0.35	0.50～0.90	≦0.050
	CFS7	S	0.20～0.30	≦0.35	1.20～1.50	≦0.050
	CFS8	S	0.40～0.55	≦0.35	0.50～0.90	≦0.050
	CFS9	S	≦0.29	≦0.35	≦1.50	≦0.050
	CFS10	S	≦0.26	≦0.35	≦0.80	≦0.050
	CFS11	S	≦0.45	≦0.35	≦1.00	≦0.050
	ERW1	E	≦0.13	−	≦0.60	≦0.050
	ERW2	E	≦0.16	−	≦0.70	≦0.050
	ERW3	E	≦0.20	≦0.35	≦0.90	≦0.050
	ERW4	E	≦0.25	≦0.35	≦1.20	≦0.050
	ERW5	E	≦0.23	≦0.50	≦1.50	≦0.050
	CEW1	E	≦0.13	−	≦0.60	≦0.050

1 N/mm² or MPa = 1.01972 × 10^{-1} kgf/mm²

化学成分 (%)						Tensile Test N/mm² 引張試験		Remarks 備考	Index No.
S	Ni	Cr	Mo	Others		Min. Yield Point 最小降伏点	Tensile Strength 引張強さ	(Similar to JIS)	索引番号
≦0.050	–	–	–	–	①	480	≧600	(STKM19)	C318
					②	380	≧550		
					③	–	–		
					④	340	≧490		
≦0.050	–	–	–	–	①	470	≧590	(STKM16)	
					②	350	≧540		
					③	305	≧440		
					④	285	≧460		
≦0.050	–	–	–	–	①	560	≧700	–	
					②	460	≧650		
≦0.050	–	–	–	–	①	580	≧720	(STKM17)	
					②	470	≧670		
					③	305	≧510		
					④	340	≧540		
≦0.050	–	–	0.15～0.25	–	①	580	≧720	–	
					②	470	≧670		
≦0.050	–	0.80～1.20	0.15～0.30	–	①	580	≧720	(SCM418TK)	
					②	470	≧670		
≦0.050	–	0.80～1.20	0.15～0.30	–	①	580	≧720	(SCM440TK)	
					②	–	≧670		
≦0.050	–	–	–	–	–	195	≧305	(STAM290GA) (STAM290GB)	
					③	150	≧275		
					④	160	≧285		
≦0.050	–	–	–	–	–	255	≧345	(STAM340G)	
					③	160	≧305		
					④	195	≧325		
≦0.050	–	–	–	–	–	305	≧400	(STAM5G)	
					③	170	≧340		
					④	205	≧360		
≦0.050	–	–	–	–	–	350	≧450	(STAM440H)	
					③	195	≧400		
					④	235	≧410		
≦0.050	–	–	–	–	–	420	≧500	(SCM470G)	
					④	340	≧490	(SCM470G)	
≦0.050	–	–	–	–	①	325	≧400	(STAM290GA) (STAM290GA)	
					②	245	≧350		
					③	150	≧275		
					④	160	≧285		

Standard No. Year Designation 規格番号 年号 規格名称	Grade グレード	Mfg. Process 製造方法	Chemical Composition C	 Si	 Mn	 P
6323－82 (Continued)	CEW2	E	≦0.16	－	≦0.70	≦0.050
	CEW3	E	≦0.20	≦0.35	≦0.90	≦0.050
	CEW4	E	≦0.25	≦0.35	≦1.20	≦0.050
	CEW5	E	≦0.23	≦0.50	≦1.50	≦0.050
	SAW4	AM	≦0.25	≦0.35	≦1.20	≦0.050
	SAW5	AM	≦0.23	≦0.50	≦1.50	≦0.050
	LW12	A	≦0.06	≦0.90	≦0.60	≦0.040
	LW13	A	≦0.08	0.20～1.00	0.50～2.00	≦0.045
	LWCF13	A				
	LW14	A	≦0.03	0.20～1.00	0.50～2.00	≦0.045
	LWCF14	A				
	LW15	A	≦0.06	0.20～1.00	0.50～2.00	≦0.045
	LWCF15	A				
	LW16	A	≦0.03	0.20～1.00	0.50～2.00	≦0.045
	LWCF16	A				
	LW17	A	≦0.07	0.20～1.00	0.50～2.00	≦0.045
	LWCF17	A				
	LW18	A	≦0.08	0.20～1.00	0.50～2.00	≦0.045
	LWCF18	A				
6363－83	34/26	E	≦0.16	－	≦1.20	≦0.050
Welded Cold Formed Steel Structural Hollow	43/36	E	≦0.20	≦0.40	≦1.20	≦0.050
Sections 溶接冷間成形構造用中空型鋼	50/45	E	≦0.23	≦0.40	≦1.50	≦0.050

化　学　成　分 (%)						Tensile Test N/mm² 引張試験		Remarks 備考	Index No.
S	Ni	Cr	Mo	Others		Min. Yield Point 最小降伏点	Tensile Strength 引張強さ	(Similar to JIS)	索引番号
≦0.050	–	–	–	–	①	350	≧420	–	C318
					②	265	≧370	–	
					③	160	≧305	–	
					④	195	≧325	(STAM340G)	
≦0.050	–	–	–	–	①	370	≧450	–	
					②	285	≧400		
					③	170	≧340		
					④	215	≧360		
≦0.050	–	–	–	–	①	410	≧520		
					②	315	≧450	(STAM440G)	
					③	195	≧400	(STAM390G)	
					④	235	≧410		
≦0.050	–	–	–	–	①	485	≧600	–	
					②	385	≧550		
					③	–	–		
					④	340	≧490		
≦0.050	–	–	–	–		235	≧410	(STK400)	
≦0.050	–	–	–	–		340	≧490	(STK500)	
≦0.020	≦0.50	11.0～13.0	–	N ≦0.025 Ti:5×C～0.70		305	≧400	–	
≦0.030	8.0～11.0	17.0～19.0	–	–	–	450	≧560	(SUS304TK)	
					⑤	205	≧510		
					⑥	205	≧510		
≦0.030	9.0～12.0	17.5～19.0	–	–	–	420	≧520	–	
					⑤	185	≧480		
					⑥	185	≧480		
≦0.030	8.0～11.0	17.5～19.0	–	–	–	450	≧560	(SUS304TK)	
					⑤	205	≧510		
					⑥	205	≧510		
≦0.030	11.0～14.0	16.5～18.5	2.25～3.0	–	–	420	≧520	–	
					⑤	185	≧480		
					⑥	185	≧480		
≦0.030	10.0～13.0	16.5～18.5	2.25～3.0	–	–	450	≧560	(SUS316TK)	
					⑤	205	≧510		
					⑥	205	≧510		
≦0.030	9.0～12.0	17.0～19.0	–	Ti 5×C～0.70	⑥	205	≧500	(SUS321TK)	
≦0.050	–	–	–	–		265	≧340	(STKR460)	C319
≦0.050	–	–	–	–		360	≧430	(STKR400)	
≦0.050	–	–	–	–		450	≧500	(STKR490)	

15. DIN Steel Tubes and Piping for General Purposes

Standard No. Year Designation 規格番号 年号 規格名称	Grade グレード	Mfg. Process 製造方法	Chemical Composition C	Si	Mn	P
1615－84 Welded Circular Unalloyed Steel Tubes Not Subject to Special Requirement 特別規定なしの非合金溶接鋼管	－	W	ref. DIN 17100	－	－	－
1626－84 Welded Circular Tubes of Non-Alloy Steels with Special Quality Requirements 特別品質要求付溶接非合金鋼鋼管	－	－	－	－	－	－
	USt37.0	W	≦0.20	－	－	≦0.040
	St37.0	W	≦0.17	－	－	≦0.040
	St44.0	W	≦0.21	－	－	≦0.040
	St52.0	W	≦0.22	≦0.55	≦1.60	≦0.040
1628－84 Welded Circular Tubes of Non-Alloy Steels with Very High Quality Requirements 高品質要求付溶接非合金鋼鋼管	St37.4	W	≦0.17	≦0.35	≦0.35	≦0.040
	St44.4	W	≦0.20	≦0.35	≦0.40	≦0.040
	St52.4	W	≦0.22	≦0.35	≦1.60	≦0.040
1629－84 Seamless Circular Tubes of Non-Alloy Steels with Special Quality Requirements 特別品質要求付継目無非合金鋼鋼管	St37.0	S	≦0.17	－	－	≦0.040
	St44.0	S	≦0.21	－	－	≦0.040
	St52.0	S	≦0.22	≦0.55	≦1.60	≦0.040
1630－84 Seamless Circular Tubes of Non-Alloy Steels with Very High Quality Requirements 高品質要求付継目無非合金鋼鋼管	St37.4	W	≦0.17	≦0.35	≦0.35	≦0.040
	St44.4	W	≦0.20	≦0.35	≦0.40	≦0.040
	St52.4	W	≦0.22	≦0.55	≦1.60	≦0.040
2440－78 Steel Tubes Medium-Weight Suitable for Screwing ねじ切用中肉鋼管	St33.2	S, W	ref. DIN 17100		－	－
2441－78 Steel Tubes Heavy-Weight for Suitable for Screwing ねじ切用厚肉鋼管	St33.2	S, W	ref. DIN 17100		－	－
2462－69 Seamless Tubes of Stainless Steel 継目無ステンレス鋼鋼管	X20Cr13	S	－	－	－	－
	X22CrNi17	S	－	－	－	－
	X7CrAl13	S	－	－	－	－

一般用鋼管

1 N/mm² or MPa = 1.01972 × 10⁻¹ kgf/mm²

化学成分 (%)					Tensile Test N/mm² 引張試験				Remarks 備考 (Similar to JIS)	Index No. 索引番号
S	Ni	Cr	Mo	Others	Min. Yield Point 最小降伏点			Tensile Strength 引張強さ		
–	–	–	–	–	–			–	(SGP)	C 401
–	–	–	–	–	1≤16	16<t ≤40		–	–	C 402
≤0.040	–	–	–	N ≤0.007	235	–		350~480	–	
≤0.040	–	–	–	N ≤0.009	235	225		350~480	(STPG370)	
≤0.040	–	–	–	N ≤0.009	275	265		420~550	(STPG410)	
≤0.035	–	–	–	N-stabilizing element (e.g. Al 0.020)	350	346		500~650	–	
≤0.040				N-stabilizing element (e.g. Al 0.020)	235	225		350~480	–	C 403
≤0.040	–	–	–		275	265		420~550	–	
≤0.035					350	340		500~650	–	
					t≤16	16<t ≤40	40<t ≤45			
≤0.040	–	–	–	N ≤0.009	235	225	215	36~49	(STPG370)	C 404
≤0.040	–	–	–	N ≤0.009	275	265	255	43~56	(STPG410)	
≤0.035	–	–	–	N-stabilizing element (e.g. Al 0.020)	350	340	334	500~650	–	
					1≤16	16<t ≤40	40<t ≤45			
≤0.040	–	–	–	N-stabilizing element (e.g. Al 0.020)	235	225	215	350~480	(STS370)	C 405
≤0.040	–	–	–		275	265	255	420~550	(STS410)	
≤0.040	–	–	–		350	340	335	500~650	(STS480)	
–	–	–	–	–	–			–	(SGP)	C 406
–	–	–	–	–	–			–	(SGP)	C 407
–	–	–	–	–	–			–	–	C 408
–	–	–	–	–	–			–	–	
–	–	–	–	–	–			–	–	

Standard No. Year Designation 規格番号 年号 規格名称	Grade グレード	Mfg. Process 製造方法	Chemical Composition			
			C	Si	Mn	P
2462−69 (Continued)	X10Cr13	S	−	−	−	−
	X8Cr17	S	−	−	−	−
	X8CrTi17	S	−	−	−	−
	X8CrNb17	S	−	−	−	−
	X8CrNi189	S	ref. DIN 17440		−	−
	X2CrNi189	S	−	−	−	−
	X5CrNiMo1810	S	−	−	−	−
	X2CrNiMo1810	S	−	−	−	−
	X2CrNiMo1812	S	−	−	−	−
	X5CrNiMo1812	S	−	−	−	−
	X10CrNiTi189	S	−	−	−	−
	X10CrNiNo189	S	−	−	−	−
	X10CrNiMoTi1810	S	−	−	−	−
	X10CrNiMoNb1810	S	−	−	−	−
	X10CrNiMoNb1812	S	−	−	−	−
2463−69	X5CrNi189	EA	−	−	−	−
Weldel Tubes of Austenitic Stainless Steel	X2CrNi189	EA	−	−	−	−
溶接オーステナイトステンレス鋼鋼管	X5CrNiMo1810	EA	−	−	−	−
	X2CrNiMo1810	EA	ref. DIN 17440		−	−
	X2CrNiMo1812	EA	−	−	−	−
	X5CrNiMo1812	EA	−	−	−	−
	X10CrNiTi189	EA	−	−	−	−
	X10CrNiMoTi1810	EA	−	−	−	−
17172−78	As Rolled or Normalized Steel					
Steel Pipes for Pipe Lines for the	StE210.7	S, E, A	≦0.17	≦0.45	0.35≦	≦0.040
Transport of Combustible Fluids and Gases	StE240.7	S, E, A	≦0.17	≦0.45	0.40≦	≦0.040
可燃性流体輸送パイプライン用鋼管	StE290.7	S, E, A	≦0.22	≦0.45	0.50〜1.10	≦0.040
	StE320.7	S, E, A	≦0.22	≦0.45	0.70〜1.30	≦0.040
	StE360.7	S, E, A	≦0.22	≦0.55	0.90〜1.50	≦0.040
	StE385.7	S, E, A	≦0.23	≦0.55	1.00〜1.50	≦0.040
	StE415.7	S, E, A	≦0.23	≦0.55	1.00〜1.50	≦0.040
	Thermo Mechanical Treated Steel					
	StE290.7TM	S, E, A	0.04〜0.12	≦0.40	0.50〜1.50	≦0.035
	StE207.7TM	S, E, A	0.04〜0.12	≦0.40	0.70〜1.50	≦0.035
	StE360.7TM	S, E, A	0.04〜0.12	≦0.45	0.90〜1.50	≦0.035
	StE385.7TM	S, E, A	0.04〜0.14	≦0.45	1.00〜1.60	≦0.035
	StE415.7TM	S, E, A	0.04〜0.14	≦0.45	1.00〜1.60	≦0.035
	StE445.7TM	S, E, A	0.04〜0.16	≦0.55	1.00〜1.60	≦0.035
	StE480.7TM	S, E, A	0.04〜0.16	≦0.55	1.10〜1.70	≦0.035

1 N/mm² or MPa = 1.01972×10^{-1} kgf/mm²

化　学　成　分　(%)					Tensile Test N/mm² 引張試験		Remarks 備考 (Similar to JIS)	Index No. 索引番号
S	Ni	Cr	Mo	Others	Min. Yield Point 最小降伏点	Tensile Strength 引張強さ		
－	－	－	－	－	－	－	(SUS410TB)	C 409
－	－	－	－	－	－	－	(SUS430TB)	
－	－	－	－	－	－	－	－	
－	－	－	－	－	－	－	－	
－	－	－	－	－	－	－	(SUS304TP, TB)	
－	－	－	－	－	－	－	(SUS304LTP, LTB)	
－	－	－	－	－	－	－	(SUS316TP, TB)	
－	－	－	－	－	－	－	(SUS316LTP, LTB)	
－	－	－	－	－	－	－	(SUS316LTP, LTB)	
－	－	－	－	－	－	－	(SUS316TP, TB)	
－	－	－	－	－	－	－	(SUS321TP, TB)	
－	－	－	－	－	－	－	(SUS347TP, TB)	
－	－	－	－	－	－	－	－	
－	－	－	－	－	－	－	－	
－	－	－	－	－	－	－	－	
－	－	－	－	－	－	－	(SUS304TP, TB)	C 410
－	－	－	－	－	－	－	(SUS304LTP, TB)	
－	－	－	－	－	－	－	(SUS316TP, TB)	
－	－	－	－	－	－	－	(SUS316LTP, TB)	
－	－	－	－	－	－	－	(SUS316LTP, TB)	
－	－	－	－	－	－	－	(SUS316TP, TB)	
－	－	－	－	－	－	－	(SUS321TP, TB)	
－	－	－	－	－	－	－	－	
製造のまままたは焼ならし鋼								C 411
≦0.035	－	－	－	－	205	325~440	(STPG370)	
≦0.035	－	－	－	－	235	370~490	－	
≦0.035	－	－	－	－	275	420~540	－	
≦0.035	－	－	－	－	325	460~580	－	
≦0.035	－	－	－	－	360	510~630	－	
≦0.035	－	－	－	－	380	530~680	－	
≦0.035	－	－	－	－	410	550~700	－	
加工熱処理鋼								
≦0.025	－	－	－	－	295	420~540	－	
≦0.025	－	－	－	－	325	460~580	－	
≦0.025	－	－	－	－	360	510~630	－	
≦0.025	－	－	－	－	380	530~680	－	
≦0.025	－	－	－	－	380	550~700	－	
≦0.035	－	－	－	－	440	560~710	－	
≦0.035	－	－	－	－	480	600~750	－	

Standard No. Year Designation 規格番号 年号 規格名称	Grade グレード	Mfg. Process 製造方法	Chemical Composition			
			C	Si	Mn	P
17173－85 Seamless Circular Tubes Made from Steels with Low Temperature Service 低温用継目無鋼管	TST35N	S	≦0.17	≦0.35	≦0.40	≦0.030
	TST35V	S	≦0.17	≦0.35	0.50～0.80	≦0.030
	26CrMo4	S	0.22～0.29	≦0.35	0.70～1.50	≦0.030
	11MnNi53	S	≦0.14	≦0.50	0.85～1.65	≦0.030
	13MnNi63	S	≦0.18	≦0.50	0.85～1.65	≦0.030
	10Ni14	S	≦0.15	≦0.35	0.30～0.80	≦0.025
	12Ni19	S	≦0.15	≦0.35	0.30～0.80	≦0.025
	X8Ni9	S	≦0.10	≦0.35	0.30～0.80	≦0.025
17174－85 Seamless Circular Tubes Made from Steels with Low Temperature Service 低温用継目無鋼管	TST35N	S	≦0.17	≦0.35	≦0.40	≦0.030
	TST35V	S	≦0.17	≦0.35	0.50～0.80	≦0.030
	26CrMo4	S	0.22～0.29	≦0.35	0.70～1.50	≦0.030
	11MnNi53	S	≦0.14	≦0.50	0.85～1.65	≦0.030

1 N/mm² or MPa = 1.01972×10⁻¹kgf/mm²

化　学　成　分　(%)					Tensile Test N/mm² 引張試験		Remarks 備考 (Similar to JIS)	Index No. 索引番号
S	Ni	Cr	Mo	Others	Min. Yield Point 最小降伏点	Tensile Strength 引張強さ		
≦0.025	−	−	−	Al ≧0.020	225	340~460	−	C412
≦0.025	−	−	−	Al ≧0.020	t≦25mm255 25mm<t ≦40mm 235	340~490	−	
≦0.025	−	0.90〜1.20	−	−	t≦25mm440 25mm<t ≦40mm 420	560~740	−	
≦0.025	0.30〜0.80	−	−	Al ≧0.020	t≦25mm285 13mm<t ≦25mm 275 25mm<t ≦40mm 265	410~530	−	
≦0.025	0.30〜0.85	−	−	−	t≦13mm355 13mm<t ≦25mm 345 25mm<t ≦40mm 335	490~610	−	
≦0.020	3.25〜3.75	−	−	−	t≦25mm345 25mm<t ≦40mm 335	470~640	(STBL46)	
≦0.020	4.50〜5.30	−	−	−	t≦25mm390 25mm<t ≦40mm 380	510~710	−	
≦0.025	8.00~10.00	−	−	−	t≦25mm490 25mm<t ≦40mm 480	640~840	(STBL70)	
≦0.025	−	−	−	Al ≧0.020	225	340~460	−	C413
≦0.025	−	−	−	Al ≧0.020	t≦25mm255 25mm<t ≦40mm 235	340~460	−	
≦0.025	−	0.90〜1.20	−	−	t≦25mm440 25mm<t ≦40mm 420	560~740	−	
≦0.025	0.30〜0.80	−	−	Al ≧0.020	t≦13mm285 13mm<t ≦25mm 275 25mm<t ≦40mm 265	410~530	−	

Standard No. Year Designation 規格番号 年号 規格名称	Grade グレード	Mfg. Process 製造方法	Chemical Composition			
			C	Si	Mn	P
17174－85 (Continued)	13MnNi63	S	≦0.18	≦0.50	0.85～1.65	≦0.030
	10Ni14	S	≦0.15	≦0.35	0.30～0.80	≦0.025
	12Ni19	S	≦0.15	≦0.35	0.30～0.80	≦0.025
	X8Ni9	S	≦0.10	≦0.35	0.30～0.80	≦0.025
17175－79 Seamless Steel Tubes for Elevated Temperatures 高温用継目無鋼管	St35.8	S	≦0.17	0.10～0.35	0.40～0.80	≦0.040
	St45.8	S	≦0.21	0.10～0.35	0.40～1.20	≦0.040
	17Mn4	S	0.14～0.20	0.20～0.40	0.90～1.20	≦0.040
	19Mn5	S	0.17～0.22	0.30～0.60	1.00～1.30	≦0.040
	15Mo3	S	0.12～0.20	0.10～0.35	0.40～0.80	≦0.035
	13CrMo44	S	0.10～0.18	0.10～0.35	0.40～0.70	≦0.035
	10CrMo910	S	0.08～0.15	≦0.50	0.40～0.70	≦0.035
	14MoV63	S	0.10～0.18	0.10～0.35	0.40～0.70	≦0.035
	X20CrMoV121	S	0.17～0.23	≦0.50	≦1.00	≦0.030
17177－79 Electrically Resistance or Induction Welded Steel Tubes for Elevated Temperatures 高温用電気溶接鋼管	St37.8	E	≦0.17	0.10～0.35	0.40～0.80	≦0.040
	St42.8	E	≦0.21	0.10～0.35	0.40～1.20	≦0.040
	15Mo3	E	0.12～0.20	0.10～0.35	0.40～0.80	≦0.035
17455－85 General Purpose Welded Circular Stainless Steel Tubes 一般用ステンレス鋼溶接鋼管	X6CrTi12	W	≦0.08	－	－	－
	X6Cr17	W	≦0.08	－	－	－
	X6CrTi17	W	≦0.08	－	－	－
	X5CrNi1810	W	≦0.07	－	－	－
	X2CrNi1911	W	≦0.03	－	－	－

$1\ N/mm^2$ or $MPa = 1.01972 \times 10^{-1} kgf/mm^2$

<table>
<tr><th colspan="5">化　学　成　分　(%)</th><th colspan="3">Tensile Test
N/mm²
引　張　試　験</th><th rowspan="2">Remarks
備　　考
(Similar to JIS)</th><th rowspan="2">Index No.
索引番号</th></tr>
<tr><th>S</th><th>Ni</th><th>Cr</th><th>Mo</th><th>Others</th><th colspan="2">Min. Yield Point
最小降伏点</th><th>Tensile Strength
引張強さ</th></tr>
<tr><td>≦0.025</td><td>0.30～0.80</td><td>–</td><td>–</td><td>–</td><td colspan="2">t≦13mm355
13mm<t
≦25mm 345
25mm<t
≦40mm 335</td><td>490~610</td><td>–</td><td rowspan="4">C414</td></tr>
<tr><td>≦0.020</td><td>3.25～3.75</td><td>–</td><td>–</td><td>–</td><td colspan="2">t≦25mm345
25mm<t
≦40mm 335</td><td>470~640</td><td>(STBL46)</td></tr>
<tr><td>≦0.020</td><td>4.50～5.30</td><td>–</td><td>–</td><td>–</td><td colspan="2">t≦25mm390
25mm<t
≦40mm 380</td><td>510~710</td><td>–</td></tr>
<tr><td>≦0.025</td><td>8.00~10.00</td><td>–</td><td>–</td><td>–</td><td colspan="2">t≦25mm490
25mm<t
≦40mm 480</td><td>640~840</td><td>(STBL46)</td></tr>
<tr><td>≦0.040</td><td>–</td><td>–</td><td>–</td><td>–</td><td>t≦16
235</td><td>16<t ≦40
225</td><td>855~480</td><td>Impact Test at Room Temp. kgf-m (STPT38)</td><td rowspan="9">C415</td></tr>
<tr><td>≦0.040</td><td>–</td><td>–</td><td>–</td><td>–</td><td>–</td><td>245</td><td>410~530</td><td>23.5(STPT410, STB410)</td></tr>
<tr><td>≦0.040</td><td>–</td><td>≦0.30</td><td>–</td><td>–</td><td>–</td><td>275</td><td>460~580</td><td>22.8</td></tr>
<tr><td>≦0.040</td><td>–</td><td>≦0.30</td><td>–</td><td>–</td><td>–</td><td>315</td><td>510~610</td><td>23.5(STB52)</td></tr>
<tr><td>≦0.035</td><td>–</td><td>–</td><td>0.25～0.35</td><td>–</td><td>275</td><td>275</td><td>450~600</td><td>23.5</td></tr>
<tr><td>≦0.035</td><td>–</td><td>0.70～1.10</td><td>0.45～0.65</td><td>–</td><td>295</td><td>295</td><td>440~590</td><td>23.5(STPA22, STBA22)</td></tr>
<tr><td>≦0.035</td><td>–</td><td>2.00～2.50</td><td>0.90～1.20</td><td>–</td><td>285</td><td>285</td><td>450~600</td><td>23.2(STPA24, STBA24)</td></tr>
<tr><td>≦0.035</td><td>–</td><td>0.30～0.60</td><td>0.50～0.70</td><td>V:0.22～0.32</td><td>325</td><td>325</td><td>460~610</td><td>24.2</td></tr>
<tr><td>≦0.030</td><td>0.30～0.80</td><td>10.00～12.50</td><td>0.80～1.20</td><td>V:0.25～0.35</td><td>490</td><td>490</td><td>690~850</td><td>23.5</td></tr>
<tr><td>≦0.040</td><td>–</td><td>–</td><td>–</td><td>–</td><td colspan="2">235</td><td>360~480</td><td>(STPT370)</td><td rowspan="4">C416</td></tr>
<tr><td>≦0.040</td><td>–</td><td>–</td><td>–</td><td>–</td><td colspan="2">255</td><td>410~530</td><td>(STPT410, STB410)</td></tr>
<tr><td>≦0.035</td><td>–</td><td>–</td><td>0.25～0.35</td><td>–</td><td colspan="2">275</td><td>450~600</td><td>–</td></tr>
<tr><td></td><td></td><td></td><td></td><td></td><td colspan="2"></td><td></td><td></td></tr>
<tr><td>–</td><td>–</td><td>10.5～12.5</td><td>–</td><td>Ti:6×%C≦1.00</td><td colspan="2">190</td><td>390~560</td><td>–</td><td rowspan="5">C417</td></tr>
<tr><td>–</td><td>–</td><td>15.5～17.5</td><td>–</td><td>–</td><td colspan="2">270</td><td>450~600</td><td>–</td></tr>
<tr><td>–</td><td>–</td><td>16.0～18.0</td><td>–</td><td>Ti:7×%C≦1.20</td><td colspan="2">270</td><td>430~600</td><td>–</td></tr>
<tr><td>–</td><td>8.5～10.5</td><td>17.0～19.0</td><td>–</td><td>–</td><td colspan="2">195</td><td>500~720</td><td>–</td></tr>
<tr><td>–</td><td>10.5～12.5</td><td>18.0～20.0</td><td>–</td><td>–</td><td colspan="2">180</td><td>460~680</td><td>–</td></tr>
</table>

Standard No. Year Designation 規格番号 年号 規格名称	Grade グレード	Mfg. Process 製造方法	Chemical Composition C	 Si	 Mn	 P
17455－85 (Continued)	X2CrNiN1810	W	≦0.03	－	－	－
	X6CrNiTi1810	W	≦0.08	－	－	－
	X5CrNiN61810	W	≦0.08	－	－	－
	X5CrNiMo 17122	W	≦0.07	－	－	－
	X2CrNiMo 17132	W	≦0.030	－	－	－
	X2CrNiMoTi 17122	W	≦0.08	－	－	－
	X6CrNiMoN 17133	W	≦0.030	－	－	－
	X2CrNiMo 18143	W	≦0.030	－	－	－
	X5CrNiMo 17133	W	≦0.07	－	－	－
	X2CrNiMoN 17135	W	≦0.030	－	－	－
17456－85 General Purpose Seamless Circular Stainless Steel Tubes 一般用ステンレス鋼継目無鋼管	X6CrTi12	S	≦0.08	－	－	－
	X6CrAl13	S	≦0.08	－	－	－
	X5CrNi1911	S	≦0.030	－	－	－
	X6Cr17	S	≦0.08	－	－	－
	X6CrTi17	S	≦0.08	－	－	－
	X5CrNi1810	S	≦0.07	－	－	－
	X2CrNi1911	S	≦0.03	－	－	－
	X2CrNiN1810	S	≦0.03	－	－	－
	X6CrNiTi1810	S	≦0.08	－	－	－
	X5CrNiNB1810	S	≦0.08	－	－	－
	X5CrNiMo 17122	S	≦0.07	－	－	－
	X2CrNiMo 17132	S	≦0.030	－	－	－

1 N/mm² or MPa = 1.01972×10⁻¹kgf/mm²

化　学　成　分　(%)					Tensile Test N/mm² 引張試験		Remarks 備考	Index No.
S	Ni	Cr	Mo	Others	Min. Yield Point 最小降伏点	Tensile Strength 引張強さ	(Similar to JIS)	索引番号
−	8.5〜10.5	17.0〜19.0	−	N:0.12〜0.22	270	550~760	−	C417
−	9.0〜12.0	17.0〜19.0	−	Ti:5×%C≦0.80	200	500~730	−	
−	9.0〜12.0	17.0〜19.0	−	Nb:10×%C≦1.00	205	510~740	−	
−	10.5〜13.5	16.5〜18.5	2.0〜2.5	−	205	510~710	SUS316TB	
−	11.0〜14.0	16.5〜18.5	2.0〜2.5	−	190	490~690	SUS316HTB	
−	10.5〜13.5	16.5〜18.5	2.0〜2.5	Ti:5×%C≦0.80	210	500~730	SUS321TB	
≦0.025	11.5〜15.0	16.5〜18.5	2.5〜3.0	N :0.14〜0.22	295	580~800	−	
≦0.025	12.5〜15.5	17.0〜18.5	2.5〜3.0	−	190	490~690	−	
≦0.025	11.0〜14.0	17.0〜18.5	2.5〜3.0	−	205	510~710	−	
≦0.025	12.5〜14.5	16.5〜18.5	4.0〜5.0	N :0.12〜0.22	285	580~800	−	
−	−	10.5〜12.5	−	Ti:6×%C≦1.00	190	390~560	−	C418
−	−	12.0〜14.0	−	Al:0.10〜0.30	250	400~600	SUS405TP	
−	8.5〜10.5	18.0〜14.0	−	−	250	450~650	−	
−	−	15.5〜17.5	−	−	270	450~600	−	
−	−	16.0〜18.0	−	Ti:7×%C≦1.20	270	430~600	−	
−	8.5〜10.5	17.0〜19.0	−	−	195	500~720	−	
−	10.5〜12.5	18.0〜20.0	−	−	180	460~680	−	
−	8.5〜10.5	17.0〜19.0	−	N :0.12〜0.22	270	550~760	−	
−	9.0〜12.0	17.0〜19.0	−	Ti:5×%C≦0.80	200	500~730	−	
−	9.0〜12.0	17.0〜19.0	−	Nb:10×%C≦1.00	205	510~740	−	
−	10.5〜13.5	16.5〜18.5	2.0〜2.5	−	205	510~710	SUS316TB	
−	11.0〜14.0	16.5〜18.5	2.0〜2.5	−	190	490~690	SUS316HTB	

Standard No. 規格番号	Year 年号	Designation 規格名称	Grade グレード	Mfg. Process 製造方法	Chemical Composition			
					C	Si	Mn	P
17456 (Continued)			X6CrNiMoTi 17122	S	≦0.08	–	–	–
			X2CrNiMoN 17133	S	≦0.030	–	–	–
			X2CrNiMo 18143	S	≦0.030	–	–	–
			X5CrNiMo 17133	S	≦0.07	–	–	–
			X2CrNiMoN 17135	S	≦0.030	–	–	–

1 N/mm² or MPa = 1.01972×10⁻¹kgf/mm²

化　学　成　分　(%)					Tensile Test N/mm² 引張試験		Remarks 備考	Index No.
S	Ni	Cr	Mo	Others	Min. Yield Point 最小降伏点	Tensile Strength 引張強さ	(Similar to JIS)	索引番号
–	10.5～13.5	16.5～18.5	2.0～2.5	Ti:5×%C≦0.80	210	500~730	SUS321TB	C418
≦0.025	11.5～14.5	16.5～18.5	2.5～3.0	N :0.14～0.22	295	580~800	–	
≦0.025	12.5～15.5	17.0～18.5	2.5～3.0	–	190	490~690	–	
≦0.025	11.0～14.0	17.0～18.5	2.5～3.0	–	205	510~710	–	
≦0.025	12.5～14.5	16.5～18.5	4.0～5.0	N :0.12～0.22	285	580~800	–	

16. DIN Precision Tubing　精密鋼管

Standard No. Year Designation 規格番号 年号 規格名称	Grade グレード	Mfg. Process 製造方法	Chemical Composition C	Si	Mn	P
2391－81 Seamless Precision Steel Tubes 継目無精密鋼管 ① Cold-finished／hard ② Cold-finished／soft ③ Annealed ④ Normalized	St30Si	S	≦0.10	≦0.30	≦0.055	≦0.040
	St30Al	S	≦0.10	≦0.05	≦0.55	≦0.040
	St35	S	≦0.17	≦0.35	≧0.40	≦0.050
	St45	S	≦0.21	≦0.35	≧0.40	≦0.050
	St52	S	≦0.22	≦0.55	≦1.60	≦0.050
2393－81 Welded Precision Steel Tubes 溶接精密鋼管 ① Cold-finished／hard ② Cold-finished／soft ③ Annealed ④ Normalized	St28 USt28 RSt28	W	≦0.13	－	－	≦0.050
	St34-2 USt34-2 RSt34-2	W	≦0.15	－	－	≦0.050
	St37-2 USt37-2 RSt37-2	W	≦0.17	－	－	≦0.050
	St44-2	W	≦0.21	－	－	≦0.050
	St52-3	W	≦0.22	≦0.55	≧1.60	≦0.040

1 N/mm² or MPa = 1.01972 × 10^{-1} kgf/mm²

化学成分(%)						Tensile Test N/mm² 引張試験		Remarks 備考	Index No.
S	Ni	Cr	Mo	Others		Min. Yield Point 最小降伏点	Tensile Strength 引張強さ	(Similar to JIS)	索引番号
≦0.040	−	−	−	−	①	−	≧400	(STKM11)	C419
					②	−	≧335		
					③	−	≧285		
					④	215	295〜420		
≦0.040	−	−	−	Al deoxydized	①	−	≧400	(STKM11)	
					②	−	≧335		
					③	−	≧285		
					④	215	295〜420		
≦0.050	−	−	−	−	①	−	≧440	(STC38)	
					②	−	≧370		
					③	−	≧315		
					④	235	340〜470		
≦0.050	−	−	−	−	①	−	≧540	(STKM13)	
					②	−	≧470		
					③	−	≧390		
					④	255	440〜570		
≦0.050	−	−	−	−	①	−	≧590	(STKM19)	
					②	−	≧540		
					③	−	≧490		
					④	350	490〜630		
≦0.050	−	−	−	−	①	−	≧400	(STKM11) (STAM80G)	C420
					②	−	≧325		
					③	−	≧265		
					④	175	275〜380		
≦0.050	−	−	−	−	①	−	≧410	−	
					②	−	≧350		
					③	−	≧305		
					④	205	315〜410		
≦0.050	−	−	−	−	①	−	≧440	(STKM12)	
					②	−	≧370		
					③	−	≧315		
					④	235	340〜470		
≦0.050	−	−	−	−	①	−	≧570	(STKM13) (STAM40G)	
					②	−	≧450		
					③	−	≧390		
					④	255	410〜540		
≦0.040	−	−	−	−	①	−	≧590	(STKM19)	
					②	−	≧540		
					③	−	≧490		

Standard No. Year Designation 規格番号 年号 規格名称	Grade グレード	Mfg. Process 製造方法	Chemical Composition			
			C	Si	Mn	P
2393－81(Continued)						
2394－81 Welded and Sized Precision Steel Tubes 溶接定径精密鋼管 ① Welded and Sized ② Annealed ③ Noemalized	St28 USt28 RSt28	W	≦0.13	－	－	≦0.050
	St34-2 USt34-2 RSt34-2	W	≦0.15	－	－	≦0.050
	St37-2 USt37-2 RSt37-2	W	≦0.17	－	－	≦0.050
	St44-2	W	≦0.21	－	－	≦0.050
	St52-3	W	≦0.25	≦0.55	≦1.60	≦0.040
17457－85 Welded Circular Austenitic Stainless Steel Tubes Subject to Special Requirements 特別規定付ステンレス鋼溶接鋼管	X5CrNi1820	W	≦0.07	－	－	－
	X2CrNi1911	W	≦0.030	－	－	－
	X2CrNiN1810	W	≦0.030	－	－	－
	X6CrNiTi 1810	W	≦0.08	－	－	－
	X6CrNiNb 1810	W	≦0.08	－	－	－
	X5CrNiMo 17122	W	≦0.07	－	－	－
	X2CrNiMo 17132	W	≦0.030	－	－	－
	X6CrNiMoTi 17122	W	≦0.08	－	－	－
	X2CrNiMoN 17133	W	≦0.030	－	－	－
	X2CrNiMo 18143	W	≦0.030	－	－	－
	X5CrNiMo 17133	W	≦0.07	－	－	－
	X2CrNiMoN 17135	W	≦0.030	－	－	－

1 N/mm² or MPa = 1.01972 × 10^{-1} kgf/mm²

化　学　成　分　(%)						Tensile Test N/mm² 引張試験		Remarks 備考 (Similar to JIS)	Index No. 索引番号
S	Ni	Cr	Mo	Others		Min. Yield Point 最小降伏点	Tensile Strength 引張強さ		
					④	350	490～630		
≦0.050	−	−	−	−	①	−	≧305	(STKM11) (STAM290G)	C421
					②	−	≧265		
					③	180	275～385		
≦0.050	−	−	−	−	①	−	≧335	−	
					②	−	≧305		
					③	205	315～410		
≦0.050	−	−	−	−	①	−	≧390	(STKM12)	
					②	−	≧315		
					③	235	340～470		
≦0.050	−	−	−	−	①	−	≧440	(STKM13) (STAM390G)	
					②	−	≧390		
					③	255	410～540		
≦0.040	−	−	−	−	①	−	≧540	(STKM19)	
					②	−	≧490		
					③	350	490～630		
−	8.5～10.5	17.0～19.0	−	−		195	500～720	−	C422
−	10.0～12.5	18.0～20.0	−	−		180	460～680	−	
−	8.5～11.5	17.0～19.0	−	No.12 ～0.22		270	550～760	−	
−	9.0～12.0	17.0～19.0	−	Ti:5×%C, ≦0.80		200	500～730	SUS321TB	
−	9.0～12.0	17.0～19.0	−	Nb:10×%C, ≦1.00		205	510～740	SUS347TB	
−	10.5～13.5	16.5～18.5	2.0 ～ 2.5	−		205	510～710	SUS316TB	
−	11.0～14.0	16.5～18.5	2.0 ～ 2.5	−		190	490～690	−	
−	10.5～13.5	16.5～18.5	2.0 ～ 2.5	Ti:5×%C, ≦0.80		210	500～730	−	
≦0.025	11.5～14.5	16.5～18.5	2.5 ～ 3.0	N:0.14～0.22		295	580～800	−	
≦0.025	12.5～15.0	17.0～18.5	2.5 ～ 3.0	−		190	490～690	−	
≦0.025	11.0～14.0	16.5～18.5	2.5 ～ 3.0	N:0.12～0.22		285	510～710	−	
≦0.025	12.5～14.5	16.5～18.5	4.0 ～ 5.0	N:0.12～0.22		315	580～800	−	

Standard No. Year Designation 規格番号 年号 規格名称	Grade グレード	Mfg. Process 製造方法	Chemical Composition			
			C	Si	Mn	P
17458－85 Seamless Circular Austenitic Stainless Steel Tubes Subject to Special Requirements 特別規定付ステンレス鋼溶接鋼管	X5CrNi1820	S	≦0.07	－	－	－
	X2CrNi1911	S	≦0.030	－	－	－
	X2CrNiN1810	S	≦0.030	－	－	－
	X6CrNiTi 1810	S	≦0.08	－	－	－
	X6CrNiNb 1810	S	≦0.08	－	－	－
	X5CrNiMo 17122	S	≦0.07	－	－	－
	X2CrNiMo 17132	S	≦0.030	－	－	－
	X6CrNiMoTi 17122	S	≦0.08	－	－	－
	X6CrNiMoNb 17122	S	≦0.08	－	－	－
	X2CrNiMoN 17133	S	≦0.030	－	－	－
	X2CrNiMo 18143	S	≦0.030	－	－	－
	X5CrNiMo 17133	S	≦0.07	－	－	－
	X2CrNiMoN 17135	S	≦0.030	－	－	－

1 N/mm^2 or MPa = 1.01972×10^{-1} kgf/mm^2

化　学　成　分 (%)					Tensile Test N/mm^2 引張試験		Remarks 備考	Index No.
S	Ni	Cr	Mo	Others	Min. Yield Point 最小降伏点	Tensile Strength 引張強さ	(Similar to JIS)	索引番号
−	8.5 ~10.5	17.0~19.0	−	−	195	500~720	−	C423
−	10.0~12.5	18.0~20.0	−	−	180	460~680	−	
−	8.5~11.5	17.0~19.0	−	N:12~0.22	270	550~760	−	
−	9.0~12.0	17.0~19.0	−	Ti : ≦0.80	200	500~730	SUS321TB	
−	9.0~12.0	17.0~19.0	−	Nb : 10×%C≦1.00	205	510~740	SUS347TB	
−	10.5~13.5	16.5~18.5	2.0 ~ 2.5	−	205	510~710	SUS316TB	
−	11.0~14.0	16.5~18.5	2.0 ~ 2.5	−	190	490~690	−	
−	10.5~13.5	16.5~18.5	2.0 ~ 2.5	Ti:5×%C, ≦0.80	210	500~730	−	
−	10.5~13.5	16.5~18.5	2.0 ~ 2.5	Nb : 10×%C≦1.00	210	510~740	−	
≦0.025	11.5~14.5	16.5~18.5	2.5 ~ 3.0	N:0.14~0.22	295	580~800	−	
≦0.025	12.5~15.0	17.0~18.5	2.5 ~ 3.0	−	190	490~690	−	
≦0.025	11.0~14.0	16.5~18.5	2.5 ~ 3.0	N:0.12~0.22	285	510~710	−	
≦0.025	12.5~14.5	16.5~18.5	4.0 ~ 5.0	N:0.12~0.22	315	580~800	−	

17. DIN Steel Tubes for Structural Purposes 構造用鋼管

Standard No. Year Designation 規格番号 年号 規格名称	Grade グレード	Mfg. Process 製造方法	Chemical Composition			
			C	Si	Mn	P
17120－84 Welded Circular Steel Tubes for Structural Steelwork 一般構造用溶接鋼管	USt37-2	W	≦0.17	–	–	≦0.050
	RSt37-2	W	≦0.17	–	–	≦0.040
	St37-2	W	≦0.17	–	–	≦0.040
	St44-2	W	≦0.21	–	–	≦0.050
	St44-3	W	≦0.20	–	–	≦0.040
	St52-3	W	≦0.22	–	–	≦0.040
17121－84 Seamless Circular Steel Tubes for Structural Steelwork 一般構造用継目無鋼管	RSt37-2	S	≦0.17	–	–	≦0.050
	St37-3	S	≦0.17	–	–	≦0.040
	St44-2	S	≦0.21	–	–	≦0.050

1 N/mm² or MPa = 1.01972×10^{-1} kgf/mm²

化　学　成　分　(%)					Tensile Test N/mm² 引張試験		Remarks 備考 (Similar to JIS)	Index No. 索引番号
S	Ni	Cr	Mo	Others	Min. Yield Point 最小降伏点	Tensile Strength 引張強さ		
≦0.050	–	–	–	N≦0.007	t≦16mm 235 16mm<t≦40mm –	340～470	–	C424
≦0.040	–	–	–	N≦0.009	t≦16mm 235 16mm<t≦40mm 225	340～470	–	
≦0.040	–	–	–	–	t≦16mm 235 16mm<t≦40mm 225	340～470	–	
≦0.050	–	–	–	N≦0.009	t≦16mm 275 16mm<t≦40mm 265	410～540	STK400	
≦0.040	–	–	–	–	t≦16mm 275 16mm<t≦40mm 265	410～540	STK400	
≦0.040	–	–	–	–	t≦16mm 355 16mm<t≦40mm 345	490～630	STK500	
≦0.050	–	–	–	N≦0.009	t≦16mm 235 16mm<t≦40mm 225 40mm<t≦65mm 215	340～470	–	C425
≦0.040	–	–	–	–	t≦16mm 275 16mm<t≦40mm 225 40mm<t≦65mm 215	340～470	–	
≦0.050	–	–	–	N≦0.009	t≦16mm 275 16mm<t≦40mm 265	410～540	STK400	

Standard No. Year Designation 規格番号 年号 規格名称	Grade グレード	Mfg. Process 製造方法	Chemical Composition C	 Si	 Mn	 P
17121－84(continued)						
	St44-3	S	≦0.20	－	－	≦0.040
	St52-3	S	≦0.23	－	－	≦0.040
17123－86 Welded Circular Fine Grain Steel Tubes for Structural Steelwork 構造用溶接細粒鋼鋼管	STE225	W	≦0.18	≦0.40	0.50～1.30	≦0.035
	TSTE255	W	≦0.16	≦0.40	0.50～1.30	≦0.030
	ESTE255	W	≦0.16	≦0.40	0.50～1.30	≦0.025

1 N/mm² or MPa = 1.01972 × 10^{-1} kgf/mm²

化　学　成　分　(%)					Tensile Test N/mm² 引張試験		Remarks 備考 (Similar to JIS)	Index No. 索引番号
S	Ni	Cr	Mo	Others	Min. Yield Point 最小降伏点	Tensile Strength 引張強さ		
					40mm<t ≦65mm 255			C 426
≦0.040	–	–	–	–	t ≦16mm 275 16mm<t ≦40mm 265 40mm<t ≦65mm 255	410～540	STK400	
≦0.040	–	–	–	–	t ≦16mm 355 16mm<t ≦40mm 345 40mm<t ≦65mm 335	490～630	STK500	
≦0.030	≦0.30	≦0.30	≦0.08	N ≦0.020 Al ≧0.020 Cu ≦0.20 Nb ≦0.03 Nb+Ti+V ≦0.05	t ≦12mm 255 12mm<t ≦20mm 255 20mm<t ≦40mm 245 40mm<t ≦50mm 235 50mm<t ≦65mm 225	360～480	–	C 427
≦0.025	≦0.30	≦0.30	≦0.08	N ≦0.020 Al ≧0.020 Cu ≦0.20 Nb ≦0.03 Nb+Ti+V ≦0.05	t ≦12mm 255 12mm<t ≦20mm 255 20mm<t ≦40mm 245 40mm<t ≦50mm 235 50mm<t ≦65mm 225	360～480	–	
≦0.015	≦0.30	≦0.30	≦0.08	N ≦0.020 Al ≧0.020 Cu ≦0.20 Nb ≦0.03	t ≦12mm 255 12mm<t ≦20mm 255 20mm<t ≦40mm 245	360～480		

Standard No. Year Designation 規格番号 年号 規格名称	Grade グレード	Mfg. Process 製造方法	Chemical Composition C	 Si	 Mn	 P
17123－86 (Continued)						
	STE285	W	≦0.18	≦0.40	0.60～1.40	≦0.035
	STE285	W	≦0.16	≦0.40	0.60～1.40	≦0.030
	STE285	W	≦0.16	≦0.040	0.60～1.40	≦0.025
	STE355	W	≦0.20	0.10～0.50	0.90～1.65	≦0.035

1 N/mm² or MPa = 1.01972 × 10^{-1} kgf/mm²

化　学　成　分 (%)					Tensile Test N/mm² 引張試験		Remarks 備　考 (Similar to JIS)	Index No. 索引番号
S	Ni	Cr	Mo	Others	Min. Yield Point 最小降伏点	Tensile Strength 引張強さ		
				Nb+Ti+V ≦0.05	40mm<t≦50mm 235 50mm<t≦65mm 225			C427
≦0.030	≦0.30	≦0.30	≦0.08	N ≦0.020 Al≧0.020 Cu≦0.20 Nb≦0.03 Nb+Ti+V ≦0.05	t ≦12mm 285 12mm<t≦20mm 285 20mm<t≦40mm 275 40mm<t≦50mm 265 50mm<t≦65mm 255	390〜510	STK400	
≦0.025	≦0.30	≦0.30	≦0.08	N ≦0.020 Al≧0.020 Cu≦0.20 Nb≦0.03 Nb+Ti+V ≦0.05	t ≦12mm 285 12mm<t≦20mm 285 20mm<t≦40mm 275 40mm<t≦50mm 265 50mm<t≦65mm 255	390〜510	STK400	
≦0.015	≦0.30	≦0.30	≦0.08	N ≦0.020 Al≧0.020 Cu≦0.20 Nb≦0.03 Nb+Ti+V ≦0.05	t ≦12mm 285 12mm<t≦20mm 285 20mm<t≦40mm 275 40mm<t≦50mm 265 50mm<t≦65mm 255	390〜510	STK400	
≦0.030	≦0.30	≦0.30	≦0.08	N ≦0.020 Al≧0.020 Cu≦0.20 Nb≦0.05	t ≦12mm 355 12mm<t≦20mm 355 20mm<t≦40mm 345	490〜639	STK500	

Standard No. Year Designation 規格番号 年号 規格名称	Grade グレード	Mfg. Process 製造方法	Chemical Composition C	 Si	 Mn	 P
17123−86(Continued)						
	TSTE355	W	≦0.18	0.10〜0.50	0.90〜1.65	≦0.030
	ESTE355	W	≦0.18	0.10〜0.50	0.90〜1.65	≦0.025
	STE420	W	≦0.20	0.10〜0.60	1.00〜1.70	≦0.035
	TSTE420	W	≦0.20	0.10〜0.60	1.00〜1.70	≦0.030

$1\ N/mm^2$ or $MPa = 1.01972 \times 10^{-1} kgf/mm^2$

化　学　成　分　(%)					Tensile Test N/mm² 引張試験		Remarks 備　考	Index No.
S	Ni	Cr	Mo	Others	Min. Yield Point 最小降伏点	Tensile Strength 引張強さ	(Similar to JIS)	索引番号
				V ≦0.10 Nb+Ti+V ≦0.12	40mm<t≦50mm 335 50mm<t≦65mm 325			C427
≦0.025	≦0.30	≦0.30	≦0.08	N ≦0.020 Al≧0.020 Cu≦0.20 Nb≦0.05 V ≦0.10 Nb+Ti+V ≦0.12	t ≦12mm 355 12mm<t≦20mm 355 20mm<t≦40mm 345 40mm<t≦50mm 335 50mm<t≦65mm 325	490～639	STK500	
≦0.015	≦0.30	≦0.30	≦0.08	N ≦0.020 Al≧0.020 Cu≦0.20 Nb≦0.05 V ≦0.10 Nb+Ti+V ≦0.12	t ≦12mm 355 12mm<t≦20mm 355 20mm<t≦40mm 345 40mm<t≦50mm 335 50mm<t≦65mm 325	490～639	STK500	
≦0.030	≦1.00	≦0.30	≦0.10	N ≦0.020 Al≧0.020 Cu≦0.20 Nb≦0.05 V ≦0.20 Nb+Ti+V ≦0.22	t ≦12mm 420 12mm<t≦20mm 410 20mm<t≦40mm 400 40mm<t≦50mm 385 50mm<t≦65mm 375	530～680	–	
≦0.025	≦1.00	≦0.30	≦0.10	N ≦0.020 Al≧0.020 Cu≦0.20 Nb≦0.05	t ≦12mm 420 12mm<t≦20mm 410 20mm<t≦40mm 400	530～680		

Standard No. 規格番号	Year 年号	Designation 規格名称	Grade グレード	Mfg. Process 製造方法	Chemical Composition C	Si	Mn	P
17123－86(Continued)								
			ESTE420	W	≦0.20	0.10～0.60	1.00～1.70	≦0.025
			STE460	W	≦0.20	0.10～0.60	1.00～1.70	≦0.035
			TSTE460	W	≦0.20	0.10～0.60	1.00～1.70	≦0.030
			ESTE460	W	≦0.20	0.10～0.60	1.00～1.70	≦0.025

1 N/mm² or MPa =1.01972×10⁻¹kgf/mm²

化　学　成　分　(%)					Tensile Test N/mm² 引張試験		Remarks 備考 (Similar to JIS)	Index No. 索引番号
S	Ni	Cr	Mo	Others	Min. Yield Point 最小降伏点	Tensile Strength 引張強さ		
				V ≦0.20 Nb+Ti+V ≦0.22	40mm<t≦50mm 385 50mm<t≦65mm 375			C427
≦0.015	≦1.00	≦0.30	≦0.10	N ≦0.020 Al≧0.020 Cu≦0.20 Nb≦0.05 V ≦0.20 Nb+Ti+V ≦0.22	t ≦12mm 420 12mm<t≦20mm 410 20mm<t≦40mm 400 40mm<t≦50mm 385 50mm<t≦65mm 375	530〜680	−	
≦0.030	≦1.00	≦0.30	≦0.10	N ≦0.020 Al≧0.020 Cu≦0.20 Nb≦0.05 V ≦0.20 Nb+Ti+V ≦0.22	t ≦12mm 460 12mm<t≦20mm 450 20mm<t≦40mm 440 40mm<t≦50mm 425 50mm<t≦65mm 410	560〜730	−	
≦0.025	≦1.00	≦0.30	≦0.10	N ≦0.020 Al≧0.020 Cu≦0.20 Nb≦0.05 V ≦0.20 Nb+Ti+V ≦0.22	t ≦12mm 460 12mm<t≦20mm 450 20mm<t≦40mm 440 40mm<t≦50mm 425 50mm<t≦65mm 410	560〜730	−	
≦0.015	≦1.00	≦0.30	≦0.10	N ≦0.020 Al≧0.020 Cu≦0.20 Nb≦0.05	t ≦12mm 460 12mm<t≦20mm 450 20mm<t≦40mm 440	560〜730		

Standard No. Year Designation 規格番号 年号 規格名称	Grade グレード	Mfg. Process 製造方法	Chemical Composition C	 Si	 Mn	 P
17123－86(Continued)						
17124－86 Seamless Circular Fine Grain Steel Tubes for Structural Steelwork 構造用溶接細粒鋼鋼管	STE255	S	≦0.18	≦0.40	0.50～1.30	≦0.035
	TSTE255	S	≦0.16	≦0.40	0.50～1.30	≦0.030
	ESTE255	S	≦0.16	≦0.40	0.50～1.30	≦0.025
	STE285	S	≦0.18	≦0.40	0.60～1.40	≦0.035

1 N/mm² or MPa = 1.01972×10^{-1} kgf/mm²

化　学　成　分（%）					Tensile Test N/mm² 引張試験		Remarks 備考	Index No.
S	Ni	Cr	Mo	Others	Min. Yield Point 最小降伏点	Tensile Strength 引張強さ	(Similar to JIS)	索引番号
				V ≦0.20 Nb+Ti+V ≦0.22	40mm<t≦50mm 425 50mm<t≦65mm 410			C427
≦0.030	≦0.30	≦0.30	≦0.08	N ≦0.020 Al≧0.020 Cu≦0.20 Nb≦0.03 Nb+Ti+V ≦0.05	t ≦12mm 255 12mm<t≦20mm 255 20mm<t≦40mm 245 40mm<t≦50mm 235 50mm<t≦65mm 225	360～480	−	C428
≦0.025	≦0.30	≦0.30	≦0.08	N ≦0.020 Al≧0.020 Cu≦0.20 Nb≦0.03 Nb+Ti+V ≦0.05	t ≦12mm 255 12mm<t≦20mm 255 20mm<t≦40mm 245 40mm<t≦50mm 235 50mm<t≦65mm 225	360～480	−	
≦0.015	≦0.30	≦0.30	≦0.08	N ≦0.020 Al≧0.020 Cu≦0.20 Nb≦0.03 Nb+Ti+V ≦0.05	t ≦12mm 255 12mm<t≦20mm 255 20mm<t≦40mm 245 40mm<t≦50mm 235 50mm<t≦65mm 225	360～480	−	
≦0.030	≦0.30	≦0.30	≦0.08	N ≦0.020 Al≧0.020 Cu≦0.20 Nb≦0.03	t ≦12mm 285 12mm<t≦20mm 285 20mm<t≦40mm 275	390～510	STK400	

Standard No. Year Designation 規格番号 年号 規格名称	Grade グレード	Mfg. Process 製造方法	Chemical Composition C	 Si	 Mn	 P
17124−86(Continued)						
	TSTE285	S	≦0.16	≦0.40	0.60〜1.40	≦0.030
	ESTE285	S	≦0.16	≦0.040	0.60〜1.40	≦0.025
	STE355	S	≦0.20	0.10〜0.50	0.90〜1.65	≦0.035
	TSTE355	S	≦0.18	0.10〜0.50	0.90〜1.65	≦0.030

1 N/mm² or MP a = 1.01972×10^{-1} kgf/mm²

化学成分（%）					Tensile Test N/mm² 引張試験		Remarks 備考	Index No.
S	Ni	Cr	Mo	Others	Min. Yield Point 最小降伏点	Tensile Strength 引張強さ	(Similar to JIS)	索引番号
				Nb+Ti+V ≦0.05	40mm<t≦50mm 265 50mm<t≦65mm 255			C428
≦0.025	≦0.30	≦0.30	≦0.08	N ≦0.020 Al≧0.020 Cu≦0.20 Nb≦0.03 Nb+Ti+V ≦0.05	t≦12mm 285 12mm<t≦20mm 285 20mm<t≦40mm 275 40mm<t≦50mm 265 50mm<t≦65mm 255	390～510	STK400	
≦0.015	≦0.30	≦0.30	≦0.08	N ≦0.020 Al≧0.020 Cu≦0.20 Nb≦0.03 Nb+Ti+V ≦0.05	t≦12mm 285 12mm<t≦20mm 285 20mm<t≦40mm 275 40mm<t≦50mm 265 50mm<t≦65mm 255	390～510	STK400	
≦0.030	≦0.30	≦0.30	≦0.08	N ≦0.020 Al≧0.020 Cu≦0.20 Nb≦0.05 V ≦0.10 Nb+Ti+V ≦0.12	t≦12mm 355 12mm<t≦20mm 355 20mm<t≦40mm 345 40mm<t≦50mm 335 50mm<t≦65mm 325	490～639	STK500	
≦0.025	≦0.30	≦0.30	≦0.08	N ≦0.020 Al≧0.020 Cu≦0.20 Nb≦0.05	t≦12mm 355 12mm<t≦20mm 355 20mm<t≦40mm 345	490～639	STK500	

Standard No. Year Designation 規格番号 年号 規格名称	Grade グレード	Mfg. Process 製造方法	Chemical Composition			
			C	Si	Mn	P
17124－86(Continued)						
	ESTE355	S	≦0.18	0.10～0.50	0.90～1.65	≦0.025
	STE420	S	≦0.20	0.10～0.60	1.00～1.70	≦0.035
	TSTE420	S	≦0.20	0.10～0.60	1.00～1.70	≦0.030
	ESTE420	S	≦0.20	0.10～0.60	1.00～1.70	≦0.025

1 N/mm² or MPa = 1.01972×10⁻¹ kgf/mm²

化　学　成　分　(%)					Tensile Test N/mm² 引張試験		Remarks 備考	Index No. 索引番号
S	Ni	Cr	Mo	Others	Min. Yield Point 最小降伏点	Tensile Strength 引張強さ	(Similar to JIS)	
				V ≦0.10 Nb+Ti+V ≦0.12	40mm<t≦50mm 335 50mm<t≦65mm 325			C428
≦0.015	≦0.30	≦0.30	≦0.08	N ≦0.020 Al≧0.020 Cu≦0.20 Nb≦0.05 V ≦0.10 Nb+Ti+V ≦0.12	t ≦12mm 355 12mm<t≦20mm 355 20mm<t≦40mm 345 40mm<t≦50mm 335 50mm<t≦65mm 325	490～639	STK500	
≦0.030	≦1.00	≦0.30	≦0.10	N ≦0.020 Al≧0.020 Cu≦0.20 Nb≦0.05 V ≦0.20 Nb+Ti+V ≦0.22	t ≦12mm 420 12mm<t≦20mm 410 20mm<t≦40mm 400 40mm<t≦50mm 385 50mm<t≦65mm 375	390～510	−	
≦0.025	≦1.00	≦0.30	≦0.10	N ≦0.020 Al≧0.020 Cu≦0.20 Nb≦0.05 V ≦0.20 Nb+Ti+V ≦0.22	t ≦12mm 420 12mm<t≦20mm 410 20mm<t≦40mm 400 40mm<t≦50mm 385 50mm<t≦65mm 375	530～680	−	
≦0.015	≦1.00	≦0.30	≦0.10	N ≦0.020 Al≧0.020 Cu≦0.20 Nb≦0.05	t ≦12mm 420 12mm<t≦20mm 410 20mm<t≦40mm 400	530～680	−	

Standard No. 規格番号	Year 年号	Designation 規格名称	Grade グレード	Mfg. Process 製造方法	Chemical Composition C	Si	Mn	P
17142−86(Continued)								
			STE460	S	≦0.20	0.10〜0.60	1.00〜1.70	≦0.035
			TSTE460	S	≦0.20	0.10〜0.60	1.00〜1.70	≦0.030
			ESTE460	S	≦0.20	0.10〜0.60	1.00〜1.70	≦0.025

1 N/mm² or MPa = 1.01972×10^{-1} kgf/mm²

化 学 成 分 (%)					Tensile Test N/mm² 引張試験		Remarks 備考	Index No.
S	Ni	Cr	Mo	Others	Min. Yield Point 最小降伏点	Tensile Strength 引張強さ	(Similar to JIS)	索引番号
				V ≦0.20 Nb+Ti+V ≦0.22	40mm<t≦50mm 385 50mm<t≦65mm 375			C428
≦0.030	≦1.00	≦0.30	≦0.10	N ≦0.020 Al≧0.020 Cu≦0.20 Nb≦0.03 V ≦0.20 Nb+Ti+V ≦0.22	t≦12mm 460 12mm<t≦20mm 450 20mm<t≦40mm 440 40mm<t≦50mm 425 50mm<t≦65mm 410	560～730	−	
≦0.025	≦1.00	≦0.30	≦0.10	N ≦0.020 Al≧0.020 Cu≦0.20 Nb≦0.05 V ≦0.20 Nb+Ti+V ≦0.22	t≦12mm 460 12mm<t≦20mm 450 20mm<t≦40mm 440 40mm<t≦50mm 425 50mm<t≦65mm 410	560～730	−	
≦0.015	≦1.00	≦0.30	≦0.10	N ≦0.020 Al≧0.020 Cu≦0.20 Nb≦0.20 V ≦0.20 Nb+Ti+V ≦0.22	t≦12mm 460 12mm<t≦20mm 450 20mm<t≦40mm 440 40mm<t≦50mm 425 50mm<t≦65mm 410	560～730	−	

18. NF Steel Pipes for Piping and Pressure Vessels

Standard No. Year Designation 規格番号 年号 規格名称	Grade グレード	Mfg. Process 製造方法	Chemical Composition			
			C	Si	Mn	P
A49－111－78 Steel Tubes-Plain End Seamless Tubes of Commercial Quality for General Purposes at Medium Pressure 常圧一般用市販品質プレンエンド継目無鋼管	TU37a	S	≦0.24	－	≦1.30	≦0.05
A49－112－87 Steel Tubes-Plain End Seamless Hot Rolled Tubes with Specical Delivery Conditions 特殊出荷条件プレンエンド熱間圧延継目無鋼管	TUE220A	S	≦0.18	≦0.35	≦0.75	≦0.040
	TUE235A	S	≦0.22	≦0.35	≦0.95	≦0.040
A49－115－78 Steel Tubes-Hot-Finished Seamless Tubes Suitable for Threading ねじ切用熱間仕上継目無鋼管	TU34-1	S	－	－	－	≦0.06
A49－115－85 Steel Tubes-Seamless Plain End Tubes for Pipe Lines and General Use- Ferritic and Austenitic Stainless Steels ﾊﾟｲﾌﾟﾗｲﾝ及び一般用ﾌﾟﾚﾝｴﾝﾄﾞ継目無ﾌｪﾗｲﾄ 系及びｵｰｽﾃﾅｲﾄ系ｽﾃﾝﾚｽ鋼鋼管	TUZ2CN 18-10	S	≦0.030	≦1.00	≦2.00	≦0.040
	TUZ6CN 18-09	S	≦0.080	≦1.00	≦2.00	≦0.040
	TUZ6CNT 18-10	S	≦0.080	≦1.00	≦2.00	≦0.040
	TUZ12CN 24-12	S	≦0.150	≦1.00	≦2.00	≦0.040
	TUZ12CN 25-20	S	≦0.150	≦1.00	≦2.00	≦0.040
	TUZ2CND 17-12	S	≦0.030	≦1.00	≦2.00	≦0.040
	TUZ6CND 17-11	S	≦0.070	≦1.00	≦2.00	≦0.040
	TUZ6CNDT 17-12	S	≦0.080	≦1.00	≦2.00	≦0.040
	TUZ12C13	S	≦0.15	≦1.00	≦1.00	≦0.040
	TUZ10C17	S	≦0.12	≦0.75	≦1.00	≦0.040
A49－141－78 Steel Tubes-Welded Plain End Tubes of Commercial Quality for General Purposes at	TS37a	E, B	Efferves-cent ≦0.22	－	≦1.30	≦0.05

配管用・圧力容器用鋼管

1 N/mm² or MPa = 1.01972 × 10^{-1} kgf/mm²

化　学　成　分　(%)					Tensile Test N/mm² 引張試験		Remarks 備　考 (Similar to JIS)	Index No. 索引番号
S	Ni	Cr	Mo	Others	Min. Yield Point 最小降伏点	Tensile Strength 引張強さ		
≦0.05	–	–	–	–	210	≧360	–	C601
≦0.040	–	–	–	–	t ≦16mm 220 t >16mm 220	360～500	(STPG370)	C602
≦0.040	–	–	–	–	t ≦16mm 230 t >16mm 410～550	410～550	(STPG410)	
≦0.05	–	–	–	–	185	≧320	–	C603
≦0.030	9～12.00	17 ～20.0	–	–	175	470～720	(SUS304LTP)	C604
≦0.030	8～11.00	17 ～20.0	–	–	200	490～740	(SUS304TP)	
≦0.030	9～12.00	17 ～20.0	–	5C≦Ti≦0.6	190	490～740	(SUS321TP)	
≦0.030	10.5～13.00	23.0～25.0	–	–	220	530～780	–	
≦0.030	19.0～22.00	24.0～26.0	–	–	220	530～780	(SUS310STP)	
≦0.030	10.5～13.00	16 ～18.0	2.00～2.40	–	175	470～720	–	
≦0.030	10.0～12.50	16 ～18.0	2.00～2.40	–	190	490～740	(SUS316TP)	
≦0.030	10.5～13.00	16 ～18.0	2.00～2.40	5C≦Ti≦0.6	190	490～740	–	
≦0.030	≦0.45	11.5～13.5	–	–	210	420～670	–	
≦0.030	≦0.45	16.0～18.0	–	–	250	420～670	–	
≦0.05	–	–		Effervescent N ≦0.009	240	≦360	(STPG370)	C605

Standard No. Year Designation 規格番号 年号 規格名称	Grade グレード	Mfg. Process 製造方法	Chemical Composition			
			C	Si	Mn	P
Medium Pressure 常圧一般用市販品質プレンエンド溶接鋼管	TS37a	E, B	Noneffer-vescent ≦0.20	–	≦1.30	≦0.05
A49-142-87 Steel Tubes-Longitudinally Pressure Welded Plain Ended and Hot Finished Tubes プレンエンド縦加圧溶接加熱仕上鋼管	TES185A	B, E	≦0.16	≦0.30	≦0.55	≦0.040
	TSE235A	B, E	≦0.15	≦0.30	≦0.70	≦0.040
	TSE250A	B, E	≦0.18	≦0.30	≦0.95	≦0.040
	TSE275A	B, E	≦0.20	≦0.30	≦1.20	≦0.040
A49-145-78 Steel Tubes-Hot-Finished Welded Tubes Suitable for Threading ねじ切用熱間仕上溶接鋼管	TS34-1	–	–	–	–	≦0.06
A49-146-75 Steel Tubes-Unthreaded, Plain Ended, Welded Tubes for Liquids 流体用ねじ無プレンエンジ溶接鋼管	TS34-a	B, E	≦0.18	–	–	≦0.062
A49-147-80 Steel Tubes-Plain End Longitudinally Welded Tubes for Pipe Lines and General Use -Austenitic Stainless Steels パイプライン及び一般用プレンエンド縦溶接オーステナイト系ステンレス鋼鋼管	TSZ2CN 18-10	E, A	≦0.030	≦1.00	≦2.00	≦0.040
	TSZ6CN 18-09	E, A	≦0.080	≦1.00	≦2.00	≦0.040
	TSZ6CNT 18-10	E, A	≦0.080	≦1.00	≦2.00	≦0.040
	TSZ2CND 17-12	E, A	≦0.030	≦1.00	≦2.00	≦0.040
	TSZ6CND 17-11	E, A	≦0.070	≦1.00	≦2.00	≦0.040
	TSZ6CNDT 17-12	E, A	≦0.080	≦1.00	≦2.00	≦0.040
A49-148-80 Steel Tubes-Plain End Longitudinally Welded Tubes for Pipe Lines and General Use-Ferritic Stainless Steels パイプライン及び一般用プレンエンド縦溶接フェライト系ステンレス 鋼鋼管	TSZ5 CT14	E, A	≦0.045	≦0.60	≦0.40	≦0.030
	TSZ6 CT12	E, A	≦0.080	≦1.00	≦1.00	≦0.040
	TSZ6 CT17	E, A	≦0.50	≦0.50	≦0.50	≦0.030
	TSZ18 C17	E, A	≦0.10	≦1.00	≦1.00	≦0.040

1 N/mm² or MPa = 1.01972 × 10⁻¹ kgf/mm²

化　学　成　分　(%)					Tensile Test N/mm² 引張試験		Remarks 備考 (Similar to JIS)	Index No. 索引番号
S	Ni	Cr	Mo	Others	Min. Yield Point 最小降伏点	Tensile Strength 引張強さ		
≦0.05	−	−		Noneffervescent N ≦0.008	240	≦360	(STPG370)	C605
≦0.035	−	−	−	−	185	330〜410	−	C606
≦0.035	−	−	−	−	235	360〜450	(STPG370)	
≦0.035	−	−	−	−	250	410〜510	(STPG410)	
≦0.035	−	−	−	−	275	470〜570	−	
≦0.05	−	−	−	−	190	≧320	(SGP)	C607
≦0.062	−	−	−	N ≦0.009	220	≧330	(SGP)	C608
≦0.030	9〜12	17〜20	−	−	175	460〜660	(SUS304LTP)	C609
≦0.030	8〜11	17〜20	−	−	200	510〜710	(SUS304TP)	
≦0.030	9〜12	17〜20	−	5×C≦Ti≦0.6	210	520〜720	(SUS321TP)	
≦0.030	10.5〜13	16〜18	2.00〜2.40	−	200	490〜690	−	
≦0.030	10〜12.5	16〜18	2.00〜2.40	−	210	510〜710	(SUS316TP)	
≦0.030	10.5〜13	16〜18	2.00〜2.40	5×C≦Ti≦0.6	220	540〜740	−	
≦0.020	−	13〜15.00	−	Ti≧10×C	230	400〜600	−	C610
≦0.030	−	11〜13.00	−	Ti≧6×C	250	420〜620	−	
≦0.020	−	16〜18.0	−	Ti≧10×C	320	510〜720	−	
≦0.030	−	16〜18.0	−	−	250	500〜700	−	

Standard No. Year Designation 規格番号 年号 規格名称	Grade グレード	Mfg. Process 製造方法	Chemical Composition			
			C	Si	Mn	P
A49−150−85 Steel Tubes-Welded Tubes Intended to Be Coated of Protocted for Use in Water Piping Systems 排水用被覆又は保護管原管用溶接鋼管	TSE235	W	≦0.16	−	≦1.10	≦0.040
A49−210−85 Steel Tubes-Seamless Cold-Drawn Tubes for Fluid Piping 流体輸送用冷間引抜継目無鋼管	TU37B	S	≦0.18	≦0.35	≦0.75	≦0.040
	TU42B	S	≦0.22	≦0.35	≦0.75	≦0.040
A49−211−89 Steel Tubes-Seamless Plain End Unalloyed Steel Tubes for Fluid Piping at Elevated Temperatures 高温流体輸送用プレンエンド継目無炭素鋼管	TUE220	S	≦0.15	≦0.30	≦0.80	≦0.025
	TUE250	S	≦0.21	≦0.30	≦1.00	≦0.025
	TUE275	S	≦0.23	≦0.30	≦1.35	≦0.025
A49−212−83 Steel Tubes-Seamless Unalloyed Steel Tubes for Use at Medium Temperatures 中温用継目無非合金鋼鋼管	TU37C	S	≦0.16	0.06〜0.30	0.35〜0.75	≦0.035
	TU42C	S	≦0.20	0.08〜0.35	0.45〜1.00	≦0.035
A49−213−84 Steel Tubes-Seamless Unalloyed and Mo or Cr-Mo Alloy Steel Tubes for Use at High Temperatures 高温用継目無非合金及びMo, Cr-Mo系合金鋼鋼管	TU37C	S	≦0.16	0.06〜0.30	0.35〜0.75	≦0.035
	TU42C	S	≦0.20	0.08〜0.35	0.45〜1.00	≦0.035
	TU42CR	S	≦0.20	0.08〜0.35	0.60〜1.15	≦0.035
	TU48C	S	≦0.22	0.10〜0.35	0.65〜1.25	≦0.035
	TU48CR	S	≦0.20	0.15〜0.50	1.00〜1.50	≦0.035
	TU52C	S	≦0.20	0.15〜0.50	1.00〜1.50	≦0.035
	TU15D3	S	0.12〜0.20	0.15〜0.35	0.50〜0.80	≦0.035
	TU15CD 2-05	S	0.10〜0.18	0.10〜0.35	0.50〜0.90	≦0.030
	TU10CD 4-04	S	0.10〜0.18	0.10〜0.35	0.40〜0.70	≦0.030
	TU10CD 5-05	S	≦0.15	0.50〜1.00	0.30〜0.60	≦0.030

1 N/mm² or MPa = 1.01972×10^{-1} kgf/mm²

化　学　成　分　(%)					Tensile Test N/mm² 引張試験		Remarks 備　考	Index No.
S	Ni	Cr	Mo	Others	Min. Yield Point 最小降伏点	Tensile Strength 引張強さ	(Similar to JIS)	索引番号
≦0.045	–	–	–	–	235	≧360	(STPG370)	C611
≦0.045	–	–	–	–	220	360～480	(STS370)	C612
≦0.045	–	–	–	–	235	410～530	(STS410)	
≦0.025	–	–	–	–	220	370～490	(STPT370)	C613
≦0.025	–	–	–	–	250	410～530	(STPT410)	
≦0.025	–	–	–	–	275	470～590	(STPT480)	
≦0.035	–	–	–	Cu≦0.25 Sn≦0.030	210	360～460	(STPG370)	C614
≦0.035	–	–	–	Cu≦0.25 Sn≦0.030	230	410～510	(STPG410)	
≦0.035	–	–	–	Cu≦0.25 Sn≦0.030	220	360～460	(STPT370)	C615
≦0.035	–	–	–	Cu≦0.25 Sn≦0.030	235	410～520	(STPT410) (STB410)	
≦0.035	–	–	–	Cu≦0.25 Sn≦0.030	235	420～510	–	
≦0.035	–	–	–	Cu≦0.25 Sn≦0.030	275	470～590	(STB480)	
≦0.035	–	–	–	Cu≦0.25 Sn≦0.030	275	490～610	–	
≦0.035	–	–	–	Cu≦0.25 Sn≦0.030	t ≦20 350 t >20 310	510～630	(STB510)	
≦0.035	≦0.30	≦0.30	0.25～0.35	Cu≦0.25 Sn≦0.030	265	430～550	–	
≦0.030	≦0.30	0.40～0.65	0.45～0.60	Cu≦0.25 Sn≦0.030	275	440～570	–	
≦0.030	≦0.30	0.70～1.10	0.45～0.65	Cu≦0.25 Sn≦0.030	290	440～590	–	
≦0.030	≦0.30	1.00～1.50	0.45～0.65	Cu≦0.25 Sn≦0.030	225	440～590	(STPA23) (STBA23)	
					325	490～640	–	

Standard No. Year Designation 規格番号 年号 規格名称	Grade グレード	Mfg. Process 製造方法	Chemical Composition			
			C	Si	Mn	P
A49-213-90 (Continued)	TU10CD 9-10	S	≦0.15	0.10～0.50	0.30～0.60	≦0.030
	TUZ12CD 05-05	S	≦0.15	0.05～0.50	0.30～0.60	≦0.030
	TUZ10CD 09	S	≦0.15	0.25～1.00	0.30～0.60	≦0.030
	TUZ10 CD/Nb 09-01	S	0.08～0.12	0.20～0.50	0.30～0.60	≦0.020
	TUZ10 CDNbV 09-02	S	≦0.15	0.20～0.65	0.80～1.30	≦0.030
A49-214-78 Steel Tubes-Seamless Austenitic Steel Tubes for Use at High Temperatures 高温用継目無オーステナイト系ステンレス鋼鋼管	Z6CN 19-10	S	0.04～0.08	≦1.0	≦2.0	≦0.035
	Z6CND 17-12B	S	0.04～0.08	≦1.0	≦2.0	≦0.035
	Z6CNT 18-12B	S	0.04～0.08	≦1.0	≦2.0	≦0.035
	Z6CNNb 18-12B	S	0.04～0.08	≦1.0	≦2.0	≦0.035
	Z8CNDT 17-13B	S	0.05～0.10	≦1.0	≦2.0	≦0.035
	Z10CNWT 17-13B	S	0.07～0.13	≦1.0	≦1.0	≦0.035
A49-230-85 Steel Tubes-Plain End Seamless Tubes for Pressure Vessels and Piping Systems Used at Low Temperatures 低温圧力容器及び配管系用プレンエンド継目鋼管	TU42BT	S	≦0.20	0.10～0.35	0.30～1.05	≦0.035
	TU17N2	S	≦0.21	≦0.35	≦1.50	≦0.040
	TU10N9	S	≦0.15	≦0.30	≦0.90	≦0.030
	TU10N14	S	≦0.15	≦0.35	≦0.65	≦0.030
	TUZ6N9	S	≦0.10	≦0.30	≦0.90	≦0.030
A49-240-83 Steel Tube Longitudinally Buttwelding Plain End for Pressure Vessels and Pipen Systems Used at Low Temperatures 低温圧力容器及び配管用プレンエンド縦鍛接鋼管	TS42BT	S, W	≦0.22	0.05～0.40	0.20～1.15	≦0.040
	TS17N2	S, W	≦0.23	≦0.40	≦1.60	≦0.045
	TS10N9	S, W	≦0.17	≦0.035	≦1.00	≦0.035

$1\ N/mm^2$ or $MPa = 1.01972 \times 10^{-1} kgf/mm^2$

化学成分（%）					Tensile Test N/mm^2 引張試験		Remarks 備考	Index No.
S	Ni	Cr	Mo	Others	Min. Yield Point 最小降伏点	Tensile Strength 引張強さ	(Similar to JIS)	索引番号
≦0.030	≦0.30	2.00～2.50	0.90～1.10	Cu≦0.25 Sn≦0.030	225	410～560	(STPA24) (STBA24)	C615
					325	490～640	－	
≦0.030	－	4.00～6.00	0.45～0.65	Cu≦0.25 Sn≦0.030	205 280	520～640	(STPA25) (STBA25)	
≦0.030	≦0.30	8.00～10.00	0.90～1.10	Cu≦0.25 Sn≦0.03	205	440～590	(STPA26) (STBA26)	
≦0.010	≦0.40	8.00～9.00	0.85～1.05	Cu ≦0.25 Sn ≦0.030 Nb 0.06～0.10 V 0.18～0.25	415	585～770	－	
≦0.030	≦0.30	8.50～10.50	1.70～2.30	Nb 0.30～0.55 V 0.20～0.40 Cu ≦0.25 Sn ≦0.03	390	590～740	－	
≦0.030	8～11	18～20	－	B0.0010 ～0.0030	190	490～740	－	C616
≦0.030	11～14	16～18	2.0～3.0	B0.0010 ～0.0030	190	490～690	－	
≦0.030	10～13	17～19	－	B0.0010 ～0.0030 4C≦Ti≦0.60	190	490～690	－	
≦0.030	10～13	17～19	－	B0.0010 ～0.0030 8C≦(Nb+Ta)1.00	190	490～690	－	
≦0.030	12～15	16～18	2.0～3.0	B0.0010 ～0.0030 4C≦Ti≦0.75	190	540～740	－	
≦0.030	12～14	16～18	－	B0.0010 ～0.0030 2.5 ≦W≦4 5C≦Ti≦0.8	190	540～740	－	
≦0.035	－	－	－	－	240	415～510	－	C617
≦0.040	0.6～0.8	－	－	－	325	490～640	－	
≦0.030	2.0～2.6	－	－	－	245	455～600	－	
≦0.030	3.2～3.8	－	－	－	245	450～600	(STPL450) (STBL450)	
≦0.030	8.5～9.6	－	－	－	520	650～800	(STPL690) (STBL690)	
≦0.040	－	－	－	Nb≦0.040	240	415～510	－	C618
≦0.045	0.6～0.8	≦0.20	－	V ≦0.050	325	490～640	－	
≦0.035	2.0～2.6	≦0.20	－	V ≦0.050	245	450～600	－	

Standard No. Year Designation 規格番号 年号 規格名称	Grade グレード	Mfg. Process 製造方法	Chemical Composition			
			C	Si	Mn	P
A49－241－86 Steel Tubes-Longitudinally Pressure Welded Tubes in Not Alloyed Steel Grades for Fiuid Piping up to 425 ℃ 425℃以下配管用非合金縦加圧溶接鋼管	TSE220	W	≦0.15	≦0.30	≦0.75	≦0.035
	TSE250	W	≦0.16	≦0.35	≦1.00	≦0.035
	TSE275	W	≦0.20	≦0.35	≦1.00	≦0.035
	TSE355	W	≦0.20	≦0.50	≦1.50	≦0.035
A49－242－85 Steel Tubes-Longitudinally Pressure Welded Tubes, D ≦168.3 mm, in Not Alloyed and Low Alloyed at Medium Elevated Temperatures 外径168.3 mm以下中温用縦溶接非合金及び低合金鋼鋼管	TS37C	W	≦0.15	0.06～0.30	0.35～0.75	≦0.035
	TS42C	W	≦0.18	0.08～0.35	0.45～1.00	≦0.035
	TSE24W3	W	≦0.13	0.10～0.40	0.20～0.60	≦0.040
	TSE36WB3	W	≦0.19	≦0.50	0.50～1.60	≦0.045
A49－243－85 Steel Tubes-Longitudinally Pressure Welded Tubes, D ≦168.3 mm, in Not Alloyed and Ferritic Alloyed Steels, Used at Elevated Temperatures 外径168.3 mm以下高温用縦溶接非合金及びフェライト合金鋼鋼管	TS37C	W	≦0.15	0.06～0.30	0.35～0.75	≦0.035
	TS42C	W	≦0.18	0.08～0.35	0.45～1.00	≦0.035
	TS48C	W	≦0.20	0.10～0.35	0.65～1.25	≦0.035
	TS52C	W	≦0.20	0.15～0.50	1.00～1.50	≦0.035
	TS15D3	W	0.12～0.20	0.15～0.35	0.50～0.90	≦0.030
	TS15CD2-05	W	0.10～0.18	0.10～0.35	0.50～0.90	≦0.030
	TS15CD4-05	W	0.08～0.20	0.10～0.40	0.40～0.85	≦0.035
A49－250－80 Steel Tubes-Welded Plain End Tubes of Commercial Quality with or without Special Delivery Conditions-D≦168.3 mm 外径168.3 mm以上特殊出荷条件付又はなしの市販品質プレンエンド溶接鋼管	TSE24a	W	≦0.18	－	≦1.20	≦0.045
	TSE26b	W	≦0.20	≦0.40	≦1.30	≦0.045

$1\ N/mm^2$ or $MPa = 1.01972 \times 10^{-1} kgf/mm^2$

化　学　成　分　(%)					Tensile Test N/mm² 引張試験		Remarks 備　考	Index No.
S	Ni	Cr	Mo	Others	Min. Yield Point 最小降伏点	Tensile Strength 引張強さ	(Similar to JIS)	索引番号
≦0.025	−	−	−	−	220	360〜480	−	C619
≦0.025	−	−	−	−	250	410〜530	−	
≦0.025	−	−	−	−	275	470〜590	−	
≦0.025	−	−	−	−	355	510〜630	−	
≦0.025	−	−	−	Cu≦0.25 Sn≦0.030	235	360〜460	(STPG370)	C620
≦0.025	−	−	−	Cu≦0.25 Sn≦0.035	255	410〜510	(STPG420)	
≦0.035	≦0.65	0.40〜0.80	−	−	255	380〜450	−	
≦0.045	≦0.65	0.40〜0.80	−	−	330	450〜520	−	
≦0.025	−	−	−	Cu≦0.25 Sn≦0.030	235	360〜460	(STPG370)	C621
≦0.025	−	−	−	Cu≦0.25 Sn≦0.030	255	410〜510	(STPG420)	
≦0.025	−	−	−	Cu≦0.25 Sn≦0.030	275	470〜570	−	
≦0.025	−	−	−	Cu≦0.25 Sn≦0.030	355	510〜630	−	
≦0.025	≦0.30	≦0.30	0.25〜0.35	Cu≦0.25 Sn≦0.030 Al≦0.025	265	430〜530	−	
≦0.025	≦0.30	0.40〜0.65	0.45〜0.60	Cu≦0.25 Sn≦0.030 Al≦0.025	275	440〜570	−	
≦0.030	≦0.30	0.75〜1.25	0.40〜0.60	Cu≦0.25 Sn≦0.030 Al≦0.025	295 mm t≦30 / mm t<30	470〜610	−	
≦0.045	−	−	−	−	230	≧360	STPG370	C622
≦0.045	−	−	−	−	250	410〜490	STPG410	

Standard No. Year Designation 規格番号 年号 規格名称	Grade グレード	Mfg. Process 製造方法	Chemical Composition			
			C	Si	Mn	P
A49-253-82 Steel Tubes-Longitudinally Fusion Welded Non Alloy and Ferritic Alloy Steel Tubes for Use at Elevated Temperatures 高温用縦溶融溶接非合金鋼及びフェライト系合金鋼鋼管	TS37CP	W	≦0.16	≦0.30	≦0.40	≦0.035
	TS42CP	W	≦0.18	≦0.30	≦0.60	≦0.035
	TS48CP	W	≦0.20	≦0.35	0.80～1.50	≦0.035
	TS52CP	W	≦0.20	≦0.55	1.0 ～1.60	≦0.040
	TS15D3	W	≦0.20	0.15～0.30	0.50～0.80	≦0.035
	TS15CD2-05	W	≦0.18	0.15～0.30	0.50～0.90	0.030～0.035
	TS10CD9-10	W	≦0.15	0.15～0.35	0.40～0.80	≦0.030
	TSZ10CD5-05	W	≦0.15	0.15～0.50	0.30～0.60	≦0.030
A49-400-82 Longitudinally Electric Resistance welded Unalloyed Steel Tubes 17.2≦D≦406.4 mm for the Transport of Pressurized Fluids 外径17.2～406.4 mmの加圧流体輸送用電機抵抗溶接非合金鋼鋼管	TSE220	E	≦0.15	≦0.22	≦0.70	≦0.030
	TSE250	E	≦0.16	≦0.30	≦1.00	≦0.030
	TSE290	E	≦0.16	≦0.30	≦1.20	≦0.030
	TSE320	E	≦0.16	≦0.35	≦1.25	≦0.030
	TSE360Nb	E	≦0.16	≦0.35	≦1.45	≦0.030
	TSE360V	E	≦0.16	≦0.35	≦1.45	≦0.030
	TSE415Nb	E	≦0.20	≦0.35	≦1.50	≦0.030
	TSE415V	E	≦0.20	≦0.35	≦1.50	≦0.03
A49-401-88 Steel Tubes-Longitudinally Fusion Welded Non-Alloy and Micro-Alloy Steel Tubes for Fluid Transporting Pipes and Pressure Vessels 液体輸送及び圧力容器用縦溶融溶接非合金及び低合金鋼鋼管	TSE220	W	≦0.16	≦0.30	≦1.00	≦0.03
	TSE250	W	≦0.17	≦0.30	≦1.10	≦0.03
	TSE290	W	≦0.17	≦0.35	≦1.20	≦0.03
	TSE320	W	≦0.17	≦0.35	≦1.25	≦0.03
	TSE360	W	≦0.17	≦0.40	0.80～1.55	≦0.03
	TSE415	W	≦0.17	≦0.45	0.80～1.55	≦0.03
	TSE450	W	≦0.17	≦0.45	0.80～1.55	≦0.03
	TSE480	W	≦0.17	≦0.45	0.80～1.55	≦0.03
A49-402-88 Steel Tubes Spiral Fusion Welded Non Alloy	TSE220	W	≦0.16	≦0.30	≦1.00	≦0.03
	TSE250	W	≦0.17	≦0.30	≦1.10	≦0.03

$1\ N/mm^2$ or $MPa = 1.01972 \times 10^{-1} kgf/mm^2$

化　学　成　分　(%)					Tensile Test N/mm² 引張試験		Remarks 備考	Index No.
S	Ni	Cr	Mo	Others	Min. Yield Point 最小降伏点	Tensile Strength 引張強さ	(Similar to JIS)	索引番号
≦0.030	–	–	–	–	225　205	360～480	–	C623
≦0.030	–	–	–	–	245　225	410～520	–	
≦0.030	–	–	–	–	285　265	470～590	–	
≦0.030	–	–	–	–	335　315	510～630	–	
≦0.030	–	≦0.30	0.25～0.35	–	265	430～550	–	
≦0.030	–	0.40～0.60	0.40～0.60	–	275	450～570	–	
≦0.030	–	2.0～2.50	0.90～1.10	–	310	t ≦20mm 540～660 t >20mm 520～640	–	
≦0.030	–	4.0～6.00	0.45～0.65	–	390	590～730	–	
≦0.030	–	–	–	Nb≦0.04	210	370～490	–	C624
≦0.025	–	–	–	Nb≦0.04	240	410～530	–	(STPG370)
≦0.025	–	–	–	Nb≦0.04	290	420～540	–	(STPG410)
≦0.025	–	–	–	Nb≦0.04	320	440～560	–	
≦0.025	–	–	–	Nb 0.01 ～0.05	360	480～620	–	
≦0.025	–	–	–	V 0.040～0.100	360	480～620	–	
≦0.025	–	–	–	Nb 0.015～0.055	410	530～680	–	
≦0.025	–	–	–	V 0.040～0.100	410	530～680	–	
≦0.025	–	–	–	–	220	370～480	–	C625
≦0.025	–	–	Nb≦0.4	–	250	420～530	–	
≦0.025	–	–	Nb≦0.4	–	290	420～530	–	
≦0.025	–	–	Nb≦0.4	–	320	440～560	–	
≦0.025	–	–	Nb≦0.4	–	360	480～620	–	
≦0.025	–	–	Nb0.01~0.05	V ≦0.06	415	550～670	–	
≦0.025	–	–	Nb0.01~0.05	V ≦0.06	450	560～680	–	
≦0.025	–	–	Nb0.02~0.06	V ≦0.10	480	580～720	–	
≦0.025	–	–	–	–	220	370～480	–	C626
≦0.025	–	–	Nb≦0.4	–	250	420～530	–	

Standard No. Year Designation 規格番号 年号 規格名称	Grade グレード	Mfg. Process 製造方法	Chemical Composition			
			C	Si	Mn	P
and Micro-Alloy Steel Tubes for Fluid Transporting Pipes and Pressure Vessels 液体輸送及び圧力容器用螺旋溶融溶接非合金及び低合金鋼鋼管	TSE290	W	≦0.17	≦0.35	≦1.20	≦0.03
	TSE320	W	≦0.17	≦0.35	≦1.20	≦0.03
	TSE360	W	≦0.17	≦0.40	0.80~1.55	≦0.03
	TSE415	W	≦0.17	≦0.45	0.80~1.55	≦0.03
	TSE450	W	≦0.17	≦0.45	0.80~1.55	≦0.03
	TSE480	W	≦0.17	≦0.45	0.80~1.55	≦0.03
A49-643-87 Steel Tubes-Round Square and Rectangular Steel Tubes for Ordinary Uses, Longitudinally Pressure Welded and Cold Formed from Flat Products 平板原料縦溶融溶接冷間成形丸・角・矩形一般用鋼管	TS30-1	W	≦0.11	≦0.40	≦0.50	≦0.050
	TS30-2	W	≦0.11	≦0.40	≦0.50	≦0.045
	TS34-2	W	≦0.14	≦0.40	≦0.50	≦0.045
	TS37-2	W	≦0.15	≦0.40	≦0.70	≦0.045
	TS42-2	W	≦0.18	≦0.40	≦1.00	≦0.045
	TS47-2	W	≦0.20	≦0.40	≦1.20	≦0.045
	TS335D	W	≦0.12	≦0.40	≦1.50	≦0.030
	TS390D	W	≦0.12	≦0.40	≦1.50	≦0.030
	TS445D	W	≦0.12	≦0.40	≦1.60	≦0.030
A49-645-87 Steel Tubes-Round Square and Rectangular Steel Tubes for Ordinary Uses, Longitudinally Pressure Welded and Cold Formed from Cold-Rolled Flat Products 冷間圧延平板原料冷間成形縦加圧溶接丸・角・矩形一般用鋼管	TS30-0	W	≦0.12	-	-	≦0.050
	TS30E	W	≦0.09	-	-	≦0.045
	TS30ES	W	≦0.09	-	-	≦0.045
	TS335D	W	≦0.100	-	-	≦0.045
	TS390D	W	0.100	-	-	≦0.045

1 N/mm² or MPa = 1.01972×10^{-1} kgf/mm²

化　学　成　分　(%)					Tensile Test N/mm² 引張試験		Remarks 備考	Index No.
S	Ni	Cr	Mo	Others	Min. Yield Point 最小降伏点	Tensile Strength 引張強さ	(Similar to JIS)	索引番号
≦0.025	−	−	Nb≦0.4	−	290	420～530	−	C626
≦0.025	−	−	Nb≦0.4	−	320	440～560	−	
≦0.025	−	−	Nb≦0.4	−	360	480～620	−	
≦0.025	−	−	Nb0.01~0.05	V ≦0.06	415	550～670	−	
≦0.025	−	−	Nb0.01~0.05	V ≦0.06	450	560～680	−	
≦0.025	−	−	Nb0.01~0.05	V ≦0.06	480	580～720	−	
≦0.050	−	−	N ≦0.009	AL≦0.08	−	−	−	C627
≦0.045	−	−	N ≦0.009	AL≦0.08	190	310	(STK290)	
≦0.045	−	−	N ≦0.009	AL≦0.08	200	330	(STK290)	
≦0.045	−	−	N ≦0.009	AL≦0.08	235	360	−	
≦0.045	−	−	N ≦0.009	AL≦0.08	255	410	(STK410)	
≦0.045	−	−	N ≦0.009	AL≦0.08	275	470	(STK490)	
≦0.030	−	−	N ≦0.009	AL≦0.08	335	370	−	
≦0.030	−	−	N ≦0.009	AL≦0.08	390	420	−	
≦0.030	−	−	N ≦0.009	AL≦0.08	445	480	−	
≦0.050	−	−	N ≦0.060	−	280	300～520	(STK290)	C628
≦0.045	−	−	N ≦0.050	−	280	300～480	(STK290)	
≦0.045	−	−	N ≦0.050	−	280	300～480	(STK290)	
≦0.040	−	−	N ≦0.007	Nb≦0.060	−	≦450	−	
≦0.040	−	−	N ≦0.007	Nb≦0.060	−	≦480	−	

19. NF Steel Tubes for Heat Exchanger and Furnace

Standard No. Year Designation 規格番号 年号 規格名称		Grade グレード	Mfg. Process 製造方法	Chemical Composition C	Si	Mn	P
A49−207−81 Steel Tubes-Seamless or Longitudinally Welded Stainless Steel Tubes for Exchangers in Aircraft Construction 航空機熱交換器用ステンレス鋼鋼管		TSZ2 CN18-10	E	≦0.030	≦1.00	≦2.00	≦0.040
		TUZ2 CN18-10	S				
		TSZ6 CNNb18-12	E	≦0.090	≦1.00	≦2.00	≦0.040
		TUZ6 CNNb18-12	S				
A49−215−81 Steel Tubes-Seamless Tubes for Ferritic Non-Alloy and Alloy Heat Exchangers 熱交換器用継目非合金及びﾌｪﾗｲﾄ 系合金鋼鋼管	高温特性規定 High Temp. Spec.	TU37C	S	≦0.18	0.05〜0.27	0.30〜0.80	≦0.045
		TU42C	S	≦0.22	0.07〜0.40	0.40〜1.05	≦0.045
		TU48C	S	≦0.24	0.09〜0.40	0.60〜1.30	≦0.045
		TU15D3	S	0.18〜0.22	0.10〜0.40	0.40〜0.90	≦0.045
		TU15 CD2-05	S	0.08〜0.20	0.05〜0.40	0.40〜1.00	≦0.035
		TU10 CD5-05	S	≦0.17	0.45〜1.05	0.20〜0.70	≦0.035
		TU10 CD9-10	S	≦0.17	0.05〜0.55	0.20〜0.70	≦0.035
		TUZ10 CD5-05	S	≦0.17	0.10〜0.55	0.30〜0.65	≦0.035
		TUZ10 CD9	S	≦0.17	0.20〜1.05	0.20〜0.70	≦0.035
	低温用 For Low Temp.	TU42BT	S	≦0.22	≧0.40	≦1.15	≦0.04
		TU17N2	S	≦0.23	≦0.40	≦1.60	≦0.045
		TU10N9	S	≦0.17	≦0.35	≦1.10	≦0.035
		TU10N14	S	≦0.17	≦0.40	≦0.75	≦0.035
		TUZ6N9	S	≦0.12	≦0.35	≦1.00	≦0.035
A49−217−87 Steel Tubes-Seamless Tubes for Heat Exchangers		TUZ12C13	S	≦0.15	≦1.00	≦1.00	≦0.040
		TUZ10C17	S	≦0.12	≦1.00	≦1.00	≦0.040
		TUZ2CN1810	S	≦0.030	≦1.00	≦2.00	≦0.040

熱交換器用・加熱炉用鋼管

$1\,N/mm^2$ or $MPa = 1.01972 \times 10^{-1} kgf/mm^2$

化学成分 (%)						Tensile Test N/mm^2 引張試験		Remarks 備考	Index No.
S	Ni	Cr	Mo	Others		Min. Yield Point 最小降伏点	Tensile Strength 引張強さ	(Similar to JIS)	索引番号
≦0.030	9.00～12.00	17.0～20.0	–	–		175	470～720	Impact Test (℃, J／cm²) −196, 120 (SUS304LTB)	C629
≦0.030	10.50～13.00	17.0～20.0	–	10C≦Nb＜1.0		185	490～740	−196, 100 (SUS347TB)	
≦0.045	–	–	–	Cu≦0.25 Sn≦0.03		210	360～450	–	C630
≦0.045	–	–	–	Cu≦0.25 Sn≦0.03		230	410～510	(STB42)	
≦0.045	–	–	–	Cu≦0.25 Sn≦0.03		270	470～570	–	
≦0.045	≦0.30	≦0.40	0.21～0.39	Cu≦0.25 Sn≦0.03		260	430～530	–	
≦0.035	≦0.30	0.30～0.75	0.41～0.64	Cu≦0.25 Sn≦0.03		270	430～570	–	
≦0.035	≦0.30	0.90～1.60	0.41～0.69	Cu≦0.25 Sn≦0.03	a	220	440～570	(STBA23)	
					b	320	490～640	–	
≦0.035	≦0.30	1.90～2.60	0.85～1.15	Cu≦0.25 Sn≦0.03	a	220	410～560	(STBA24)	
					b	320	490～640	–	
≦0.035	–	3.90～6.10	0.40～0.65	–		390	590～710	–	
≦0.035	≦0.30	7.90～10.15	0.85～1.15	Cu≦0.25 Sn≦0.03		200	440～590	(STBA26)	
≦0.04	–	–	–	Cu≦0.30		230	410～510	Impact Test (℃ J／cm²) −46, 35	
≦0.045	0.6～0.8	≦0.20	≦0.10	Cu≦0.30 V ≦0.05		320	≧490	−60, 50	
≦0.035	2.0～2.6	≦0.20	≦0.10	Cu≦0.30 V ≦0.05		240	≧450	−80, 50	
≦0.035	3.2～3.8	≦0.20	≦0.10	Cu≦0.30 V ≦0.05		240	≧450	(STBL46) −100, 50	
≦0.035	8.5～9.6	≦0.20	≦0.10	Cu≦0.30 V ≦0.05		530	≧650	(STBL70) −196, 60	
≦0.030	≦0.50	11.50～13.50	–	–		210	420～670	Ferritic Stain- less Steels	C631
≦0.030	≦0.50	16.00～18.00	–	–		250	420～670		
≦0.030	9.00～12.00	17.00～20.00	–	–		175	470～720		

Standard No. Year Designation 規格番号 年号 規格名称	Grade グレード	Mfg. Process 製造方法	Chemical Composition			
			C	Si	Mn	P
熱交換器用継目無鋼管	TUZ2CN 1810AZ	S	≦0.030	≦1.00	≦2.00	≦0.040
	TUZ6CN 1809	S	≦0.080	≦2.00	≦1.00	≦0.040
	TUZ1CNS 1815	S	≦0.15	3.50～4.50	≦2.00	≦0.030
	TUZ6CNT 1810	S	≦0.080	≦1.00	≦2.00	≦0.040
	TUZ2CNNb 2520	S	≦0.030	≦0.40	≦1.00	≦0.030
	TUZ2CND 1712	S	≦0.030	≦1.00	≦2.00	≦0.040
	TUZ2CND 1814	S	≦0.030	≦1.00	≦2.00	≦0.020
	TUZ2CND 1712AZ	S	≦0.030	≦1.00	≦2.00	≦0.040
	TUZ6CND 1711	S	≦0.070	≦1.00	≦2.00	≦0.040
	TUZ1CND 2522AZ	S	≦0.020	≦1.00	1.50～2.00	≦0.020
	TUZ2CNDU 1716	S	≦0.030	≦1.00	≦1.00	≦0.040
	TUZ1NCDU 201806AZ	S	≦0.020	≦0.80	≦2.00	≦0.030
	TUZ1NCDU 252004	S	≦0.020	≦0.40	≦2.00	≦0.030
	TUZ1NCDU 312703	S	≦0.020	≦1.00	≦2.00	≦0.020
	TUZ2CND 180503	S	≦0.030	1.40～2.00	1.20～2.00	≦0.030
	TUZ2CND 220503	S	≦0.030	≦1.00	≦2.00	≦0.030
	TUZ2CND 250703	S	≦0.030	≦0.70	≦1.70	≦0.030
	TUZ5CNDU 210802	S	≦0.060	≦1.00	≦2.00	≦0.030
A49−218−79 Steel Pipes-Seamless Pipes For Furnaces-Austenitic Stainless Steels 加熱炉用継目無ｵｰｽﾃﾅｲﾄ系ｽﾃﾝﾚｽ鋼鋼管	TUZ6CN 18-09	S	≦0.08	≦1.0	≦2.0	≦0.04
	TUZ2CN 18-10	S	≦0.03	≦1.0	≦2.0	≦0.04

$1\,N/mm^2$ or $MPa = 1.01972 \times 10^{-1} kgf/mm^2$

化学成分 (%)					Tensile Test N/mm^2 引張試験		Remarks 備考	Index No.
S	Ni	Cr	Mo	Others	Min. Yield Point 最小降伏点	Tensile Strength 引張強さ	(Similar to JIS)	索引番号
≦0.030	9.00～11.00	17.00～20.00	－	N 0.10～0.20	240	500～800	Austenitic Stainless Steels オーステナイト系ステンレス鋼	C631
≦0.030	8.00～11.00	17.00～20.00	－	－	200	490～740		
≦0.020	13.80～16.00	16.50～18.50	≦0.50	－	220	540～740		
≦0.030	9.00～12.00	17.00～20.00	－	Ti 5C ～0.6	190	490～740		
≦0.020	23.00～26.00	19.00～22.00	－	N6≦0.25	215	490～690		
≦0.030	10.50～13.00	16.00～18.00	2.00～2.40	－	175	470～720		
≦0.015	13.00～16.00	17.00～18.50	2.20～3.00	－	210	490～690		
≦0.030	11.00～13.50	16.00～18.00	2.00～2.40	－	280	600～800		
≦0.030	10.00～12.50	16.00～18.00	2.00～2.40	－	190	490～740		
≦0.015	21.50～23.00	24.50～26.00	1.90～2.40	N 0.10～0.15	260	540～740		
≦0.030	15.00～17.00	16.50～18.50	5.00～6.00	Cu 2.50 ～3.50	255	590～790		
≦0.010	17.50～18.50	19.50～20.50	6.00～6.50	Cu 0.50 ～1.00	300	650～850		
≦0.010	24.00～27.00	19.00～22.00	4.00～5.00	Cu 1.00 ～2.00	230	550～700		
≦0.015	29.00～32.00	25.00～29.00	3.00～4.00	Cu 0.70 ～1.20	210	500～750		
≦0.030	4.25～5.25	18.00～19.00	2.50～3.00	－	440	630～880	Austenitic and Ferritic Stainless Steels オーステナイトフェライト系ステンレス鋼	
≦0.020	4.50～6.50	21.00～23.00	2.50～3.50	N 0.08 ～0.20	450	680～880		
≦0.020	5.50～7.50	23.50～25.50	2.50～3.50	N 0.15 ～0.25	450	700～900		
≦0.030	6.00～9.00	20.00～22.00	2.00～3.00	Cu 1.00 ～2.00	380	635～835		
≦0.03	8～11	18～20	－	－	190	490～690	(SUS304TF)	C632
≦0.03	9～12	18～20	－	－	190	460～660	－	

Standard No. Year Designation 規格番号 年号 規格名称	Grade グレード	Mfg. Process 製造方法	Chemical Composition			
			C	Si	Mn	P
A49−218−79 (Continued)	TUZ6CNT 18-10	S	≦0.08	≦1.0	≦2.0	≦0.04
	TUZ6 CNNb 18-10	S	≦0.08	≦1.0	≦2.0	≦0.04
	TUZ6 CND 17-11	S	≦0.08	≦1.0	≦2.0	≦0.04
	TUZ2 CND 17-12	S	≦0.03	≦1.0	≦2.0	≦0.04
A49−245−86 Steel Tubes-Longitudinally Welded Tubes for Ferric Non-Alloy and Alloy Steel Heat Exchangers 熱交換器用縦溶接フェライト系非合金鋼及び合金鋼鋼管	TS34C	E	≦0.14	0.06〜0.30	0.30〜0.60	≦0.035
	TS37C	E	≦0.15	0.06〜0.30	0.35〜0.75	≦0.035
	TS42C	E	≦0.18	0.08〜0.35	0.45〜1.00	≦0.035
	TS48C	E	≦0.20	0.10〜0.35	0.65〜1.20	≦0.035
	TS24W3	E	≦0.13	0.10〜0.40	0.20〜0.60	≦0.040
	TSE36WB3	E	≦0.19	≦0.50	0.50〜1.50	≦0.040
	TS34BT	E	≦0.14	0.06〜0.30	0.30〜0.60	≦0.035
	TS42BT	E	≦0.18	0.08〜0.35	0.45〜1.00	≦0.035
	TS17N2	E	≦0.21	≦0.35	≦1.50	≦0.030
	TS10N9	E	≦0.15	≦0.30	≦0.90	≦0.030
A49−249−80 Steel Tubes-Longitudinally Welded Plain End Tubes for Food Industry-Austenitic Stainless Steels 食品工業用プレンエンド縦溶接オーステナイト系ステンレス鋼鋼管	TSZ6 CN18-09	E	≦0.09	≦1.05	≦2.04	≦0.045
	TSZ2 CN18-10	E	≦0.030	≦1.05	≦2.04	≦0.045
	TSZ2 CND17-12	E	≦0.030	≦1.05	≦2.04	≦0.045
	TSZ2 CND17-11	E	≦0.08	≦1.05	≦2.04	≦0.045

1 N/mm² or MPa = 1.01972 × 10^{-1} kgf/mm²

化学成分 (%)					Tensile Test N/mm² 引張試験		Remarks 備考	Index No.
S	Ni	Cr	Mo	Others	Min. Yield Point 最小降伏点	Tensile Strength 引張強さ	(Similar to JIS)	索引番号
≦0.03	10～13	17～19	−	4C≦Ti≦0.6	190	490～690	(SUS321TF)	C632
≦0.03	10～13	17～19	−	8C≦Nb+Ta≦1.0	190	490～690	(SUS347TF)	
≦0.03	11～13	16～18	2.0～3.0	−	190	490～690	(SUS316TF)	
≦0.03	12～14	16～18	2.0～3.0	−	190	460～660	−	
≦0.025	−	−	−	Cu≦0.25 Sn≦0.030	185	330～410	−	C633
≦0.025	−	−	−	Cu≦0.25 Sn≦0.030	235	360～450	−	
≦0.025	−	−	−	Cu≦0.25 Sn≦0.030	255	410～510	−	
≦0.025	−	−	−	Cu≦0.25 Sn≦0.030	275	470～570	−	
≦0.035	≦0.65	0.40～0.80	−	Cu0.20～0.50	255	380～450	−	
≦0.040	≦0.65	0.40～0.80	−	Cu0.20～0.50	330	450～520	−	
≦0.025	−	−	−	Cu≦0.25 Sn≦0.030	185	330～410	Normalization 920～ 960℃	
≦0.025	−	−	−	Cu≦0.25 Sn≦0.030	235	410～510	Normalization 875～ 925℃	
≦0.035	0.6～0.8	−	−	−	325	450～520	Normalization 850～ 900℃plus	
≦0.030	2.0～2.6	−	−	−	245	450～520	Tempering 650～ 700℃	
≦0.035	9～12.5	17～20.2	−	−	190	≧510	(SUS304TBS)	C634
≦0.035	9～12.5	17～20.2	−	−	175	≧460	(SUS304LTBS)	
≦0.035	10.5～13.15	16～18.2	1.90～2.50	−	190	≧490	(SUS316LTBS)	
≦0.035	10～12.65	16～18.2	1.90～2.50	−	200	≧510	(SUS316LTBS)	

20. NF Steel Pipes for Structural Purposes 構造用鋼管

Standard No. Year Designation 規格番号 年号 規格名称	Grade グレード	Mfg. Process 製造方法	Chemical Composition			
			C	Si	Mn	P
A49－310－94 Steel Tubes-Seamless Precision Tubes for Mechanical Application 機械用継目無精密鋼管	TU37-b	S	≦0.18	≦0.35	≦0.75	≦0.040
	TU52-b	S	≦0.20	≦0.50	≦1.50	≦0.040
	TU20MV6	S	≦0.22	≦0.50	≦1.70	≦0.040
	S100	S	≦0.12	0.10～0.35	0.70～1.10	≦0.030
	18MF6	S	0.14～0.20	0.10～0.35	1.30～1.60	≦0.030
	37MF6	S	0.32～0.39	0.10～0.35	1.35～1.65	≦0.030
A49－311－74 Steel Tubes-Seamless for Mechanical Application 機械用継目無鋼管	TU37-b	S	≦0.18	≦0.35	≦0.75	≦0.04
	TU52-b	S	≦0.20	≦0.50	≦1.5	≦0.04
	TU56-b	S	≦0.42	≦0.35	≦0.95	≦0.04
	TUXC35	S	0.38～0.38	0.15～0.40	0.50～0.80	≦0.035
A49－312－93 Steel Tubes-Seamless Tubes with Improved Machinability for Mechanical Machined Parts 機械用継目無鋼管（補完用）	S470M	S	0.15～0.22	≦0.50	1.00～1.70	≦0.030
	S450MG2	S	0.15～0.22	≦0.50	1.00～1.70	≦0.030
	S650MG2	S	0.15～0.22	≦0.50	1.00～1.70	≦0.030
A49－317－80 Steel Tubes-Seamless Plain End Tubes for Ergineering Use, Austenitic Stainless Steels 機械用プレンエンド継目無オーステナイト系ステンレス鋼鋼管	TUZ2 CN18-10	S	≦0.030	≦1.00	≦2.00	≦0.040
	TUZ2 CND17-12	S	≦0.030	≦2.00	≦0.030	≦0.040
A49－321－78 Steel Tubes-Jacks for Hydraulic Transmissions-Cold-Rolled or Cold-Drawn Seamless Tubes, Type "Standard"	TU37b	S	≦0.18	≦0.35	≦0.75	≦0.04

1 N/mm² or MPa = 1.01972×10^{-1} kgf/mm²

化　学　成　分　(%)					Tensile Test N/mm² 引張試験		Remarks 備考 (Similar to JIS)	Index No. 索引番号
S	Ni	Cr	Mo	Others	Min. Yield Point 最小降伏点	Tensile Strength 引張強さ		
≦0.040	−	−	−	−	240	360~500	Normalized	C635
≦0.040	−	−	−	−	350	510~650	Normalized	
≦0.040	−	−	−	V ≦0.15	410	550~700	Normalized	
0.080~0.130	−	−	−	−	240	360~500	Normalized	
0.080~0.130	−	−	−	−	360	520~650	Normalized	
0.080~0.130	−	−	−	−	420	650~720	Normalized	
≦0.04	−	−	−	−	t <16 210	≧360	−	C636
					t ≧16 190	≧340		
≦0.04	−	−	−	−	t <16 340	≧520	(STKM18)	
					t ≧16 320	≧490		
≦0.04	−	−	−	−	t <16 340	≧510	−	
					t ≧16 320	≧490		
≦0.035	−	−	−	−	t <16 320	≧540	(STKM16)	
					t ≧16 300	≧520		
0.020~0.040	−	−	−	Cu ≦0.30 V 0.08~0.15	≦16 470	≧620	(STKM18)	C637
					≦20 460	≧610		
					≦25 460	≧610		
					≦30 430	≧550		
					≦40 420	≧550		
					≦52.5 410	≧550		
0.020~0.040	−	−	−	Cu ≦0.30 V 0.08~0.15	≦16 450	550~720	(STKM16)	
					≦20 440	550~720		
					≦25 440	550~720		
					≦30 430	550~720		
					≦40 420	550~720		
					≦52.5 410	550~720		
0.020~0.040	−	−	−	Cu ≦0.30 V 0.08~0.15	≦16 650	≧750		
					≦20 650	≧750		
					≦25 620	≧700		
					≦30 570	≧650		
≦0.030	9~12.00	17~20.0	−	−	175	≧470	−	C638
≦1.00	10.5~13.0	16.0~18.0	2.00~2.40	−	175	≧470	−	
≦0.04	−	−	−	−	(Normalized) 240	360~480	Impact Test (C, J／cm²) −20 , 35	C639

Standard No. Year Designation 規格番号 年号 規格名称	Grade グレード	Mfg. Process 製造方法	Chemical Composition			
			C	Si	Mn	P
水圧伝達装置用ジャッキ，冷間圧延または冷間引抜継目無鋼管（標準形）	TU52b	S	≦0.20	≦0.50	≦1.5	≦0.04
	TU52BT	S	≦0.20	≦0.50	≦1.5	≦0.04
	TU18 MDV5	S	≦0.20	≦0.45	≦1.55	≦0.035
	TU17 MV5	S	≦0.21	≦0.45	≦1.50	≦0.040
A49－322－78 Steel Tubes-Jack for Hydraulic Transmissions-Cold-Rolled or Cold-Drawn Seamless Tubes, Type "Suitable for Grinding in " 水圧伝達装置用ｼﾞｬｯｷ，冷間圧延または冷間引抜継目無鋼管（すりあわせ形）	TU37b	S	≦0.18	≦0.35	≦0.75	≦0.04
	TU52b	S	≦0.20	≦0.50	≦1.5	≦0.04
	TU17 MV5	S	≦0.21	≦0.45	≦1.50	≦0.040
A49－323－78 Steel Tubes-Jacks for Hydraulic Transmissions-Cold-Rolled or Cold-Drawn Seamless Tubes, Type "Ready for Use" 水圧伝達装置用ｼﾞｬｯｷ，冷間圧延または冷間引抜継目無鋼管（直接適用形）	TU37b	S	≦0.18	≦0.35	≦0.75	≦0.04
	TU52b	S	≦0.20	≦0.50	≦1.50	≦0.040
	TU52BT	S	≦0.20	≦0.50	≦1.50	≦0.40
	TU17 MV5	S	≦0.21	≦0.45	≦1.50	≦0.040
A49－326－75 Steel Tubes-Jacks for Pneumatic Transmissions-Cold Drawn Seamless Tubes, Type "Suitable for Grinding in " 空気圧伝達用ｼﾞｬｯｷ，冷間引抜継目無鋼管（すりあわせ形）	TU37b	S	≦0.18	≦0.35	≦0.75	≦0.04
	TU52b	S	≦0.20	≦0.50	≦1.5	≦0.04
	TU52BT	S	≦0.20	≦1.5	≦0.50	≦0.04
A49－327－75 Steel Tubes-Jacks for Pneumatic Transmissions-Cold Drawn Seamless Tubes, Tvpe "Ready for Use " 空気圧伝達用ｼﾞｬｯｷ，冷間引抜継目無鋼管（直接適用形）	TU37b	S	≦0.18	≦0.35	≦0.75	≦0.04
	TU52b	S	≦0.20	≦0.50	≦1.5	≦0.04
A49－330－85 Steel Tubes-Seamless Cold Drawn Tubes, for Hydraulic and Pneumatic Power Systems	TU37b	S	≦0.18	≦0.35	≦0.75	≦0.040
	TU52b	S	≦0.20	≦0.50	≦1.50	≦0.040

1 N/mm² or MPa = 1.01972 × 10⁻¹ kgf/mm²

化学成分 (%)					Tensile Test N/mm² 引張試験		Remarks 備考	Index No.
S	Ni	Cr	Mo	Others	Min. Yield Point 最小降伏点	Tensile Strength 引張強さ	(Similar to JIS)	索引番号
≦0.04	−	−	−	−	(Normalized) 350	510〜630	− , − (STKM18)	C639
≦0.04	−	−	−	−	(〃) 290	510〜630	−20 , 50 −46 , 35	
≦0.035	−	−	≦0.35	V ≦0.14	(〃) 410	510〜630	−20 , 50	
≦0.040	−	−	≦0.35	V ≦0.14	(〃) 480	590	−20 , 50	
≦0.04	−	−	−	−	350	≧440	(STKM12)	C640
≦0.04	−	−	−	−	490	≧600	−	
≦0.040	−	−	−	V ≦0.15	510	≧640	−	
≦0.040	−	−	−	−	(Normalized) 240	360〜480	Impact Test (℃, J/cm²) −20 , 35	C641
≦0.040	−	−	−	−	(〃) 350	510〜630	− , − (STKM18)	
≦0.040	−	−	−	−	(〃) 350	510〜630	−20 , 50 −46 , 35	
≦0.040	−	−	−	V ≦0.15	(〃) 410	510〜630	−20 , 50	
≦0.04	−	−	−	−	210	360〜480	(STKM13)	C642
≦0.04	−	−	−	−	350	510〜630	(STKM18)	
≦0.04	−	−	−	−	350	510〜630	Impact Test −46℃ 35J/cm²	
≦0.04	−	−	−	−	−	≧440	(STKM12)	C643
≦0.04	−	−	−	−	−	≧650		
≦0.040	−	−	−	−	220	360〜480	(STKM13)	C644
≦0.040	−	−	−	−	350	510〜630	(STKM18) Impact Test	

Standard No. Year Designation 規格番号 年号 規格名称	Grade グレード	Mfg. Process 製造方法	Chemical Composition C	 Si	 Mn	 P
水圧及び空気圧用冷間引抜継目無鋼管	TU42BT	S	≦0.20	≦0.35	≦1.05	≦0.035
	TU17N2	S	≦0.21	≦0.35	≦1.50	≦0.040
	TU10N14	S	≦0.15	≦0.35	≦0.65	≦0.030
A49−341−75 Steel Tubes Precision Welded Tubes for Mechanical Application 機械用精密溶接鋼管	TS30o	B, E	≦0.15	−	−	≦0.075
	TS30a	B, E	≦0.15	−	−	≦0.062
	TS34a	B, E	≦0.18	−	−	≦0.062
	TS37a	B, E	≦0.22	−	−	≦0.062
	TS42a	B, E	≦0.24	≦0.45	≦1.40	≦0.055
	TS47a	B, E	≦0.24	≦0.60	≦1.60	≦0.055
A49−343−80 Steel Tubes-Longitudinally Welded Tubes D≦168.3 mm for Engineering Use 外径168.3 mm以下の機械用縦溶接鋼管	TS37b	E	≦0.18	≦0.38	≦0.81	≦0.045
	TS18M5	E	≦0.22	≦0.55	≦1.6	≦0.045
	TS18 MDV5	E	≦0.22	≦0.30	≦1.5	≦0.045
A49−501−86 Steel Tubes-Seamless or Welded Hot Finished Structural Hollow Sections 継目無または溶接熱間仕上中空形鋼	TUTSE 235-2	Sorw	≦0.20	−	−	≦0.050
	TUTSE 235-3	Sorw	≦0.20	−	−	≦0.045
	TUTSE 235-4	Sorw	≦0.20	−	−	≦0.040
	TUTSE 275-2	Sorw	≦0.22	−	−	≦0.050
	TUTSE 275-3	Sorw	≦0.22	−	−	≦0.045
	TUTSE 275-4	Sorw	≦0.22	−	−	≦0.040
	TSE 295-2	Sorw	≦0.24	−	−	≦0.040
	TSE 295-3	Sorw	≦0.22	−	−	≦0.040
	TSE 295-4	Sorw	≦0.22	−	−	≦0.040
	TUTSE 355-2	Sorw	≦0.24	−	−	≦0.040

1 N/mm² or MPa = 1.01972 × 10^{-1} kgf/mm²

化　学　成　分　(%)					Tensile Test N/mm² 引張試験		Remarks 備考	Index No.
S	Ni	Cr	Mo	Others	Min. Yield Point 最小降伏点	Tensile Strength 引張強さ	(Similar to JIS)	索引番号
≦0.035					240	415～510	(℃, J／cm²) −45， 35	C644
≦0.040	0.6 ～0.8				325	490～640	− 60， 50	
≦0.030	3.2 ～3.8				245	450～600	−100， 50	
≦0.062	−	−	−	−	(Normalized) 175	≧310	−	C645
≦0.062	−	−	−	N ≦0.009	(〃) 185	≧310	−	
≦0.062	−	−	−	N ≦0.009	(〃) 210	≧330	−	
≦0.062	−	−	−	N ≦0.009	(〃) 240	≧380	−	
≦0.055	−	−	−	−	(〃) 270	≧430	(STKM18)	
≦0.055	−	−	−	−	(〃) 300	≧480	(STKM18)	
≦0.045	−	−	−	−	230	≧360	(STKM13)	C646
≦0.045	−	−	−	−	340	≧510	(STKM18)	
≦0.045	−	−	≦0.3	V ≦0.12	470	≧610	−	
≦0.050	−	−	−	−	235	340～480	−	C647
≦0.045	−	−	−	−	235	340～480	−	
≦0.040	−	−	−	−	235	340～480	−	
≦0.050	−	−	−	−	275	410～550	−	
≦0.050	−	−	−	−	275	410～550	−	
≦0.040	−	−	−	−	275	410～550	−	
≦0.040	−	−	−	−	295	420～560	−	
≦0.040	−	−	−	−	295	420～560	−	
≦0.040	−	−	−	−	295	420～560	−	
≦0.040	−	−	−	−	355	490～640	−	

Standard No. Year Designation 規格番号 年号 規格名称	Grade グレード	Mfg. Process 製造方法	Chemical Composition			
			C	Si	Mn	P
A49−501−86 (Continued)	TUTSE 355-3	Sorw	≦0.24	−	−	≦0.040
	TUTSE 355-4	Sorw	≦0.24	−	−	≦0.040
	TUTSE 450-2	Sorw	≦0.25	−	−	≦0.040
	TUTSE 450-3	Sorw	≦0.25	−	−	≦0.040
	TUTSE 450-4	Sorw	≦0.25	−	−	≦0.040
A49−541−86 Steel Tubes-Cold Finished Welded Structural Hollow Sections 構造用冷間仕上溶接中空形鋼	TSE 235-2	W	≦0.20	≦0.40	≦1.10	≦0.050
	TSE 235-3	W	≦0.18	≦0.40	≦1.20	≦0.045
	TSE 235-4	W	≦0.18	≦0.40	≦1.20	≦0.040
	TSE 275-2	W	≦0.22	≦0.40	≦1.20	≦0.050
	TSE 275-3	W	≦0.20	≦0.40	≦1.20	≦0.045
	TSE 275-4	W	≦0.20	≦0.40	≦1.20	≦0.040
A49−647−79 Steel Tubes-Structural Welded Tubes, Circular, Square Rectangular or Oval, in Ferritic or Austenitic Stainless Steels 構造用円形，方形，矩形又は楕円形フェライト系，オーステナイト系ステンレス鋼鋼管	TSZ2CN 18-10	W	≦0.030	≦1.0	≦2.0	≦0.040
	TSZ6CN 18-09	W	≦0.07	≦1.0	≦2.0	≦0.040
	TSZ6CND 17-11	W	≦0.08	≦1.0	≦2.0	≦0.040

1 N/mm² or MPa = 1.01972×10^{-1} kgf/mm²

化 学 成 分 (%)					Tensile Test N/mm² 引張試験		Remarks 備考 (Similar to JIS)	Index No. 索引番号
S	Ni	Cr	Mo	Others	Min. Yield Point 最小降伏点	Tensile Strength 引張強さ		
≦0.040	−	−	−	−	355	490〜640	−	C 647
≦0.040	−	−	−	−	355	490〜640	−	
≦0.040	−	−	−	−	450	550〜750	−	
≦0.040	−	−	−	−	450	550〜720	−	
≦0.040	−	−	−	−	450	550〜720	−	
≦0.050	−	−	−	N ≦0.009	t ≦3.2 mm 250 t >3.2 mm 235	360〜480	Impact Test +20℃, 35	C 648
≦0.045	−	−	−	N ≦0.009	t ≦3.2 mm 250 t >3.2 mm 230	360〜480	Impact Test 0℃, 35	
≦0.040	−	−	−	N ≦0.009	t ≦3.2 mm 250 t >3.2 mm 230	360〜480	Impact Test −20℃, 35	
≦0.050	−	−	−	N ≦0.009	t ≦3.2 mm 290 t >3.2 mm 275	410〜550	Impact Test +20℃, 35	
≦0.045	−	−	−	N ≦0.009	t ≦3.2 mm 290 t >3.2 mm 275	410〜550	Impact Test 0℃, 35	
≦0.040	−	−	−	N ≦0.009	t ≦3.2 mm 290 t >3.2 mm 275	410〜550	Impact Test −20℃, 35	
≦0.030	8〜12	18〜20	−	−	400	≧600	−	C 649
≦0.030	8〜12	18〜20	−	−	400	≧600	−	
≦0.030	10〜12.5	16〜18	2〜2.5	−	400	≧600	−	

Standard No. Year Designation 規格番号 年号 規格名称	Grade グレード	Mfg. Process 製造方法	Chemical Composition			
			C	Si	Mn	P
A49－647－79 (Continued)	TSZ12CN 17-07	W	≦0.15	≦1.0	≦2.0	≦0.040

1 N/mm² or MPa = 1.01972×10^{-1} kgf/mm²

化　学　成　分　(%)					Tensile Test N/mm² 引　張　試　験		Remarks 備　考 (Similar to JIS)	Index No. 索引番号
S	Ni	Cr	Mo	Others	Min. Yield Point 最小降伏点	Tensile Strength 引張強さ		
≦0.030	6～8.5	16～18	－	－	500	≧800	－	C649

21. ISO Steel Pipes for General Purposes　一般用鋼管

Standard No. Year Designation 規格番号 年号 規格名称	Grade グレード	Mfg. Process 製造方法	Chemical Composition			
			C	Si	Mn	P
65－81 Carbon Steel Tubes Suitable for Screwing in Accordance with ISO 7/1 ねじ付炭素鋼鋼管	TS	S	－	－	－	≦0.06
	TW	W	－	－	－	≦0.06
559－77 Welded or Seamless Steel Tubes for Water, Sewage and Gas 給排水・ガス用溶接又は継目無鋼管	TS0	S	－	－		≦0.060
	TW0	W	－	－		≦0.060
	TS1	S	≦0.16	－	0.30～0.70	≦0.050
	TW1	W	≦0.16	－	0.30～0.70	≦0.050
	TS4	S	≦0.17	≦0.35	0.40～0.80	≦0.045
	TW4	W	≦0.17	≦0.35	0.40～0.80	≦0.045
	TS9	S	≦0.21	≦0.35	0.40～1.20	≦0.045
	TW9	W	≦0.21	≦0.35	0.40～1.20	≦0.045

1 N/mm² or MPa = 1.01972×10^{-1} kgf/mm²

化 学 成 分 (%)					Tensile Test N/mm² 引 張 試 験		Remarks 備 考	Index No.
S	Ni	Cr	Mo	Others	Min. Yield Point 最小降伏点	Tensile Strength 引張強さ	(Similar to JIS)	索引番号
≦0.06	–	–	–	–	–	320～520	–	C701
≦0.06	–	–	–	–	–	320～520	(SGP)	
≦0.060	–	–	–	–	–	320～520	–	C702
≦0.060	–	–	–	–	–	320～520	(SGP)	
≦0.050	–	–	–	–	195	320～440	–	
≦0.050	–	–	–	–	195	320～440	–	
≦0.045	–	–	–	–	215	360～480	(STPG370)	
≦0.045	–	–	–	–	215	360～480	(STPG370)	
≦0.045	–	–	–	–	235	410～530	(STPG410)	
≦0.045	–	–	–	–	235	410～530	(STPG410)	

22. ISO Steel Tubes for Pressure Purposes　圧力容器用鋼管

Standard No. Year Designation 規格番号 年号 規格名称	Grade グレード	Mfg. Process 製造方法	Chemical Composition C	Si	Mn	P
2604／2－75 Steel Products for Pressure Purposes Part2:Wrought Seamless Tubes 圧力容器用継目無鋼管	TS1	S	≦0.16	－	0.30～0.70	≦0.050
	TS2	S	≦0.16	－	0.40～0.70	≦0.050
	TS4	S	≦0.17	≦0.35	0.40～0.80	≦0.045
	TS5	S	≦0.17	≦0.35	0.40～0.80	≦0.045
	TS6	S	≦0.17	≦0.35	0.40～1.00	≦0.045
	TS9	S	≦0.21	≦0.35	0.40～1.20	≦0.045
	TS9H	S	≦0.21	≦0.35	0.40～1.20	≦0.045
	TS10	S	≦0.19	≦0.35	0.60～1.20	≦0.045
	TS13	S	≦0.22	≦0.35	0.60～1.40	≦0.045
	TS14	S	≦0.22	≦0.35	0.80～1.40	≦0.045
	TS15	S	≦0.20	≦0.35	0.80～1.40	≦0.045
	TS18	S	≦0.23	≦0.35	0.80～1.50	≦0.045
	TS26	S	0.12～0.20	0.10～0.35	0.40～0.80	≦0.040
	TS32	S	0.10～0.18	0.10～0.35	0.40～0.70	≦0.040
	TS33	S	0.10～0.18	0.10～0.35	0.40～0.70	≦0.040
	TS34	S	0.08～0.15	≦0.50	0.40～0.70	≦0.040
	TS37	S	≦0.15	≦0.50	0.30～0.60	≦0.030
	TS38	S	≦0.15	0.25～1.00	0.30～0.60	≦0.030
	TS39	S	≦0.08	≦1.00	≦1.00	≦0.040
	TS40	S	0.17～0.23	≦0.50	≦1.00	≦0.030
	TS43	S	≦0.15	0.15～0.35	0.30～0.80	≦0.040
	TS45	S	≦0.13	0.15～0.30	0.30～0.80	≦0.040
	TS46	S	≦0.03	≦1.00	≦2.00	≦0.045
	TS47	S	≦0.07	≦1.00	≦2.00	≦0.045

1 N/mm² or MPa = 1.01972 × 10⁻¹ kgf/mm²

化学成分（%）					Tensile Test N/mm² 引張試験		Remarks 備考	Index No.
S	Ni	Cr	Mo	Others	Min. Yield Point 最小降伏点	Tensile Strength 引張強さ	(Similar to JIS)	索引番号
≦0.050	−	−	−	−	195	320～440	−	C703
≦0.050	−	−	−	−	195	320～440	−	
≦0.045	−	−	−	−	215	360～480	(STPG370) (STS 370)	
≦0.045	−	−	−	−	215	360～480	(STPT370) (STF 370)	
≦0.045	−	−	−	Al＞0.015	215	360～480	(STPL380) (STBL380)	
≦0.045	−	−	−	−	235	410～530	(STS 410)	
≦0.045	−	−	−	−	235	410～530	(STS 410) (STB 410) (STF 410)	
≦0.045	−	−	−	Al＞0.015	235	410～530	−	
≦0.045	−	−	−	−	265	460～580	(STS 480)	
≦0.045	−	−	−	−	265	460～580	(STS 480)	
≦0.045	−	−	−	Al＞0.015	265	460～580	−	
≦0.045	−	−	−	−	285	490～610	(STB 510)	
≦0.040	−	−	0.25～0.35	Al≦0.012	255	450～600	−	
≦0.040	−	0.70～1.10	0.45～0.65	Al≦0.02	275	440～590	−	
≦0.040	−	0.30～0.80	0.50～0.70	V 0.22～0.32 Al≦0.02	275	460～610	−	
≦0.045	−	2.00～2.50	0.90～1.20	Al≦0.02	A 140 N+T 275	410～560 490～640	−	
≦0.030	−	4.00～6.00	0.45～0.65	Al≦0.02	205	410～560	(STPA25) (STBA25) (STFA25)	
≦0.030	−	8.00～10.00	0.90～1.10	Al≦0.02	A 140 N+T 390	410～560 590～	−	
≦0.030	≦0.50	11.5～14.0	−	−	A 245 Q+T 390	440～590 590～750	−	
≦0.030	0.30～0.80	10.00～12.50	0.80～1.20	V 0.25～0.35	430	～	−	
≦0.040	3.25～3.75	−	−	−	245	440～590	(STPL450) (STBL450)	
≦0.040	8.50～9.50	−	−	−	510	690～850	(STPL690) (STBL690)	
≦0.030	9.00～13.00	17.00～19.00	−	−	180	490～690	(SUS304LTP) (SUS304LT)	
≦0.030	8.00～12.00	17.00～19.00	−	−	195	490～690	(SUS304TP) (SUS304TB) (SUS304TF)	

StandardNo. Year Designation 規格番号 年号 規格名称	Grade グレード	Mfg. Process 製造方法	Chemical Composition C	 Si	 Mn	 P
2604/2-75 (Continued)	TS48	S	0.04~0.09	≦0.75	≦2.00	≦0.045
	TS50	S	≦0.08	≦1.00	≦2.00	≦0.045
	TS53	S	≦0.08	≦1.00	≦2.00	≦0.045
	TS54	S	0.04~0.10	0.20~0.80	≦2.00	≦0.045
	TS56	S	0.04~0.10	0.20~0.80	≦2.00	≦0.045
	TS57	S	≦0.03	≦1.00	≦2.00	≦0.045
	TS58	S	≦0.03	≦1.00	≦2.00	≦0.045
	TS60	S	≦0.07	≦1.00	≦2.00	≦0.045
	TS61	S	≦0.07	≦1.00	≦2.00	≦0.045
	TS63	S	0.04~0.09	≦0.75	1.00~2.00	≦0.045
	TS67	S	0.04~0.10	0.30~0.60	1.00~1.50	≦0.045
	TS68	S	≦0.15	≦0.75	≦2.00	≦0.045
	TS69	S	≦0.12	≦1.00	≦1.50	≦0.045
2604/3-75 Steel Products for Pressure Purposes Part3:Electric Resistance and Induction-Welded Tubes 圧力容器用電気抵抗溶接鋼管	TW1	E	≦0.16	–	0.30~0.70	≦0.050
	TW2	E	≦0.16	–	0.30~0.70	≦0.050
	TW4	E	≦0.17	≦0.35	0.40~0.80	≦0.045
	TW5	E	≦0.17	≦0.35	0.40~0.80	≦0.045
	TW6	E	≦0.17	≦0.35	0.40~1.00	≦0.045
	TW9	E	≦0.21	≦0.35	0.40~1.20	≦0.045

$1\,N/mm^2$ or $MPa = 1.01972 \times 10^{-1}\,kgf/mm^2$

化　学　成　分　(%)					Tensile Test N/mm² 引張試験		Remarks 備考	Index No.
S	Ni	Cr	Mo	Others	Min. Yield Point 最小降伏点	Tensile Strength 引張強さ	(Similar to JIS)	索引番号
≦0.030	8.00〜12.00	17.00〜20.00	−	−	195	490〜690	(SUS304HTP) (SUS304HTB) (SUS347HTF)	C703
≦0.030	9.00〜13.00	17.00〜19.00	−	Nb10×C〜1.00	205	510〜720	(SUS347HTP) (SUS347HTB) (SUS347HTF)	
≦0.030	9.00〜13.00	17.00〜19.00	−	Ti5 ×C〜0.80	235	510〜720	(SUS321TP) (SUS321TB) (SUS321TC)	
≦0.030	9.00〜13.00	17.00〜20.00	−	Ti4 ×C〜0.60	160 195	490〜690 500〜720	Q1070〜1140℃ Q 950〜1070℃	
≦0.030	11.00〜14.00	16.00〜20.00	−	Nb10×C〜1.4	205	510〜720	(SUS347HTP) (SUS347HTB)	
≦0.030	11.00〜14.00	16.00〜18.50	2.00〜2.50	−	185	490〜690	(SUS316LTP) (SUS316LTB)	
≦0.030	11.50〜14.50	16.00〜18.50	2.00〜3.00	−	185	490〜690	(SUS316LTP) (SUS316LTB)	
≦0.030	11.00〜14.00	16.00〜18.50	2.00〜2.50	−	205	510〜720	(SUS316TP) (SUS316TB) (SUS316TF)	
≦0.030	11.00〜14.50	16.00〜18.50	2.50〜3.00	−	205	510〜720	(SUS316TP) (SUS316TB) (SUS316TF)	
≦0.030	12.00〜14.00	16.00〜18.00	2.00〜2.75	−	205	510〜720	(SUS316HTP) (SUS316HTB) (SUS316HTF)	
≦0.030	15.50〜17.50	15.50〜17.50	1.60〜2.00	Nb : 10×C〜 10×C+0.4	215	510〜720	−	
≦0.030	19.00〜22.00	24.00〜26.00	−	−	205	510〜720	(SUS316TP) (SUS310TB) (SUS310TF)	
≦0.030	30.00〜35.00	19.00〜23.00	−	Ti : 0.15〜0.60 Al : 0.15〜0.60	205	480〜690	−	
≦0.050	−	−	−	−	195	320〜440	−	C704
≦0.050	−	−	−	−	195	320〜440	−	
≦0.045	−	−	−	−	215	360〜480	(STPG370)	
≦0.045	−	−	−	−	215	360〜480	(STPT370)	
≦0.045	−	−	−	Al≦0.015	215	360〜480	(STPL380) (STBL380)	
≦0.045	−	−	−	−	235	410〜530	(STPG410)	

Standard No. Year Designation 規格番号 年号 規格名称	Grade グレード	Mfg. Process 製造方法	Chemical Composition			
			C	Si	Mn	P
2604／3－75 (Continued)	TW9H	E	≦0.21	≦0.35	0.40～1.20	≦0.045
	TW10	E	≦0.19	≦0.35	0.60～1.20	≦0.045
	TW13	E	≦0.22	≦0.35	0.60～1.40	≦0.045
	TW14	E	≦0.22	≦0.35	0.80～1.40	≦0.045
	TW15	E	≦0.20	≦0.35	0.80～1.40	≦0.045
	TW26	E	0.12～0.20	0.10～0.35	0.40～0.80	≦0.040
	TW32	E	0.10～0.18	0.10～0.35	0.40～0.70	≦0.040
2604／5－78 Steel Products for Pressure Purposes Part5:Longitudinally Welded Austenitic Stainless Steel Tubes 圧力容器用縦溶接オーステナイト系ステンレス鋼鋼管	TW46	A	≦0.03	≦1.00	≦2.00	≦0.045
	TW47	A	≦0.07	≦1.00	≦2.00	≦0.045
	TW50	A	≦0.08	≦1.00	≦2.00	≦0.045
	TW53	A	≦0.08	≦1.00	≦2.00	≦0.045
	TW57	A	≦0.03	≦1.00	≦2.00	≦0.045
	TW58	A	≦0.03	≦1.00	≦2.00	≦0.045
	TW60	A	≦0.07	≦1.00	≦2.00	≦0.045
	TW61	A	≦0.07	≦1.00	≦2.00	≦0.045
	TW69	A	≦0.10	≦1.00	≦1.50	≦0.045

1 N/mm² or MPa = 1.01972×10^{-1} kgf/mm²

化学成分 (%)					Tensile Test N/mm² 引張試験		Remarks 備考	Index No.
S	Ni	Cr	Mo	Others	Min. Yield Point 最小降伏点	Tensile Strength 引張強さ	(Similar to JIS)	索引番号
≦0.045	–	–	–	–	235	410～530	(STPT410) (STPB410)	C704
≦0.045	–	–	–	Al≦0.015	235	410～530	–	
≦0.045	–	–	–	–	235	460～580	–	
≦0.045	–	–	–	–	265	460～580	–	
≦0.045	–	–	–	Al≦0.015	265	460～580	–	
≦0.040	–	–	0.25～0.35	Al≦0.012	255	450～600	–	
≦0.040	–	0.70～1.10	0.45～0.65	Al≦0.020	275	440～590	–	
≦0.030	9.00～12.00	17.00～19.00	–	–	175	490～690	(SUS304LTPY)	C705
≦0.030	8.00～11.00	17.00～19.00	–	–	195	490～690	(SUS304TPY)	
≦0.030	9.00～12.00	17.00～19.00	–	Nb : 10×c～1.00	205	510～710	(SUS347TPY)	
≦0.030	9.00～12.00	17.00～19.00	–	Ti : 5 ×c～0.80	195	510～710	(SUS321TPY)	
≦0.030	11.00～14.00	16.00～18.50	2.00～2.50	–	185	490～690	(SUS316LTPY)	
≦0.030	11.50～14.50	16.00～18.50	2.50～3.00	–	185	490～690	(SUS316LTPY)	
≦0.030	10.50～14.00	16.00～18.50	2.00～2.50	–	205	510～710	(SUS316TPY)	
≦0.030	11.00～14.50	16.00～18.50	2.50～3.00	–	205	510～710	(SUS316TPY)	
≦0.030	30.00～35.00	19.00～23.00	–	Ti : 0.15～0.60 Al : 0.15～0.60	195	480～680	–	

23. ISO Steel Tuaes for Mechnical Application 機械用鋼管

Standard No. Year Designation 規格番号 年号 規格名称	Grade グレード	Mfg. Process 製造方法	Chemical Composition			
			C	Si	Mn	P
2937－74 Plain End Seamless Steel Tubes for Mechanical Application 機械用プレンドエンド継目無鋼管	TS1	S	≦0.16	–	0.30～0.70	≦0.050
	TS4	S	≦0.17	≦0.35	0.40～0.80	≦0.045
	TS9	S	≦0.21	≦0.35	0.40～1.20	≦0.045
	TS18	S	≦0.23	≦0.35	0.80～1.50	≦0.045
	C35	S	0.32～0.39	0.15～0.40	0.50～0.80	≦0.050
2938－74 Hollow Steel Bars for Machining 機械用中空鋼棒 ① Hot-finished ② Normalized	Gr1	S	≦0.20	≦0.50	≦1.6	≦0.045
	Gr2	S	0.32～0.39	0.15～0.40	0.50～0.80	≦0.035
3304－85 Plain End Seamless Precision Steel Tubes プレーンエンド継目無精密鋼管 ① Cold-finished as drawn ② Lightly Cold-Worked ③ Annealed ④ Normalized	R28	S	≦0.13	–	≦0.060	≦0.050
	R33	S	≦0.16	–	≦0.070	≦0.050
	R37	S	≦0.17	≦0.35	≦0.8	≦0.050
	R42	S	≦0.21	≦0.35	≦1.2	≦0.050
	R50	S	≦0.23	≦0.35	≦1.5	≦0.050
3305－85 Plain End Welded Precision Steel Tubes プレーンエンド溶接精密鋼管 ① Cold-finished as drawn	R28	E	≦0.13	–	≦0.60	≦0.050

$1 N/mm^2$ or $MPa = 1.01972 \times 10^{-1} kgf/mm^2$

化　学　成　分　(%)					Tensile Test N/mm^2 引張試験		Remarks 備考	Index No.
S	Ni	Cr	Mo	Others	Min. Yield Point 最小降伏点	Tensile Strength 引張強さ	(Similar to JIS)	索引番号
≦0.050	–	–	–	–	195	320～440	–	C706
≦0.045	–	–	–	–	215	360～480	(STKM13) (STC38)	
≦0.045	–	–	–	–	235	410～530	(STKM14)	
≦0.045	–	–	–	–	285	490～610	(STKM19)	
≦0.050	–	–	–	–	275	540～660	–	
≦0.045	–	–	–	① t ≦16 16< t ≦30 30< t	330 325 305	490～610	(STKM19)	C707
				② t ≦16 16< t ≦30 30< t	340 225 315	490～610		
≦0.035	–	–	–	① t ≦16 t >16	275 265	490～640	–	
				② t ≦16 t >16	275 265	490～610		
≦0.050	–	–	–	–	① – ② – ③ – ④ 155	≧400 ≧350 ≧275 ≧285	(STKM11)	C708
≦0.050	–	–	–	–	① – ② – ③ – ④ 195	≧420 ≧370 ≧325 ≧325	(STKM12)	
≦0.050	–	–	–	–	① – ② – ③ – ④ 215	≧450 ≧400 ≧340 ≧360	(STKM13) (STC38)	
≦0.050	–	–	–	–	① – ② – ③ – ④ 235	≧520 ≧450 ≧400 ≧410	(STKM14)	
≦0.050	–	–	–	–	① – ② – ③ – ④ 285	≧600 ≧550 ≧480 ≧490	(STKM19)	
≦0.050	–	–	–	–	① – ② – ③ – ④ 155	≧400 ≧350 ≧270 ≧280	(STKM11) (STAM30G)	C709

Standard No. Year Designation 規格番号 年号 規格名称	Grade グレード	Mfg. Process 製造方法	Chemical Composition			
			C	Si	Mn	P
② Lightly Cold-Worked ③ Annealed ④ Normalized	R33	E	≦0.16	−	≦0.70	≦0.050
	R37	E	≦0.17	≦0.35	≦0.8	≦0.050
	R44	E	≦0.21	≦0.35	≦1.2	≦0.050
	R50	E	≦0.23	≦0.35	≦1.5	≦0.050
3306−85 Plain End as Welded and Sized Precision Steel Tubes プレーンエンド溶接定径精密鋼管 ① As-Welded and Sized ② Annealed ③ Normalized	R28	E	≦0.13	−	≦0.60	≦0.050
	R33	E	≦0.16	−	≦0.70	≦0.050
	R37	E	≦0.17	≦0.35	≦0.8	≦0.050
	R44	E	≦0.21	≦0.35	≦1.2	≦0.050
	R50	E	≦0.23	≦0.35	≦1.5	≦0.050

1 N/mm² or MPa = 1.01972 × 10⁻¹ kgf/mm²

化学成分（%）					Tensile Test N/mm² 引張試験		Remarks 備考	Index No.
S	Ni	Cr	Mo	Others	Min. Yield Point 最小降伏点	Tensile Strength 引張強さ	(Similar to JIS)	索引番号
≦0.050	–	–	–	–	① – ② – ③ – ④ 190	≧420 ≧370 ≧320 ≧320	(STKM12) (STAM340G)	C709
≦0.050	–	–	–	–	① – ② – ③ – ④ 210	≧450 ≧400 ≧340 ≧360	(STKM13) (STAM390G)	
≦0.050	–	–	–	–	① – ② – ③ – ④ 230	≧520 ≧450 ≧400 ≧410	(STKM14)	
≦0.050	–	–	–	–	① – ② – ③ – ④ 280	≧600 ≧550 ≧480 ≧490	(STKM19)	
≦0.050	–	–	–	–	① – ② – ③ 150	≧300 ≧270 ≧280	(STKM11) (STAM290G)	C710
≦0.050	–	–	–	–	① – ② – ③ 195	≧330 ≧320 ≧320	(STKM12) (STAM340G)	
≦0.050	–	–	–	–	① – ② – ③ 215	≧400 ≧340 ≧360	(STKM12) (STAM340G) (STAM390G)	
≦0.050	–	–	–	–	① – ② – ③ 235	≧460 ≧400 ≧410	(STKM14)	
≦0.050	–	–	–	–	① – ② – ③ 285	≧520 ≧480 ≧490	(STKM19)	

24. ISO Steel Tubes for Petroleum and Natural Gas Industries

Standard No. Year Designation 規格番号 年号 規格名称	Grade グレード	Mfg. Process 製造方法	Chemical Composition			
			C	Si	Mn	P
3183－80 Oil and Natural Gas Industries-Steel Line Pipe 石油及び天然ガス用ラインパイプ	E17	S, W	≦0.21	－	0.30～0.60	≦0.045
	E21	S, W	≦0.22	－	≦2.0	≦0.04
	E24-1	S, W	≦0.22	－	≦1.15	≦0.04
	E24-2	S, W	≦0.27	－	≦1.15	≦0.04

石油天然ガス用鋼管

1 N/mm² or MPa = 1.01972 × 10^{-1} kgf/mm²

化学成分 (%)					Tensile Test N/mm² 引張試験		Remarks 備考 (Similar to JIS)	Index No. 索引番号
S	Ni	Cr	Mo	Others	Min. Yield Point 最小降伏点	Tensile Strength 引張強さ		
≦0.06	−	−	−	−	18	≧315	−	C711
≦0.05	−	−	−	−	21	≧330	−	
≦0.05	−	−	−	−	25	≧410	(STPG410)	
≦0.05	−	−	−	−	25	≧410	(STPG410)	

INDEX OF CHAPTER III

STEEL PIPES AND TUBES

第3部　鋼管

索　　引

"Index No." stands for the
"Index No." in the last column
of each page.

「索引番号」は本文各頁の右端の
「索引番号」の欄の数字をあらわす。

Standard No. 規格番号	Grade グレード	Index No. 索引番号	Standard No. 規格番号	Grade グレード	Index No. 索引番号
〔JIS〕			JIS-G3446	SUS304TKA	C019, C123
JIS-C8305	—	C130		SUS304TKC	C019, C123
				SUS316TKA	C019, C123
JIS-G3429	STH11	C127		SUS316TKC	C019, C123
	STH12	C127		SUS321TKA	C019, C123
	STH21	C127		SUS321TKC	C019, C123
	STH22	C127		SUS347TKA	C019, C123
	STH31	C127		SUS347CKC	C019, C123
				SUS410TKA	C019, C123
JIS-G3439	—	C128		SUS410TKC	C019, C123
				SUS420J1TKA	C019, C123
JIS-G3441	SCM415TK	C018, C120		SUS420J2TKA	C019, C123
	SCM418TK	C018, C120		SUS430TKA	C019, C123
	SCM420TK	C018, C120		SUS430TKC	C019, C123
	SCM430TK	C018, C120			
	SCM435TK	C018, C120	JIS-G3447	SUS304LTBS	C015, C101
	SCM440TK	C018, C120		SUS304TBS	C015, C101
	SCY420TK	C018, C120		SUS316STBS	C015, C101
				SUS316TBS	C015, C101
JIS-G3442	SGPW	C100			
			JIS-G3448	SUS304TPD	C008, C102
JIS-G3444	STK290(STK30)	C016, C121		SUS316TPD	C008, C102
	STK400(STK41)	C016, C121			
	STK490(STK50)	C016, C121	JIS-G3452	SGP	C001, C103
	STK500(STK51)	C016, C121			
	STK540(STK55)	C016, C121	JIS-G3454	STPG370(STPG38)	C002, C103
				STPG410(STPG42)	C002, C103
JIS-G3445	STKM11A	C017, C122			
	STKM12A	C017, C122	JIS-G3455	STS370(STS38)	C003, C105
	STKM12B	C017, C122		STS410(STS42)	C003, C105
	STKM12C	C017, C122			
	STKM13A	C017, C122	JIS-G3456	STPT370(STPT38)	C004, C106
	STKM13B	C017, C122		STPT410(STPT42)	C004, C106
	STKM13C	C017, C122		STPT480(STPT49)	C004, C106
	STKM14A	C017, C122			
	STKM14B	C017, C122	JIS-G3457	STPY400(STPY41)	C107
	STKM14C	C017, C122			
	STKM15A	C017, C122	JIS-G3458	STPA12	C006, C108
	STKM15C	C017, C122		STPA20	C006, C108
	STKM16A	C017, C122		STPA23	C006, C108
	STKM16C	C017, C122		STPA24	C006, C108
	STKM17A	C017, C122		STPA25	C006, C108
	STKM17C	C017, C122		STPA26	C006, C108
	STKM18A	C017, C122			
	STKM18B	C017, C122	JIS-G3459	SUS304TP	C008, C109
	STKM18C	C017, C122		SUS304HTP	C008, C109
	STKM19A	C017, C122		SUS304LTP	C008, C109
	STKM19C	C017, C122		SUS309STP	C008, C109
	STKM20A	C017, C122		SUS309TP	C008, C109
				SUS310STP	C008, C109

Standard No. 規格番号	Grade グレード	Index No. 索引番号
	SUS310TP	C008, C109
	SUS316HTP	C008, C109
	SUS316LTP	C008, C109
	SUS316TP	C008, C109
	SUS317LTP	C008, C109
	SUS321HTP	C008, C109
	SUS321TP	C008, C109
	SUS329J1TP	C008, C109
	SUS329J2LTP	C008, C109
	SUS347HTP	C008, C109
	SUS347TP	C008, C109
	SUS405TP	C008, C109
JIS-G3460	STPL380(STPL39)	C007, C110
	STPL450(STPL46)	C007, C110
	STPL690(STPL70)	C007, C110
JIS-G3461	STB340(STB35)	C010, C114
	STB410(STB42)	C010, C114
	STB510(STB52)	C010, C114
JIS-G3462	STBA12	C011, C115
	STBA13	C011, C115
	STBA20	C011, C115
	STBA22	C011, C115
	STBA23	C011, C115
	STBA24	C011, C115
	STBA25	C011, C115
	STBA26	C011, C115
JIS-G3463	SUS304HTB	C012, C116
	SUS304LTB	C012, C116
	SUS304TB	C012, C116
	SUS309STB	C012, C116
	SUS309TB	C012, C116
	SUS310STB	C012, C116
	SUS310TB	C012, C116
	SUS316HTB	C012, C116
	SUS316LTB	C012, C116
	SUS316TB	C012, C116
	SUS317LTB	C012, C116
	SUS317TB	C012, C116
	SUS321HTB	C012, C116
	SUS321TB	C012, C116
	SUS329J1TB	C012, C116
	SUS329J2LTB	C012, C116
	SUS347HTB	C012, C116
	SUS347TB	C012, C116
	SUS405TB	C012, C116
	SUS409TB	C012, C116

Standard No. 規格番号	Grade グレード	Index No. 索引番号
	SUS410TB	C012, C116
	SUS410TiTB	C012, C116
	SUS430TB	C012, C116
	SUS444TB	C012, C116
	SUSXM8TB	C012, C116
	SUSXM15TB	C012, C116
	SUSXM27TB	C012, C116
JIS-G3464	STBL380(STBL39)	C014, C117
	STBL450(STBL46)	C014, C117
	STBL690(STBL70)	C014, C117
JIS-3465	STM-C540(STM-C55)	C129
	STM-C640(STM-C65)	C129
	STM-R590(STM-R60)	C129
	STM-R690(STM-R70)	C129
	STM-R790(STM-R80)	C129
	STM-R840(STM-R85)	C129
JIS-3466	STKR400(STKR41)	C020, C124
	STKR490(STKR50)	C020, C124
JIS-G3467	NCF2HTF	C015, C118
	NCF2TF	C015, C118
	STF38	C015, C118
	STF42	C015, C118
	STFA12	C015, C118
	STFA22	C015, C118
	STFA23	C015, C118
	STFA24	C015, C118
	STFA25	C015, C118
	STFA26	C015, C118
	SUS304HTF	C015, C118
	SUS304TF	C015, C118
	SUS309STF	C015, C118
	SUS316HTF	C015, C118
	SUS321HTF	C015, C118
	SUS321TF	C015, C118
	SUS347HTF	C015, C118
	SUS347TF	C015, C118
JIS-G3468	SUS304TLPY	C009, C111
	SUS304TPY	C009, C111
	SUS309STPY	C009, C111
	SUS310STPY	C009, C111
	SUS310STPY	C009, C111
	SUS316LTPY	C009, C111
	SUS316TPY	C009, C111
	SUS317TPY	C009, C111
	SUS317TPY	C009, C111

Standard No. 規 格 番 号	Grade グレード	Index No. 索引番号	Standard No. 規 格 番 号	Grade グレード	Index No. 索引番号
	SUS321TPY	C009, C111		Gr. B	C002, C205
	SUS3129J1TPY	C009, C111			
	SUS347TPY	C009, C111	ASTM-A139	Gr. A	C206
				Gr. B	C005, C206
JIS-G3469	P1H	C112		Gr. C	C206
	P2S	C112		Gr. D	C206
	P1F	C112		Gr. E	C206
JIS-G3472	STAM290GA(STAM30GA)	C125	ASTM-A161	LC	C013, C233
	STAM290GB(STAM30GB)	C125		Gr. T1	C233
	STAM340G(STAM35G)	C125			
	STAM390G(STAM40G)	C125	ASTM-A178M	Gr. A	C010, C234
	STAM440H(STAM45H)	C125		Gr. C	C010, C234
	STAM470G(STAM48G)	C125		Gr. D	C234
	STAM500G(STAM51G)	C125			
	STAM500H(STAM51H)	C125	ASTM-A179M	—	C235
	STAM540H(STAM55H)	C125			
			ASTM-A181M	Cl. 60	C208
JIS-G3473	STC370(STC38)	C126		Cl. 60	C208
	STC440(STC45)	C126			
	STC510A(STC52A)	C126	ASTM-A192M	—	C010, C236
	STC510B(STC52B)	C126			
	STC540(STC55)	C126	ASTM-A199M	T4	C237
	STC590A(STC60A)	C126		T5	C011, C237
	STC590B(STC60B)	C126		T7	C237
				T9	C011, C237
JIS-G4903	NCF600TP	C113		T11	C011, C237
	NCF800TP	C113		T21	C237
	NCF800HTP	C113		T22	C011, C237
	NCF825TP	C113		T91	C237
JIS-G4904	NCF600TB	C119	ASTM-A200M	T4	C238
	NCF800TB	C119		T5	C013, C238
	NCF800HTB	C119		T7	C238
	NCF825TB	C119		T9	C013, C238
				T11	C013, C238
〔ASTM〕				T21	C238
ASTM-A53	TypeE	C201		T22	C013, C238
	TypeF	C001, C201		T91	C238
	TypeS	C201			
			ASTM-A209M	T1	C011, C239
ASTM-A105M	—	C002, C202		T1a	C011, C239
				T1b	C239
ASTM-A106M	Gr. A	C004, C203			
	Gr. B	C004, C203	ASTM-A210M	Gr. A-1	C010, C240
	Gr. C	C004, C203		Gr. C	C240
ASTM-A134		C204	ASTM-A211	A570	C209
ASTM-A135	Gr. A	C002, C205	ASTM-A213M	18Cr2Mo	C012, C241

Standard No. 規格番号	Grade グレード	Index No. 索引番号
	S21500	C241
	S30815	C241
	S31725	C241
	S31726	C241
	S32615	C241
	T2	C012, C241
	T5	C012, C241
	T56	C241
	T5C	C241
	T7	C241
	T9	C011, C241
	T11	C011, C241
	T12	C011, C241
	Y22	C011, C241
	T91	C241
	TP201	C241
	TP202	C241
	TP304	C012, C241
	TP304H	C012, C241
	TP304L	C012, C241
	TP304LN	C241
	TP304N	C241
	TP309Cb	C241
	TP309H	C241
	TP309HCb	C241
	TP309S	C012, C241
	TP310Cb	C241
	TP310H	C241
	TP310HCb	C241
	TP316	C012, C241
	T9316H	C012, C241
	TP316L	C241
	TP316LN	C241
	TP316N	C241
	TP321	C012, C241
	TP321H	C012, C241
	TP347	C012, C241
	TP347H	C012, C241
	TP348	C241
	TP348H	C241
	XM-15	C241
ASTM-A214M	–	C242
ASTM-A226M	–	C010, C243
ASTM-A249M	S30815	C243
	S31050	C243
	S31254	C243
	S31725	C243

Standard No. 規格番号	Grade グレード	Index No. 索引番号
	S31726	C243
	TP201	C243
	TP202	C243
	TP202	C243
	TP202	C243
	TP304	C012, C243
	TP304H	C012, C243
	TP304L	C012, C243
	TP304LN	C243
	TP304N	C243
	TP305	C243
	TP309Cb	C243
	TP309H	C243
	TP309HCb	C243
	TP309S	C243
	TP310Cb	C243
	TP310H	C243
	TP310HCb	C243
	TP310S	C243
	TP316	C012, C243
	TP316H	C243
	TP316L	C243
	TP316LN	C243
	TP316N	C243
	TP317	C012, C243
	TP317L	C012, C243
	TP321	C012, C243
	TP321H	C012, C243
	TP347	C012, C243
	TP347H	C012, C243
	TP348	C243
	TP348H	C243
	XM-15	C243
	XM-19	C243
	XM-29	C243
ASTM-A250M	T1a	C011, C245
	T1b	C245
	T2	C245
	T11	C245
	T12	C245
	T22	C245
ASTM-A252	Gr. 1	C016, C261
	Gr. 2	C016, C261
	Gr. 3	C261
ASTM-A266M	Cl. 1	C246
	Cl. 2	C246
	Cl. 3	C246

Standard No. 規格番号	Grade グレード	Index No. 索引番号	Standard No. 規格番号	Grade グレード	Index No. 索引番号
	Cl.4	C246		TP316	C013, C249
				TP316H	C013, C249
ASTM-A268	25-4-4			TP321	C013, C249
	26-3-3			TP321H	C013, C249
	29-4			TP347	C013, C249
	29-4-2			TP347H	C008, C249
	18Cr-2Mo	C247			
	S40800	C247	ASTM-312M	S30600	C211
	S41500	C247		S30815	C211
	S44735	C247		S31050	C211
	TP405	C012, C247		S31254	C211
	TP409	C012, C247		S31725	C211
	TP410	C247		S31726	C211
	TP429	C247		TP304	C008, C211
	TP430	C012, C247		TP304H	C008, C211
	TP430Ti	C247		TP304L	C008, C211
	TP439	C247		TP304LN	C211
	TP443	C247		TP304N	C211
	TP446-1	C247		TP309Cb	C211
	TP446-2	C247		TP309H	C211
	TPXM-27	C247		TP309HCb	C211
	XM33	C247		TP309S	C211
				TP310Cb	C211
ASTM-A269	S30600	C248		TP310H	C211
	S31725	C248		TP310HCb	C211
	S31726	C248		TP310S	C211
	TP304	C012, C248		TP316	C008, C211
	TP304L	C012, C248		TP316H	C008, C211
	TP304LN	C248		TP316L	C008, C211
	TP316	C012, C248		TP316LN	C211
	TP316LN	C248		TP316N	C211
	TP316N	C248		TP317	C008, C211
	TP317	C248		TP317L	C008, C211
	TP321	C012, C248		TP321	C008, C211
	TP347	C012, C248		TP321H	C008, C211
	TP348	C248		TP347	C008, C211
	TPS31254	C248		TP347H	C211
	XM-10	C248		TP348	C211
	XM-11	C248		TP348H	C211
	XM-15	C248		TPXM-10	C211
	XM-19	C248		TPXM-11	C211
	XM-29	C248		TPXM-15	C211
				TPXM-19	C211
ASTM-A270	TP304	C015, C210		XPXM-29	C211
	TP304L	C015, C210			
	TP316	C015, C210	ASTM-A333M	Gr.1	C007, C212
	TP316L	C015, C210		Gr.3	C007, C212
				Gr.4	C212
ASTM-A271	TP304	C013, C249		Gr.6	C212
	TP304H	C013, C249		Gr.7	C212

Standard No. 規 格 番 号	Grade グレード	Index No. 索引番号
	Gr. 8	C007, C212
	Gr. 9	C212
ASTM-A334M	Gr. 1	C014, C250
	Gr. 3	C014, C250
	Gr. 6	C250
	Gr. 7	C250
	Gr. 8	C014, C250
	Gr. 9	C250
ASTM-A335M	P1	C006, C213
	P2	C006, C213
	P5	C213
	P5b	C213
	P5c	C213
	P7	C213
	P9	C006, C213
	P11	C006, C213
	P12	C006, C213
	P15	C213
	P21	C213
	P22	C006, C213
	P91	C213
ASTM-A336	Cl. F1	C251
	Cl. F3V	C251
	Cl. F5	C251
	Cl. F5A	C251
	Cl. F6	C251
	Cl. F9	C251
	Cl. F11	C251
	Cl. F11A	C251
	Cl. F11B	C251
	Cl. F12	C251
	Cl. F21	C251
	Cl. F21A	C251
	Cl. F22	C251
	Cl. F22A	C251
	Cl. F46	C251
	Cl. F91	C251
	Cl. F304	C251
	Cl. F304H	C251
	Cl. F304L	C251
	Cl. F304LN	C251
	Cl. F304N	C251
	Cl. F309H	C251
	Cl. F310	C251
	Cl. F310H	C251
	Cl. F316	C251
	Cl. F316H	C251

Standard No. 規 格 番 号	Grade グレード	Index No. 索引番号
	Cl. F316L	C251
	Cl. F316LN	C251
	Cl. F316N	C251
	Cl. F321	C251
	Cl. F321H	C251
	Cl. F347	C251
	Cl. F347H	C251
	Cl. F348	C251
	Cl. F348H	C251
	Cl. FXM-11	C251
	Cl. FXM-19	C251
ASTM-A358M	304	C009, C215
	304H	C215
	304L	C009, C215
	304LN	C215
	309	C009, C215
	309Cb	C215
	309S	C215
	310	C009, C215
	310Cb	C215
	316	C215
	316H	C215
	316L	C009, C215
	316LN	C215
	316N	C215
ASTM-A376M	16-8-2H	C216
	S31725	C216
	S31726	C216
	TP304	C008, C216
	TP304H	C008, C216
	TP304LN	C216
	TP304N	C216
	TP316	C008, C216
	TP316H	C008, C216
	TP316LN	C216
	TP316N	C216
	TP321	C008, C216
	TP321H	C008, C216
	TP347	C008, C216
	TP347H	C008, C216
	TP348	C216
ASTM-A381	Cl. Y35	C217
	Cl. Y42	C217
	Cl. Y46	C217
	Cl. Y48	C217
	Cl. Y50	C217
	Cl. Y52	C217

Standard No. 規格番号	Grade グレード	Index No. 索引番号	Standard No. 規格番号	Grade グレード	Index No. 索引番号
	Cl.Y56	C217	ASTM-A511	29-4	C264
	Cl.Y60	C217		29-4-2	C264
	Cl.Y65	C217			
				MT302	C264
ASTM-A405	P24	C218		MT303Se	C264
				MT304	C017, C264
ASTM-A409M	S30815	C219		MT304L	C264
	S31254	C219		MT305	C264
	S31725	C219		MT309S	C264
	S31726	C219		MT310S	C264
	TP304	C009, C219		MT316	C019, C264
	TP304L	C009, C219		MT316L	C264
	TP309Cb	C219		MT317	C264
	TP309S	C219		MT321	C019, C264
	TP310Cb	C219		MT347	C019, C264
	TP310S	C219		MT403	C264
	TP316	C219		MT410	C019, C264
	TP316L	C009, C219		MT414	C264
	TP317	C219		MT416Se	C264
	TP321	C219		MT431	C264
	TP347	C219		MT440A	C264
	TP348	C219		MT405	C264
				MT429	C264
ASTM-A422	—	C252		MT430	C019, C264
				MT443	C264
ASTM-A423M	Gr.1	C253		MT446-1	C264
	Gr.2	C253		MT446-2	C264
ASTM-A430M	FP16-8-2H	C220	ASTM-A512	1008	C265
	FP304	C220		1010	C265
	FP304H	C008, C220		1012	C265
	FP304N	C220		1015	C265
	FP316	C220		1016	C265
	FP316H	C220		1018	C265
	FP316N	C220		1019	C265
	FP321	C220		1020	C265
	FP321H	C220		1021	C265
	FP347	C220		1025	C265
	FP347H	C220		1026	C265
				1030	C265
ASTM-A452	Gr.TP304H	C271		1035	C265
	Gr.TP347H	C271		1110	C265
	Gr.TP316H	C271		1115	C265
				1117	C265
ASTM-A500	Gr.A	C016, C262		MT1010	C017, C265
	Gr.B	C016, C262		MT1015	C017, C265
	Gr.C	C262		MT1017	C265
				MT1020	C265
ASTM-A501	—	C016, C263		MTX1015	C265
				MTX1020	C017, C265

Standard No. 規格番号	Grade グレード	Index No. 索引番号	Standard No. 規格番号	Grade グレード	Index No. 索引番号
ASTM-A513	1008	C266		1012	C267
	1010	C266		1015	C267
	1015	C266		1016	C267
	1016	C266		1017	C267
	1017	C266		1018	C267
	1018	C266		1019	C267
	1019	C266		1020	C267
	1020	C266		1021	C267
	1021	C266		1022	C267
	1022	C266		1025	C267
	1023	C266		1026	C267
	1024	C266		1030	C017, C267
	1025	C266		1035	C267
	1026	C266		1040	C017, C267
	1027	C266		1045	C267
	1030	C017, C266		1050	C017, C267
	1033	C266		1118	C267
	1035	C266		1132	C267
	1040	C266		1137	C267
	1050	C266		1141	C267
	1060	C266		1144	C267
	1340	C266		1213	C267
	1524	C266		1215	C267
	4118	C266		1330	C267
	4130	C266		1335	C267
	4140	C266		1340	C267
	5130	C266		1345	C267
	8620	C266		1518	C017, C267
	8630	C266		1524	C017, C267
	MT1010	C017, C266		1541	C267
	MT1015	C017, C266		3140	C267
	MT1020	C017, C266		4012	C267
	MTX1015	C266		4023	C267
	MTX1020	C017, C266		4024	C267
				4027	C267
ASTM-A519	11L18	C267		4028	C267
	12L14	C267		4037	C267
	50B40	C267		4042	C267
	50B46	C267		4047	C267
	50B50	C267		4063	C267
	50B60	C267		4118	C267
	51B60	C267		4130	C017, C267
	81B45	C267		4135	C017, C267
	86B45	C267		4137	C267
	94B15	C267		4140	C017, C267
	94B17	C267		4142	C267
	94B30	C267		4115	C267
	94B40	C267		4117	C267
	1008	C267		4150	C267
	1010	C267		4320	C267

Standard No. 規格番号	Grade グレード	Index No. 索引番号	Standard No. 規格番号	Grade グレード	Index No. 索引番号
	4337	C267		9840	C267
	4340	C267		9850	C267
	4422	C267			
	4427	C267	ASTM-A523	Gr. A	C272
	4520	C267		Gr. B	C272
	4615	C267			
	4617	C267	ASTM-A524	—	C002, C221
	4620	C267			
	4621	C267	ASTM-A539	—	C222
	4718	C267			
	4720	C267	ASTM-A554	MT301	C268
	4815	C267		MT302	C268
	5015	C267		MT304	C017, C268
	5046	C267		MT304L	C268
	5115	C267		MT305	C268
	5120	C017, C267		MT309S	C268
	5130	C267		MT309S-Cb	C268
	5132	C267		MT310S	C268
	5135	C267		MT316	C019, C268
	5145	C267		MT316L	C268
	5147	C267		MT317	C268
	5150	C267		MT321	C019, C268
	5155	C267		MT330	C268
	5160	C267		MT347	C019, C268
	6118	C267		MT429	C268
	6120	C267		MT430	C019, C268
	6150	C267		MT430Ti	C268
	8115	C267			
	8615	C267	ASTM-A556M	Gr. A2	C010, C254
	8617	C267		Gr. B2	C010, C254
	8620	C267		Gr. C2	C254
	8622	C267			
	8625	C267	ASTM-A557M	Gr. A2	C010, C255
	8627	C267		Gr. B2	C255
	8630	C267		Gr. C2	C010, C255
	8637	C267			
	8640	C267	ASTM-A587	—	C002, C223
	8642	C267			
	8645	C267	ASTM-A589	ButW	C273
	8650	C267		Gr. A	C273
	8655	C267		Gr. B	C273
	8660	C267			
	8720	C267	ASTM-A595	Gr. A	C269
	8735	C267		Gr. B	C269
	8740	C267		Gr. C	C269
	8742	C267			
	8822	C267	ASTM-A608	HC30	C275
	9255	C267		HD50	C275
	9260	C267		HE35	C275
	9262	C267		HF30	C275

Standard No. 規格番号	Grade グレード	Index No. 索引番号
	HH30	C275
	HH33	C275
	HI35	C275
	HK30	C275
	HK40	C275
	HL30	C275
	HL40	C275
	HN40	C275
	HT50	C275
	HU50	C275
	HW50	C275
	HX50	C275
ASTM-A618	Gr. Ⅰa	C269
	Gr. Ⅰb	C269
	Gr. Ⅱ	C269
	Gr. Ⅲ	C269
ASTM-A632	TP304	C012, C256
	TP304L	C012, C256
	TP310	C012, C256
	TP316	C012, C256
	TP316L	C256
	TP317	C012, C256
	TP321	C012, C256
	TP347	C012, C256
	TP348	C256
ASTM-A660	WCA	C276
	WCB	C276
	WCC	C276
ASTM-A671	CA55	C224
	CB60	C224
	CB65	C224
	CB70	C224
	CC70	C224
	CD70	C224
	CD80	C224
	CE55	C224
	CE60	C224
	CF65	C224
	CF66	C224
	CF70	C224
	CF71	C224
	CJ101	C224
	CJ102	C224
	CJ103	C224
	CJ104	C224
	CJ105	C224

Standard No. 規格番号	Grade グレード	Index No. 索引番号
	CJ106	C224
	CJ107	C224
	CJ108	C224
	CJ109	C224
	CJ110	C224
	CJ111	C224
	CJ112	C224
	CJ113	C224
	CK75	C224
	CP65	C224
	CP75	C224
ASTM-A672	A45	C225
	A50	C225
	A55	C225
	B55	C225
	B60	C225
	B65	C225
	B70	C225
	C55	C225
	C60	C225
	C65	C225
	D70	C225
	D80	C225
	E55	C225
	E60	C225
	H75	C225
	H80	C225
	J80	C225
	J90	C224
	J100	C224
	K75	C224
	K85	C224
	L65	C224
	L70	C224
	L75	C224
	N75	C224
ASTM-A688M	TP304	C012, C257
	TP304L	C012, C257
	TP304LN	C257
	TP304N	C012, C257
	TP316	C257
	TP316L	C257
	TP316LN	C257
	TP316N	C257
	TPXM-29	C257
ASTM-A691	½CR	C226
	1CR	C226

Standard No. 規格番号	Grade グレード	Index No. 索引番号
	1¼CR	C226
	2¼CR	C226
	3CR	C226
	5CR	C226
	9CR	C226
	CM-65	C226
	CM-70	C226
	CM-75	C226
	CMSH-70	C226
	CMSH-75	C226
	CMSH-80	C226
ASTM-A692	—	C011, C258
ASTM-A714	Gr. I	C227
	Gr. II	C227
	Gr. III	C227
	Gr. IV	C227
	Gr. V	C227
	Gr. VI	C227
	Gr. VII	C227
	Gr. VIII	C227
ASTM-A727N	—	C228
ASTM-A731M	18Cr2Mo	C012, C229
	26-3-3	C229
	29-4	C229
	29-4-2	C229
	S41500	C229
	TP439	C012, C229
	TPXM27	C229
	TPXM33	C229
ASTM-A766M	—	C230
ASTM-A771	S38660	C274
	Type316	C274
ASTM-A789	S31200	C277
	S31260	C277
	S31500	C277
	S31803	C277
	S32304	C277
	S32500	C277
	S32750	C277
	S32900	C277
	S32950	C277
ASTM-A790	S31200	C278
	S31260	C278
	S31500	C278
	S31803	C278
	S32304	C278
	S32500	C278
	S32750	C278
	S32900	C278
	S32950	C278
ASTM-A791	S31200	C279
	S31260	C279
	S31500	C279
	S31803	C279
	S32304	C279
	S32500	C279
	S32750	C279
	S32900	C279
	S32950	C279
ASTM-A803	S44400	C259
	S44635	C259
	S44660	C259
	S44700	C259
	S44735	C259
	S44800	C259
	TP409	C259
	TP439	C259
	TPXM-27	C259
	TPXM-33	C259
ASTM-A813M	S30815	C280
	S31254	C280
	TP304	C280
	TP304Cb	C280
	TP304L	C280
	TP304LN	C280
	TP304N	C280
	TP309S	C280
	TP310Cb	C280
	TP310S	C280
	TP316H	C280
	TP316L	C280
	TP316LN	C280
	TP316N	C280
	TP317	C280
	TP317L	C280
	TP321	C280
	TP321H	C280
	TP347	C280
	TP347H	C280

Standard No. 規格番号	Grade グレード	Index No. 索引番号
	TP348	C280
	TP348H	C280
	TPXM10	C280
	TPXM11	C280
	TPXM15	C280
	TPXM19	C280
	TPXM29	C280
ASTM-A814M	S30815	C281
	S31254	C281
	TP304	C281
	TP304Cb	C281
	TP304H	C281
	TP304L	C281
	TP304LN	C281
	TP304N	C281
	TP309S	C281
	TP310Cb	C281
	TP310S	C281
	TP316	C281
	TP316H	C281
	TP316L	C281
	TP316LN	C281
	TP316N	C281
	TP317	C281
	TP317L	C281
	TP321	C281
	TP321H	C281
	TP347	C281
	TP347H	C281
	TP348	C281
	TP348H	C281
	TPXM10	C281
	TPXM11	C281
	TPXM15	C281
	TPXM15	C281
	TPXM19	C281
	TPXM29	C281
ASTM-A822	−	C232
ASTM-A826	S38660	C282
	S42100	C282
	T91	C282
	TP316	C282
ASTM-A847	−	C270
ASTM-A851	TP304	C260
	TP304L	C260
ASTM-A872	JP93183	C284
	JP93350	C284
ASTM-B167	No6600	C010, C285
	No6090	C285
ASTM-B407	No8800	C010, C286
	No8810	C286
	No8811	C286
ASTM-B423	No8221	C284, C287
	No8825	C287
〔API〕		
API-Spec-5CT	C-75-1	C288
	C-75-2	C288
	C-75-3	C288
	C-75-9Cr	C288
	C-75-13Cr	C288
	C-90-1	C288
	C-90-2	C288
	C-95	C288
	H-40	C288
	J-55	C288
	L-80-1	C288
	L-80-9Cr	C288
	L-80-13Cr	C288
	N-80	C288
	P-105	C288
	P-110	C288
	Q-125-1	C288
	Q-125-2	C288
	Q-125-3	C288
	Q-125-4	C288
	T-95-1	C288
	T-95-2	C288
API-Spec-5D	E-75	C289
	G-105	C289
	S-135	C289
	X-95	C289
API-Spec-5L	A	C300
	A25Cl Ⅰ	C300
	A25Cl Ⅱ	C300
	B	C300
	X42	C300
	X46	C300
	X52	C300
	X56	C300

Standard No. 規 格 番 号	Grade グレード	Index No. 索引番号
	X60	C300
	X65	C300
	X70	C300
	X80	C300
〔BS〕		
BS534	—	C301
BS778	—	C301
BS879	ERW26	C314
	ERW27	C314
	HFS27	C314
	HFS35	C314
BS1139	—	C315
BS1339	—	C303
BS1717	CEWC1	C316
	CEWC2	C017, C316
	CEWC3	C316
	CFSC3	C017, C316
	CFSC4	C017, C316
	CFSC6	C316
	ERWC1	C017, C316
	ERWC2	C017, C316
	ERWC3	C017, C316
	ERWC5	C017, C316
BS1864	Gr. 1	C015, C304
	Gr. 2	C304
	Gr. 3	C304
	Gr. 4	C304
	Gr. 5	C015, C304
	Gr. 6	C015, C304
	Gr. 7	C304
	Gr. 8	C304
	Gr. 9	C304
BS3059 Part1	CEW320	C010, C310
	CFS320	C010, C310
	ERW320	C010, C310
	HFS320	C010, C310
BS3059 Part2	CEW243	C310
	CEW360	C010, C310
	CEW440	C310
	CEW620	C011, C310
	CFS304S59	C310

Standard No. 規 格 番 号	Grade グレード	Index No. 索引番号
	CFS316S59	C012, C310
	CFS321S59	C012, C310
	CFS347S59	C012, C310
	CFS1250	C310
	ERW243	C310
	ERW360	C010, C310
	ERW440	C310
	S1243	C310
	S1360	C010, C310
	S1440	C310
	S1620	C310
	S1622-440	C010, C310
	S1629-440	C011, C310
	S1629-470	C011, C310
	S1629-590	C011, C310
	S1762	C310
	S2243	C310
	S2360	C010, C310
	S2440	C310
	S2620	C010, C310
	S2622-440	C012, C310
	S2622-470	C012, C310
	S2629-490	C012, C310
	S2672	C310
BS3601	BW320	C305
	ERW320	C305
	ERW360	C002, C305
	ERW430	C305
	S360	C002, C305
	S430	C305
	SAW430	C305
BS3602	500Nb	C306
	CEW360	C004, C306
	CEW430	C306
	CEW460	C306
	CFS360	C004, C306
	CFS430	C306
	CFS460	C004, C306
	ERW430	C306
	ERW460	C004, C306
	HFS360	C004, C306
	HFS430	C306
	HFS460	C004, C306
	SAW410	C306
	SAW460	C306
BS3603	CEW410LT50	C014, C307
	CFS410LT50	C014, C307

Standard No. 規格番号	Grade グレード	Index No. 索引番号	Standard No. 規格番号	Grade グレード	Index No. 索引番号
	CFS503LT100	C014, C307		CFS261	C311
	CFS509LT196	C307		CFS304S22	C012, C311
	ERW410LT50	C014, C307		CFS304S24	C311
	HFS410LT50	C014, C307		CFS316S24	C012, C311
	HFS503LT100	C014, C307		CFS316S25	C011, C012, C311
	HFS509LT196	C014, C307		CFS316S29	C012, C311
				CFS316S30	C012, C311
BS3604	CEW620	C006, C308		CFS320	C311
	CEW621	C006, C308		CFS321S22	C012, C311
	CFS620	C006, C308		CFS347S17	C012, C311
	CFS621	C006, C308		CFS440	C010, C311
	CFS622	C006, C308		CFS620	C010, C311
	CFS625	C006, C011, C308		CFS621	C010, C311
	CFS660	C011, C308		CFS622	C311
	CFS762	C308		CFS625	C311
	ERW620	C006, C308		ERW243	C311
	ERW621	C006, C308		ERW245	C010, C311
	HFS620	C006, C308		ERW320	C311
	HFS621	C006, C308		ERW620	C311
	HFS622	C006, C308		LWBC304S22	C012, C311
	HFS625	C006, C011, C308		LWBC304S25	C012, C311
	HFS629	C006, C011, C308		LWBC316S24	C311
	HFS660	C011, C308		LWBC316S25	C311
	HFS762	C308		LWBC316S29	C012, C311
				LWBC316S30	C012, C311
BS3605	304S14	C309		LWBC321S22	C012, C311
	304S18	C008, C012, C309		LWBC347S17	C012, C311
	304S22	C309		LWBF304S22	C012, C311
	304S25	C008, C012, C309		LWBF304S25	C311
	304S59	C008, C012, C309		LWBF316S24	C012, C311
	316S14	C008, C012, C309		LWBF316S25	C012, C311
	316S18	C008, C012, C309		LWBF316S29	C012, C311
	316S22	C008, C012, C309		LWBF316S30	C012, C311
	316S59	C309		LWBF321S22	C311
	321S18	C008, C012, C309		LWBF347S17	C012, C311
	321S22	C008, C012, C309		LWHT304S22	C012, C311
	321S59	C008, C012, C309		LWHT304S25	C012, C311
	347S17	C012, C309		LWHT316S24	C012, C311
	347S18	C008, C012, C309		LWHT316S25	C012, C311
	347S59	C008, C012, C309		LWHT316S29	C311
				LWHT316S30	C012, C311
BS3606	CEW243	C311		LWHT321S22	C311
	CEW245	C010, C311		LWHT347S17	C311
	CEW261	C311			
	CEW320	C311	BS4127	−	C312
	CEW440	C010, C311			
	CEW620	C006, C311	BS4825	−	C312
	CEW621	C012, C311			
	CFS243	C311	BS5242	HP1	C022, C317
	CFS245	C010, C311		HP2	C022, C317

Standard No. 規格番号	Grade グレード	Index No. 索引番号	Standard No. 規格番号	Grade グレード	Index No. 索引番号
	HP3	C317	BS6363	34/26	C020, C319
	HP4	C022, C317		43/36	C020, C319
	HP5	C022, C317		50/45	C020, C319
	HP6	C317			
	HP7	C317	〔DIN〕		
	HP8	C317	DIN1615	—	C401
BS6323	CEW1	C021, C318	DIN1626	St37.0	C002, C402
	CEW2	C021, C318		St44.0	C002, C402
	CEW3	C021, C318		St52.0	C402
	CEW4	C021, C318		USt37.0	C402
	CEW5	C021, C318			
	CFS3	C017, C318	DIN1628	St37.4	C403
	CFS3A	C017, C318		St44.4	C403
	CFS4	C017, C318		St52.4	C403
	CFS5	C017, C318			
	CFS6	C318	DIN1629	St00	C404
	CFS7	C318		St35	C404
	CFS8	C017, C318		St45	C404
	CFS9	C318		St52	C404
	CFS10	C018, C318		St55	C404
	CFS11	C318		St354	C404
	ERW1	C021, C318		St454	C404
	ERW2	C318		St524	C404
	ERW3	C318		St524	C404
	ERW4	C021, C318		St554	C404
	ERW5	C021, C318			
	HFS3	C017, C318	DIN1630	St37.4	C003, C405
	HFS4	C017, C318		St44.4	C003, C405
	HFS5	C017, C318		St52.4	C003, C405
	HFS8	C017, C318			
	HFW2	C021, C318	DIN2391	St30Al	C017, C419
	HFW3	C021, C318		St30Si	C017, C419
	HFW4	C021, C318		St35	C017, C419
	HFW5	C021, C318		St45	C017, C419
	LW12	C318		St52	C017, C419
	LW13	C019, C318			
	LW14	C318	DIN2393	RSt28	C017, C420
	LW15	C019, C318		RSt34-2	C420
	LW16	C318		RSt37-2	C017, C420
	LW17	C318		St28	C017, C420
	LW18	C318		St44-2	C420
	LWCF13	C019, C318		St52-3	C017, C420
	LWCF14	C318		Ust34-2	C420
	LWCF15	C318		Ust37-2	C420
	LWCF16	C318			
	LWCF17	C019, C318	DIN2394	RSt28	C017, C420
	LWCF18	C019, C318		RSt34-2	C420
	SAW4	C014, C318		RSt37-2	C017, C420
	SAW5	C014, C318		St28	C017, C420

Standard No. 規格番号	Grade グレード	Index No. 索引番号	Standard No. 規格番号	Grade グレード	Index No. 索引番号
	St34-2	C420		ESTE285	C427
	St37-2	C017, C420		ESTE355	C427
				ESTE420	C427
DIN2440	St33	C001, C406		ESTE460	C427
		C406		STE255	C427
DIN2441	St33	C001, C407		STE285	C427
				STE355	C427
DIN2462	X2CrNi189	C012, C408		STE420	C427
	X2CrNiMo1810	C008, C408		STE460	C427
	X2CrNiMo1812	C008, C408		TSTE255	C427
	X5CrNi189	C008, C408		TSTE285	C427
	X5CrNiMo1810	C008, C408		TSTE355	C427
	X5CrNiMo1812	C408		TSTE420	C427
	X7CrAlT3	C408		TSTE460	C427
	X8Cr7	C408			
	X8CrNb17	C408	DIN17124	ESTE255	C428
	X8CrTi17	C408		ESTE285	C428
	X10Cr13	C408		ESTE355	C428
	X10CrNiMoNb1810	C408		ESTE420	C428
	X10CrNiMoNb1812	C408		ESTE460	C428
	X10CrNiMoTi1810	C408		STE255	C428
	X10CrNiNb189	C008, C408		STE285	C428
	X10CrNiTi189	C008, C408		STE355	C428
	X20Cr13	C012, C408		STE420	C428
	X22CrNi17	C408		STE460	C428
	X22CrNi17	C408		TSTE255	C428
				TSTE285	C428
DIN2463	X2CrNi189	C008, C410		TSTE355	C428
	X2CrNiMo1810	C008, C410		TSTE420	C428
	X2CrNiMo1812	C410		TSTE460	C428
	X5CrNi189	C008, C410			
	X5CrNiMo1810	C008, C410	DIN17172	STE210.7	C411
	X5CrNiMo1812	C008, C410		STE240.7	C411
	X10CrNiMoTi1810	C410		STE290.7	C411
	X10CrNiTi189	C012, C410		STE290.7TM	C411
				STE320.7	C411
DIN17120	RST37-2	C424		STE360.7	C411
	ST37-2	C424		STE360.7TM	C411
	ST44-2	C424		STE385.7	C411
	ST44-3	C424		ETE415.7	C411
	ST52-3	C424		STE415.7TM	C411
	USt37-2	C424		ETE445.7TM	C411
				STE480.7TM	C411
DIN17121	RST37-2	C425			
	ST37-2	C425	DIN17173	10Ni14	C412
	ST44-2	C425		11MnNi53	C412
	ST44-3	C425		12Ni19	C412
	ST52-3	C425		13MnNi63	C412
				26CrMo4	C412
DIN17123	ESTE255	C427		TST35N	C412

Standard No. 規格番号	Grade グレード	Index No. 索引番号
	TST35V	C412
	X8Ni9	C412
DIN17174	10Ni14	C413
	11MnNi53	C413
	12Ni19	C413
	13MnNi63	C413
	26CrMo4	C413
	TST35N	C413
	TST35V	C413
	X8Ni9	C413
DIN17175	10CrMo910	C004, C415
	13CrMo44	C004, C415
	14MoV63	C415
	15Mo3	C415
	17Mn4	C415
	19Mn5	C015, C415
	St35.8	C004, C415
	St45.8	C004, C415
DIN17177	15Mo3	C416
	St37.8	C416
	St42.8	C416
DIN17455	X2CrNi1911	C417
	X2CrNiMo17132	C417
	X2CrNiMo18143	C417
	X2CrNiMoN17133	C417
	X2CrNiMoN17135	C417
	X2CrNiMoTi17122	C417
	X2CrNiN1810	C417
	X5CrNi1810	C417
	X5CrNiMo17122	C417
	X5CrNiMo17133	C417
	X5CrNiNb1810	C417
	X6CrNiMoTi17122	C417
	X6CrNiTi1810	C417
DIN17456	X2CrNi1911	C418
	X2CrNiMo17132	C418
	X2CrNiMo18143	C418
	X2CrNiMoN17133	C418
	X2CrNiMoTi17122	C418
	X2CrNiN1810	C418
	X5CrNi1810	C418
	X5CrNi1911	C418
	X5CrNiMo17133	C418
	X5CrNiNb1810	C418
	X5CrNi1911	C416

Standard No. 規格番号	Grade グレード	Index No. 索引番号
	X5CrNiMo17122	C418
	X5CrNiNb1810	C418
	X6CrNiA113	C418
	X6CrNiMoTi17122	C418
	X6CrNiTi1810	C418
DIN17457	X2CrNi1911	C422
	X2CrNiMo17122	C422
	X2CrNiMo18143	C422
	X2CrNiMoN17122	C422
	X2CrNiMoN17133	C422
	X5CrNi1820	C422
	X5CrNiMo17122	C422
	X5CrNiMo17133	C422
	X6CrNiMoTi17122	C422
	X6CrNiNb1810	C422
	X6CrNiTi1810	C422
DIN17458	X2CrNi1911	C423
	X2CrNiMo17122	C423
	X2CrNiMo18143	C423
	X2CrNiMoN17122	C423
	X2CrNiMoN17133	C423
	X5CrNi1820	C423
	X5CrNiMo17133	C423
	X5CrNiMb17122	C423
	X6CrNiMoTb17122	C423
	X6CrNiNb1810	C423
	X6CrNiTi1810	C423
〔NF〕		
NF-A49-111	TU37a	C601
NF-A49-112	TUE220A	C602
	TUE235A	C602
NF-A49-115	TU34-1	C603
NF-A49-117	TUZ2CN18-10	C604
	TUZ2CND17-12	C604
	TUZ6CN18-09	C604
	TUZ6CND17-11	C604
	TUZ6CND17-12	C604
	TUZ6CNT18-10	C604
	TUZ10C17	C604
	TUZ2C13	C604
	TUZ12CN24-12	C604
	TUZ12CN25-20	C604
NF-A49-141	TS37a	C605

Standard No. 規 格 番 号	Grade グレード	Index No. 索引番号
NF-A49-142	TSE185A	C606
	TSE235A	C606
	TSE250A	C606
	TSE275A	C606
NF-A49-145	TS34-1	C607
NF-A49-146	TS34-a	C608
NF-A49-147	TSZ2CN18-10	C609
	TSZ2CND17-12	C609
	TSZ2CN18-09	C609
	TSZ2CND17-11	C609
	TSZ2CNDT17-12	C609
	TSZ6CNT	C609
NF-A49-148	TSZ5CT14	C610
	TSZ6CT12	C610
	TSZ6CT17	C610
	TSZ18C17	C610
NF-A49-150	TSE235	C611
NF-A49-207	TSZ2CN18-10	C629
	TUZ2CN18-10	C629
	TSZ6CNNb18-12	C629
	TUZ6CNNb18-12	C629
NF-A49-210	TU37B	C612
	TU42B	C612
NF-A49-211	TUE220	C613
	TUE250	C613
	TUE275	C613
NF-A49-212	TU37C	C614
	TU42C	C614
NF-A49-213	TU10CD5-05	C615
	TU10CD9-10	C615
	TU15D3	C615
	TU15CD5-05	C615
	TUZ10CD5-05	C615
	TUZ10CD9	C615
	TUZ10CDNbV9-2	C615
	TU37C	C615
	TU42C	C615
	TU42CR	C615
	TU48C	C615
	TU48CR	C615

Standard No. 規 格 番 号	Grade グレード	Index No. 索引番号
	TU52C	C615
NF-A49-214	Z6CN19-10	C616
	Z6CND17-12B	C616
	Z6CNDNb18-12B	C616
	Z6CNT18-12B	C616
	Z8CNDT17-13B	C616
	Z10CNWT17-13B	C616
NF-A49-215	TU10CD5-05	C630
	TU10CD9-10	C630
	TU10N9	C630
	TU10N14	C630
	TU15CD2-05	C630
	TU15D3	C630
	TU17N2	C630
	TU42BT	C630
	TU37C	C630
	TU42C	C630
	TU48C	C630
	TUZ6N9	C630
	TUZ10CD5-05	C630
	TUZ10CD9	C630
NF-A49-217	TUZ1CND2522AZ	C631
	TUZ1NCDU20180AZ	C631
	TUZ1NCDU312703	C631
	TUZ1NCDU312703	C631
	TUZ2CN1810	C631
	TUZ2CN1810AZ	C631
	TUZ2CND1712	C631
	TUZ2CND1814	C631
	TUZ2CND1712AZ	C631
	TUZ2CND180503	C631
	TUZ2CND220503	C631
	TUZ2CND250703	C631
	TUZ2CNNb2520	C631
	TUZ5CNDU210802	C631
	TUZC6CN1809	C631
	TUZ6CND1711	C631
	TUZ6CNT1810	C631
	TUZ10C17	C631
	TUZ12C13	C631
NF-A49-218	TUZ2CN18-10	C632
	TUZ2CND17-12	C632
	TUZ6CN18-09	C632
	TUZ6CND17-11	C632
	TUZ6CNNb18-10	C632
	TUZ6CNT18-10	C632

Standard No. 規格番号	Grade グレード	Index No. 索引番号
NF-A49-230	TU10N9	C617
	TU10N14	C617
	TU17N2	C617
	TU42BT	C617
	TUZ6N9	C617
NF-A49-240	TS10N9	C618
	TS17N2	C618
	TS42BT	C618
NF-A49-241	TSE220	C619
	TDE250	C619
	TDE275	C619
	TDE350	C619
NF-A49-242	TS37C	C620
	TS42C	C620
	TSE24W3	C620
	TSE36WB3	C620
NF-A49-243	TS15CD2-05	C621
	TS15D3	C621
	TS15CD4-05	C621
	TS37C	C621
	TS42C	C621
	TS48C	C621
	TS52C	C621
NF-A49-245	TS10N9	C633
	TS17N2	C633
	TS24W3	C633
	TS34BT	C633
	TS34C	C633
	TS36WB3	C633
	TS37C	C633
	TS42C	C633
	TS48C	C633
NF-A49-249	TSZ2CN18-10	C634
	TSZ2CND17-11	C634
	TSZ2CND17-12	C634
	TSZ2CN18-09	C634
NF-A49-250	TSE24a	C622
	TSE26b	C622
NF-A49-253	TS10CD9-10	C623
	TS15CD2-05	C623
	TS15D3	C623
	TS37CP	C623

Standard No. 規格番号	Grade グレード	Index No. 索引番号
	TS42CP	C623
	TS48CP	C623
	TS52CP	C623
NF-A49-310	TU37-b	C635
	TU52-b	C635
	TU56-b	C635
NF-A49-311	TU37-b	C636
	TU52-b	C636
	TU56-b	C636
	TUXC35	C636
NF-A49-312	TU56-b	C637
	TUXC35	C637
NF-A49-321	TU17MV5	C639
	TU18MDV5	C639
	TU37b	C639
	TU52b	C639
	TU52BT	C639
NF-A49-322	TU17MV5	C640
	TU37b	C640
	TU52b	C640
NF-A49-323	TU17MV5	C641
	TU37b	C641
	TU52b	C641
	TU52BT	C641
NF-A49-326	TU37b	C642
	TU52b	C642
	TU52BT	C642
NF-A49-327	TU37b	C643
	TU52b	C643
NF-A49-330	TU10N14	C644
	TU17N2	C644
NF-A49-341	TS30-0	C645
	TS30a	C645
	TS34a	C645
	TS37a	C645
	TS42a	C645
	TS47a	C645
NF-A49-343	TS18M5	C646
	TS18MDV5	C646

Standard No. 規格番号	Grade グレード	Index No. 索引番号
	TS37b	C646
NF-A49-400	TSE220	C624
	TSE250	C624
	TSE290	C624
	TSE320	C624
	TSE360Nb	C624
	TSE360V	C624
	TSE415Nb	C624
	TSE415V	C624
NF-A49-401	TSE220	C625
	TSE250	C625
	TSE290	C625
	TSE320	C625
	TSE360	C625
	TSE415	C625
	TSE450	C625
	TSE480	C625
NF-A49-402	TSE220	C626
	TSE250	C626
	TSE290	C626
	TSE320	C626
	TSE360	C626
	TSE415	C626
	TSE450	C626
	TSE480	C626
NF-A49-501	TSE295-2	C647
	TSE295-3	C647
	TSE295-4	C647
	TUTSE235-2	C647
	TUTSE235-3	C647
	TUTSE235-4	C647
	TUTSE275-2	C647
	TUTSE275-3	C647
	TUTSE355-2	C647
	TUTSE355-3	C647
	TUTSE355-4	C647
	TUTSE450-2	C647
	TUTSE450-3	C647
	TUTSE450-4	C647
NF-A49-541	TSE235-2	C648
	TSE235-3	C648
	TSE235-4	C648
	TSE275-2	C648
	TSE275-3	C648
	TSE275-4	C648
NF-A49-643	TS30-1	C627
	TS30-2	C627
	TS34-2	C627
	TS37-2	C627
	TS42-2	C627
	TS47-2	C627
	TS335D	C627
	TS390D	C627
	TS445D	C627
NF-A49-645	TS30-0	C628
	TS30E	C628
	TS30ES	C628
	TS335D	C628
	TS390D	C628
NF-A49-647	TSZ2CN18-10	C649
	TSZ6CN18-09	C649
	TSZ6CND17-11	C649
	TS12CN17-07	C649
〔ISO〕		
ISO-65	TS	C701
	TW	C001, C701
ISO-559	TS0	C702
	TS1	C702
	TS4	C002, C702
	TS9	C002, C702
ISO-2604/2	TS1	C703
	TS2	C003, C703
	TS4	C002, C703
	TS5	C004, C703
	TS6	C007, C703
	TS9	C003, C703
	TS9H	C004, C703
	TS10	C703
	TS13	C003, C703
	TS14	C004, C703
	TS15	C703
	TS18	C010, C703
	TS26	C703
	TS32	C703
	TS33	C703
	TS34	C703
	TS37	C004, C703
	TS34	C703
	TS37	C013, C703
	TS38	C703

Standard No. 規格番号	Grade グレード	Index No. 索引番号	Standard No. 規格番号	Grade グレード	Index No. 索引番号
	TS39	C703	ISO-2938	Gr.1	C017, C707
	TS40	C703		Gr.2	C707
	TS43	C007, C703			
	TS45	C007, C703	ISO-3183	E17	C711
	TS46	C012, C703		E21	C711
	TS47	C008, C703		E24-1	C002, C711
	TS48	C008, C703		E24-2	C002, C711
	TS50	C008, C703			
	TS53	C008, C703	ISO-3304	R28	C708
	TS54	C703		R33	C017, C708
	TS56	C008, C703		R37	C017, C708
	TS57	C008, C703		R42	C708
	TS58	C008, C703		R50	C017, C708
	TS60	C008, C703			
	TS61	C008, C703	ISO-3305	R28	C018, C709
	TS63	C008, C703		R33	C017, C709
	TS67	C703		R37	C017, C709
	TS68	C008, C703		R42	C709
	TS69	C703		R50	C017, C709
ISO-2604/3	TW1	C704	ISO-3306	R28	C710
	TW2	C704		R33	C017, C710
	TW4	C002, C704		R37	C017, C710
	TW5	C704		R42	C710
	TW6	C007, C704		R50	C017, C710
	TW9	C002, C704			
	TW9H	C004, C704			
	TW10	C704			
	TW13	C704			
	TW14	C704			
	TW15	C704			
	TW26	C704			
	TW32	C704			
ISO-2604/5	TW46	C008, C705			
	TW47	C008, C705			
	TW50	C009, C705			
	TW53	C705			
	TW57	C009, C705			
	TW58	C705			
	TW60	C009, C705			
	TW61	C009, C705			
	TW69	C706			
ISO-2937	C35	C706			
	TS1	C705			
	TS4	C017, C706			
	TS9	C706			
	TS18	C017, C706			

CHAPTER IV

STEELS FOR MACHINE STRUCTURAL USE

第4部

機械構造用鋼

CHAPTER IV CONTENTS　第4部　目次

1. STANDARDS LIST 収録各国規格リスト

1-1 JIS（Japan 日 本）

G4051-79 Carbon Steels for Machine Structural Use
機械構造用炭素鋼鋼材

G4052-79 Structural Steels with Specified Hardenability Bands
焼入性を保証した構造用鋼鋼材（H鋼）

G4102-79 Nickel Chromium Steels
ニッケルクロム鋼鋼材

G4103-79 Nickel Chromium Molybdenum Steels
ニッケルクロムモリブデン鋼鋼材

G4104-79 Chromium Steels
クロム鋼鋼材

G4105-79 Chromium Molybdenum Steels
クロムモリブデン鋼鋼材

G4106-79 Manganese Steels and Manganese Chromium Steels for Machine Structural Use
機械構造用マンガン鋼鋼材及びマンガンクロム鋼鋼材

G4107-94 Alloy Steel Bolting Materials for High Temperature Services
高温用合金鋼ボルト材

G4108-94 Alloy Steel Bars for Special Application Bolting Materials
特殊用途合金鋼ボルト用棒鋼

G4202-79 Aluminum Chromium Molybdenum Steels
アルミニウムクロムモリブデン鋼鋼材

1-2 ASTM（U.S.A. アメリカ）

A193M-94c Alloy-Steel and Stainless Bolting Materials for High Temperature Service
高温用ステンレス合金鋼ボルト材

A540M-93 Alloy-Steel Bolting Materials for Special Applications
特殊用途用合金鋼ボルト材

1-3 AISI（U.S.A. アメリカ）

MN06/210-86 Alloy, Carbon and High Strength Low Alloy Steels, Semifinished for Forging; Hot Rolled Bars; Cold Finished Steel Bars; Hot Rolled Deformed and Plain Concrete Reinforcing Bar
合金鋼，炭素鋼及び高強度低合金鋼，鍛造用半製品；熱間及び冷間仕上鋼棒
熱間圧延棒鋼鉄筋コンクリート用棒鋼

MN06/211-81 Hot Rolled Floor Plates: Carbon, High Strength Low Alloy and Alloy
鋼板，床用鋼板；炭素鋼，高強度低合金鋼及び合金鋼

MN06/214-88 Sheet Steel; Carbon, High Strength low Alloy and Alloy Coils and Cutlength
薄鋼板；炭素鋼，高強度鋼及び合金鋼コイル及び寸法

MN06/215-88 Strip Steel
鋼帯

MN06/217-84 Wire and Rods, Carbon Steel
鋼棒及び線材；炭素鋼

MN06/221-75 Steel Speciality Tubular Products
特殊鋼管

MN06/224-75 Wire and Rods, Alloy Steel
鋼棒及び線材；合金鋼

1-4 BS（U.K.　イギリス）

970Part1-91 General Inspection and Testing Procedures and Specific Requirements for Carbon, Carbon Manganese and Stainless Steels
炭素鋼，炭素マンガン鋼およびステンレス鋼鋼材

1506-90 Specification for Carbon, Low Alloy and Stainless Steel Bars and Billets for Bolting Material to Be Used in Pressure Retaining Applications
高圧用ボルト材用炭素鋼，低合金鋼，ステンレス鋼鋼棒及び鋼片

1-5 DIN（West Germany 西ドイツ）

17200-88 Direct Hardening Steels Superseded by European Standard EN 10083 - 1&-2, 1991
調質鋼鋼材

17210-86 Case Hardening Steels
はだ焼鋼鋼材

17211-87 Nitriding Steels
窒化鋼鋼材

17240-76 Heat-Resisting Materials for Bolts and Nuts
ボルト・ナット用耐熱鋼鋼材

1-6 NF（France　フランス）

A35-551 Unalloyed and Alloyed Steel Bars and Wires for Machine Structural Use
機械構造用非合金及び合金特殊鋼鋼棒及び鋼線

A35-552 Heat-Treated Unalloyed and Alloyed Steels for Machine Structural Use
機械構造用熱処理非合金及び合金特殊鋼鋼材

A35-553 Unalloyed and Alloyed Steel Strips for Structural Use
構造用非合金及び合金鋼鋼帯

A35-554 Unalloyed and Alloyed Steel Plates and Sheets for Structural Use
構造用非合金及び合金鋼鋼板

A35-558 Steels for Fastening
固定用鋼材

1-7 ГОСТ（U.S.S.R. ソ 連）

1050-74 High Quality Structural Carbon Steels
高品質構造用炭素鋼鋼材

4543-71 Alloyed Structural Steels
合金構造用鋼鋼材

1. Carbon Steels for Machine Structural Use

JIS	AISI	BS	DIN	NF	ГОСТ		
G4051	MN06- 210, 211, 214 215, 217	970 Part 1	17200 17210 17211	A35- 551～554	1050	C	Si
–	–					–	–
	1008					≦0.10	–
		040A04				≦0.08	–
S10C						0.08～0.13	0.15～0.35
	1010					0.08～0.13	by agreement
		040A10				0.08～0.13	–
			CK10			0.07～0.13	0.15～0.35
			C10			0.07～0.13	0.15～0.35
				XC10		0.06～0.12	0.05～0.30
					10	0.07～0.14	0.17～0.37
S12C						0.10～0.15	0.15～0.35
	1012						by agreement
		040A12				0.10～0.15	–
			CK12	CK12		0.10～0.16	0.15～0.35
S15C						0.13～0.18	0.15～0.35
	1015					0.13～0.18	by agreement
		055M15				≦0.20	–
			CK15			0.12～0.18	0.15～0.35
			C15			0.12～0.18	0.15～0.35
					15	0.12～0.19	0.17～0.37
–						–	–
	1016					0.13～0.18	by agreement
						–	–
						–	–
						–	–
						–	–

機械構造用炭素鋼

Chemical Composition 化学成分 (%)								Index Number 索引番号
Mn	P	S	Ni	Cr	Mo	Cu	Others	
－	－	－	－	－	－	－	－	D001
0.30～0.50	≦0.040	≦0.050	－	－	－	－	－	
0.30～0.50	≦0.050	0.025～0.050	≦0.40	≦0.30	≦0.15	－	－	
0.30～0.60	≦0.030	≦0.035	≦0.020	≦0.020	－	≦0.30	Ni+Cr≦0.35	D002
0.30～0.60	≦0.040	≦0.050	－	－	－	－	－	
0.30～0.50	≦0.050	0.025～0.050	≦0.40	≦0.30	≦0.15	－	－	
0.30～0.60	≦0.035	≦0.035	－	－	－	－	－	
0.30～0.60	≦0.045	≦0.045	－	－	－	－	－	
0.30～0.60	≦0.035	≦0.035	－	－	－	－	－	
0.35～0.65	≦0.035	≦0.040	≦0.25	≦0.15	≦0.15	－	－	
0.30～0.60	≦0.030	≦0.035	≦0.20	≦0.20	－	≦0.30	Ni+Cr≦0.35	D003
0.30～0.60	≦0.040	≦0.050	－	－	－	－	－	
0.30～0.50	≦0.050	0.025～0.050	≦0.40	≦0.30	≦0.15	－	－	
0.30～0.60	≦0.035	≦0.035	－	－	－	－	－	
0.30～0.60	≦0.030	≦0.035	≦0.20	≦0.20	－	≦0.30	Ni+Cr≦0.35	D004
0.30～0.60	≦0.040	≦0.050	－	－	－	－	－	
≦0.80	≦0.050	0.025～0.050	≦0.40	≦0.30	≦0.15	－	－	
0.30～0.60	≦0.035	≦0.035	－	－	－	－	－	
0.30～0.60	≦0.045	≦0.045	－	－	－	－	－	
0.35～0.65	≦0.035	≦0.040	≦0.25	≦0.25	－	－	－	
－	－	－	－	－	－	－	－	D005
0.60～0.90	≦0.040	≦0.050	－	－	－	－	－	
－	－	－	－	－	－	－	－	
－	－	－	－	－	－	－	－	
－	－	－	－	－	－	－	－	
－	－	－	－	－	－	－	－	

JIS	AISI	BS	DIN	NF	ГОСТ		
G4051	MN06－ 210, 211, 214 215, 217	970 Part 1 －91	17200 17210 17211	A35－ 551～554	1050	C	Si
S17C						0.15～0.20	0.15～0.35
	1017					0.15～0.20	by agreement
				XC18		0.16～0.22	0.15～0.35
				XC18S		0.16～0.22	0.15～0.35
S20C						0.18～0.23	0.15～0.35
	1020					0.18～0.23	by agreement
		070M20	Replaced by European Standard EN 10083 - 1: 91, Grade 2C22			0.16～0.24	–
			CK20			0.17～0.23	0.15～0.35
			C20			0.17～0.23	0.15～0.35
					20	0.17～0.24	0.17～0.37
S22C						0.20～0.25	0.15～0.35
	1023					0.20～0.25	by agreement
			CK22	Replaced by EN 10083-1:91,Grade 2C22		0.18～0.25	0.15～0.35
			C22			0.18～0.25	0.15～0.35
S25C						0.22～0.28	0.15～0.35
	1025					0.22～0.28	by agreement
			CK25	Replaced by EN 10083-1:91,Grade 2C25		0.22～0.28	0.15～0.35
			C25			0.22～0.28	0.15～0.35
				XC25		0.23～0.29	0.15～0.35
					25	0.22～0.30	–
–						–	–
	1026					0.22～0.28	by agreement
		070M26	Replaced by European Standard EN 10083 - 1: 91, Grade 2C25			0.22～0.30	–
S28C						0.25～0.31	0.15～0.35
	1029					0.22～0.30	by agreement
S30C						0.27～0.33	0.15～0.33
	1030					0.28～0.34	by agreement
		080A30	Not listed in BS970 Part 1- 91			0.26～0.34	–
		080M30	Replaced by European Standard EN 10083 - 1: 91, Grade 2C30			0.26～0.34	–
			CK30	Replaced by EN 10083-1:91, Grade 2C30		0.27～0.33	0.15～0.35
			C30			0.27～0.33	0.15～0.35
					30	0.27～0.35	0.17～0.35
S33C						0.27～0.33	0.15～0.33
		060A32	Not listed in BS970 Part 1- 91			0.28～0.36	–
		080A32	Not listed in BS970 Part 1- 91			0.28～0.36	–
				XC32	–	0.30～0.36	0.15～0.35
S35C						0.32～0.38	0.15～0.33
	1035					0.32～0.38	by agreement
		080A35	Not listed in BS970 Part 1- 91			0.33～0.38	–
		080M36	Replaced by European Standard EN 10083 - 1: 91, Grade 2C35			0.32～0.40	–

Chemical Composition 化学成分 (%)								Index Number 索引番号
Mn	P	S	Ni	Cr	Mo	Cu	Others	
0.30～0.60	≦0.030	≦0.035	≦0.20	≦0.20	−	≦0.30	Ni+Cr≦0.35	D006
≦0.040	≦0.050	−	−	−	−	−	−	
0.40～0.70	≦0.035	≦0.035	−	−	−	−	−	
0.40～0.70	≦0.035	≦0.035	−	−	−	−	−	
0.30～0.60	≦0.030	≦0.035	−	−	−	−	−	D007
0.30～0.60	≦0.040	≦0.050	−	−	−	−	−	
0.50～0.90	≦0.050	0.025～0.050	−	−	−	−	−	
0.30～0.60	≦0.035	≦0.035	−	−	−	−	−	
0.30～0.60	≦0.045	≦0.045	−	−	−	−	−	
0.35～0.65	≦0.035	≦0.035	≦0.25	≦0.25	−	−	−	
0.30～0.60	≦0.030	≦0.035	−	−	−	−	−	D008
0.30～0.60	≦0.040	≦0.050	−	−	−	−	−	
0.30～0.60	≦0.035	≦0.035	−	−	−	−	−	
0.30～0.60	≦0.045	≦0.045	−	−	−	−	−	
0.30～0.60	≦0.030	≦0.035	≦0.20	≦0.20	−	≦0.30	Ni+Cr≦0.35	D009
0.30～0.60	≦0.040	≦0.050	−	−	−	−	−	
0.30～0.60	≦0.035	≦0.035	−	−	−	−	−	
0.30～0.60	≦0.045	≦0.045	−	−	−	−	−	
0.40～0.70	≦0.035	≦0.035	−	−	−	−	−	
0.50～0.90	≦0.050	≦0.050	≦0.40	≦0.30	≦0.15	−	−	
−	−	−	−	−	−	−	−	D010
0.50～0.90	≦0.050	≦0.050	≦0.40	≦0.30	≦0.15	−	−	
0.50～0.90	≦0.050	0.025～0.050	−	−	−	−	−	
0.60～0.90	≦0.030	≦0.035	−	−	−	−	−	D011
0.60～0.90	≦0.040	≦0.050	−	−	−	−	−	
0.60～0.90	≦0.030	≦0.035	≦0.20	≦0.20	−	≦0.30	Ni+Cr≦0.35	D012
0.70～1.00	≦0.040	≦0.050	−	−	−	−	−	
0.70～0.90	≦0.050	0.025～0.050	≦0.40	≦0.30	≦0.15	−	−	
0.60～1.00	≦0.050	0.025～0.050	−	−	−	−	−	
0.60～0.90	≦0.035	≦0.035	−	−	−	−	−	
0.60～0.90	≦0.045	≦0.045	−	−	−	−	−	
0.50～0.80	≦0.035	≦0.040	≦0.25	≦0.25	−	−	−	
0.60～0.90	≦0.030	≦0.035	≦0.20	≦0.20	−	≦0.30	Ni+Cr≦0.35	D013
0.50～0.70	≦0.050	0.025～0.050	−	−	−	−	−	
0.60～1.00	≦0.050	0.025～0.050	−	−	−	−	−	
0.40～0.70	≦0.035	≦0.035	−	−	−	−	−	
0.60～0.90	≦0.030	≦0.035	−	−	−	≦0.30	Ni+Cr≦0.35	D014
0.60～0.90	≦0.040	≦0.050	−	−	−	−	−	
0.70～0.90	≦0.050	0.025～0.050	−	−	−	−	−	
0.60～1.00	≦0.050	0.025～0.050	−	−	−	−	−	

JIS	AISI	BS	DIN	NF	ГОСТ		
G4051	MN06－ 210, 211, 214 215, 217	970 Part 1 －91	17200 17210 17211	A35－ 551～554	1050	C	Si
			CK35	Replaced by EN 10083-1:91, Grade 2C35		0.32～0.38	0.15～0.35
			C35			0.32～0.39	0.15～0.35
					35	0.32～0.40	0.17～0.37
－						－	－
	1037					0.32～0.38	by agreement
		080A37	Not listed in BS970 Part 1- 91			0.32～0.40	0.17～0.37
S38C						0.35～0.41	0.15～0.35
	1038					0.32～0.42	by agreement
				XC38		0.36～0.41	0.15～0.35
				XC38H1		0.36～0.41	0.15～0.35
				XC38H2		0.36～0.41	0.15～0.35
S40C						0.37～0.43	0.15～0.35
	1040					0.37～0.44	by agreement
		060A40	Not listed in BS970 Part 1- 91			0.36～0.44	－
		080A40	Not listed in BS970 Part 1- 91			0.36～0.44	－
		080M40	Replaced by European Standard EN 10083 - 1: 91, Grade 2C40			0.36～0.44	－
			CK40	Replaced by EN 10083-1:91, Grade 2C40		0.37～0.44	0.15～0.35
			C40			0.37～0.44	0.15～0.35
					40	0.37～0.45	0.17～0.37
－						－	－
	1042					0.40～0.47	by agreement
		080A42	Not listed in BS970 Part 1- 91			0.40～0.45	－
				XC42H1		0.40～0.44	0.15～0.30
				XC42H2		0.40～0.44	0.15～0.30

Chemical Composition 化 学 成 分 (%)								Index Number 索引番号
Mn	P	S	Ni	Cr	Mo	Cu	Others	
0.50～0.80	≦0.035	≦0.035	–	–	–	–	–	
0.50～0.80	≦0.045	≦0.045	–	–	–	–	–	
0.50～0.80	≦0.035	≦0.035	≦0.25	≦0.25	–	–	–	
–	–	–	–	–	–	–	–	D015
0.70～1.00	≦0.040	≦0.050	–	–	–	–	–	
0.50～0.80	≦0.050	0.025～0.050	≦0.25	≦0.25	–	–	–	
0.60～0.90	≦0.030	≦0.035	≦0.20	≦0.20	–	≦0.30	Ni+Cr≦0.35	D016
0.60～0.90	≦0.040	≦0.050	–	–	–	–	–	
0.30	≦0.035	≦0.035	–	–	–	–	–	
0.60～0.90	≦0.035	≦0.035	–	–	–	–	–	
0.60～0.90	≦0.035	≦0.035	–	–	–	–	–	
0.60～0.90	≦0.030	≦0.035	≦0.20	≦0.20	–	≦0.30	–	D017
0.70～1.00	≦0.040	≦0.050	–	–	–	–	–	
0.50～0.70	≦0.050	0.025～0.050	–	–	–	–	–	
0.70～0.90	≦0.050	0.025～0.050	–	–	–	–	–	
0.60～1.00	≦0.050	0.025～0.050	–	–	–	–	–	
0.50～0.80	≦0.035	≦0.035	–	–				
0.50～0.80	≦0.045	≦0.045	–	–				
8.50～0.80	≦0.035	≦0.040	≦0.25	≦0.25	–	–	–	
–	–	–	–	–	–	–	–	D018
0.60～0.90	≦0.040	≦0.050	–	–	–	–	–	
0.70～0.90	≦0.050	0.025～0.050	≦0.040	≦0.30	≦0.15	–	–	
0.70～0.90	≦0.035	≦0.035	–	–	–	–	–	
0.70～0.90	≦0.035	≦0.035	–	–	–	–	–	

J I S	A I S I	B S	D I N	N F	ГОСТ		
G4051	MN06－ 210, 211, 214 215, 217	970 Part 1 －91	17200 17210 17211	A35－ 551～554	1050	C	S i
S43C						0. 40～0. 46	0. 15～0. 35
	1043					0. 40～0. 47	by agreement
S45C						0. 42～0. 48	0. 15～0. 35
	1045					0. 43～0. 50	by agreement
		060A45	Not listed in BS970 Part 1- 91			0. 41～0. 49	－
			CK45	Replaced by EN 10083-1:91, Grade 2C45		0. 42～0. 50	0. 15～0. 35
			C45			0. 42～0. 50	0. 15～0. 35
				XC45		0. 42～0. 48	0. 15～0. 35
					45	0. 42～0. 50	0. 17～0. 37
－						－	－
	1046					0. 43～0. 50	by agreement
		080M46				0. 42～0. 50	－
S48C						0. 45～0. 51	0. 15～0. 35
		060A47	Not listed in BS970 Part 1- 91			0. 44～0. 50	－
		080A47	Not listed in BS970 Part 1- 91			0. 44～0. 50	－
				XC48		0. 46～0. 50	0. 15～0. 35
				XC8H1		0. 46～0. 50	0. 15～0. 35
				XC8H2		0. 46～0. 50	0. 15～0. 35
S50C						0. 47～0. 53	0. 15～0. 35
	1050					0. 48～0. 55	by agreement
		080M50	Replaced by European Standard EN 10083 - 1: 91, Grade 2C50			0. 47～0. 54	－
			CK50	Replaced by EN 10083-1:91, Grade 2C50		0. 47～0. 54	0. 15～0. 35
			C50			0. 47～0. 54	0. 15～0. 35
				XC50		0. 47～0. 54	0. 15～0. 35
					50	0. 47～0. 55	0. 17～0. 35

Chemical Composition 化学成分 (%)								Index Number 索引番号
Mn	P	S	Ni	Cr	Mo	Cu	Others	
0.60～0.90	≦0.030	≦0.035	≦0.20	≦0.20	–	≦0.30	Ni+Cr≦0.35	D019
0.70～1.00	≦0.040	≦0.050	–	–	–	–	–	
0.60～0.90	≦0.030	≦0.035	≦0.020	≦0.20	–	≦0.30	Ni+Cr≦0.35	D020
0.60～0.90	≦0.040	≦0.050	–	–	–	–	–	
0.50～0.70	≦0.050	0.025～0.050	–	–	–	–	–	
0.50～0.80	≦0.035	≦0.035	–	–	–	–	–	
0.50～0.80	≦0.045	≦0.045	–	–	–	–	–	
0.70～0.90	≦0.035	≦0.035	–	–	–	–	–	
0.50～0.80	≦0.035	≦0.040	≦0.25	≦0.25	–	–	–	
–	–	–	–	–	–	–	–	D021
0.60～0.90	≦0.040	≦0.050	–	–	–	–	–	
0.60～1.00	≦0.050	0.025～0.050	–	–	–	–	–	
0.60～0.90	≦0.030	≦0.035	≦0.020	≦0.20	–	≦0.30	Ni+Cr≦0.35	D022
0.50～0.70	≦0.050	0.025～0.050	–	–	–	–	–	
0.60～1.00	≦0.050	0.025～0.050	–	–	–	–	–	
0.60～1.00	≦0.035	≦0.035	–	–	–	–	–	
0.60～1.00	≦0.035	≦0.035	–	–	–	–	–	
0.60～1.00	≦0.035	≦0.035	–	–	–	–	–	
0.60～0.90	≦0.030	≦0.035	≦0.020	≦0.20	–	≦0.30	Ni+Cr≦0.35	D023
0.60～0.90	≦0.040	≦0.050	–	–	–	–	–	
0.60～1.00	≦0.050	0.025～0.050	–	–	–	–	–	
0.60～0.90	≦0.035	≦0.035	–	–	–	–	–	
0.60～0.90	≦0.045	≦0.045	–	–	–	–	–	
0.60～1.00	≦0.035	≦0.035	–	–	–	–	–	
0.50～0.80	≦0.040	≦0.040	≦0.025	≦0.25	–	–	–	

JIS	AISI	BS	DIN	NF	ГОСТ		
G4051	MN06－ 210, 211, 214 215, 217	970 Part 1 －91	17200 17210 17211	A35－ 551～554	1050	C	Si
S53C						0.50～0.50	0.15～0.35
	1053					0.48～0.55	by agreement
				XC54		0.51～0.56	0.15～0.35
S55C						0.52～0.58	0.15～0.35
	1055					0.50～0.60	by agreement
		070M50	Not listed in BS970 Part 1-91			0.50～0.60	－
			CK55	Replaced by EN 10083-1:91, Grade 2C55		0.52～0.60	0.15～0.35
			C55			0.52～0.60	0.15～0.35
				XC55H1		0.51～0.59	0.15～0.35
					55	0.52～0.60	0.17～0.37
S58C						0.55～0.61	0.15～0.35
	1060					0.55～0.65	by agreement
		060A57	Not listed in BS970 Part 1-91			0.54～0.60	－
		080A57	Not listed in BS970 Part 1-91			0.54～0.60	－
			CK60	Replaced by EN 10083-1:91, Grade 2C60		0.57～0.65	0.15～0.35
			C60			0.57～0.65	0.15～0.35
				XC60		0.58～0.62	0.15～0.35
					58	0.55～0.63	0.17～0.35
					60	0.57～0.65	0.17～0.35
－						－	－
	1070					0.65～0.75	by agreement
		060A72				0.70～0.75	－
S09CK						0.07～0.12	0.10～0.35
			CK10			0.07～0.13	≦0.40
				XC10		0.06～0.12	0.05～0.30
					10	0.07～0.14	0.17～0.37
S15CK						0.13～0.18	0.15～0.35
			CK15			0.12～0.18	≦0.40
				XC12		0.12～0.18	0.05～0.30
					15	0.12～0.19	0.17～0.37
CSK20	－	－				－	－
			CK22	Replaced by EN 10083-1:91, Grade 2C22		0.18～0.25	0.15～0.30
				XC18		0.16～0.20	0.17～0.30
					20	0.17～0.24	0.17～0.37

Chemical Composition 化学成分 (%)								Index Number 索引番号
Mn	P	S	Ni	Cr	Mo	Cu	Others	
0.60～0.90	≦0.030	≦0.035	≦0.020	≦0.20	–	≦0.30	Ni+Cr≦0.35	D024
0.60～1.00	≦0.040	≦0.050	–	–	–	–	–	
0.60～1.00	≦0.035	≦0.035	–	–	–	–	–	
0.60～0.90	≦0.030	≦0.035	≦0.020	≦0.20	–	≦0.30	Ni+Cr≦0.35	D025
0.60～0.90	≦0.040	≦0.050	–	–	–	–	–	
0.60～0.90	≦0.050	0.025～0.050	–	–	–	–	–	
0.60～0.90	≦0.035	≦0.035	–	–	–	–	–	
0.60～0.90	≦0.045	≦0.045	–	–	–	–	–	
0.60～1.00	≦0.035	≦0.035	–	–	–	–	–	
0.50～0.80	≦0.035	≦0.040	≦0.25	≦0.25	–	–	–	
0.60～0.90	≦0.030	≦0.035	≦0.020	≦0.20	–	≦0.30	Ni+Cr≦0.35	D026
0.50～0.90	≦0.040	≦0.050	–	–	–	–	–	
0.50～0.90	≦0.050	≦0.050	–	–	–	–	–	
0.60～1.00	≦0.050	≦0.050	–	–	–	–	–	
0.60～0.90	≦0.035	≦0.035	–	–	–	–	–	
0.60～0.90	≦0.045	≦0.045	–	–	–	–	–	
0.60～0.90	≦0.035	≦0.035	–	–	–	–	–	
0.50～0.80	≦0.035	≦0.035	≦0.25	≦0.25	–	–	–	
0.70～1.00	≦0.040	≦0.040	≦0.25	≦0.25	–	–	–	
–	–	–	–	–	–	–	–	D027
0.50～0.70	≦0.050	≦0.050	≦0.040	≦0.30	≦0.15	–	–	
0.50～0.70	≦0.050	0.025～0.050	≦0.40	≦0.30	≦0.15	–	–	
0.30～0.60	≦0.025	≦0.025	≦0.20	≦0.20	–	≦0.25	Ni+Cr≦0.30	D028
0.30～0.60	≦0.035	≦0.035	–	–	–	–	–	
0.30～0.60	≦0.035	≦0.035	–	–	–	–	–	
0.35～0.65	≦0.035	≦0.040	≦0.25	≦0.15	–	–	–	
0.30～0.60	≦0.025	≦0.025	≦0.20	≦0.20	–	–	Ni+Cr≦0.30	D029
0.30～0.60	≦0.035	≦0.035	–	–	–	–	–	
0.30～0.60	≦0.035	≦0.035	–	–	–	–	–	
0.35～0.65	≦0.035	≦0.040	≦0.25	≦0.15	–	–	–	
–	–	–	≦0.25	≦0.25	–	–	Ni+Cr≦0.30	D030
0.30～0.60	≦0.035	≦0.030	–	–	–	–	–	
0.30～0.60	≦0.035	≦0.035	–	–	–	–	–	
0.35～0.65	≦0.035	≦0.040	≦0.25	≦0.25	–	–	Ni+Cr≦0.30	

JIS	AISI	BS	DIN	NF	ГОСТ		
G4051	MN06- 210, 211, 214 215, 217	970 Part 1 -91	17200 17210 17211	A35- 551～554	4543	C	Si
	1019					0.15～0.20	by agreement
	1021					0.18～0.23	by agreement
	1022					0.18～0.23	by agreement
	1039					0.37～0.44	by agreement
	1049					0.46～0.53	by agreement
	1078					0.18～0.75	by agreement
	1080					0.75～0.88	by agreement
	1084					0.85～0.83	by agreement
	1090					0.85～0.98	–
	1095					0.90～0.03	–
		060A78				0.75～0.82	–
		060A81				0.78～0.85	–

Chemical Composition 化学成分 (%)								Index Number 索引番号
Mn	P	S	Ni	Cr	Mo	Cu	Others	
0.70~1.00	≦0.040	≦0.050	–	–	–	–	–	D031
0.30~0.60	≦0.040	≦0.050	–	–	–	–	–	
0.70~1.00	≦0.040	≦0.050	–	–	–	–	–	
0.70~1.00	≦0.040	≦0.050	–	–	–	–	–	
0.60~0.90	≦0.040	≦0.050	–	–	–	–	–	
0.30~0.60	≦0.040	≦0.050	–	–	–	–	–	
0.60~0.90	≦0.040	≦0.050	–	–	–	–	–	
0.60~0.90	≦0.040	≦0.050	–	–	–	–	–	
0.60~0.90	≦0.040	≦0.050	–	–	–	–	–	
0.60~0.90	≦0.040	≦0.050	–	–	–	–	–	
0.50~0.70	≦0.050	0.025~0.050	≤0.40	≤0.30	≤0.15	–	–	D032
0.50~0.70	≦0.050	0.025~0.050	≤0.40	≤0.30	≤0.15	–	–	

JIS	AISI	BS	DIN	NF	ГОСТ		
G4102~ G4106 G4202	MN06- 210, 211, 214 215, 217	970 Part 1 -91	17200 17210 17211	A35- 551~554	4543	C	Si
Ni-Cr Steel							
SNC236						0.32~40	0.15~0.35
				35NC6		0.32~0.39	0.10~0.40
SNC415						0.12~0.18	0.15~0.35
SN631						0.27~0.35	0.15~0.35
SNC815						0.12~0.18	0.15~0.35
		655M13				0.10~0.16	–
				14NC11		0.11~0.17	0.10~0.40
SNC836						0.32~0.40	0.15~0.35
Ni-Cr-Mo Steel							
SNCM220						0.17~0.23	0.15~0.35
	8615					0.13~0.18	0.15~0.30
	8617					0.15~0.20	0.15~0.30
	8620					0.18~0.23	0.15~0.30
	8622					0.20~0.25	0.15~0.30
				20NCD2		0.17~0.23	0.10~0.40
SNCM240						0.38~0.43	0.15~0.35
	8637					0.35~0.40	0.15~0.35
	8640					0.38~0.43	0.15~0.35
SNCM415						0.12~0.18	0.15~0.35
SNCM420						0.17~0.23	0.15~0.35
	4320					0.17~0.22	0.15~0.30
SNCM431						0.27~0.35	0.15~0.35
SNCM439						0.36~0.43	0.15~0.35
	4340					0.38~0.43	0.15~0.35
SNCM447						0.44~0.50	0.15~0.35
SNCM616						0.13~0.20	0.15~0.35
SNCM625						0.20~0.30	0.15~0.35
SNCM630						0.25~0.35	0.15~0.35
SNCM815						0.12~0.18	0.15~0.35
Cr Steel							
SCr415						0.13~0.18	0.15~0.35
					15X	0.12~0.18	0.17~0.37
					15XA	0.12~0.17	0.17~0.37
SCr420						0.18~0.23	0.15~0.35
	5120					0.28~0.33	0.15~0.35
					20X	0.17~0.23	0.17~0.37
SCr430						0.28~0.33	0.15~0.35

機械構造用合金鋼

Chemical Composition 化学成分 (%)								Index Number 索引番号
Mn	P	S	Ni	Cr	Mo	Cu	Others	
0.50～0.80	≦0.030	≦0.030	1.00～1.50	0.50～0.90	–	≦0.30	–	D033
0.60～0.90	≦0.035	≦0.035	1.20～1.60	0.85～1.15	–	≦0.30	–	
0.35～0.65	≦0.030	≦0.030	2.00～2.50	0.20～0.50	–	≦0.30	–	D034
0.35～0.65	≦0.030	≦0.030	2.00～2.50	0.60～1.00	–	≦0.30	–	D035
0.35～0.65	≦0.030	≦0.030	3.00～3.50	0.70～1.00	–	≦0.30	–	D036
0.35～0.60	≦0.035	≦0.040	3.00～3.75	0.70～1.00	≦0.15	–	–	
0.25～0.60	≦0.035	≦0.035	2.50～3.00	0.60～0.90	–	–	–	
0.35～0.65	≦0.035	≦0.030	3.00～3.50	0.60～1.00	–	≦0.30	–	D037
0.60～0.90	≦0.030	≦0.030	0.40～0.70	0.40～0.65	0.15～0.30	≦0.30	–	D038
0.70～0.90	≦0.035	≦0.040	0.40～0.70	0.40～0.60	0.15～0.25	–	–	
0.70～0.90	≦0.035	≦0.040	0.40～0.70	0.40～0.60	0.15～0.25	–	–	
0.70～0.90	≦0.035	≦0.040	0.40～0.70	0.40～0.60	0.15～0.25	–	–	
0.70～0.90	≦0.035	≦0.040	0.40～0.70	0.40～0.60	0.15～0.25	–	–	
0.65～0.95	≦0.035	≦0.035	0.40～0.70	0.40～0.65	0.15～0.25	–	–	
0.70～1.00	≦0.030	≦0.030	0.40～0.70	0.40～0.65	0.15～0.30	≦0.30	–	D039
0.75～1.00	≦0.035	≦0.040	0.40～0.70	0.40～0.60	0.15～0.25	–	–	
0.75～1.00	≦0.035	≦0.040	0.40～0.70	0.40～0.60	0.15～0.25	–	–	
0.40～0.70	≦0.030	≦0.030	1.60～2.00	0.40～0.65	0.15～0.30	≦0.30	–	D040
0.40～0.70	≦0.030	≦0.030	1.60～2.00	0.40～0.65	0.15～0.30	≦0.30	–	D041
0.45～0.65	≦0.035	≦0.040	1.65～2.00	0.40～0.60	0.20～0.30	–	–	
0.60～0.90	≦0.030	≦0.030	1.60～2.00	0.60～1.00	0.15～0.30	≦0.30	–	D042
0.60～0.90	≦0.030	≦0.030	1.60～2.00	0.60～1.00	0.15～0.30	≦0.30	–	D043
0.60～0.80	≦0.035	≦0.040	1.50～2.00	0.70～0.90	0.20～0.30	–	–	
0.60～0.90	≦0.030	≦0.030	1.60～2.00	0.60～1.00	0.15～0.30	≦0.30	–	D044
0.80～1.20	≦0.030	≦0.030	2.80～3.20	1.40～1.80	0.40～0.60	≦0.30	–	D045
0.35～0.60	≦0.030	≦0.030	3.00～3.50	1.00～1.50	0.15～0.30	≦0.30	–	D046
0.35～0.60	≦0.030	≦0.030	2.50～3.50	2.50～3.50	0.15～0.70	≦0.30	–	D047
0.30～0.60	≦0.030	≦0.030	4.00～4.50	0.70～1.00	0.15～0.30	≦0.30	–	D048
0.60～0.85	≦0.030	≦0.030	≦0.25	0.90～1.20	–	≦0.30	–	D049
0.40～0.70	–	–	–	0.70～1.00	–	–	–	
0.40～0.70	–	–	–	0.70～1.00	–	–	–	
0.60～0.85	≦0.030	≦0.030	≦0.25	0.90～1.20	–	≦0.30	–	D050
0.70～0.90	≦0.035	≦0.040	–	0.80～1.10	–	–	–	
0.50～0.80	–	–	–	0.70～1.10	–	–	–	
0.60～0.85	≦0.030	≦0.030	≦0.25	0.90～1.20	–	≦0.30	–	D051

JIS	AISI	BS	DIN	NF	ГОСТ		
G4102~ G4106 G4202	MN06－ 210, 211, 214 215, 217	970 Part 1 -91	17200 17210 17211	A35－ 551~554	4543	C	Si
	5130					0.28~0.33	0.15~0.30
	5132					0.30~0.35	0.15~0.30
SCr430		530A30	Not listed in BS970 Part 1-91			0.28~0.33	–
		530A32	Not listed in BS970 Part 1-91			0.30~0.35	–
			34Cr4	Replaced by EN 10083-1:91, Grade 34 Cr 4		0.24~0.32	0.17~0.37
					30X	0.24~0.32	0.17~0.37
SCr435						0.33~0.38	0.15~0.35
	5135					0.33~0.38	0.15~0.30
		530A36	Not listed in BS970 Part 1-91			0.34~0.39	–
			37Cr4	Replaced by EN 10083-1:91, Grade 37 Cr 4		0.34~0.41	0.15~0.40
					35X	0.31~0.39	0.17~0.37
SCr440						0.38~0.43	0.15~0.35
	5140					0.38~0.43	0.15~0.30
		530A40	Not listed in BS970 Part 1-91			0.38~0.43	–
			41Cr4	Replaced by EN 10083-1:91, Grade 41 Cr 4		0.38~0.45	0.15~0.40
					40X	0.36~0.44	0.50~0.80
SCr445						0.43~0.48	0.15~0.35
	5147					0.46~0.51	0.15~0.35
					45X	0.41~0.49	0.17~0.37
Cr－Mo Steel							
SCM415						0.13~0.18	0.15~0.35
SCM418						0.16~0.21	0.15~0.35
				18CD4		0.16~0.22	0.10~0.40
SCM420						0.18~0.23	0.15~0.35
		708M20				0.17~0.23	0.10~0.35
SCM421						0.77~0.23	0.15~0.35
SCM430						0.28~0.33	0.15~0.35
	4130					0.28~0.33	0.15~0.30
		708A30	Not listed in BS970 Part 1-91			0.28~0.33	–
				30CD4		0.28~0.44	0.10~0.40
					30XM	0.26~0.34	0.17~0.37
					30XMA	0.26~0.34	0.17~0.37
SCM432						0.27~0.37	0.15~0.35
SCM435						0.33~0.38	0.15~0.35
	4135					0.32~0.39	0.15~0.30
	4137					0.35~0.40	0.15~0.30
		708A37	Not listed in BS970 Part 1-91			0.35~0.40	–
		709A37	Not listed in BS970 Part 1-91			0.35~0.40	–
			34CrMo4	Replaced by EN 10083-1:91, Grade 34 Cr Mo 4		0.30~0.37	0.15~0.30

Chemical Composition 化学成分 (%)								Index Number 索引番号
Mn	P	S	Ni	Cr	Mo	Cu	Others	
0.70～0.90	≦0.035	≦0.040	−	0.80～1.10	−	−	−	D051
0.60～0.80	≦0.035	≦0.040	−	0.75～1.00	−	−	−	
0.60～0.80	≦0.050	0.025～0.050	−	0.90～1.20	−	−	−	
0.60～0.80	≦0.050	0.025～0.050	−	0.90～1.20	−	−	−	
0.50～0.80	−	−	−	0.80～1.10	−	−	−	
0.50～0.80	−	−	−	0.80～1.10	−	−	−	
0.60～0.85	≦0.030	≦0.030	≦0.25	0.90～1.20	−	≦0.35	−	D052
0.60～0.80	≦0.035	≦0.040	−	0.80～1.05	−	−	−	
0.60～0.80	≦0.050	0.025～0.050	−	0.90～1.20	−	−	−	
0.60～0.90	≦0.035	≦0.035	−	0.90～1.20	−	−	−	
0.50～0.80	−	−	−	0.80～1.10	−	−	−	
0.60～0.85	≦0.030	≦0.00	≦0.25	0.90～1.20	−	≦0.30	−	D053
0.70～0.90	≦0.035	≦0.040		0.70～0.90	−	−	−	
0.60～0.80	≦0.050	0.025～0.050		0.90～1.20	−	−	−	
0.50～0.80	≦0.035	≦0.035		0.90～1.20	−	−	−	
−	−	−		0.80～1.10	−	−	−	
0.60～0.85	≦0.030	≦0.030	≦0.25	0.90～1.20	−	≦0.35	−	D054
0.70～0.95	≦0.035	≦0.040	−	0.85～1.15	−	−	−	
0.50～0.80	−	−	−	0.80～1.00	−	−	−	
0.60～0.85	≦0.030	≦0.030	≦0.25	0.90～1.20	0.15～0.30	≦0.30	−	D055
0.60～0.85	≦0.030	≦0.030	≦0.25	0.90～1.20	0.15～0.30	≦0.30	−	D056
0.60～0.90	≦0.035	≦0.035	−	0.90～1.20	0.15～0.25	−	−	
0.60～0.90	≦0.030	≦0.030	≦0.25	0.85～1.15	0.15～0.25	≦0.30	−	D057
0.50～0.60	≦0.050	0.025～0.050	−	0.90～1.20	0.15～0.25	−	−	
0.70～1.00	≦0.030	≦0.030	≦0.25	0.90～1.20	0.15～0.30	≦0.30	−	D058
0.60～0.85	≦0.030	≦0.030	≦0.25	0.90～1.20	0.15～0.30	≦0.30	−	D059
0.40～0.60	≦0.035	≦0.040	−	0.80～1.10	0.15～0.25	−	−	
0.40～0.60	≦0.050	0.025～0.050	−	0.90～1.20	0.15～0.25	−	−	
0.40～0.60	≦0.035	≦0.035	−	0.90～1.20	0.15～0.25	−	−	
0.40～0.60	−	−	−	0.80～1.15	0.15～0.30	−	−	
0.40～0.60	−	−	−	0.80～1.15	0.15～0.30	−	−	
0.30～0.60	≦0.030	≦0.030	≦0.25	1.00～1.50	0.15～0.30	−	−	D060
0.60～0.85	≦0.030	≦0.030	−	0.90～1.20	0.15～0.30	≦0.30	−	D061
0.65～0.95	≦0.035	≦0.040	−	0.80～1.15	0.15～0.25	−	−	
0.70～0.90	≦0.035	≦0.040	−	0.80～1.10	0.15～0.25	−	−	
0.75～1.00	≦0.035	≦0.040	−	0.90～1.20	0.15～0.25	−	−	
0.75～1.00	≦0.050	0.025～0.050	−	0.90～1.20	0.25～0.35	−	−	
0.50～0.80	≦0.035	≦0.035	−	0.90～1.20	0.15～0.30	−	−	

JIS	AISI	BS	DIN	NF	ГОСТ		
G4102~ G4106 G4202	MN06- 210, 211, 214 215, 217	970 Part 1 -91	17200 17210 17211	A35- 551~554	4543	C	Si
				34CD4		0.32~0.38	0.15~0.30
					35XM	0.32~0.40	0.17~0.37
SCM440						0.38~0.43	0.15~0.35
	4140					0.38~0.43	0.15~0.30
	4142					0.40~0.45	0.15~0.30
		708A40	Not listed in BS970 Part 1-91			0.38~0.43	-
		708A42	Not listed in BS970 Part 1-91			0.40~0.45	-
		709A40	Not listed in BS970 Part 1-91			0.38~0.43	-
		709A42	Not listed in BS970 Part 1-91			0.40~0.45	-
			42CrMo4	Replaced by EN 10083-1:91, Grade 42 Cr Mo 4		0.38~0.45	0.15~0.40
				42CD4		0.40~0.44	0.15~0.30
SCM445						0.43~0.48	0.15~0.30
	4145					0.43~0.48	0.15~0.30
	4147					0.45~0.50	0.15~0.30
		708A47	Not listed in BS970 Part 1-91			0.45~0.50	-
SCM822						0.20~0.25	0.15~0.35
Mn-Steel, Mn-Cr Steel							
SMn420						0.17~0.23	0.15~0.35
	1522					0.18~0.24	-
		120M19	Not listed in BS970 Part 1-91			0.15~0.23	0.10~0.40
				20M5		0.7~0.23	0.1~0.4
SMn433						0.30~0.36	0.15~0.35
		120M36	Not listed in BS970 Part 1-91			0.32~0.40	-
				35M5		0.32~0.37	0.1~0.4
					35Γ2	0.31~0.39	0.17~0.37
SMn438						0.35~0.41	0.15~0.35
	1541					0.36~0.44	-
		150M36	Not listed in BS970 Part 1-91			0.32~0.40	-
				40M6		0.35~0.44	0.1~0.4
					40Γ2	0.36~0.44	0.17~0.37
SMn443						0.40~0.46	-
		135M44	Not listed in BS970 Part 1-91			0.40~0.48	-
					45Γ2	0.41~0.49	0.17~0.37
SMnC420						0.17~0.23	0.15~0.35
SMnC443						0.40~0.46	0.15~0.35
Al-Cr-Mo-Steel							
SACM645						0.38~0.45	0.20~0.50
		905M39				0.35~0.43	-
			41CrAlMo7	No longer listed		0.35~0.45	0.20~0.50
				40CSD6.12		0.38~0.44	0.2~0.5

Chemical Composition 化学成分 (%)								Index Number 索引番号
Mn	P	S	Ni	Cr	Mo	Cu	Others	
0.60～0.90	≦0.035	≦0.035	－	0.90～1.20	0.15～0.25			D061
0.40～0.70	－	－	－	0.80～1.10	0.15～0.25			
0.60～0.85	≦0.030	≦0.030	≦0.025	0.90～1.20	0.15～0.30	－	－	D062
0.75～1.00	≦0.035	≦0.035	≦0.040	0.80～1.10	0.15～0.25	－	－	
0.75～1.00	≦0.035	≦0.040	－	0.80～1.10	0.15～0.25	－	－	
0.75～1.00	≦0.050	0.025～0.050	－	0.90～1.20	0.15～0.25	－	－	
0.75～1.00	≦0.050	0.025～0.050	－	0.90～1.20	0.15～0.25	－	－	
0.70～1.00	≦0.035	0.025～0.050	－	0.90～1.20	0.25～0.30	－	－	
0.75～1.00	≦0.050	0.025～0.050	－	0.90～1.20	0.25～0.35	－	－	
0.50～0.80	≦0.035	≦0.035	－	0.90～1.20	0.15～0.30	－	－	
0.75～1.00	≦0.035	≦0.035	－	0.90～1.20	0.15～0.25	－	－	
0.60～0.85	≦0.030	≦0.030	≦0.25	0.90～1.20	0.15～0.30	≦0.30	－	D063
0.75～1.00	≦0.035	≦0.040	－	0.80～1.10	0.15～0.35	－	－	
0.75～1.00	≦0.035	≦0.040	－	0.80～1.10	0.15～0.25	－	－	
0.75～1.00	≦0.050	0.025～0.050	－	0.90～1.20	0.15～0.25	－	－	
0.60～0.85	≦0.030	≦0.030	≦0.25	0.90～1.25	0.30～0.45	－	－	D064
1.20～1.50	≦0.030	≦0.030	≦0.25	≦0.35	－	≦0.30	－	D065
1.10～1.40	≦0.040	≦0.050	－	－	－	－	－	
1.00～1.40	≦0.050	0.025～0.050	－	－	－	－	－	
1.00～1.4	≦0.040	≦0.035	－	－	－	－	－	
1.20～1.50	≦0.030	≦0.030	≦0.25	≦0.35	－	≦0.30	－	D066
1.30～1.70	≦0.05	0.025～0.050	－	－	－	－	－	
1.3～1.7	≦0.040	≦0.035	－	－	－	－	－	
1.40～1.80	－	－	－	－	－	－	－	
1.35～1.65	≦0.030	≦0.030	≦0.25	≦0.35	－	≦0.30	－	D067
1.35～1.65	≦0.040	≦0.030	－	－	－	－	－	
1.30～1.70	≦0.050	0.025～0.050	－	－	－	－	－	
1.3～1.8	≦0.040	≦0.035	－	－	－	－	－	
1.40～1.80	－	－	－	－	－	－	－	
1.35～1.65	≦0.040	≦0.050	－	－	－	－	－	D068
1.35～1.50	≦0.050	0.025～0.050	－	－	－	－	－	
1.40～1.80	－	－	－	－	－	－	－	
1.35～1.65	≦0.030	≦0.030	≦0.25	0.35～0.70	0.15～0.35	≦0.30	－	D069
1.20～1.50	≦0.030	≦0.030	≦0.25	0.35～0.70	－	≦0.30	－	D070
0.50～0.80	≦0.030	≦0.035	－	1.50～1.80	0.25～0.40	－	Al 0.80~1.20	D071
0.40～0.65	≦0.025	≦0.025	－	1.40～1.80	0.15～0.25	4XP+Sn≦0.10	Al 0.90~1.30	
0.50～0.80	≦0.030	≦0.035	－	1.50～1.80	0.25～0.40	－	Al 0.80~1.20	
0.5～0.8	≦0.035	≦0.03	－	1.5～1.8	0.2～0.3	－	Al 0.7~1.1	

JIS	ASTM	BS	DIN	NF	ГОСТ		
G4107 G4108	A193M A540M	1506 -90	17240	A35－558	－	C	Si
Alloy Steels for Bolting Materials ボルト用合金鋼							
SNB5						≧0.10	≦1.00
	B5					≧0.10	≦1.00
SNB7						0.38～0.48	0.20～0.35
	B7					0.37～0.49	0.15～0.25
		630-790				0.37～0.49	0.15～0.35
			42CrMo4	Replaced by EN 10083-1:91,Grade 42 Cr Mo 4		0.38～0.45	0.15～0.40
SNB16						0.36～0.44	0.20～0.35
SNB21						0.36～0.44	0.20～0.35
	B16					0.36～0.47	0.15～0.35
		670-860				0.36～0.44	0.15～0.35
			40CrMoV47	Not listed		0.35～0.45	0.2～0.35
				42CDV4		0.36～0.44	0.2～0.35
SNB22						0.39～0.46	0.20～0.35
	B22					0.36～0.44	0.15～0.35
		630-790				0.37～0.49	0.15～0.35
			42CrMo4	Replaced by EN 10083-1:91,Grade 42 Cr Mo 4		0.38～0.45	0.15～0.40
SNB23						0.37～0.44	0.20～0.35
	B23					0.37～0.44	0.15～0.35
SNB24						0.37～0.44	0.20～0.35
	B24					0.37～0.44	0.15～0.35

Chemical Composition 化学成分 (%)								Index Number 索引番号
Mn	P	S	Ni	Cr	Mo	Cu	Others	
≦1.00	≦0.040	≦0.030	–	4.00～6.00	0.45～0.65	–	–	D072
≦1.00	≦0.040	≦0.030	–	4.00～6.00	0.40～0.65	–	–	
0.75～1.00	≦0.040	≦0.040	–	0.80～1.10	0.15～0.25	–	–	D073
0.65～1.00	≦0.035	≦0.040	–	0.75～1.20	0.15～0.25	–	–	
0.65～1.10	≦0.035	≦0.040	–	0.75～1.20	0.15～0.25	–	–	
0.50～0.80	≦0.035	≦0.035	–	0.90～1.20	0.15～0.30	–	–	
0.45～0.70	≦0.040	≦0.040	–	0.80～1.15	0.50～0.65	–	V0.25~0.35	D074
0.45～0.70	≦0.025	≦0.025	–	0.80～1.50	0.50～0.65	–	V0.25~0.35	
0.45～0.70	≦0.035	≦0.040	–	0.80～1.15	0.50～0.65	–	V0.25~0.35	
0.4～0.7	≦0.04	≦0.04	–	0.8～1.2	0.5～0.7	–	V0.25~0.35	
0.45～0.7	≦0.035	≦0.035	–	0.8～1.1	0.5～0.6	–	V0.25~0.35	
0.65～1.10	≦0.025	≦0.025	–	0.75～1.20	0.15～0.25	–	–	D075
0.65～1.10	≦0.025	≦0.025	–	0.75～1.20	0.15～0.25	–	–	
0.65～1.10	≦0.035	≦0.040	–	0.75～1.20	0.15～0.25	–	–	
0.50～0.80	≦0.035	≦0.035	–	0.90～1.20	0.15～0.30	–	V 0.2~0.3	
0.60～0.95	≦0.025	≦0.025	1.55～2.00	0.65～0.95	0.20～0.30	–	–	D076
0.65～1.10	≦0.025	≦0.025	1.55～2.00	0.65～0.95	0.20～0.30	–	–	
0.70～0.90	≦0.025	≦0.025	1.65～2.00	0.70～0.95	0.30～0.40	–	–	D077
0.70～0.90	≦0.025	≦0.025	1.65～2.00	0.70～0.95	0.30～0.40	–	–	
0.45～0.70	≦0.035	≦0.040	–	0.80～1.15	0.50～0.65	–	V0.25~0.35	

3. Structural Steels with Specified Hardenability Bands

JIS	AISI	BS	DIN	NF	ГОСТ		
G4052	MN06 – 210, 217, 224	970 Part 1 -91	17200 17210 17211	–	–	C	Si
Mn – Steel							
SMn420H						0.16～0.23	0.15～0.35
SMn433H						0.29～0.34	0.15～0.35
	1330H					0.27～0.35	0.15～0.35
SMn438H						0.34～0.41	0.15～0.35
	1335H					0.32～0.38	0.15～0.35
SMn443H						0.39～0.40	0.15～0.35
Mn – Cr – Steel							
SMnC420H						0.16～0.23	0.15～0.35
SMnC443H						0.39～0.40	0.15～0.35
Cr – Steel							
SCr415H						0.12～0.18	0.15～0.35
SCr420H						0.17～0.23	0.15～0.35
	5120H					0.17～0.23	0.15～0.35
SCr430H						0.27～0.34	0.15～0.35
	5130H					0.27～0.33	0.15～0.30
	5132H					0.29～0.35	0.15～0.30
		530H32	Replaced by European Standard EN 10083 - 1: 91, Grade 34 Cr 4			0.29～0.35	–
			34Cr4	–	–	0.30～0.37	0.15～0.40
SCr435H						0.32～0.39	0.15～0.35
	5135H					0.32～0.38	0.15～0.35
		530H36	Replaced by European Standard EN 10083 - 1: 91, Grade 37 Cr 4			0.33～0.40	–
			37Cr4			0.34～0.41	0.15～0.40
SCr440H						0.37～0.44	0.15～0.35
	5140H					0.37～0.44	0.15～0.35
		530H40	Replaced by European Standard EN 10083 - 1: 91, Grade 41 Cr 4			0.37～0.44	–
			41Cr4			0.38～0.45	0.15～0.40
Cr – Mo – Steel							
SCM415H						0.12～0.18	0.15～0.35
SCM418H						0.15～0.21	0.15～0.35
SCM420H						0.17～0.23	0.15～0.35
		708M20				0.17～0.23	0.10～0.35
SCM435H						0.32～0.39	0.15～0.35
	4135H					0.32～0.38	0.15～0.30
	4137H					0.34～0.41	0.15～0.35
		708H37	Not listed in BS970 Part 1- 91			0.39～0.46	–
			34CrMo4	Replaced by EN 10083-1:91,Grade 34Cr Mo 4		0.32～0.34	0.15～0.40
SCM440H						0.37～0.44	0.15～0.35

焼入れ性を保証した構造用鋼

Chemical Composition 化学成分 (%)								Index Number 索引番号
Mn	P	S	Ni	Cr	Mo	Cu	Others	
1.15~1.55	≦0.030	≦0.030	≦0.25	≦0.35	−	≦0.30	−	D078
1.15~1.55	≦0.030	≦0.030	≦0.25	≦0.35	−	≦0.30	−	D079
1.45~2.05	−	−	−	−	−	−	−	
1.30~1.70	≦0.030	≦0.030	≦0.25	≦0.35	−	≦0.30	−	D080
1.45~2.05	−	−	−	−	−	−	−	
1.30~1.70	≦0.030	≦0.030	≦0.25	0.35~0.70	−	≦0.30	−	D081
1.15~1.55	≦0.030	≦0.030	≦0.25	0.35~0.70	−	≦0.30	−	D082
1.30~1.70	≦0.030	≦0.030	≦0.25	≦0.35~0.70	−	≦0.30	−	D083
0.55~0.90	≦0.030	≦0.030	≦0.25	0.85~1.25	−	≦0.30	−	D084
0.60~1.00	≦0.030	≦0.030	≦0.25	0.85~1.20	−	≦0.30	−	D085
0.60~1.00	≦0.035	≦0.040	−	0.60~1.00	−	−	−	
0.55~0.90	≦0.030	≦0.030	≦0.25	0.85~1.25	−	≦0.30	−	D086
0.60~1.10	≦0.035	≦0.040	−	0.75~1.20	−	−	−	
0.50~0.90	≦0.035	≦0.040	−	0.65~1.10	−	−	−	
0.50~0.90	≦0.050	0.025~0.050	−	0.80~1.25	−	−	−	
0.60~0.90	≦0.035	≦0.035	−	0.90~1.20	−	−		
0.55~0.90	≦0.030	≦0.030	≦0.25	0.85~1.20	−	≦0.30	−	D087
0.50~0.90	≦0.030	≦0.030	≦0.25	0.85~1.25	−	−	−	
0.50~0.90	≦0.050	0.025~0.050	−	0.50~1.25	−	−	−	
0.60~0.90	≦0.035	≦0.035	−	0.90~1.20	−	−	−	
0.55~0.90	≦0.030	≦0.030	≦0.25	0.85~1.25	−	≦0.30	−	D088
0.60~1.00	≦0.035	≦0.040	−	0.60~1.00	−	−	−	
0.50~0.90	≦0.050	0.025~0.050	−	0.80~1.25	−	−	−	
0.50~0.80	≦0.035	≦0.0355	−	0.90~1.25	−	−	−	
0.35~0.90	≦0.030	≦0.030	≦0.25	0.85~1.25	0.15~0.35	≦0.30	−	D089
0.55~0.90	≦0.030	≦0.030	≦0.25	0.85~1.25	0.15~0.35	≦0.30	−	D090
0.55~0.90	≦0.030	≦0.030	≦0.25	0.85~1.25	0.15~0.35	≦0.30	−	D091
0.60~0.90	≦0.035	≦0.040	≦0.40	0.85~1.15	0.15~0.25	−		
0.55~0.90	≦0.030	≦0.030	≦0.25	0.85~1.25	0.15~0.35	−	−	D092
0.60~1.00	≦0.030	≦0.040	−	0.75~1.20	0.15~0.25	−	−	
0.60~1.00	−	−	−	0.75~1.20	0.15~0.25	−	−	
0.65~1.05	≦0.050	0.025~0.050	−	0.80~1.25	0.15~0.25	−	−	
0.50~0.80	≦0.035	≦0.035	−	0.90~1.20	0.15~0.30	−	−	
0.50~0.90	≦0.030	≦0.030	≦0.25	0.85~1.25	0.15~0.35	≦0.30	−	D093

JIS	AISI	BS	DIN	NF	ГОСТ		
G4051	MN06－ 210, 217, 244	970 Part 1	17200 17210 17211	－	－	C	Si
	4140H					0.37～0.44	0.15～0.30
	4142H					0.39～0.46	0.15～0.30
		708H42	Not listed in BS970 Part 1-91			0.39～0.46	－
			42CrMo4	Replaced by EN 10083-1:91,Grade 42Cr Mo 4		0.38～0.45	0.15～0.40
SCM445H						0.42～0.49	0.15～0.35
	4145H					0.42～0.49	0.15～0.30
	4145H					0.44～0.51	0.15～0.30
		708H45	Not listed in BS970 Part 1-91			0.42～0.49	－
SCM822H						0.19～0.25	0.15～0.35
Ni－Cr－Steel							
SNC415H						0.11～0.18	0.15～0.35
SNC631H						0.26～0.35	0.15～0.35
SNC815H						0.11～0.18	0.15～0.35
		655H13				0.10～0.16	0.10～0.35
Ni－Cr－Mo－Steel							
SNCM220H						0.17～0.23	0.15～0.35
	8617H					0.14～0.20	0.15～0.35
	8620H					0.17～0.23	0.15～0.30
	8622H					0.19～0.25	0.15～0.30
		805H20				0.17～0.23	0.10～0.35
SNCM420H						0.17～0.23	0.15～0.35
	4320H					0.17～0.23	0.15～0.35

Chemical Composition 化学成分 (%)								Index Number 索引番号
Mn	P	S	Ni	Cr	Mo	Cu	Others	
0.65～1.10	≦0.035	≦0.040	–	0.75～1.20	0.15～0.25	–	–	D093
0.65～1.10	≦0.035	≦0.040	–	0.75～1.20	0.15～0.25	–	–	
0.65～1.05	≦0.050	0.025～0.050	–	0.80～1.25	0.15～0.25	–	–	
0.60～0.80	≦0.035	≦0.035	–	0.90～1.20	0.15～0.30	–		
0.55～0.90	≦0.030	≦0.030	≦0.25	0.85～1.25	0.15～0.35	≦0.30	–	D094
0.65～1.10	≦0.035	≦0.040	–	0.75～1.20	0.15～0.25	–	–	
0.65～1.10	≦0.035	≦0.040	–	0.75～1.20	0.15～0.25	–	–	
0.65～1.05	≦0.050	0.025～0.050	–	0.80～1.25	0.15～0.25	–	–	
0.55～0.90	≦0.030	≦0.030	≦0.25	0.85～1.25	0.35～0.45	≦0.30	–	D095
0.30～0.70	≦0.030	≦0.030	2.95～3.50	0.65～1.05	–	–	–	D096
0.35～0.60	≦0.030	≦0.030	2.95～3.50	0.65～1.05	–	–	–	D097
0.30～0.70	≦0.030	≦0.030	2.95～3.50	0.65～1.05	–	–	–	D098
0.35～0.60	≦0.035	≦0.040	3.00～3.75	0.70～1.00	–	–	–	
0.60～0.95	≦0.030	≦0.030	0.35～0.75	0.35～0.65	0.15～0.30	–	–	D099
0.60～0.95	≦0.035	≦0.040	0.35～0.75	0.35～0.65	0.15～0.25	–	–	
0.60～0.95	≦0.035	≦0.040	0.35～0.75	0.35～0.65	0.15～0.25	–	–	
0.60～0.95	≦0.035	≦0.040	0.35～0.75	0.35～0.65	0.15～0.25	–	–	
0.60～0.95	≦0.035	≦0.040	–	–	–	–	–	
0.40～0.70	≦0.030	≦0.030	1.55～2.00	0.35～0.65	0.15～0.30	–	–	D100
0.40～0.70	≦0.035	≦0.040	1.55～2.00	0.35～0.65	0.15～0.30	–	–	

INDEX OF CHAPTER IV

STEELS
FOR MACHINE STRUCTURAL USE

第4部　機械構造用鋼

索　　引

"Index No." stands for the
"Index No." in the last column
of each page.

「索引番号」は本文各頁の右端の
「索引番号」の欄の数字をあらわす。

Standard No. 規格番号	Grade グレード	Index No. 索引番号	Standard No. 規格番号	Grade グレード	Index No. 索引番号
〔JIS〕			JIS-G4104	SCr415	D049
JIS-G4051	S09CK	D028		SCr420	D050
	S10C	D002		SCr430	D051
	S12C	D003		SCr435	D052
	S15C	D004		SCr440	D053
	S15CK	D029		SCr445	D054
	S17C	D006			
	S20C	D007	JIS-G4105	SCM415	D055
	S20CK	D030		SCM418	D056
	S22C	D008		SCM420	D057
	S25C	D009		SCM421	D058
	S28C	D0011		SCM430	D059
	S30C	D012		SCM432	D060
	S33C	D013		SCM435	D061
	S35C	D014		SCM440	D062
	S38C	D016		SCM445	D063
	S40C	D017		SCM822	D064
	S43C	D019			
	S45C	D020	JIS-G4106	SMn420	D065
	S48C	D022		SMn433	D066
	S50C	D023		SMn438	D067
	S53C	D024		SMn443	D068
	S55C	D025		SMnC420	D069
	S58C	D026		SMnC443	D070
JIS-G4052	SCM415H	D089	JIS-G4202	SACM645	D071
	SCM418H	D090			
	SCM420H	D091	JIS-G4107	SNB5	D072
	SCM435H	D092		SNB7	D073
	SCM440H	D093		SNB16	D074
	SCM445H	D094			
	SCM822H	D095	JIS-G4108	SNB21	D075
				SNB22	D075
JIS-G4102	SNC236	D033		SNB23	D076
	SNC415	D034		SNB24	D077
	SNC631	D035			
	SNC815	D036	〔ASTM〕		
	SNC836	D037	ASTM-A193	B5	D072
				B7	D073
JIS-G4103	SNCM220	D038		B16	D074
	SNCM240	D039			
	SNCM415	D040	ASTM-A540	B22	D075
	SNCM420	D041		B23	D076
	SNCM431	D042		B24	D077
	SNCM439	D043			
	SNCM447	D044	〔AISI〕		
	SNCM616	D045	AISI-MN06-	1008	D001
	SNCM625	D046	210, 211, 214,	1010	D002
	SNCM630	D047	215, 217, 224	1012	D003
	SNCM815	D048		1015	D004

Standard No. 規 格 番 号	Grade グレード	Index No. 索引番号
	1016	D005
	1017	D006
	1019	D031
	1020	D007
	1021	D031
	1022	D031
	1023	D008
	1025	D009
	1026	D010
	1029	D011
	1030	D012
	1035	D014
	1037	D015
	1038	D016
	1039	D031
	1040	D017
	1042	D018
	1043	D019
	1045	D020
	1046	D021
	1049	D031
	1050	D023
	1053	D024
	1055	D025
	1060	D026
	1070	D027
	1078	D031
	1080	D031
	1084	D031
	1090	D031
	1095	D031
	1330H	D079
	1335H	D080
	1522	D065
	1541	D067
	4135H	D092
	4137H	D092
	4140H	D093
	4142H	D093
	4145H	D094
	4147H	D094
	4320	D041
	4320H	D100
	4340	D043
	51201	D050
	5120H	D085
	5130	D051
	5130H	D086
	5132	D051
	5132H	D086
	5135	D052
	5135H	D087
	5140	D053
	5140H	D089
	5145H	D094
	5147	D054
	5147H	D094
	8615	D038
	8617	D038
	8617H	D093
	8620	D038
	8620H	D093
	8622	D038
	8622H	D093
	8637	D039
	8640	D039
〔BS〕		
BS-970 Part 1	040A04	D001
	040A10	D002
	040A12	D003
	055M15	D005
	060A30	D012
	060M30	D012
	060A40	D017
	060A45	D020
	060A47	D022
	060A57	D026
	060A72	D027
	060A78	D031
	060A81	D031
	070M20	D007
	070M26	D010
	070M55	D025
	080A30	D012
	080M30	D012
	080A37	D015
	080A40	D017
	080M40	D017
	080A42	D018
	080M46	D021
	080M50	D023
	080A57	D026
	120M19	D065
	120M36	D066
	135M44	D068
	150M36	D067
	530A30	D051
	530A32	D051
	530A36	D052

Standard No. 規格番号	Grade グレード	Index No. 索引番号	Standard No. 規格番号	Grade グレード	Index No. 索引番号
	530A40	D058		CK30	D012
	530H32	D080		CK35	D014
	530H36	D087		CK40	D017
	530H40	D088		CK45	D020
	655H13	D098		CK50	D023
	655M13	D036		CK55	D025
	708A30	D059		CK60	D026
	708M30	D057			
	708A37	D061	DIN-17240	42CrMo4	D073
	708A40	D062		40CrMoV47	D074
	708A42	D062		42CrMo4	D075
	708H45	D094			
	708A47	D063	〔NF〕		
	708H20	D091	NF-A35-551~554	14NC11	D036
	708H37	D092		18CD4	D056
	708H42	D093		20NCD2	D038
	708H45	D094		20M5	D065
	708H20	D057		30CD4	D059
	709A37	D061		34CD4	D061
	709A40	D062		35M5	D066
	805H20	D099		35NC6	D033
	905M39	D071		40CAD6.12	D071
				40M6	D067
BS-1506	708A40	D073		42CDV4	D074
	708A42	D073		XC10	D002, D028
				XC12	D003, D029
〔DIN〕				XC18	D006, D030
DIN-17200	34Cr4	D051, D080		XC18S	D006
DIN-17210	37CrMo4	D086		XC25	D009
DIN-17211	37Cr4	D052, D087		XC32	D013
	40CrMoV47	D074		XC38	D016
	41Cr4	D058, D089		XC38H1	D016
	41CrAlMo7	D071		XC38H2	D016
	42CrMo4	D073, D075, D093		XC42H1	D018
	C10	D002		XC42H2	D018
	C15	D004		XC45	D020
	C20	D007		XC48	D022
	C22	D008		XC48H1	D022
	C25	D009		XC48H2	D022
	C30	D012		XC50	D023
	C35	D014		XC55	D024
	C40	D017		XC55H1	D025
	C45	D020		XC60	D026
	C55	D025			
	C60	D026	NF-A35-558	42CDV4	D074
	CK10	D002, D028			
	CK15	D004, D029	〔ГOCT〕		
	CK20	D007	ГOCT-1050	10	D001, D028
	CK22	D008, D030		15	D004, D029
	CK25	D009		20	D007, D030

Standard No. 規格番号	Grade グレード	Index No. 索引番号	Standard No. 規格番号	Grade グレード	INdex No. 索引番号
	25	D009			
	30	D012			
	35	D014			
	40	D018			
	45	D020			
	50	D023			
	58	D026			
	60	D026			
ΓOCT-4543	15X	D049			
	15XA	D049			
	20X	D050			
	30X	D051			
	30XM	D059			
	30XMA	D059			
	35Γ2	D066			
	35X	D052			
	35XM	D061			
	40Γ2	D067			
	40X	D058			
	45Γ2	D068			
	45X	D053			

CHAPTER V

STEELS FOR SPECIAL PURPOSES

第 5 部

特 殊 用 途 鋼

CHAPTER V CONTENTS 第5部 目次

1. STANDARDS LIST 収録各国規格リスト

1-1 JIS (Japan 日 本)

G3320-91 Protected Stainless Steel Sheets
塗装ステンレス鋼板

G4303-91 Stainless Steel Bars
ステンレス鋼棒

G4304-91 Hot Rolled Stainless Sheets and Plates
熱間圧延ステンレス鋼板

G4305-91 Cold Rolled Stainless Sheets and Plates
冷間圧延ステンレス鋼帯

G4306-88 Hot Rolled Stainless Steel Strips
熱間圧延ステンレス鋼帯

G4307-87 Cold Rolled Stainless Steel Strips
冷間圧延ステンレス鋼帯

G4308-91 Stainless Steel Wire Rods
ステンレス鋼線材

G4309-94 Stainless Steel Wires
ステンレス鋼線

G4311-91 Heat-Resisting Steel Bars
耐熱鋼棒

G4312-91 Heat-Resisting Steel Sheets and Plates
耐熱鋼板

G4313-88 Cold Rolled Stainless Steel Strips for Springs
ばね用ステンレス鋼帯

G4314-94 Stainless Steel Wires for Spring
ばね用ステンレス鋼線

G4315-94 Stainless Steel Wires for Cold Heading and Cold Forging
冷間圧造用ステンレス鋼線

G4317-91 Hot Rolled Stainless Steel Equal Leg Angles
熱間圧延ステンレス鋼等辺山形鋼

G4318-91 Cold Finished Stainless Steel Bars
冷間仕上ステンレス鋼棒

G4320-91 Cold-rolled Formed Stainless Steel Equal Leg Angles
冷間成形ステンレス鋼等辺山形鋼

G4401-83 Carbon Tool Steels
炭素工具鋼鋼材

G4403-83 High Speed Tool Steels
高速度工具鋼鋼材

G4404-83 Alloy Tool Steels
合金工具鋼鋼材

G4410-84 Hollow Drill Steels
中空鋼鋼材

G4801-84 Spring Steels
ばね鋼鋼材

G4804-83 Free Cutting Carbon Steels
硫黄及び硫黄複合快削鋼鋼材

G4805-90 High Carbon Chromium Bearing Steels
高炭素クロム軸受鋼鋼材

G4901-91 Corrosion-Resisting and Heat-Resisting Superalloy Bars
耐食耐熱超合金棒

G4902-91 Corrosion-Resisting and Heat-Resisting Superalloy Sheets and Plates
耐食耐熱超合金板

1-2 ASTM (U.S.A. アメリカ)

A167-94a Stainless and Heat-Resisting Chromium-Nickel Steel Plate, Sheet and Strip
ステンレス及び耐熱Cr-Ni 鋼板, 薄板, 鋼帯

A176-94 Stainless and Heat-Resisting Chromium Steel Plate, Sheet and Strip
ステンレス及び耐熱Cr鋼板, 薄板, 鋼帯

A240-94a Heat-Resisting Chromium and Chromium-Nickel Stainless Steel, Plate, Sheet and Strip for Fusion-Welded Unfired Pressure Vessels
圧力容器用耐熱Cr及びCr-Ni ステンレス鋼板, 薄板, 鋼帯

A276-95 Stainless and Heating Steel Bars and Shapes
ステンレス及び耐熱鋼棒及び型鋼

A295-94 High Carbon Ball and Roller Bearing Steel
球及び転がり軸受用高炭素鋼鋼材

A313-95 Chromium-Nickel Stainless and Heat-Resisting Steel Spring Wire
Cr-Ni ステンレス及び耐熱鋼ばね線

A314-95 Stainless and Heat-Resisting Steel Billets and Bars for Forging
鍛造用ステンレス及び耐熱鋼棒及び鋼塊

A478-95a Chromium-Nickel Stainless and Heat-Resisting Steel Weaving Wire
ステンレス及び耐熱Cr-Ni 鋼編み用線

A479-95 Stainless and Heat-Resisting Steel Wires, Bars and Shapes for Use in Boilers and Other Pressure Vessels
ボイラ・圧力容器用ステンレス及び耐熱鋼棒及び型鋼

A485-94 High Hardenability Bearing Steels
高硬度軸受鋼鋼材

A492-95　Stainless and Heat-Resisting Steel Rope Wire
ステンレス及び耐熱鋼ロープ用線材

A493-95　Stainless and Heat-Resisting Steel for Cold Heading and Cold Forging-Bar and Wire
冷間圧造・冷間鍛造用ステンレス及び耐熱鋼鋼棒及び線材

A535-85　Special Quality Ball and Roller Bearing Steel
特殊品質球及び転がり軸受用鋼材

A580-95　Stainless and Heat-Resisting Steel Wire
ステンレス及び耐熱鋼鋼線材

A581-95a　Free-Machining Stainless and Heat-Resisting Steel Wire
快削ステンレス及び耐熱鋼鋼線

A582-95a　Free-Machining Stainless and Heat-Resisting Steel Bars, Hot-Rolled or Cold-Finished
熱間圧延または冷間仕上快削ステンレス及び耐熱鋼鋼棒

A600-92a　High Speed Tool Steels
高速度工具鋼鋼材

A666-94a　Austenitic Stainless Steel, Sheet, Strip, Plate and Flat Bar for Structural Applications
構造用オーステナイト系ステンレス鋼薄板, 鋼帯, 鋼板及び扁平鋼棒

A681-94　Alloy Tool Steels
合金工具鋼鋼材

A686-92　Carbon Tool Steels
炭素工具鋼鋼材

1-3 AISI（U.S.A.　アメリカ）

MN06/222-74 Stainless and Heat Resisting Steels
ステンレス鋼及び耐熱鋼鋼材

MN06/224-75 Wires and Rods, Alloy Steel
合金鋼線材及び鋼棒

1-4 BS（U.K.　イギリス）

970-Part1-91　General Inspection and Testing Procedures and Specific Requirements for Carbon, Carbon Manganese and Stainless Steels
炭素鋼, 炭素マンガン鋼, ステンレス鋼鋼材

970-Part4-70　Stainless, Heat Resisting and Valve Steels
ステンレス鋼, 耐熱鋼及びバルブ鋼

970-Part5-72　Carbon and Alloy Spring Steels for the Manufacture of Hot-Formed Springs
熱間成形炭素鋼及び合金鋼ばね用鋼鋼材

1449-Part2-83　Stainless and Heat-Resisting Plate, Sheet and Strip
ステンレス及び耐熱鋼板シート

4659-89 Tool Steels
工具鋼鋼材

1-5 DIN (West Germany 西ドイツ)

1651-88 Free-Machining Steels
快削鋼

17221-88 Hot-Rolled Spring Steels for Hardening and Tempering
調質ばね用熱間圧延鋼材

17224-82 Stainless Steel Wire and Strip for Spring
ばね用ステンレス鋼線及び帯

17230-80 Bearing Steels
軸受鋼

17350-80 Tool Steels
工具鋼

17440-85 Stainless Steels
ステンレス鋼鋼材

17441-85 Stainless Steels
ステンレス鋼鋼材

1-6 NF (France フランス)

A35-565-94 Bearing Steels
軸受鋼鋼材

A35-571-84 Spring Steels for Hot Forming
熱間成形用ばね鋼鋼材

A35-573-81 Stainless Steels for General Purposes, II
一般用ステンレス鋼種類II

A35-574-81 Stainless Steels for General Purposes, III
一般用ステンレス鋼種類III

A35-575-81 Stainless Steels for General Purposes
一般用ステンレス鋼

A35-576-81 Stainless Steels, Chemical and Mechanical Characteristics
ステンレス鋼化学的機械的性質

A35-578-91 Heat-Resisting Steels
耐熱鋼

A35-581-84 Precipitation-Hardened Stainless Steels
析出硬化ステンレス鋼

A35-582-79 Nitrided Stainless Steels
窒素添加ステンレス鋼

A35-584-81 Corrosion-Resisting Stainless Steels
特殊腐食用ステンレス鋼

A35-585-91 Stainless Steels
ステンレス鋼

A35-586-81 Codification of Stainless Steels
ステンレス鋼コード

A35-590-92 Tool Steels
工具鋼

1-7 ГОСТ（U.S.S.R. ソ 連）

1435-74 Carbon Tool Steels
炭素工具鋼

5632-72 Highly Alloyed Steels and Corrosion-Proof, Heat-Resisting Alloys
耐食耐熱高合金鋼

5950-70 Alloy Tool Steels
合金工具鋼鋼材

14959-79 Spring-Type Steels
ばね鋼鋼材

20072-74 Heat-Resisting Steels
耐熱鋼

21022-75 Chromic Steels for Precise Bearing
精密軸受用クロム鋼鋼材

2. Stainless Steels　ステンレス鋼

JIS	AISI	BS	DIN	NF	ГОСТ		
G4303～4309 4313～4320	MN06－222	970 Part1 1449 Part2	17224 17440 17441	A35－573 ～584 586	5632	C	Si
SUS201						≦0.15	≦1.00
	201					≦0.15	≦1.00
SUS202						≦0.15	≦1.00
	202					≦0.15	≦1.00
		284S16				≦0.07	≦1.00
					12X17Г9AH4	≦0.12	≦0.80
SUS301						≦0.15	≦1.00
SUS301J1						0.08～0.12	≦1.00
	301					≦0.15	≦1.00
		301S21				≦0.15	≦1.00
			X12CrNi177			≦0.12	≦1.5
				Z12CN17.07		0.08～0.15	≦1.0
SUS302						≦0.15	≦1.00
SUS302B						≦0.15	2.00～3.00
	302					≦0.15	≦1.00
	302B					≦0.15	2.00～3.00
		302S31				≦0.12	≦1.00
			X10CrNiS189			≦0.12	－
				Z10CN18.09		≦0.12	≦0.75
SUS303						≦0.15	≦1.00
SUS303Se						≦0.15	≦1.00
	303					≦0.15	≦1.00
	303Se					≦0.15	≦1.00
		303S31				≦0.12	≦1.00
		303S42				≦0.12	≦1.00
			X10CrNiS189			≦0.12	≦1.00
				Z10CNF18.09		≦0.12	≦1.00
					12X18H10E	≦0.12	≦0.80
SUS304						≦0.08	≦1.00
SUS304L						≦0.03	≦1.00
SUS304N1						≦0.08	≦1.00
SUS304N2						≦0.08	≦1.00
304LN						≦0.03	≦1.00
	304					≦0.08	≦1.00
	304L					≦0.03	≦1.00
	304N					≦0.08	≦1.00
	XM21(ASTM)					≦0.08	≦1.00
	304LN(ASTM)					≦0.03	≦1.00

Chemical Composition 化学成分 (%)								Index Number 索引番号
Mn	P	S	Ni	Cr	Mo	N	Others	
5.50～7.50	≦0.060	≦0.030	3.50～5.50	16.00～18.00	–	≦0.25	–	E001
6.50～7.50	≦0.060	≦0.030	3.50～5.50	16.00～18.00	–	≦0.25	–	
7.50～10.00	≦0.060	≦0.030	4.00～6.00	17.00～19.00	–	≦0.25	–	E002
7.50～10.00	≦0.060	≦0.030	4.00～6.00	17.00～19.00	–	≦0.25	–	
7.00～10.00	≦0.060	≦0.030	4.00～6.50	16.50～18.50	–	0.15～0.25	–	
8.0 ～10.5	≦0.035	≦0.035	3.5～4.5	16.00～18.00	–	0.15～0.25	–	
≦2.00	≦0.045	≦0.030	6.00～8.00	16.00～18.00	–	–	–	E003
≦2.00	≦0.045	≦0.035	7.00～9.00	16.00～18.00	–	–	–	
≦2.00	≦0.045	≦0.030	6.00～8.00	16.00～18.00	–	–	–	
≦2.00	≦0.045	≦0.030	6.00～8.00	16.0～18.0	–	–	–	
≦2.00	–	–	6.0～9.0	16.0～18.0	≦0.8	–	–	
≦2.00	≦0.040	≦0.030	6.0－8.0	16.0～18.0	–	–	–	
≦2.00	≦0.045	≦0.030	8.00～10.00	17.00～19.00	–	–	–	E004
≦2.00	≦0.045	≦0.030	8.00～10.00	17.00～19.00	–	–	–	
≦2.00	≦0.045	≦0.030	8.00～10.00	17.00～19.00	–	–	–	
≦2.00	≦0.045	≦0.030	8.0～10.0	17.0～19.0	–	–	–	
≦2.00	≦0.045	≦0.030	8.0～10.0	17.0～19.0	–	–	–	
–	≦0.060	0.15～0.35	8.0～10.0	17.0～19.0	–	–	–	
≦2.00	≦0.040	≦0.030	7.50～9.50	17.0～19.0	–	–	–	
≦2.00	≦0.200	≦0.150	8.00～10.00	17.00～19.00	(≦0.60)	–	–	E005
≦2.00	≦0.200	≦0.060	8.00～10.00	17.00～19.00	–	–	Se≧0.15	
≦2.00	≦0.200	≦0.150	8.00～10.00	17.00～19.00	(≦0.60)	–	–	
≦2.00	≦0.200	≦0.060	8.00～10.00	17.00～19.00	–	–	Se≧0.15	
≦2.00	≦0.060	0.15～0.35	8.0～10.0	17.00～19.00	(1.00)	–	–	
≦2.00	≦0.060	≦0.060	8.0～10.0	17.0～19.0	(1.00)	–		
≦2.00	≦0.060	0.15～0.35	8.0～10.0	17.0～19.0	–	–	–	
≦2.00	≦0.05	≦0.06	8.0～11.0	17.0～19.0	–	–	–	
≦2.00	–	–	9.0～11.0	17.0～19.0	–	–	–	
≦2.00	≦0.045	≦0.030	8.00～10.50	18.00～20.00	–	–	–	E006
≦2.00	≦0.045	≦0.030	9.00～13.00	18.00～20.00	–	–	–	
≦2.50	≦0.045	≦0.030	7.00～10.50	18.00～20.00	–	0.10～0.25	–	
≦2.50	≦0.045	≦0.030	7.50～10.50	18.00～20.00	–	0.15～0.30		
≦2.00	≦0.045	≦0.030	8.50～11.5	17.00～19.0	–	–	–	
≦2.00	≦0 045	≦0 030	8.00～10 50	18 00～20.00	–	–	–	
≦2.00	≦0.045	≦0.030	8.00～12.00	18.00～20.00	–	–	–	
≦2.00	≦0.045	≦0.030	8.00～10.50	18.00～20.00	–	0.10～0.16	–	
≦2.00	≦0.045	≦0.030	7.00～10.50	18.00～20.00	–	0.10～0.30	–	
≦2.00	≦0.045	≦0.030	8.00～12.00	18.00～20.00	–	0.10～0.16	–	

JIS	AISI	BS	DIN	NF	ГОСТ		
G4303～4309 4313～4320	MN06－222	970 Part 1 1449 Part2	17224 17440 17441	A35－573 ～584 586	5632	C	Si
		304S11				≦0.03	≦1.00
		304S15				≦0.06	≦1.0
		304S31				≦0.07	≦1.0
			X5CrNi1810			≦0.03	≦1.00
			X2CrNi1911			≦0.03	≦1.00
				Z6CN18.09		≦0.07	≦1.00
				Z2CN18.10		≦0.03	≦1.00
				Z5CN18.09AZ		≦0.06	≦1.00
				Z2CN18.10AZ		≦0.05	≦1.00
					08X18H10	≦0.04	≦0.80
					03X18H11	≦0.04	≦0.80
SUS305						≦0.12	≦1.00
SUS305J1						≦0.08	≦1.00
	305					≦0.12	≦1.00
		305S19				≦0.10	≦1.00
			X5CrNi1911			≦0.07	－
				Z8CN19-12		≦0.08	≦1.0
SUS309S						≦0.08	≦1.00
	309S					≦0.08	≦1.00
				Z10CN24-13		≦0.100	≦1.00
SUS310S						≦0.08	≦1.50
	310S					≦0.08	≦1.50
				Z12CN25-20		≦0.08	≦1.20
					10X23H18	≦0.10	≦1.0
					20X23H18	≦0.20	≦1.0
SUS316						≦0.08	≦1.00
SUS316J1						≦0.08	≦1.00
SUS316L						≦0.03	≦1.00
SUS316J1L						≦0.03	≦1.00
SUS316N						≦0.08	≦1.00
SUS316LN						≦0.03	≦1.00
	316					≦0.08	≦1.00
	316L					≦0.03	≦1.00
	316N					≦0.08	≦1.00
	316LN					≦0.03	≦1.00
		316S33				≦0.07	≦1.0
		316S31				≦0.07	≦1.00
		316S11				≦0.03	≦1.00
		316S13				≦0.030	≦1.0
			X5CrNiMo 17122			≦0.07	－
			X5CrNiMo1810			≦0.07	≦1.0

Chemical Composition 化 学 成 分 (%)								Index Number 索引番号
Mn	P	S	Ni	Cr	Mo	N	Others	
≦2.00	≦0.045	≦0.030	9.00～12.0	17.0～19.0	–	–	–	E006
≦2.0	≦0.045	≦0.030	8.0～11.0	17.5～19.0	–	–	–	
≦2.0	≦0.045	≦0.030	8.0～11.0	17.0～19.0	–	–	–	
≦2.00	≦0.045	≦0.030	8.00～12.0	17.00～19.0	–	–	–	
≦2.00	≦0.045	≦0.030	9.00～12.0	17.00～19.00	–	–	–	
≦2.00	≦0.040	≦0.030	8～11	17～19	–	–	–	
≦2.00	≦0.040	≦0.030	9～11	17～19	–	–	–	
≦2.00	≦0.040	≦0.030	8～11	17～20	–	0.12～0.20	–	
≦2.00	≦0.040	≦0.030	9～12	17～19	–	0.12～0.20	–	
≦2.00	–	–	9.00～11.00	17.00～19.0	–	–	–	
≦2.00	–	–	10.00～12.0	17.00～19.0	–	–	–	
≦2.00	≦0.045	≦0.030	10.50～13.00	17.00～19.00	–	–	–	E007
≦2.00	≦0.045	≦0.030	11.00～13.50	16.50～19.00	–	–	–	
≦2.00	≦0.045	≦0.030	10.50～13.00	17.00～19.00	–	–	–	
≦2.00	≦0.045	≦0.030	11.00～13.00	17.00～19.00	–	–	–	
–	–	–	10.5～12.0	17.00～19.00	–	–	–	
≦2.00	≦0.040	≦0.030	11～13	17～19	–	–	–	
≦2.00	≦0.045	≦0.030	12.00～15.00	22.00～24.00	–	–	–	E008
≦2.00	≦0.045	≦0.030	12.00～15.00	22.00～24.00	–	–	–	
≦2.0	≦0.040	≦0.030	12～14	22～24	–	–	–	
≦2.00	≦0.045	≦0.030	19.00～22.00	24.00～26.00	–	–	–	E009
≦2.00	≦0.045	≦0.030	19.00～22.00	24.00～26.00	–	–	–	
≦2.00	≦0.040	≦0.030	19.00～21.00	24.00～26.00	–	–	–	
≦2.00	≦0.035	≦0.020	17.0～20.0	22.0～25.0	–	–	–	
≦2.0	≦0.035	≦0.020	17.0～20.0	22.0～25.0	–	–	–	
≦2.00	≦0.045	≦0.030	10.00～14.00	16.00～18.00	2.00～3.00	–	–	E010
≦2.00	≦0.045	≦0.030	10.00～14.00	17.00～19.00	1.20～2.75	–	Cu1.00～2.50	
≦2.00	≦0.045	≦0.030	12.00～15.00	16.00～18.00	2.00～3.00	–	–	
≦2.00	≦0.045	≦0.030	12.00～13.00	17.00～19.00	1.20～2.75	–	Cu1.00～2.50	
≦2.00	≦0.045	≦0.030	10.00～14.00	16.00～18.00	2.00～3.00	0.12～0.22	–	
≦2.00	≦0.045	≦0.030	10.00～14.00	16.00～18.00	2.00～3.00	0.12～0.22	–	
≦2.00	≦0.045	≦0.030	10.00～14.00	16.00～18.00	2.00～3.00	–	–	
≦2.00	≦0.045	≦0.030	10.00～14.00	16.00～18.00	1.20～2.75	–	–	
≦2.00	≦0.045	≦0.030	10.00～14.00	16.00～18.00	2.00～3.00	0.16～0.16	–	
≦2.00	≦0.045	≦0.030	10.50～14.50	16.50～18.50	2.00～3.00	0.12～0.22	–	
≦2.0	≦0.045	≦0.030	11.0～14.0	16.5～18.5	2.50～3.00	–	–	
≦2.00	≦0.045	≦0.030	10.5～13.5	16.5～18.5	2.00～2.50	–	–	
≦2.00	≦0.045	≦0.030	11.0～14.0	16.5～18.5	2.00～2.50	–	–	
≦2.0	≦0.045	≦0.030	11.5～14.0	16.5～18.5	2.50～3.00	–	–	
–	–	–	10.5～13.5	16.5～18.5	2.0～2.5	–	–	
≦2.0	–	–	10.5～13.5	16.5～18.5	2.0～2.5	–	–	

J I S	A I S I	B S	D I N	N F	ГОСТ		
G4303～4309 4313～4320	MN06－222	970 Part1 1449 Part2	17224 17440 17441	A35－573～584 586	5632	C	S i
			X2CrNiMo 17132			≦0.03	－
				Z6CND17.11		≦0.07	≦1.00
				Z6CND17.12		≦0.03	≦1.00
				Z2CND17.12		≦0.03	≦1.00
					03X17H14M2	≦0.03	≦0.8
SUS317						≦0.08	≦1.00
SUS317L						≦0.03	≦1.00
SUS317J1						≦0.04	≦1.00
	317					≦0.08	≦1.00
	317L					≦0.03	≦1.00
		317S16				≦0.06	≦1.00
		317S12				≦0.03	≦1.00
SUS321						≦0.08	≦1.00
	321					≦0.08	≦1.00
		321S31				≦0.08	≦1.00
			X6CrNiTi1810			≦0.08	≦1.00
				Z6CNT18.10		≦0.08	≦1.00
					08X18H10T	≦0.08	≦0.80
SUS329J1						≦0.08	≦1.00
SUS329J2						≦0.030	≦1.00
	329					≦0.10	≦1.00
SUS347						≦0.08	≦1.00
	347					≦0.08	≦1.00
		347S31				≦0.08	≦1.00
			X6CrNiNb1810			≦0.08	≦1.00
				Z6CNNb18.10		≦0.08	≦1.00
					08X18H125	≦0.08	≦0.80
SUS384						≦0.08	≦1.00
	384					≦0.08	≦1.00
				Z6NC18.16		≦0.08	≦1.00
SUS403						≦0.15	≦0.50
	403					≦0.15	≦0.50
		403S17				≦0.08	≦1.00
			X6Cr13			≦0.08	－
				Z6C13		≦0.08	≦1.00
					08X13	≦0.08	≦0.8
					12X13	0.09～0.15	≦0.8

Chemical Composition 化学成分 (%)								Index Number 索引番号
Mn	P	S	Ni	Cr	Mo	N	Others	
–	–	–	11.0～14.0	16.5～18.5	2.0～2.5	–	–	E010
≦2.00	≦0.040	≦0.030	10.0～12.5	16.0～18.0	2.0～2.5	–	–	
≦2.00	≦0.040	≦0.030	10.5～13.0	16.0～18.0	2.0～2.5	0.10～0.20	–	
≦2.00	≦0.040	≦0.030	10.5～13.0	16.0～18.0	2.0～2.5	–	–	
1.0～2.0	≦0.035	≦0.020	13.0～15.0	16.0～18.0	2.0～2.3	–	–	
≦2.00	≦0.045	≦0.030	11.00～15.00	18.00～20.00	3.00～4.00	–	–	E011
≦2.00	≦0.045	≦0.030	11.00～15.00	18.00～20.00	3.00～4.00	–	–	
≦2.50	≦0.045	≦0.030	15.00～17.00	16.00～19.00	3.00～6.00	–	–	
≦2.00	≦0.045	≦0.030	11.50～15.00	18.00～20.00	3.00～4.00	–	–	
≦2.00	≦0.045	≦0.030	11.00～15.00	18.00～20.00	3.00～4.00	–	–	
≦2.00	≦0.045	≦0.030	12.0～15.0	17.5～19.0	3.0～4.0	–	–	
≦2.00	≦0.045	≦0.030	14.0～17.0	17.5～19.0	3.0～4.0	–	–	
≦2.00	≦0.045	≦0.030	9.00～13.00	17.00～19.00	–	–	Ti≧5C	E012
≦2.00	≦0.045	≦0.030	9.00～12.00	17.00～19.00	–	–	Ti≧5C	
≦2.00	≦0.045	≦0.030	9.0～12.0	17.0～19.0	–	–	Ti=5C～0.8	
≦2.00	≦0.045	≦0.030	9.0～11.0	17.0～19.0	–	–	Ti=5C～0.8	
≦2.00	≦0.045	≦0.030	9.0～11.0	17.0～19.0	–	–	Ti=5C～0.6	
≦2.00	≦0.035	≦0.020	9.0～11.0	17.0～19.0	–	–	Ti=5C～0.8	
≦1.50	≦0.040	≦0.030	3.00～6.00	23.00～28.00	1.00～3.00	–	–	E013
≦1.50	≦0.040	≦0.030	4.50～7.50	22.00～26.00	2.50～4.00	0.08～0.20	–	
≦2.00	≦0.040	≦0.020	3.00～6.00	25.00～30.00	1.00～2.00	–	–	
≦2.00	≦0.045	≦0.030	9.00～13.00	17.00～19.00	–	–	Nb≧10C	E014
≦2.00	≦0.045	≦0.030	9.00～13.00	17.00～19.00	–	–	Cb+Ta≧10C	
≦2.00	≦0.045	≦0.030	9.0～12.0	17.0～19.0	–	–	Nb10C～1.00	
≦2.00	≦0.035	≦0.020	9.0～11.0	17.0～19.0	–	–	Nb≧10C ～1.00	
≦2.00	≦0.040	≦0.030	9.0～11.0	17.0～19.0	–	–	Nb+Ta=10C ～1.0	
≦2.00	≦0.020	≦0.035	11.0～13.0	17.0～19.0	–	–	Nb≧10S～1.1	
≦2.00	≦0.045	≦0.030	15.00～17.00	17.00～19.00	–	–		E015
≦2.00	≦0.045	≦0.030	15.00～17.00	17.00～19.00	–	–	–	
≦2.00	≦0.040	≦0.030	15.0～17.0	17.0～19.0	–	–	–	
≦1.00	≦0.040	≦0.030	(≦0.60)	11.50～13.00	–	–	–	E016
≦1.00	≦0.040	≦0.030	–	11.50～13.00	–	–	–	
≦1.00	≦0.040	≦0.030	≦0.50	12.0～14.0	–	–	–	
–	–	–	–	12.0～14.0	–	–	–	
≦1.00	≦0.040	≦0.030	–	11.5～13.5	–	–	–	
≦0.8	≦0.030	≦0.025	–	12.0～13.0	–	–	–	
≦0.8	≦0.030	≦0.025	–	12.0～14.0	–	–	–	

JIS	AISI	BS	DIN	NF	ГОСТ		
G4303〜4309 4313〜4320	MN06－ 222	970 Part1 1449 Part2	17224 17440 17441	A35－573 〜584 586	5632	C	Si
SUS405						≦0.08	≦1.00
	405					≦0.08	≦1.00
		405S17				≦0.08	≦1.00
			X6CrAl13			≦0.08	－
				Z6CA13		≦0.08	≦1.00
SUS410						≦0.15	≦1.00
SUS410L						≦0.03	≦1.00
SUS410S						≦0.08	≦1.00
SUS410J1						0.08〜0.18	≦1.00
	410					≦0.15	≦1.00
	410S (ASTM-A176)					≦0.08	≦1.00
		410S21				0.09〜0.15	≦1.00
			X10Cr13			0.08〜0.12	－
				X12C13		0.08〜0.15	≦1.00
				X6C13		≦0.08	≦1.00
					12X13	0.09〜0.15	≦0.80
					08X13	≦0.08	≦0.80
SUS416						≦0.15	≦1.00
	416					≦0.15	≦1.00
	416Se					≦0.15	≦1.00
		416S21				0.09〜0.15	≦1.00
		416S29				0.14〜0.20	≦1.00
		416S41				0.09〜0.15	≦1.00
		416S37				0.20〜0.28	≦1.0
SUS420J1						0.16〜0.25	≦1.00
SUS420J2						0.26〜0.40	≦1.00
SUS420F						0.26〜0.40	≦1.00
	420					≧0.15	≦1.00
	420F					≧0.15	≦1.00
		420S29				0.14〜0.20	≦0.80
		420S45				0.28〜0.36	≦1.00
			X20Cr13			0.17〜0.22	－
				Z20C13		0.15〜0.24	≦1.00
					20X13	0.16〜0.25	≦0.80
					30X13	0.26〜0.35	≦0.80
SUS429						≦0.12	≦1.00
SUS429J1						0.25〜0.40	≦1.00
	429					≦0.12	≦1.00
SUS430						≦0.12	≦1.00
SUS430F						≦0.12	≦1.00

Chemical Composition 化 学 成 分 (%)								Index Number 索引番号
Mn	P	S	Ni	Cr	Mo	N	Others	
≦1.00	≦0.040	≦0.030	–	11.50～14.50	–	–	Al 0.10~0.30	E017
≦1.00	≦0.040	≦0.030	≦1.00	11.50～14.50	–	–	Al 0.10~0.30	
≦1.00	≦0.040	≦0.030	≦1.00	12.0～14.0	–	–	Al 0.10~0.30	
–	–	–	–	12.0～14.0	–	–	Al 0.10~0.30	
≦1.00	≦0.040	≦0.030	–	11.5～13.5	–	–	Al 0.10~0.30	
≦1.00	≦0.040	≦0.030	–	11.50～13.50	–	–	–	E018
≦1.00	≦0.040	≦0.030	–	11.50～13.50	–	–	–	
≦1.00	≦0.040	≦0.030	≦1.00	11.50～13.50	–	–	–	
≦1.00	≦0.040	≦0.030	(≦0.60)	11.50～14.00	0.30～0.60	–	–	
≦1.00	≦0.040	≦0.030	–	11.50～13.50	–	–	–	
≦1.00	≦0.040	≦0.030	≦1.00	11.50～13.50	–	–	–	
≦1.00	≦0.040	≦0.030	≦1.00	11.5～13.5	–	–	–	
–	–	–	–	12.0～14.0	–	–	–	
≦1.00	≦0.045	≦0.030	–	11.5～13.5	–	–	–	
≦1.00	≦0.040	≦0.030	–	11.5～13.5	–	–	–	
≦0.80	≦0.025	≦0.030	–	12.0～14.0	–	–	–	
≦0.80	≦0.025	≦0.030	–	12.0～14.0	–	–	–	
≦1.25	≦0.060	≧0.150	(≦0.60)	12.00～14.00	(≦0.60)	–	–	E019
≦1.25	≦0.060	≧0.150	–	12.00～14.00	≦0.60	–	–	
≦1.25	≦0.060	≦0.060	–	12.00～14.00	–	–	Se≧0.15	
≦1.50	≦0.060	0.15～0.35	≦1.00	11.5～13.0	≦0.60	–	–	
≦1.50	≦0.060	0.15～0.35	≦1.00	11.5～13.0	≦0.60	–	–	
≦1.50	≦0.060	≦0.060	≦1.00	11.5～13.5	≦0.60	–	Se 0.15~0.30	
≦1.5	≦0.060	0.15～0.35	≦1.00	12.0～14.0	–	–		
≦1.00	≦0.040	≦0.030	(≦0.60)	12.00～14.00	–	–	–	E020
≦1.00	≦0.040	≦0.030	(≦0.60)	12.00～14.00	–	–	–	
≦1.25	≦0.060	≧0.150	(≦0.60)	12.00～14.00	≦0.60	–	–	
≦1.00	≦0.040	≦0.150	–	12.00～14.00	–	–	–	
≦1.25	≦0.060	≦0.060	(≦0.60)	–	–	–	–	
≦1.00	≦0.040	≦0.030	≦1.0	11.5～13.5	–	–	–	
≦1.00	≦0.040	≦0.030	≦1.0	12.0～14.0	–	–	–	
–	–	–	–	12.0～14.0	–	–	–	E021
≦1.00	≦0.040	≦0.030	–	12.0～14.0	–	–	–	
≦0.80	≦0.025	≦0.030	–	12.0～14.0	–	–	–	
≦0.80	≦0.025	≦0.030	–	12.0～14.0	–	–	–	
≦1.00	≦0.040	≦0.030	–	14.0～16.0	–	–	–	E022
≦1.00	≦0.040	≦0.030	–	15.00～17.00	–	–	–	
≦1.00	≦0.040	≦0.030	–	14.0～16.0	–	–	–	
≦1.00	≦0.040	≦0.030	–	16.00～18.00	–	–	–	E023
≦1.25	≦0.060	≧0.150	–	16.00～18.00	–	–	–	

J I S	A I S I	B S	D I N	N F	ГОСТ		
G4303～4309 4313～4320	MN06－222	970 Part 1 1449 Part2	17224 17440 17441	A35－573 ～584 586	5632	C	S i
SUS430LX						≦0.03	≦0.75
	430					≦0.12	≦1.00
	430F					≦0.12	≦1.00
	430FSe					≦0.12	≦1.00
		430S17				≦0.08	≦1.00
			X6CrTi17			≦0.08	－
			X6CrTb17			≦0.08	－
				Z8C17		≦0.08	≦1.00
				Z8CT17		≦0.08	≦1.00
				Z8CNb17		≦0.08	≦1.00
					12X17	≦0.12	≦0.8
					10X18C10	≦0.15	1.0～1.5
					08X17T	≦0.08	≦0.08
SUS431						≦0.20	≦1.00
	431					≦0.20	≦1.00
		431S29				0.12～0.20	≦1.00
			X20CrNi172			0.14～0.23	－
			X22CrNi17			0.15～0.23	－
					14X17H2	0.11～0.17	≦0.8
					20X17H2	0.17～0.25	≦0.8
SUS434						≦0.12	≦1.00
	434					≦0.08	≦1.00
		434S17				≦0.08	≦1.00
			X6CrMo171			≦0.08	－
				Z8CD17.01		≦0.08	≦1.0
SUS436L						≦0.25	≦1.00
SUS440A						0.60～0.75	≦1.00
SUS440B						0.75～0.95	≦1.00
SUS440C						0.95～1.20	≦1.00
SUS440F						0.95～1.20	≦1.00
	440F					0.60～0.75	≦1.00
	440B					0.75～0.95	≦1.00
	440C					0.95～1.20	≦1.00
SUS444						≦0.25	≦1.00
	444 (ASTM-A176)					≦0.25	≦1.00

Chemical Composition 化学成分 (%)								Index Number 索引番号
Mn	P	S	Ni	Cr	Mo	N	Others	
≦1.00	≦0.040	≦0.030	–	16.00～19.00	–	–	Ti or Nb 0.10～1.00	E023
≦1.00	≦0.040	≦0.030	–	16.00～18.00	–	–	–	
≦1.25	≦0.060	≦0.150	–	16.00～18.00	≦0.60	–	–	
≦1.25	≦0.060	≦0.060	1.25～2.50	15.00～17.00	–	–	Se≧0.15	
≦1.00	≦0.040	≦0.030	≦1.00	16.0～18.0	–	–	–	
–	–	–	–	16.0～18.0	–	–	Ti≧7C	
–	–	–	–	16.0～18.0	–	–	Nb≦12C	
≦1.0	≦0.040	≦0.030	–	16.0～18.0	0.9～1.3	–	–	
≦1.0	≦0.040	≦0.030	–	16.0～18.0	–	–	Ti7c～1.2	
≦1.0	≦0.040	≦0.030	–	16.0～18.0	–	–	Nb7C～1.2	
≦0.8	≦0.035	≦0.025	–	16.0～18.0	–	–	–	
≦0.8	≦0.035	≦0.025	–	17.0～20.0	–	–	Ti≦7c	
≦0.8	≦0.035	≦0.025	–	16.0～18.0	–	–	Ti5C～0.8	
≦1.00	≦0.040	≦0.030	1.25～2.50	15.00～17.00	–	–	–	E024
≦1.00	≦0.040	≦0.030	1.25～2.50	15.00～17.00	–	–	–	
≦1.0	≦0.040	≦0.030	2.0～3.0	15.0～18.0	–	–	–	
–	–	–	1.5～2.5	15.5～17.5	–	–	–	
–	–	–	1.5～2.5	16.0～18.0	–	–	–	
≦0.8	≦0.030	≦0.025	1.5～2.5	16.0～18.0	–	–	–	
≦0.8	≦0.035	≦0.025	1.5～2.5	16.0～18.0	–	–	–	
≦1.00	≦0.040	≦0.030	(≦0.60)	16.00～18.00	0.75～1.25	–	–	E025
≦1.00	≦0.040	≦0.030	≦1.00	16.00～18.00	0.90～1.30	–	–	
≦1.00	≦0.004	≦0.03	≦1.0	16.0～18.0	0.90～1.30	–	–	
–	–	–	–	16.0～18.0	0.9～1.3	–	–	
≦1.0	≦0.004	≦0.03	≦1.0	16.0～18.0	0.9～1.2	–	–	
≦1.00	≦0.040	≦0.030	–	16.00～19.00	0.75～1.25	≦0.025	Ti+Nb+Zr= 8(C+N)～0.80	E026
≦1.00	≦0.040	≦0.040	(≦0.60)	16.00～18.00	≦0.75	–	–	E027
≦1.00	≦0.035	≦0.030	(≦0.60)	16.00～18.00	≦0.75	–	–	
≦1.00	≦0.040	≦0.030	(≦0.60)	16.00～18.00	≦0.75	–	–	
≦1.25	≦0.060	≧0.150	(≦0.60)	16.00～18.00	≦0.75	–	–	
≦1.00	≦0.040	≦0.030	–	16.00～18.00	≦0.75	–	–	
≦1.00	≦0.035	≦0.030	–	16.00～18.00	≦0.75	–	–	
≦1.00	≦0.030	≦0.030	–	16.00～18.00	≦0.75	–	–	
≦1.00	≦0.040	≦0.030	–	17.00～20.00	1.75～2.50	≦0.025	Ti+Nb+Zr= 8(C+N)～0.80	E028
≦1.00	≦0.040	≦0.030	≦1.00	17.50～19.50	1.75～2.50	≦1.00	Ti+Cb=0.20+ 4(C+N)～0.80	

JIS	AISI	BS	DIN	NF	ГОСТ		
G4303～4309 4313～4320	MN06－222	970 Part1 1449 Part2	17224 17440 17441	A35－573 ～584 586	5632	C	Si
SUS447J1						≦0.010	≦0.40
SUS630						≦0.07	≦1.00
				Z6CNU17.04		≦0.06	≦1.0
SUS631						≦0.09	≦1.00
SUS631JI						≦0.09	≦1.00
	631					≦0.09	≦1.00
			X7CrNiAL17.7			≦0.09	－
				Z8CNiAL177		≦0.09	≦1.0
					09X17H101	≦0.09	≦0.80
SUSXM7						≦0.08	≦1.00
	XM7 (ASTMA276)					≦0.08	≦1.00
				Z6CNU18.10		≦0.06	≦1.0
SUSXM15						≦0.08	3.00～5.00
	XM15 (ASTMA167, 240)					≦0.10	1.50～2.50
SUSXM27						≦0.010	≦0.40
	XM27 (ASTMA176)					≦0.010	≦0.40
					15X25T	≦0.15	≦1.0

Chemical Composition 化学成分 (%)								Index Number 索引番号
Mn	P	S	Ni	Cr	Mo	N	Others	
≦0.40	≦0.040	≦0.030	≦0.50	28.50～32.50	1.50～2.50	≦0.015	Cu≦0.20 Ni+Cu≦950	E029
≦1.00	≦0.040	≦0.035	3.00～5.00	15.50～17.50	–	–	Cu3.00～5.00 Nb0.15～0.45	E030
≦1.0	≦0.040	≦0.035	3.0～5.0	16.0～18.0	–	–	Cu3-5	
≦1.00	≦0.040	≦0.030	6.50～7.75	16.00～18.00	–	–	AL0.75～1.50	E031
≦1.00	≦0.040	≦0.030	7.00～8.50	16.00～18.00			AL0.75～1.50	
≦1.00	≦0.040	≦0.030	6.50～7.75	16.00～18.00	–	–	AL0.75～1.50	
–	–	–	6.50～7.75	16.0～18.0	–	–	AL0.75～1.50	
≦1.0	≦0.035	≦0.025	6.50～7.75	16.0～18.0	–	–	AL0.75～1.50	
≦0.80	≦0.025	≦0.035	6.5 ～7.5	16.5～18.0	–	–	AL0.75～1.50	
≦2.00	≦0.045	≦0.030	8.50～10.50	17.00～19.00	–	–	Cu3.00～4.00	E032
≦2.00	≦0.045	≦0.030	8.00～10.00	17.00～19.00	–	–	Cu3.00～4.00	
0.30～0.60	≦0.035	≦0.035	9.0 ～11.0	17.0～19.0	–	–	Cu3.0～4.0	
≦2.00	≦0.045	≦0.035	11.00～15.00	15.00～20.00	–	–	–	E033
≦2.00	≦0.030	≦0.030	17.50～18.50	17.00～19.00	–	–	–	
≦0.40	≦0.020	≦0.020	≦0.75	25.00～27.00	–	≦0.25	–	E034
≦0.40	≦0.020	≦0.020	≦0.75	25.00～27.00	Cb0.05～0.20 Cu≦0.20	≦0.015	Ni+Cu≦ 0.50	
≦0.8	≦0.035	≦0.035	–	24.0～27.0	–	–	Ti5C～0.90	

Appendix 1. JIS Stainless Steels List, Classified According

Designalion 名 称	Bar 棒	Hot Plate 熱 板	Cold Plate 冷 板	Hot Strip 熱 帯	Cold Strip 冷 帯
	G4303 G4318	G4304	G4305	G4306	G4307
SUS 201	○	○	○	○	○
SUS 202	○	○	○	○	○
SUS 301	○	○	○	○	○
SUS 301 J1	–	○	○	○	○
SUS 302	○	○	○	○	○
SUS 302 B	–	○	○	○	○
SUS 303	○	–	–	–	–
SUS 303 Se	○	–	–	–	–
SUS 304	○	○	○	○	○
SUS 304 L	○	○	○	○	○
SUS 304 N1	○	○	○	○	○
SUS 304 N2	○	○	○	○	○
SUS 304 LN	○	○	○	○	○
SUS 305	○	○	○	○	○
SUS 305 J1	○	–	–	–	–
SUS 309 S	○	○	○	○	○
SUS 310 S	○	○	○	○	○
SUS 316	○	○	○	○	○
SUS 316 L	○	○	○	○	○
SUS 316 J1	○	○	○	○	○
SUS 316 J1L	○	○	○	○	○
SUS 316 N	○	○	○	○	○
SUS 316 LN	○	○	○	○	○
SUS 317	○	○	○	○	○
SUS 317 L	○	○	○	○	○
SUS 317 J1	○	○	○	○	○
SUS 321	○	○	○	○	○
SUS 329 J1	○	○	○	○	○
SUS 329 J2L	–	○	○	○	○
SUS 347	○	○	○	○	○
SUS 384	–	–	–	–	–
SUS 403	○	○	○	○	○
SUS 405	○	○	○	○	○
SUS 410	○	○	○	○	○
SUS 410 L	○	○	○	○	○
SUS 410 S	○	○	○	○	○
SUS 410 J1	○	–	–	–	–
SUS 416	○	–	–	–	–
SUS 420 J1	○	○	○	○	○
SUS 420 J2	○	○	○	○	○

to Shape JIS ステンレス鋼形状別分類

Spring Strip ばね用帯	Spring Wire ばね用線	Wire 線	Cold Wire 冷　線	Protected Sheet 塗装鋼板	Leg Angle 山形鋼	Index Number 索引番号
G4313	G4314	G4308 G4309	G4315	G3320	G4317 G4320	
–	–	○	–	–	–	E 035
–	○	–	–	–	–	
○	○	–	–	–	–	
–	–	–	–	–	–	
–	○	○	–	–	○	
–	–	–	–	–	–	
–	–	○	–	–	–	
–	–	○	–	–	–	
○	○	○	○	○	○	
–	–	○	–	–	○	
–	○	○	–	–	–	
–	–	–	–	–	–	
–	–	–	–	–	–	
–	–	○	○	–	–	
–	–	○	○	–		
–	–	○	–	–	–	
–	–	○	–	–	–	
–	○	○	○	–	○	
–	–	○	–	–	○	
–	–	–	–	–	–	
–	–	–	–	–	–	
–	–	–	–	–	–	
–	–	–	–	–	–	
–	–	–	–	–	–	
–	–	–	–	–	–	
–	–	–	–	–	–	
–	–	○	–	–	○	
–	–	–	–	–	–	
–	–	–	–	–	–	
–	–	○	–	–	○	
–	○	○	○	–	–	
–	–	–	–	–	–	
–	–	–	–	–	–	
–	○	○	○	–	–	
–	–	–	–	–	–	
–	–	–	–	–	–	
–	–	–	–	–	–	
–	–	○	–	–	–	
–	–	○	–	–	–	
○	○	○	–	–	–	

Designalion 名　　称	Bar 棒	Hot Plate 熱　板	Cold Plate 冷　板	Hot Strip 熱　帯	Cold Strip 冷　帯
	G4303 G4318	G4304	G4305	G4306	G4307
SUS 420 F	○	−	−	−	−
SUS 429	−	○	○	○	○
SUS 429 J1	−	○	○	○	−
SUS 430	○	○	○	○	○
SUS 430 F	○	−	−	−	−
SUS 430 LX	−	○	○	○	○
SUS 431	○	−	−	−	−
SUS 434	○	○	○	○	○
SUS 436 L	−	−	○	○	○
SUS 440 A	○	○	○	○	○
SUS 440 B	○	−	−	−	−
SUS 440 C	○	−	−	−	−
SUS 440 F	○	−	−	−	−
SUS 444	−	○	○	○	○
SUS 447 JI	○	○	○	○	○
SUS 630	○	−	−	−	−
SUS 631	○	○	○	○	○
SUS 631 J1	−	−	−	−	−
SUSXM 7	○	−	−	−	−
SUSXM 15 JI	○	○	○	○	○
SUSXM 27	○	○	○	○	○

Spring Strip ばね用帯	Spring Wire ばね用線	Wire 線	Cold Wire 冷　　線	Protected Sheet 塗装鋼板	Leg Angle 山 形 鋼	Index Number 索引番号
G4313	G4314	G4308 G4309	G4315	G3320	G4317 G4320	
–	–	○	–	–	–	E 035
–	–	–	–	–	–	
–	–	–	–	–	–	
–	–	○	○	○	○	
–	–	○	–	–	–	
–	–	–	–	–	–	
–	–	–	–	–	–	
–	–	○	○	–	–	
–	–	–	–	–	–	
–	–	–	–	–	–	
–	–	–	–	–	–	
–	–	○	–	–	–	
–	–	–	–	–	–	
–	–	–	–	–	–	
–	–	–	–	–	–	
–	–	–	–	–	–	
○	○	–	–	–	–	
–	○	○	–	–	–	
–	○	○	○	–	–	
–	–	–	–	–	–	
–	–	–	–	–	–	

Appendix 2. ASTM Stainless Steels List, Classified According

Designation 名　称	Bar 棒				Plate, Sheet and Strip 板，シ		
	A276	A314	A479	A582	A167	A176	A240
9	○	–	–	–	–	–	–
201	○	–	–	–	–	–	○
202	○	○	–	–	–	–	○
205	○	–	–	–	–	–	–
218	–	–	–	–	–	–	–
301	–	–	–	–	○	–	–
302	○	○	○	–	○	–	○
302 B	○	○	–	–	○	–	–
303	–	○	–	○	–	–	–
303 Se	–	○	–	○	–	–	–
304	○	○	○	–	○	–	○
304 H	–	–	○	–	–	–	○
304 L	○	○	○	–	○	–	○
304 N	○	–	○	–	–	–	○
304 LN	○	–	○	–	○	–	○
305	○	○	–	–	○	–	○
308	○	○	–	–	○	–	–
ER308	–	–	○	–	–	–	–
309	○	○	–	–	○	–	–
309 S	○	○	○	–	○	–	○
309 Cb	○	○	○	–	○	–	○
310	○	○	–	–	○	–	–
310MLN	–	–	–	–	–	–	○
310 S	○	○	○	–	○	–	○
310 Cb	○	○	○	–	○	–	○
310 H	–	–	–	–	–	–	○
310HCb	–	–	–	–	–	–	○
314	○	○	○	–	–	–	–
316	○	○	○	–	○	–	○
316 Cb	○	○	○	–	○	–	○
316 H	–	–	–	–	–	–	○
316 L	○	○	○	–	○	–	○
316 N	○	–	–	–	–	–	○
316 LN	○	–	–	–	○	–	○
316 Ti	○	○	○	–	○	–	○
317	○	○	–	–	○	–	○
317 L	–	–	–	–	○	–	○
317 LN	–	–	–	–	○	–	○
321	○	○	○	–	○	–	○
321 H	–	–	○	–	○	–	○

to Shape　ASTM ステンレス鋼形状別分類

ート，帯	Wire 線						Index Number 索引番号
A666	A313	A478	A492	A493	A580	A581	
–	–	–	–	–	–	–	E036
○	–	–	–	–	–	–	
	–	–	–	–	–	–	
○	–	–	–	–	–	–	
○	–	–	–	–	–	–	
○	–	–	–	–	–	–	
○	○	○	○	○	○	–	
–	–	–	–	–	○	–	
–	–	–	–	–	–	○	
–	–	–	–	–	–	○	
○	○	○	○	○	○	–	
–	–	–	–	–	–	–	
○	–	○	–	–	○	–	
○	–	–	–	–	–	–	
○	–	–	–	–	–	–	
–	○	○	○	○	○	–	
–	–	–	–	–	○	–	
–	–	–	–	–	–	–	
–	–	–	–	–	○	–	
–	–	–	–	–	○	–	
–	–	○	–	–	○	–	
–	–	–	–	–	○	–	
–	–	–	–	–	–	–	
–	–		–	–	○	–	
–	–	–	–	–	–	–	
–	–	–	–	–	–	–	
–	–	–	–	–	–	–	
–	–	–	–	–	○	–	
○	○	○	○	○	○	–	
–	–	○	–	–	–	–	
–	–	–	–	–	–	–	
○	–	○	–	–	○	–	
○	–	–	–	–	–	–	
–	–	–	–	–	–	–	
–	–	○	–	–	–	–	
–	–	○	–	–	○	–	
–	–	–	–	–	–	–	
–	–	–	–	–	–	–	
–	○	–	–	–	○	–	
–	–	–	–	–	–	–	

Designation 名　称	Bar　棒				Plate, Sheet and Strip 板，シ		
	A276	A314	A479	A582	A167	A176	A240
329	–	–	–	–	–	–	○
347	○	○	○	–	○	–	○
347 H	–	–	○	–	–	–	○
348	○	○	○	–	○	–	○
348 H	–	–	○	–	–	–	○
384	–	–	–	–	–	–	–
403	○	○	○	–	–	○	–
403 F	–	–	–	–	–	–	–
405	○	–	○	–	–	○	○
409	–	–	–	–	–	○	○
410	○	○	○	–	–	○	○
410 S	–	–	–	–	–	○	○
414	○	○	○	–	–	–	–
416	–	○	–	○	–	–	–
416 Se	–	○	–	○	–	–	–
420	–	○	–	–	–	○	–
420 F	–	–	–	○	–	–	–
420 Se	–	–	–	○	–	–	–
420FSe	–	–	–	○	–	–	–
429	○	○	–	○	–	○	○
430	○	○	○	–	–	○	○
430 F	–	○	–	○	–	–	–
430FSe	–	○	–	○	–	–	–
431	○	○	–	–	–	–	–
439	–	–	○	–	–	–	–
440 A	○	○	–	–	–	–	–
440 B	○	○	–	–	–	–	–
440 C	○	○	–	–	–	–	–
440 F	–	–	–	○	–	–	–
440 Se	–	–	–	○	–	–	–
442	–	–	–	–	–	○	–
446	○	○	–	–	–	○	–
501	–	○	–	–	–	–	–
502	–	○	–	–	–	–	–
631	–	–	–	–	–	–	–
XM-1	–	–	–	○	–	–	–
XM-2	–	–	–	○	–	–	–
XM-3	–	–	–	○	–	–	–
XM-5	–	–	–	○	–	–	–
XM-6	–	–	–	○	–	–	–

ート，帯	Wire			線			Index Number 索引番号
A666	A313	A478	A492	A493	A580	A581	
—	—	—	—	—	—	—	E036
—	○	—	—	—	○	—	
—	—	—	—	—	—	—	
—	—	—	—	—	○	—	
—	—	—	—	—	—	—	
—	—	—	—	○	—	—	
—	—	—	—	—	—	—	
—	—	—	—	—	○	—	
—	—	—	—	—	○	—	
—	—	—	—	—	—	—	
—	—	—	—	○	○	—	
—	—	—	—	—	—	—	
—	—	—	—	—	○	—	
—	—	—	—	—	—	○	
—	—	—	—	—	○	○	
—	—	—	—	—	—	—	
—	—	—	—	—	—	—	
—	—	—	—	—	—	—	
—	—	—	—	—	—	—	
—	—	—	—	○	—	—	
—	—	—	—	○	—	—	
—	—	—	—	—	—	○	
—	—	—	—	—	—	○	
—	—	—	—	—	—	○	
—	—	—	—	—	—	—	
—	—	—	—	—	○	—	
—	—	—	—	—	○	—	
—	—	—	—	—	○	—	
—	—	—	—	—	—	—	
—	—	—	—	—	—	—	
—	—	—	—	—	—	—	
—	—	—	—	—	○	—	
—	—	—	—	—	—	—	
—	—	—	—	—	—	—	
—	○	—	—	—	—	—	
—	—	—	—	—	—	○	
—	—	—	—	—	—	○	
—	—	—	—	—	—	○	
—	—	—	—	—	—	○	
—	—	—	—	—	—	○	

Designation 名　称	Bar 棒				Plate, Sheet and Strip 板, シ		
	A276	A314	A479	A582	A167	A176	A240
XM-7	○	–	–	–	–	–	–
XM-8	○	–	–	–	–	–	–
XM-10	○	○	–	–	–	–	–
XM-11	○	○	○	–	–	–	–
XM-15	–	–	–	–	○	–	○
XM-16	–	–	–	–	–	–	–
XM-17	–	–	○	–	–	–	○
XM-18	–	–	○	–	–	–	○
XM-19	○	○	○	–	–	–	○
XM-21	○	–	–	–	–	–	○
XM-26	○	–	–	–	–	–	–
XM-27	○	○	○	–	–	○	–
XM-28	○	○	–	–	–	–	–
XM-29	○	○	○	–	–	–	○
XM-30	○	–	○	–	–	–	○
XM-31	–	–	–	–	–	–	○
XM-33	–	–	–	–	–	○	○
XM-34	–	–	–	○	–	–	–
S 018235	–	–	–	○	–	–	–
S 21800	○	○	○	–	–	–	–
S 28200	○	○	–	–	–	–	–
S 30454	○	–	–	–	–	–	–
S 30600	–	–	–	–	○	–	○
S 30815	○	–	○	–	○	–	○
S 31200	–	–	–	–	–	–	○
S 31254	○	–	○	–	○	–	○
S 31260	–	–	–	–	–	–	○
S 31654	○	–	–	–	–	–	–
S 31725	○	–	–	–	○	–	○
S 31726	○	–	○	–	○	–	○
S 31803	○	–	–	–	–	–	–
S 32550	–	–	○	–	–	–	○
S 32615	–	–	○	–	–	–	○
S 32950	–	–	○	–	–	–	○
S 41050	–	–	–	–	–	–	○
S 41500	○	–	○	–	–	–	○
S 42010	○	○	–	–	–	–	–
S 44400	○	–	○	–	–	–	○
S 44401	–	–	–	–	–	–	–
S 44625	–	–	–	–	–	–	–

ート，帯	Wire 線						Index Number 索引番号
A666	A313	A478	A492	A493	A580	A581	
－	－	－	－	－	－	○	E036
－	－	－	－	－	－	－	
－	－	－	－	－	○	－	
－	－	－	－	－	○	－	
－	－	－	－	－	－	－	
－	○	－	－	－	－	－	
－	－	－	○	－	－	－	
－	－	－	○	－	－	－	
－	－	－	－	－	－	－	
－	－	－	－	－	－	－	
－	－	－	－	－	－	－	
－	○	－	－	○	－	－	
－	－	－	－	－	○	－	
－	－	－	－	－	○	－	
－	－	－	－	－	－	－	
－	－	－	－	－	○	－	
－	－	－	－	－	－	－	
－	－	－	－	－	－	○	
－	－	－	－	－	－	○	
－	－	－	－	－	○	－	
－	－	－	－	○	○	－	
－	－	－	－	－	－	－	
－	－	－	－	－	－	－	
－	－	－	－	－	－	－	
－	－	－	－	－	－	－	
－	－	－	－	－	－	－	
－	－	－	－	－	－	－	
－	－	－	－	－	－	－	
－	－	－	－	－	－	－	
－	－	－	－	－	－	－	
－	－	－	－	－	－	－	
－	－	－	－	－	－	－	
－	－	－	－	－	－	－	
－	－	－	－	－	－	－	
－	－	－	－	－	－	－	
－	－	－	－	－	－	－	
－	－	－	－	－	－	－	
－	－	－	－	－	－	－	
－	－	－	－	－	－	－	
－	－	－	－	○	○	－	
－	－	－	－	○	－	－	

Designation 名　称	Bar　棒				Plate, Sheet and Strip 板, シ		
	A276	A314	A479	A582	A167	A176	A240
S 44635	−	−	−	−	−	○	○
S 44660	−	−	−	−	−	○	○
S 44700	○	−	○	−	−	○	○
S 44735	−	−	−	−	−	−	○
S 44800	○	−	○	−	−	○	○

ート，帯	Wire 線						Index Number 索引番号
A 6 6 6	A 3 1 3	A 4 7 8	A 4 9 2	A 4 9 3	A 5 8 0	A 5 8 1	
−	−	−	−	−	−	−	E 0 3 6
−	−	−	−	−	−	−	
−	−	−	−	○	○	−	
−	−	−	−	−	−	−	
−	−	−	−	○	○	−	

3. Heat-Resisting Steels 耐熱鋼

JIS	AISI	BS	DIN	NF	ГОСТ		
G4311 G4312	MN06－222	970 Part 4	17440	A35－578	5632	C	Si
SUH31						0.35～0.45	0.50～2.50
		331S42				0.37～0.47	1.00～2.00
SUH35						0.48～0.58	≦0.25
			349S52			0.48～0.58	≦0.25
					5X20Г9AH4	0.50～0.60	≦0.45
SUH36						0.48～0.58	≦0.25
		349S54				0.48～0.58	≦0.25
SUH37						0.15～0.25	≦1.00
		381S34				0.15～0.25	0.75～1.25
SUH38						0.25～0.35	≦1.00
SUH309						≦0.20	≦1.00
SUH309S						≦0.20	≦1.00
	309					≦0.20	≦1.00
		309S24				≦0.15	≦1.00
				Z15CN23-13		≦0.15	≦0.75
SUH310						≦0.25	≦1.50
SUH310S						≦0.25	≦1.50
	310					≦0.25	≦1.50
		310S24				≦0.25	≦1.00
			CrNi2520			≦0.25	≦1.0
				Z15CNS25-20		≦0.15	1.50～2.50
					20X25H20C2	≦0.20	2.0～3.0
SUH330						≦0.15	≦1.50
	330					≦0.08	0.75～1.50
				Z20NCS33-16		≦0.20	1.00～2.00
SUH660						≦0.08	≦1.00
				Z5NCTDV26-15B		≦0.05	≦0.50
SUH661						0.08～0.16	≦1.00
SUH21						≦0.10	≦1.50
SUH409						≦0.08	≦1.00
	409					≦0.08	≦1.00
		409S19				≦0.08	≦1.0
			X6CrTi12			≦0.08	≦1.0
				Z36CT12		≦0.03	≦0.75

Chemical Composition 化学成分 (%)								Index Number 索引番号
Mn	P	S	Ni	Cr	Mo	N	Others	
≦0.60	≦0.040	≦0.030	13.00～15.00	14.00～16.00	–	–	W2.00～3.00	E037
0.50～1.00	≦0.040	≦0.030	13.0～15.0	13.0～15.0	0.40～0.70	–	W2.00～3.00	
8.00～10.00	≦0.040	≦0.030	3.25～4.50	20.00～22.00	–	0.35～0.50	–	E038
8.0～10.0	≦0.040	≦0.035	3.25～4.50	20.0～22.0	–	0.38～0.50	–	
8.0～10.0	≦0.030	≦0.030	–	–	–	0.30～0.60	–	
8.00～10.00	≦0.040	0.040～0.090	3.25～4.50	20.00～22.00	–	0.38～0.50	–	E039
1.00～1.60	≦0.040	≦0.030	10.5～12.5	20.0～22.0	–	0.15～0.30	–	
1.00～1.60	≦0.040	≦0.030	10.00～12.00	20.50～22.50	–	0.15～0.30	–	E040
≦1.50	≦0.040	≦0.030	1.05～1.25	2.00～2.20	–	0.15～0.30		
≦1.20	0.18～0.25	≦0.030	10.00～12.00	19.00～21.00	1.80～2.50	–	B0.001～0.010	E041
≦2.00	≦0.040	≦0.030	12.00～15.00	22.00～24.00	–	–	–	E042
≦2.00	≦0.040	≦0.030	12.00～15.00	22.00～24.00	–	–	–	
≦2.00	≦0.045	≦0.030	12.00～15.00	20.20～20.40	–	–	–	
≦2.00	≦0.045	≦0.030	13.0～16.0	2.20～2.50	–	–	–	
≦2.00	≦0.035	≦0.015	12.0～14.0	22.0～24.0	–	–	–	
≦2.00	≦0.040	≦0.030	19.00～22.00	24.00～26.00	–	–	–	E043
≦2.00	≦0.040	≦0.030	19.00～22.00	24.00～26.00	–	–	–	
≦2.00	≦0.045	≦0.030	19.00～22.00	24.00～26.00	–	–	–	
≦2.00	≦0.045	≦0.030	19.0～22.0	23.0～26.0	–	–	–	
≦2.0	≦0.040	≦0.035	19～22	23～26	–	–	–	
≦2.00	≦0.035	≦0.015	9～21	24～26	–	–	–	
≦1.5	≦0.020	≦0.035	18.0～21.0	24.0～27.0	–	–	–	
≦2.00	≦0.040	≦0.030	33.00～37.00	14.00～17.00	–	–	–	E044
≦2.00	≦0.040	≦0.030	34.00～37.00	17.00～20.00	–	–	–	
≦0.50	≦0.035	≦0.015	32.0～34.0	14.0～17.0	–	–	–	
≦2.00	≦0.040	≦0.030	24.00～27.00	13.50～16.00	1.00～1.50	≦0.35 Al≦0.35	V0.10～0.50 B0.001~0.010 Ti1.90～2.35	E045
≦2.00	≦0.020	≦0.015	24.0～27.0	13.5～16.0	1.00～1.50	Al≦0.35 V01.0~0.50	Ti 1.90~2.30 B0.005～0.08	
1.00～2.00	≦0.040	≦0.030	19.00～21.00	20.00～22.50	2.50～3.50 W2.00～3.00	0.10～0.20		E046
≦1.00	0.040	≦0.030	(≦0.60)	17.00～21.00	–	–	AL2.00～4.00	E047
≦1.00	≦0.040	≦0.030	(≦0.60)	10.50～11.75	–	–	Ti6C～0.75 Cu≦0.30	E048
≦1.00	≦0.040	≦0.030	(≦0.60)	10.50～11.75	–	–	Ti6C～0.75	
≦1.0	≦0.045	≦0.030	–	10.5～12.5	–	–	Ti6C～1.0	
≦1.0	≦0.045	≦0.030	–	10.5～12.5	–	–	Ti6C～1.0	
≦0.75	≦0.035	≦0.015	–	10.5～12.5	–	–	Ti6(C+N)～0.60	

JIS	AISI	BS	DIN	NF	ГОСТ		
G4311- G4312	MN06－222	970 Part 4	17440	A35－578	5632	C	Si
SUH446						≦0.20	≦1.00
	446					≦0.20	≦1.00
				Z10C5		≦0.12	≦0.75
SUH1						0.40～0.50	3.00～3.50
		401S45				0.40～0.50	–
					40X9C2	0.35～0.45	2.0～3.0
SUH3						0.35～0.45	1.80～2.50
					40X10C2M	0.35～0.45	2.0～3.0
SUH4						0.75～0.85	1.75～2.25
		443S65				0.75～0.85	1.75～2.25
SUH11						0.45～0.55	1.00～2.00
				Z45CS9		0.4～0.5	1.0～2.0
SUH600						0.15～0.20	≦0.50
				Z21CDNbV11		0.16～0.25	0.10～0.50
					20X12BHMФ	0.17～0.23	≦0.6
				Z3CAT18		≦0.03	≦0.75

Chemical Composition 化学成分 (%)								Index Number 索引番号
Mn	P	S	Ni	Cr	Mo	N	Others	
≦1.50	≦0.040	≦0.030	(≦0.40)	23.00～27.00	–	≦0.25	–	E049
≦1.50	≦0.040	≦0.030	–	23.00～27.00	–	≦0.25	–	
0.75	≦0.035	≦0.015	–	23.0 ～ 26.0	–	–	–	
≦0.60	≦0.030	≦0.030	(≦0.60)	7.50～9.50	–	–	–	E050
–	–	–	–	–	–	–	–	
≦0.8	≦0.025	≦0.030	–	8.0～10.0	–	–	–	
≦0.60	≦0.030	≦0.030	(≦0.60)	10.00～12.00	0.70～1.30	–	–	E051
≦0.8	≦0.025	≦0.035	–	8.0～10.0	–	–	–	
0.20～0.60	≦0.030	≦0.030	1.15～1.65	19.00～20.50	–	–	–	E052
0.30～0.75	≦0.030	≦0.030	1.20～1.70	19.0～21.0	–	–	–	
≦0.60	≦0.030	≦0.030	1.20～1.70	7.50～9.50	–	–	–	E053
≦1.0	≦0.035	≦0.030	1.0～1.5	8.5～9.5	–	–	–	
0.50～1.00	≦0.040	≦0.030	(≦0.60)	10.0～13.00	0.30～0.90	0.05～0.10	V0.10～0.40 Nb0.20～0.60	E054
0.30～0.80	≦0.030	≦0.015	≦1.00	10.0～12.0	0.50～1.00	0.05～0.10	Nb0.25～0.55 V0.10～0.30	
0.50～1.00	≦0.040	≦0.030	0.50～1.10	11.00～13.00	0.75～1.25	W0.75～1.25	V0.20～0.30	E055
0.5～0.9	≦0.025	≦0.035	0.5～0.9	10.5～12.9	0.5 ～0.7	–	V0.18～0.30	

4. Tool Steels 工具鋼

4-1. Carbon Tool Steels 炭素工具鋼

JIS	ASTM	BS	DIN	NF	ГОСТ			
G4401	A686	4659	17350	A35-590	1435	C	Si	Mn
SK1						1.30～1.50	≦0.35	≦0.50
	W2A-13					1.30～1.50	0.10～0.40	0.10～0.40
	W2C-13					1.30～1.50	0.10～0.40	0.10～0.40
				C140E3U		1.30～1.50	0.10～0.30	0.10～0.40
					Y13	1.25～1.35	0.15～0.35	0.15～0.35
SK2						1.10～1.30	≦0.35	≦0.35
	W1A-11½					1.15～1.25	0.10～0.40	0.10～0.40
	W1C-11½					1.15～1.25	0.10～0.40	0.10～0.40
		BW1C	No longer listed			1.1～1.3	0.10～0.40	≦0.35
				C120E3U		1.10～1.29	0.10～0.30	0.10～0.40
					Y12	1.15～1.24	0.15～0.35	0.15～0.35
SK3						1.00～1.10	≦0.35	≦0.50
	W1A-10					1.00～1.10	0.10～0.40	0.10～0.40
	W1C-10					1.00～1.10	0.10～0.40	0.10～0.40
	W1A-9½					0.95～1.05	0.10～0.40	0.10～0.40
	W1C-9½					0.95～1.05	0.10～0.30	0.10～0.40
	W2A-9½					0.95～1.10	0.10～0.40	0.10～0.40
	W2C-9½					0.95～1.10	0.10～0.40	0.10～0.40
		BW1C	No longer listed			0.95～1.10	≦0.030	≦0.35
			C105W1			1.00～1.10	0.10～0.25	0.10～0.25
				C105E2U		0.95～1.09	0.10～0.30	0.10～0.40
					Y11	1.05～1.14	0.15～0.35	0.15～0.35
SK4						0.90～1.00	≦0.35	≦0.50
	W1A-9					0.90～1.00	0.10～0.40	0.10～0.40
	W1C-9					0.90～1.00	0.10～0.40	0.10～0.40
	W2A-9					0.90～1.00	0.10～0.40	0.10～0.40
	W2C-9					0.90～1.00	0.10～0.40	0.10～0.40
		BW1A	No longer listed			0.85～0.95	≦0.30	≦0.35
				C90E2U		0.85～0.94	0.10～0.30	0.10～0.40
					Y10	0.95～1.04	0.15～0.35	0.15～0.35
SK5						0.80～0.90	≦0.35	≦0.50
	W1A-8					0.80～0.90	0.10～0.40	0.10～0.40
	W1C-8					0.80～0.90	0.10～0.40	0.10～0.40
		BW1A	No longer listed			0.85～0.95	≦0.30	≦0.35
			C80W1			0.75～0.85	0.10～0.25	0.10～0.25
				C80E2U		0.75～0.84	0.10～0.30	0.10～0.40
					Y8Г	0.80～0.90	0.35～0.60	≦0.035
					Y9	0.85～0.94	0.15～0.35	≦0.035
SK6						0.70～0.90	≦0.35	≦0.50
			C70W2			0.65～0.75	0.10～0.25	0.10～0.25

Chemical Composition 化学成分 (%)									Index Number 索引番号
P	S	Ni	Cr	Mo	W	V	Cu	Others	
≦0.030	≦0.030	≦0.25	≦0.30	–	–	–	≦0.25	–	E056
≦0.030	≦0.030	≦0.20	≦0.15	≦0.10	≦0.15	0.15～0.35	≦0.20	–	
≦0.030	≦0.030	≦0.20	≦0.30	≦0.10	≦0.15	0.15～0.35	≦0.20	–	
≦0.025	≦0.025	–	–	–	–	0.20～0.50	–	–	
≦0.035	≦0.030	–	–	–	–	–	–	–	
≦0.050	≦0.030	≦0.25	≦0.30	–	–	–	≦0.25	–	E057
≦0.030	≦0.030	≦0.20	≦0.15	≦0.10	≦0.15	≦0.10	≦0.20		
≦0.030	≦0.030	≦0.20	≦0.15	≦0.10	≦0.15	≦0.10	≦0.20	–	
≦0.035	≦0.035	≦0.20	≦0.15	≦0.10	≦0.15	–	–	Sn≦0.05	
≦0.025	≦0.025	–	0.20~0.50	–	–	–	–	–	
≦0.035	≦0.030	–	–	–	–	–	–	–	
≦0.030	≦0.030	≦0.25	≦0.30	–	–	–	≦0.25	–	E058
≦0.030	≦0.030	≦0.20	≦0.15	≦0.10	≦0.15	≦0.10	≦0.20		
≦0.030	≦0.030	≦0.20	≦0.30	≦0.10	≦0.15	≦0.10	≦0.20		
≦0.030	≦0.030	≦0.20	≦0.15	≦0.10	≦0.15	≦0.10	≦0.20		
≦0.030	≦0.030	≦0.20	≦0.30	≦0.10	≦0.15	≦0.10	≦0.20		
≦0.030	≦0.030	≦0.20	≦0.15	≦0.10	≦0.15	0.15～0.35	≦0.20		
≦0.030	≦0.030	≦0.20	≦0.30	≦0.10	≦0.15	0.15～0.35	≦0.20		
≦0.035	≦0.035	≦0.20	≦0.15	≦0.10	–	–	≦0.20	Sn≦0.05	
≦0.020	≦0.020	≦0.20	–	–	–	–	–	–	
≦0.020	≦0.020	–	–	–	–	–	–	–	
≦0.020	≦0.020	–	–	–	–	–	–	–	
≦0.030	≦0.030	≦0.25	≦0.30	–	–	–	≦0.25	–	E059
≦0.030	≦0.030	≦0.20	≦0.15	≦0.10	≦0.15	≦0.10	≦0.20	–	
≦0.030	≦0.030	≦0.20	≦0.30	≦0.10	≦0.15	≦0.10	≦0.20	–	
≦0.030	≦0.030	≦0.20	≦0.15	≦0.10	≦0.15	0.15～0.35	≦0.20	–	
≦0.030	≦0.030	≦0.20	≦0.30	≦0.10	≦0.15	0.15～0.35	≦0.20	Sn≦0.05	
≦0.035	≦0.035	≦0.20	≦0.15	≦0.10	–	–	≦0.20	–	
≦0.020	≦0.020	–	–	–	–	–	–	–	
≦0.035	≦0.035	–	–	–	–	–	–	–	
≦0.030	≦0.030	≦0.25	≦0.30	–	–	–	≦0.25	–	E060
≦0.030	≦0.030	≦0.20	≦0.15	≦0.10	≦0.15	≦0.10	≦0.20	–	
≦0.030	≦0.030	≦0.20	≦0.30	≦0.10	≦0.15	≦0.10	≦0.20	–	
≦0.035	≦0.035	≦0.20	≦0.15	≦0.10	–	–	≦0.20	Sn≦0.05	
≦0.020	≦0.020	–	–	–	–	–	–	–	
≦0.020	≦0.020	–	–	–	–	–	–	–	
≦0.035	≦0.030	–	–	–	–	–	–	–	
≦0.035	≦0.030	–	–	–	–	–	–	–	
≦0.030	≦0.030	–	≦0.30	–	–	–	–	–	E061
≦0.020	≦0.020	–	–	–	–	–	–	–	

JIS	ASTM	BS	DIN	NF	ГОСТ			
G4401	A686	4659	17350	A35-590	1435	C	Si	Mn
SK6				C70E2U		0.65～0.74	0.10～0.30	0.10～0.40
					Y8	0.75～0.84	0.15～0.35	0.20～0.40
SK7						0.60～0.70	≦0.35	≦0.50
			C70W2			0.65～0.74	0.10～0.30	0.10～0.35
				C65E4U		0.60～0.69	0.10～0.40	0.50～0.80
					Y7	0.65～0.74	0.15～0.35	0.20～0.40

Chemical Composition 化学成分 (%)									Index Number 索引番号
P	S	Ni	Cr	Mo	W	V	Cu	Others	
≦0.020	≦0.020	—	—	—	—	—	—	—	E061
≦0.035	≦0.030	—	—	—	—	—	—	—	
≦0.030	≦0.030	≦0.25	≦0.30	—	—	—	≦0.25	—	E062
≦0.030	≦0.030	≦0.25	≦0.20	—	—	—	—	—	
≦0.020	≦0.020	—	—	—	—	—	—	—	
≦0.025	≦0.035	≦0.030	—	—	—	—	—	—	

4-2. Alloy Tool Steels　合金工具鋼

JIS	ASTM	BS	DIN	NF	ГОСТ			
G4404	A681	4659	17350	A35-590	5950	C	Si	Mn
Cutting Tool Alloy Steels　切削工具用合金鋼								
SKS11						1.20～1.30	≦0.35	≦0.50
SKS2						1.00～1.10	≦0.35	≦0.80
			105WCr6			1.00～1.10	0.10～0.40	0.80～1.10
SKS21						1.00～1.10	≦0.35	≦0.50
SKS5						0.75～0.85	≦0.35	≦0.50
SKS51						0.75～0.85	≦0.35	≦0.50
SKS7						1.10～1.20	≦0.35	≦0.50
				C120E3UCr4		1.10～1.29	0.10～0.30	0.10～0.40
SKS8						1.30～1.50	≦0.35	≦0.50
				C140E3UCr4		1.30～1.50	0.10～0.30	0.10～0.40
					13X	1.25～1.40	0.15～0.35	0.30～0.60
	F1					0.95～1.25	0.10～0.50	≦0.50
	F2					1.20～1.40	0.10～0.50	0.10～0.50
Impact－Resistant Tool Alloy Steels　耐衝撃工具用合金鋼								
SKS4						0.45～0.55	≦0.35	≦0.50
SKS41						0.35～0.45	≦0.35	≦0.50
	S1					0.40～0.55	0.55～0.20	0.10～0.40
SKS43						1.00～1.10	≦0.25	≦0.35
		BW2				0.95～1.1	≦0.30	≦0.35
				C105E2UV-1		0.95～1.09	0.10～0.30	0.10～0.40
SKS44						0.80～0.90	≦0.25	≦0.30
	S2					0.40～0.55	0.15～1.20	0.10～0.40
	S4					0.50～0.65	0.75～2.25	0.60～0.95
	S5					0.50～0.65	1.75～2.25	0.60～1.00
	S6					0.40～0.50	2.00～2.50	1.20～1.50
	S7					0.45～0.55	0.20～1.00	0.20～0.90
Cold Work Tool and Die Steels　冷間工具及び金型用鋼								
SKS3						0.90～1.00	≦0.35	0.90～1.20
SKS31						0.95～1.05	≦0.35	0.90～1.20
			105WCr6			1.00～1.10	0.10～0.40	0.80～1.10
					XBГ	0.90～1.05	0.15～0.35	0.80～1.10
SKS93						1.00～1.10	≦0.50	0.80～1.10
SKS94						1.00～1.10	≦0.50	0.80～1.10
SKS95						0.80～0.90	≦0.50	0.80～1.10
SKD1						1.80～2.40	≦0.40	≦0.60
	D3					2.00～2.35	0.10～0.60	0.10～0.60
		BD3				1.90～2.30	≦0.60	≦0.60
			X210Cr12			1.90～2.30	0.10～0.50	0.10～0.50
				X200Cr12		1.90～2.20	0.10～0.40	0.15～0.45

Chemical Composition 化学成分 (%)									Index Number 索引番号
P	S	Ni	Cr	Mo	W	V	Cu	Others	
≦0.030	≦0.030	≦0.25	0.20~0.50	–	0.10~0.30	–	≦0.25	–	E 063
≦0.030	≦0.030	≦0.25	0.50~1.00	–	1.00~1.50	(≦0.20)	(≦0.20)	–	E 064
≦0.030	≦0.030	–	0.90~1.10	–	1.00~1.30	–	≦0.25	–	
≦0.030	≦0.030	≦0.25	0.20~0.50	–	0.50~1.00	0.10～0.25	≦0.25	–	E 065
≦0.030	≦0.030	0.70~1.30	0.20~0.50	–	–	–	≦0.25	–	E 066
≦0.030	≦0.030	1.30~2.00	0.20~0.50	–	–	–	–	–	
≦0.030	≦0.030	≦0.25	0.20~0.50	–	–	–	≦0.25	–	E 067
≦0.025	≦0.025		0.20~0.50						E 068
≦0.030	≦0.030	≦0.25	0.20~0.50	–	–	–	≦0.25	–	
≦0.025	≦0.025	–	0.20~0.50	–	–	–	–	–	
–	–	–	0.40~0.7	–	–	–	–	–	
≦0.030	≦0.030	–	–	–	1.00~1.75	–	–	–	E 069
≦0.030	≦0.030	–	0.20~0.40	–	3.00~4.50	–	–	–	
≦0.030	≦0.030	≦0.25	0.50~1.00	–	0.50~1.00	–	≦0.25	–	E 070
≦0.030	≦0.030	≦0.25	1.00~1.50	–	2.50~3.50	0.15～0.30	≦0.25	–	E 071
≦0.030	≦0.030	–	1.00~1.80	≦0.50	1.50~3.60	–	–	–	
≦0.030	≦0.030	≦0.25	1.00~1.50	–	–	–	≦0.25	–	E 072
≦0.035	≦0.035	≦0.20	≦0.15	≦0.10	–	0.15～0.35	≦0.20	Sn≦0.05	
≦0.020	≦0.020	–	–	–	–	0.05～0.25	–	–	
≦0.030	≦0.030	≦0.25	≦0.20	–	–	0.15～0.25	≦0.25	–	E 073
≦0.030	≦0.030	–	–	0.30~0.60	–	≦0.50	–	–	E 074
≦0.030	≦0.030	–	0.10~0.50	–	–	0.15～0.35	–	–	
≦0.030	≦0.030	–	0.10~0.50	–	–	0.15～0.35	–	–	
≦0.030	≦0.030	–	1.20~1.50	0.30~0.50	–	0.20～0.40	–	–	
≦0.030	≦0.030	–	3.00~3.50	1.30~1.80	–	≦0.035	–	–	
≦0.030	≦0.030	0.50~1.00	≦0.25	–	0.50~1.00	–	≦0.25	–	E 075
≦0.030	≦0.030	≦0.25	0.90~1.10	–	1.00~1.30	–	≦0.25	–	E 076
≦0.030	≦0.030	0.90~1.10	–	–	1.00~1.30	–		–	
–	–	–	0.90~1.20	–	1.20~1.60	–	–	–	
≦0.030	≦0.030	≦0.25	0.20~0.60	–	–	–	≦0.25	–	E 077
≦0.030	≦0.030	≦0.25	0.20~0.60	–	–	–	≦0.25	–	E 078
≦0.030	≦0.030	≦0.20	0.20~0.60	–	–	–	≦0.25	–	E 079
≦0.030	≦0.030	(≦0.25)	12.0~15.0	–	–	(≦0.30)	≦0.25	–	E 080
≦0.030	≦0.030	–	11.0~13.0	0.70~1.20	–	–	–	–	
≦0.035	≦0.035	≦0.40	12.0~13.0	–	–	≦0.50	≦0.20	Sn≦0.05	
≦0.030	≦0.030	–	11.0~13.0	–	–	–	–	–	
≦0.025	≦0.025	–	11.0~13.0	–	–	–	–	–	

JIS	ASTM	BS	DIN	NF	ГОСТ			
G4404	A681	4659	17350	A35-590	5950	C	Si	Mn
					X12	2.00～2.20	0.15～0.35	0.15～0.40
SKD11						1.40～1.60	≦0.40	≦0.60
	D2					1.40～1.60	0.10～0.60	0.10～0.60
		BD2				1.40～1.60	≦0.60	≦0.60
				X160CrMoV12		1.45～1.70	0.10～0.40	0.15～0.45
SKD12						0.95～1.05	≦0.40	0.60～0.90
	A2					0.95～1.05	0.10～0.50	0.40～1.00
		BA2				0.90～1.05	≦0.40	0.30～0.70
				X100CrMoV5		0.90～1.05	0.10～0.40	0.50～0.80
	A3					1.20～1.30	0.10～0.70	0.40～0.60
	A4					0.95～1.05	0.10～0.70	1.80～2.20
	A5					0.95～1.05	0.10～0.70	2.80～3.20
	A6					0.65～0.75	0.10～0.70	1.80～2.50
	A7					2.00～2.85	0.10～0.70	0.20～0.80
	A8					0.50～0.60	0.75～1.10	0.20～0.50
	A9					0.45～0.55	0.95～1.15	0.20～0.50
	A10					1.25～1.50	1.00～1.50	1.60～2.10
	D4					2.05～2.40	0.10～0.60	0.10～0.60
	D5					1.40～1.60	0.10～0.60	0.10～0.60
	D7					2.15～2.50	0.10～0.60	0.10～0.60
	P2					≦0.10	0.10～0.40	0.10～0.40
	P3					≦0.10	≦0.40	0.20～0.60
	P4					≦0.12	0.10～0.40	0.20～0.60
	P5					0.06～0.10	0.10～0.40	0.20～0.60
	P6					0.05～0.15	0.10～0.40	0.35～0.70
	P20					0.28～0.40	0.20～0.80	0.60～1.00
	P21					0.18～0.22	0.20～0.40	0.20～0.40
	01					0.85～1.00	0.10～0.50	1.00～1.40
	02					0.85～0.95	≦0.50	1.40～1.80
	06					1.25～1.55	0.55～1.50	0.30～1.10
	07					1.10～1.30	0.10～0.60	0.20～1.00
Hot Work Tool and Die Steels 熱間工具及び金型鋼								
SKD4						0.25～0.35	≦0.40	≦0.40
		BH21				0.25～0.35	≦0.40	≦0.40
				X32WCrV5		0.28～0.35	0.10～0.40	0.15～0.45
SKD5						0.25～0.35	≦0.40	≦0.60
				X30WCrV9		0.25～0.32	0.10～0.40	0.15～0.45
SKD6						0.32～0.42	0.80～1.20	≦0.50
	H11					0.35～0.43	0.80～1.25	0.20～0.60
		BH11				0.32～0.42	0.85～1.15	≦0.40

Chemical Composition 化学成分 (%)									Index Number 索引番号
P	S	Ni	Cr	Mo	W	V	Cu	Others	
–	–	–	11.5~13.0	–	–	–	–	–	E081
≦0.030	≦0.030	(≦0.30)	11.0~13.0	0.80~1.20	–	0.25～0.50	≦0.25	–	
≦0.030	≦0.030	–	11.0~13.0	0.70~1.20	–	–	–	–	
≦0.035	≦0.035	≦0.40	11.5~12.5	0.70~1.20	–	0.25～1.00	≦0.20	Sn≦0.05	
≦0.025	≦0.025	–	11.0~13.0	0.70~1.10	–	0.70～1.00	–	–	
≦0.030	≦0.030	(≦0.50)	4.50~5.50	0.80~1.20	–	0.25～0.50	≦0.25	–	E082
≦0.030	≦0.030	–	4.75~5.50	0.90~1.40	–	–	–	–	
≦0.035	≦0.035	≦0.40	4.75~5.25	0.90~1.10	–	0.15～0.40	≦0.20	Sn≦0.05	
–	–	–	4.80~5.50	0.90~1.80	–	0.15～0.35	–	–	
≦0.030	≦0.030	–	4.75~5.50	0.90~1.40	–	0.80～1.40	–	–	E083
≦0.030	≦0.030	–	0.90~2.20	0.90~1.40	–	–	–	–	
≦0.030	≦0.030	–	0.90~1.20	0.90~1.40	–	–	–	–	
≦0.030	≦0.030	–	0.80~1.40	0.90~1.40	–	–	–	–	
≦0.030	≦0.030	–	5.00~5.75	0.90~1.40	0.50~1.50	3.90～5.15	–	–	
≦0.030	≦0.030	–	4.75~5.50	0.90~1.40	1.00~1.50	–	–	–	
≦0.030	≦0.030	1.25~1.75	4.75~5.50	1.30~1.80	–	0.80～1.40	–	–	
≦0.030	≦0.030	1.55~2.00	–	1.25~1.75	–	–	–	–	
≦0.030	≦0.030	–	11.0~13.0	0.70~1.20	–	0.15～1.00	–	–	
≦0.030	≦0.030	–	11.0~13.0	0.70~1.20	–	≦1.00	2.50~3.50	–	
≦0.030	≦0.030	–	11.5~13.5	0.70~1.20	–	3.80～4.40	–	–	
≦0.030	≦0.030	0.10~0.50	0.75~1.25	0.15~0.40	–	–	–	–	
≦0.030	≦0.030	1.00~1.50	0.40~0.75	–	–	–	–	–	
≦0.030	≦0.030	–	4.00~5.25	0.40~1.00	–	–	–	–	
≦0.030	≦0.030	≦0.25	–	–	–	–	–	–	
≦0.030	≦0.030	3.25~3.75	1.25~1.75	–	–	–	–	–	
≦0.030	≦0.030	–	1.40~2.00	0.30~0.55	–	–	–	–	
≦0.030	≦0.030	3.90~4.25	0.20~0.30	–	–	0.15～0.25	–	–	
≦0.030	≦0.030	–	0.40~0.70	–	0.40~0.60	≦0.30	–	–	
≦0.030	≦0.030	–	≦0.50	≦0.30	–	≦0.30	–	–	
≦0.030	≦0.030	–	≦0.30	0.20~0.30	–	–	–	–	
≦0.030	≦0.030	–	0.35~0.85	≦0.30	1.00~2.00	–	–	–	
≦0.030	≦0.030	≦0.25	2.00~3.00	–	5.00~6.00	0.35～0.50	≦0.25	–	E084
≦0.035	≦0.035	≦0.40	2.25~3.25	≦0.60	8.5~10.0	≦0.40	≦0.20	Sn≦0.05	
≦0.025	≦0.025	–	2.00~3.00	–	4.5 ~ 5.1	0.40～0.70		–	
≦0.030	≦0.030	≦0.45	2.00~3.00	–	–	0.30～0.50	–	–	E085
≦0.025	≦0.025		2.50~3.50		8.50 ~ 9.50	0.30～0.50		–	
≦0.030	≦0.030	≦0.25	4.50~5.50	1.00~1.50	–	0.30～0.50	≦0.25	–	E086
≦0.030	≦0.030	–	4.75~5.50	1.00~1.60	–	0.30～0.60	–	–	
≦0.035	≦0.035	≦0.40	4.75~5.25	1.25~1.75	–	0.30～0.50	≦0.20	Sn≦0.05	

JIS	ASTM	BS	DIN	NF	ГОСТ			
G4404	A681	4659	17350	A35-590	5950	C	Si	Mn
			X38CrMoV51			0.36～0.42	0.90～1.20	0.30～0.50
				X38CrMoV5-3		0.34～0.42	0.30～0.50	0.20～0.50
					4X5MФC	0.36～0.42	0.80～1.10	0.20～0.40
SKD61						0.32～0.42	0.80～1.20	≦0.50
	H13					0.32～0.45	0.80～1.25	0.20～0.60
		BH13				0.32～0.42	0.85～1.15	≦0.40
			X40CrMoV51			0.37～0.43	0.90～1.20	0.30～0.50
				X40CrMoV5		0.36～0.44	0.80～1.20	0.20～0.50
					4X5MФ1C	0.35～0.43	0.80～1.10	0.20～0.40
SKD62						0.32～0.42	0.80～1.20	≦0.50
	H12					0.30～0.40	0.80～1.25	0.20～0.60
		BH12				0.30～0.40	0.85～1.15	≦0.40
				X35CrWMoV5		0.32～0.40	0.80～1.20	0.20～0.50
					3X3M3Ф	0.30～0.40	0.80～1.10	0.20～0.40
SKD7						0.28～0.38	≦0.50	≦0.60
	H10					0.35～0.45	0.80～1.25	0.20～0.70
		BH10				0.30～0.40	0.75～1.10	≦0.40
			X32CrMoV3			0.28～0.35	0.10～0.40	0.15～0.45
				32CrMoV12-28		0.28～0.35	0.10～0.40	0.20～0.50
SKD8						0.35～0.45	≦0.50	≦0.60
	H19					0.32～0.45	0.15～0.50	0.20～0.50
		BH19				0.35～0.45	≦0.40	≦0.40
SKT3						0.50～0.60	≦0.35	0.60～1.00
SKT4						0.50～0.60	≦0.035	0.60～1.00
			55NiCrMoV6			0.50～0.60	0.10～0.40	0.65～0.95
				55NiCrMoV7		0.50～0.60	0.10～0.40	0.50～0.80
					5XГM	0.50～0.60	0.25～0.60	1.20～1.60
	H21					0.26～0.36	0.15～0.50	0.15～0.40
	H22					0.30～0.60	0.15～0.40	0.15～0.40
	H23					0.25～0.35	0.15～0.60	0.15～0.40
	H24					0.42～0.53	0.15～0.40	0.15～0.40
	H25					0.22～0.32	0.15～0.40	0.15～0.40
	H26					0.45～0.55	0.15～0.40	0.15～0.40
	H14					0.35～0.45	0.80～1.25	0.20～0.60

Chemical Composition 化学成分 (%)									Index Number 索引番号
P	S	Ni	Cr	Mo	W	V	Cu	Others	
≦0.030	≦0.030	≦0.40	4.80~5.50	1.10~1.40	–	0.25～0.50	–	–	E086
≦0.025	≦0.025	–	4.80~5.50	2.8 ~3.2	–	0.30～0.50	–	–	
–	–	–	4.80~5.50	1.10~1.60	–	0.20～0.50	–	–	
≦0.030	≦0.030	≦0.25	4.50~5.50	1.00~1.50	–	0.80～1.20	≦0.25	–	E087
≦0.030	≦0.030	–	4.75~5.50	1.10~1.75	–	0.80～1.20	–	–	
≦0.035	≦0.035	≦0.30	4.75~5.25	1.25~1.75	–	0.90～1.10	≦0.20	Sn≦0.05	
≦0.035	≦0.035	–	4.80~5.50	1.20~1.50	–	0.90～1.10	–	–	
≦0.025	≦0.025	–	4.80~5.50	1.10~1.50	–	0.85～1.15	–	–	
–	–	–	4.70~5.60	1.10~1.60	–	0.20～0.50	–	–	
≦0.030	≦0.030	≦0.25	4.50~5.50	1.00~1.50	1.00~1.50	0.20～0.60	≦0.25	–	E088
≦0.035	≦0.035	≦0.40	4.75~5.25	1.25~1.75	1.00~1.70	0.20～0.50	–	–	
≦0.035	≦0.035	≦0.40	4.75~5.25	1.25~1.75	1.25~1.75	≦0.50	≦0.20	Sn≦0.05	
≦0.025	≦0.025	–	4.80~5.50	1.20~1.50	1.10~1.60	0.30～0.50	–	–	
–	–	–	4.70~5.30	1.10~1.60	–	0.20～0.50	–	–	
≦0.030	≦0.030	≦0.25	2.50~3.50	2.50~3.50	–	0.40～0.70	≦0.25	–	E089
≦0.030	≦0.030	–	3.00~3.75	2.00~3.00	–	0.25～0.75	–	–	
≦0.035	≦0.035	≦0.40	2.8~3.2	2.65~2.95	–	0.30～0.50	≦0.20	Sn≦0.05	
≦0.030	≦0.030	–	2.70~3.20	2.60~3.00	–	0.40～0.70	–	–	
≦0.025	≦0.025	–	2.60~3.30	2.50~3.00	–	0.40～0.70	–	–	
≦0.030	≦0.030	≦0.25	4.00~4.70	0.30~0.50	3.80~4.50	1.70～2.20	≦0.25	–	E090
≦0.030	≦0.030	–	4.00~4.75	0.30~0.55	3.75~4.50	1.75～2.20	–	Co 4.0~4.5	
≦0.035	≦0.035	≦0.40	4.0~4.5	≦0.45	4.0~4.5	2.0～2.4	≦0.20	Sn≦0.05	
0.60~1.00	≦0.030	≦0.30	0.90~1.20	0.30~0.50	–	(≦0.20)	≦0.25	–	E091
≦0.030	≦0.030	1.30~2.00	0.70~1.00	0.20~0.50	–	(≦0.20)	≦0.25	–	E092
≦0.030	≦0.030	1.50~1.80	0.60~0.80	0.25~0.35	–	0.07～0.12	–	–	
≦0.025	≦0.025	1.50~2.00	0.70~1.00	0.30~0.50	–	0.05～0.15	–	–	
–	–	–	0.60~0.90	–	1.20~1.60	–	–	–	
≦0.030	≦0.030	–	3.00~3.75	–	8.50~10.0	0.30～0.60	–	–	E093
≦0.030	≦0.030	–	1.75~3.75	–	10.0~11.75	0.25 ～0.50	–	–	
≦0.030	≦0.030	–	11.0~12.75	–	11.0~12.75	0.75 ～1.25	–	–	
≦0.030	≦0.030	–	2.50~3.50	–	14.0~16.0	0.40～0.60	–	–	
≦0.030	≦0.030	–	3.75~4.50	–	14.0~16.0	0.40～0.60	–	–	
≦0.030	≦0.030	–	3.75~4.50	–	17.25~19.0	0.75 ～1.25	–	–	
≦0.030	≦0.030	–	4.75~5.50	–	4.00~5.25	–	–	–	

4-3. High-Speed Tool Steels 高速度工具鋼

JIS	ASTM	BS	DIN	NF	ГОСТ			
G4403	A600	4659	17350	A35-590	–	C	Si	Mn
Tungsten Type High Speed Tool Steels								
SKH2						0.73～0.83	≦0.40	≦0.40
	T1					0.65～0.80	0.20～0.40	0.10～0.40
		BT1				0.70～0.80	≦0.40	≦0.40
				Z80WKCV 18-04-01		0.75 0.83	≦0.50	≦0.40
–						–	–	–
	T2					0.80～0.90	0.20～0.40	0.20～0.40
		BT2	No longer listed			0.75～0.85	≦0.40	≦0.40
		No longer listed		Z80WKCV 18-04-02		0.85～0.95	≦0.40	≦0.40
SKH3						0.73～0.83	≦0.40	≦0.40
	T4					0.70～0.80	0.20～0.40	0.10～0.40
		BT4				0.70～0.80	≦0.40	≦0.40
			S18-1-2-5			0.75～0.83	≦0.45	≦0.030
		No longer listed		Z80WKCV 18-15-04-01		0.75～0.85	≦0.40	≦0.40
SKH4						0.73～0.83	≦0.40	≦0.40
	T15					0.75～0.85	0.20～0.40	0.20～0.40
		BT5				0.75～0.85	≦0.40	≦0.40
		No longer listed		Z85WCV 18-15-04-02		–	≦0.40	≦0.40
–						–	–	–
	T6					0.75～0.85	0.20～0.40	0.20～0.40
		BT6				0.75～0.85	≦0.40	≦0.40
SKH10						1.45～1.60	≦0.40	≦0.40
	T5					1.50～1.60	0.15～0.40	0.15～0.40
		BT15				1.40～1.60	≦0.40	≦0.40
			S12-1-4-5			1.30～1.45	≦0.45	≦0.30
		No longer listed		Z160WKCV 02-05-05-04		1.40～1.60	≦0.40	≦0.40
		BT20	No longer listed			0.75～0.85	≦0.40	≦0.40
		BT21				0.60～0.70	≦0.40	≦0.40
		BT42				1.25～1.40	≦0.40	≦0.40
Molybdenum Type High-Speed Tool Steels								
–						–	–	–

Chemical Composition 化学成分 (%)									Index Number 索引番号
P	S	Ni	Cr	Mo	W	V	Co	Others	
W 系高速度工具鋼									
≦0.030	≦0.030	–	3.80~4.50	–	17.0~19.0	0.80～1.20	–	–	E 094
≦0.030	≦0.030	–	3.75~4.50	–	17.25~18.75	0.90～1.30	–	–	
≦0.035	≦0.035	–	3.75~4.5	≦0.7	17.5~18.5	1.0～1.25	≦1.0	–	
–	–	–	3.50~4.50	–	17.2~18.7	1.00～1.30	–	–	
–	–	–	–	–	17.5~19.0	–	–	–	E 095
≦0.030	≦0.030	–	3.75~4.50	≦1.00	17.5~18.5	1.80～2.40	–	–	
≦0.035	≦0.035	–	3.75~4.5	≦0.7	–	1.75～2.05	–	–	
≦0.030	≦0.030	–	3.8~4.4	≦1.0	17.0~18.5	1.8～2.4	–	–	
≦0.030	≦0.030	≦0.25	3.80~4.50	–	17.0~19.0	0.80～1.20	4.50~5.50	Cu≦0.25	E 096
≦0.035	≦0.035	–	3.75~4.50	0.40~1.00	17.5~19.0	0.80～1.20	4.25~5.75	–	
≦0.035	≦0.035	≦0.40	3.75~4.5	≦1.0	17.5~18.5	1.0～1.25	4.5~5.5	Cu≦0.25 Sn≦0.05	
–	–	–	3.8~4.5	0.5~0.8	17.5~18.5	1.4～1.7	4.50~5.00	–	
≦0.030	≦0.030	–	3.8~4.4	≦1.0	17.5~18.5	1.0～1.2	4.5~5.0	–	
≦0.030	≦0.030	≦0.25	3.80~4.50	–	17.0~19.0	1.00～1.50	9.00~11.0	Cu≦0.25	E 097
≦0.030	≦0.030	–	3.75~5.00	0.40~1.00	17.5~19.0	1.80～2.40	7.00~9.50	–	
≦0.035	≦0.035	≦0.40	4.25~5.0	≦1.0	12.0~13.0	4.75～5.25	4.5~5.5	Cu≦0.20 Sn≦0.05	
≦0.030	≦0.030	–	3.8~4.4	≦1.0	17.5~19	1.0～1.2	8.0~9.0	–	
–	–	–	–	–	–	–	–	–	E 098
≦0.030	≦0.030	–	4.00~4.75	0.40~1.00	18.5~21.0	1.50～2.10	11.0~13.0	–	
≦0.035	≦0.035	–	3.75~4.5	≦1.0	20.0~21.0	1.25～1.75	11.25~12.25	Cu≦0.20 Sn≦0.05	
≦0.030	≦0.030	≦0.25	3.80~4.50	–	11.5~13.5	4.20～5.20	4.20~5.20	Cu≦0.25	E 099
≦0.030	≦0.030	–	3.75~5.00	≦1.00	11.75~13.0	4.50～5.25	4.75~5.25	–	
≦0.035	≦0.035	≦0.40	3.75~4.5	≦1.0	18.5～19.5	1.75～2.05	9.0～10.0	Cu≦0.25 Sn≦0.05	
–	–	≦0.40	3.8~4.5	0.70~1.00	11.5~12.5	3.50～4.00	4.50~5.00	–	
≦0.035	≦0.035	≦0.30	3.75~4.5	≦1.0	18.5~19.5	1.75～2.05	9.0~10.0	–	
≦0.035	≦0.035	–	4.25~5.0	≦1.0	21.0~22.5	1.4～1.6	≦0.60	–	E 100
≦0.035	≦0.035	–	3.5~4.25	≦0.7	13.5~14.5	0.40～0.60	≦1.0	–	
≦0.035	≦0.035	≦0.40	3.75~4.5	2.75~3.5	8.5~9.5	2.75～3.25	9.0~10.0	Cu≦0.20 Sn≦0.05	
Mo 系高速度工具鋼									
–	–	–	–	–	–	–	–	–	E 101

J I S	A S T M	B S	D I N	N F	ГОСТ			
G4403	A600	4659	17350	A35-590	–	C	S i	M n
	M1					0.78～0.88	0.20～0.50	0.15～0.40
		BM1				0.75～0.85	≦0.40	≦0.40
SKH51						0.80～0.90	≦0.40	≦0.40
	M2regular					0.78～0.88	0.20～0.45	0.15～0.40
	M2highC					0.95～1.05	0.20～0.45	0.15～0.40
		BM2				0.82～0.92	≦0.40	≦0.40
			S-6-5-2			0.86～0.94	≦0.45	≦0.40
				HS6-5-2		0.80～0.87	≦0.50	≦0.40
SKH52						1.00～1.10	≦0.40	≦0.40
	M3Class1					1.00～1.10	0.20～0.45	0.15～0.40
SKH53						1.10～1.25	≦0.40	≦0.40
	M3Class2					1.15～1.25	0.20～0.45	0.15～0.40
			S-6-5-3			1.17～1.27	≦0.45	≦0.45
				HS6-5-3		1.15～1.25	≦0.50	≦0.040
SKH54						1.25～1.40	≦0.40	≦0.40
	M4					1.25～1.40	0.20～0.45	0.15～0.40
		BM4				1.25～1.40	≦0.40	≦0.40
			HS6-5-2-5			1.25～1.40	≦0.50	≦0.040
SKH55						0.85～0.95	≦0.40	≦0.40
			S-6-5-2-5			0.88～0.96	≦0.45	≦0.40
				HS-5-2-5		0.80～0.87	≦0.50	≦0.40
SKH56						0.85～0.95	≦0.40	≦0.40
	M36					0.80～0.90	0.20～0.45	0.15～0.40
SKH57						1.20～1.35	≦0.40	≦0.40
		–				–	–	–
			S-10-4-3-10			1.20～1.35	≦0.45	≦0.40
				HS10-4-3-10		1.25～1.35	≦0.50	≦0.40
SKH58						0.95～1.05	≦0.50	≦0.40
	M7					0.95～1.05	0.20～0.45	0.15～0.40
				HS-2-9-2		0.95～1.05	≦0.50	≦0.40
SKH59						1.00～1.15	≦0.50	≦0.40
	M42					1.05～1.15	0.15～0.65	0.15～0.40
		BM42				1.00～1.10	≦0.40	≦0.40

Chemical Composition 化学成分 (%)									Index Number 索引番号
P	S	Ni	Cr	Mo	W	V	Co	Others	
≦0.030	≦0.030	≦0.25	3.50~4.00	8.20~9.20	1.40~2.10	1.00~1.35	≦0.25	–	E 101
≦0.035	≦0.035	≦0.40	3.75~4.5	8.0~9.0	1.0~2.0	1.0~1.25	≦1.0	Co≦0.20 Sn≦0.05	
≦0.030	≦0.030	≦0.25	3.80~4.50	4.80~6.20	5.50~6.70	2.30~2.80	–		E 102
≦0.03	≦0.03	–	3.75~4.50	4.50~5.50	5.50~6.75	1.75~2.20	–	–	
≦0.03	≦0.03	–	3.75~4.50	4.50~5.50	6.50~6.75	1.75~2.20	–	–	
≦0.035	≦0.035	≦0.40	3.75~4.5	4.75~5.5	6.0~6.75	1.75~2.05	≦1.0	Cu≦0.20 Sn≦0.05	
≦0.030	≦0.030	–	3.80~4.50	4.70~5.20	6.00~6.70	1.70~2.00	–	–	
≦0.030	≦0.030	–	3.50~4.50	4.60~5.30	5.7~6.70	1.7 ~2.20	–	–	
≦0.030	≦0.030	≦0.25	3.80~4.50	4.80~6.20	5.50~6.70	2.30~2.80	≦0.25	–	E 103
≦0.03	≦0.03	–	3.75~4.50	4.75~6.50	5.50~6.75	2.25~2.75	–	–	
≦0.030	≦0.030	≦0.25	3.80~4.50	4.70~5.20	6.00~6.70	2.70~3.20	≦0.25	–	E 104
≦0.03	≦0.03	–	3.75~4.50	4.75~6.50	5.00~6.75	2.75~3.25	–	–	
≦0.030	≦0.030	–	3.80~4.50	4.70~5.20	6.00~6.70	2.70~3.20	–	–	
≦0.030	≦0.030	–	3.50~4.50	4.60~5.30	5.7~6.70	2.70~3.20	–	–	
≦0.030	≦0.030	≦0.25	3.80~4.50	4.50~5.50	5.30~6.50	3.90~4.50	≦0.25	–	E 105
≦0.030	≦0.030	–	3.75~4.50	4.25~5.50	5.25~6.50	3.75~4.50	–	–	
≦0.035	≦0.035	≦0.40	3.75~4.5	4.25~5.0	5.75~6.5	3.75~4.25	≦1.0	Cu≦0.20 Sn≦0.05	
≦0.030	≦0.030	–	4.00~5.00	4.20~5.00	5.00~6.00	3.60~4.20	–	–	
≦0.030	≦0.030	≦0.25	3.80~4.50	4.60~5.30	5.70~6.70	1.70~2.20	≦0.25	–	E 106
≦0.03	≦0.03	–	3.80~4.50	4.70~5.20	6.00~6.70	1.70~2.20	≦0.25	–	
≦0.030	≦0.030		3.50~4.50	4.60~5.30	5.70~6.70	1.70~2.20	4.50~5.20		
≦0.030	≦0.030	≦0.25	3.80~4.50	4.60~5.30	5.70~6.70	1.70~2.20	7.00~9.00	Cu≦0.25	E 107
≦0.03	≦0.03	–	3.75~4.00	4.50~5.50	5.50~6.50	1.75~2.25	7.75~8.75	Co 7.75~8.75	
≦0.030	≦0.030	≦0.25	3.80~4.50	3.00~4.00	9.00~11.0	3.00~3.70	≦0.25	Cu≦0.25	E 108
–	–	–	–	–	–	–	9.00~11.0	–	
≦0.035	≦0.035	–	3.80~4.50	3.20~3.90	9.00~10.0	3.00~3.50	9.50~10.5	–	
≦0.030	≦0.030	–	3.20~4.50	3.20~3.90	9.00~10.0	3.00~3.50	9.50~10.5	–	
≦0.030	≦0.030	≦0.25	3.50~4.50	8.20~9.20	1.50~2.10	1.70~2.20	–	Cu≦0.25	E 109
≦0.03	≦0.03	–	3.50~4.00	8.20~9.20	1.40~2.10	1.75~2.25	–	–	
≦0.030	≦0.030	–	3.50~4.50	8.20~9.20	1.50~2.10	1.70~2.20	–	–	
≦0.030	≦0.030	≦0.25	3.50~4.25	9.00~10.0	1.20~1.90	0.90~1.40	7.50~8.50	Cu≦0.25	E 110
≦0.03	≦0.03	–	3.50~4.25	9.00~10.0	1.15~1.85	0.95~1.35	7.75~8.75	–	
≦0.035	≦0.035	≦0.40	3.5~4.25	9.00~10.0	1.0~2.0	1.0~1.3	7.5~8.5	Cu≦0.20 Sn≦0.05	

JIS	ASTM	BS	DIN	NF	ГОСТ			
G4403	A600	4659	17350	A35-590	—	C	Si	Mn
			S-2-10-1-8			1.00～1.15	≦0.40	≦0.40
				HS2-9-1-8		1.05～1.15	≦0.50	≦0.40
	M6					0.75～0.85	0.20～0.45	0.15～0.40
	M10regular C					0.84～0.94	0.20～0.45	0.10～0.40
	M10high C					0.95～1.05	0.20～0.45	0.10～0.40
	M30					0.75～0.80	0.20～0.45	0.15～0.40
	M33					0.85～0.92	0.15～0.50	0.15～0.40
	M34					0.85～0.92	0.20～0.45	0.15～0.40
	M41					1.05～1.15	0.15～0.50	0.20～0.60
	M43					1.15～1.25	0.15～0.65	0.20～0.40
	M46					1.22～1.30	0.40～0.65	0.20～0.40
	M47					1.05～1.15	0.20～0.45	0.15～0.40
	M48					1.42～1.52	0.15～0.40	0.15～0.40
	M50					0.78～0.88	0.20～0.60	0.15～0.45
	M52					0.85～0.95	0.20～0.60	0.15～0.45
	M62					1.25～1.35	0.15～0.40	0.15～0.40
	(ASTMA681)							
	H41					0.60～0.75	0.20～0.45	0.15～0.40
	H42					0.55～0.70	0.20～0.45	0.15～0.40
	H43					0.50～0.65	0.20～0.45	0.15～0.40

Chemical Composition 化学成分 (%)									Index Number 索引番号
P	S	Ni	Cr	Mo	W	V	Co	Others	
≦0.035	≦0.035	–	3.50~4.20	9.00~10.	1.1~1.9	1.0～1.35	7.7~8.8	–	
≦0.030	≦0.030	–	3.50~4.50	9.00~10.0	1.30~1.90	1.00～1.30	7.50~8.50	–	E110
≦0.030	≦0.03	–	3.75~4.50	4.50~5.50	3.75~4.7	1.30～1.70	11.0~13.0	–	E111
≦0.03	≦0.03	–	3.75~4.50	7.75~8.50	–	1.80～2.20	–	–	
≦0.03	≦0.03	–	3.75~4.50	7.75~8.50	–	1.80～2.20	–	–	
≦0.03	≦0.03	–	3.50~4.25	7.75~9.00	1.30~2.30	1.00～1.40	4.50~5.50	–	
≦0.03	≦0.03	–	3.50~4.00	9.00~10.0	1.30~2.10	1.00～1.35	7.75~8.75	–	
≦0.03	≦0.03	–	3.50~4.00	7.75~9.20	1.40~2.10	1.90～2.30	7.75~8.75	–	
≦0.03	≦0.03	–	3.75~4.50	3.25~4.25	6.25~7.00	1.75～2.25	4.75~5.75	–	
≦0.03	≦0.03	–	3.50~4.25	7.50~8.50	2.25~3.00	1.50～1.75	7.75~8.75	–	
≦0.03	≦0.03	–	3.70~4.20	8.00~8.50	1.90~2.20	3.00～3.30	7.80~8.80	–	
≦0.03	≦0.03	–	3.50~4.00	9.25~10.0	1.30~1.80	11.5～13.5	4.75~5.20	–	
≦0.03	≦0.03	–	3.50~4.00	4.75~5.50	9.50~10.5	2.75～3.25	8.00~10.0	–	
≦0.03	≦0.03	–	3.75~4.50	3.90~4.75	–	0.80～1.25	–	–	
≦0.03	≦0.03	–	3.50~4.30	4.00~4.90	0.75~1.5	1.65～2.25	–	–	
≦0.03	≦0.03	–	3.50~4.00	10.0~11.0	5.75~6.00	1.80～2.10	–	–	
≦0.030	≦0.030	–	3.50~4.00	8.20~8.90	1.40~2.10	1.00～1.30	–	–	
≦0.03	≦0.03	–	3.75~4.50	4.50~5.50	5.50~6.75	1.15～2.20	–	–	
≦0.03	≦0.03	–	3.75~4.50	7.75~8.50	–	1.80～2.20	–	–	

5. Spring Steels ばね鋼

JIS	AISI	B S	DIN	N F	ΓOCT		
G4801	MN06－224	970 Part 5	17211	A35－571	14959	C	Si
SUP3						0.75～0.90	0.15～0.35
	1075					0.70～0.80	by agreement
	1078					0.72～0.85	by agreement
					75	0.72～0.80	0.17～0.37
					85	0.82～0.90	0.17～0.37
SUP6						0.56～0.64	1.50～1.80
		250A53				0.50～0.57	1.70～2.10
				56S7		0.50～0.57	1.70～2.10
					60C2	0.57～0.65	1.50～2.00
SUP7						0.56～0.64	1.80～2.00
	9260					0.56～0.64	1.80～2.20
		250A58				0.55～0.62	1.70～2.10
		250A61				0.58～0.65	1.70～2.10
				61S7		0.57～0.64	1.60～2.00
					60C2Γ	0.58～0.63	1.60～2.00
SUP9						0.52～0.60	0.15～0.35
SUP9A						0.56～0.64	0.15～0.35
		527A60				0.55～0.65	0.10～0.55
			55Cr3			0.52～0.59	0.15～0.40
SUP10						0.47～0.55	0.15～0.35
	6150					0.48～0.53	0.15～0.30
		735A50				0.46～0.54	0.10～0.35
			50CrV4			0.47～0.55	0.15～0.40
				50CV4		0.47～0.55	0.15～0.40
					50XΓϕA	0.48～0.55	0.17～0.37

Chemical Composition 化学成分 (%)								Index Number 索引番号
Mn	P	S	Ni	Cr	Mo	Cu	Others	
0.30～0.60	≦0.035	≦0.035	−	−	−	≦0.30	−	E112
0.40～0.70	≦0.040	≦0.040	−	−	−	−	−	
0.30～0.60	≦0.040	≦0.050	−	−	−	−	−	
0.50～0.80	≦0.035	≦0.035	≦0.250	−	−	≦0.20	−	
0.50～0.80	≦0.035	≦0.035	≦0.250	−	−	≦0.20	−	
0.70～1.00	≦0.035	≦0.035	−	−	−	−	−	E113
0.70～1.00	by agreement	by agreement	−	−	−	−	−	
0.70～1.00	−	−	−	−	−	−	−	
0.60～0.90	≦0.035	≦0.035	≦0.025	≦0.030	−	≦0.20	−	
0.70～1.00	≦0.035	≦0.035	−	−	−	≦0.30	−	E114
0.75～1.10	≦0.035	≦0.040	−	−	−	≦0.30	−	
0.70～1.00	by agreement	by agreement	−	−	−	−	−	
0.70～1.00	by agreement	by agreement	−	−	−	−	−	
0.60～0.90	≦0.035	≦0.035	−	≦0.45	−	−	−	
0.60～0.90	−	−	−	≦0.30	−	−	−	
0.65～0.95	≦0.035	≦0.035	−	0.65～0.95	−	≦0.30	−	E115
0.70～1.00	≦0.035	≦0.035	−	0.70～1.00	−	≦0.30	−	
0.70～1.10	by agreement	by agreement	−	0.60～0.90	−	−	−	
0.70～1.00	≦0.035	≦0.035	−	0.60～0.90	−	−	−	
0.65～0.95	≦0.035	≦0.035	−	0.80～1.10	−	≦03.0	V0.15～0.25	E116
0.70～0.90	≦0.035	≦0.040	−	0.80～1.10	−	≦0.15	−	
0.60～0.90	by agreement	by agreement	−	0.80～1.10	−	≦0.15	−	
0.70～1.00	≦0.035	≦0.035	−	0.90～1.20	−	0.10～0.20	−	
0.70～1.00	≦0.035	≦0.035	−	0.90～1.20	−	0.10～0.20	−	
0.80～1.00	−	−	−	0.95～1.20	−	0.15～0.25		

JIS	AISI	B S	DIN	N F	ГОСТ		
G4801	MN06－224	970 Part 1	17211	A35－51	14959	C	Si
SUP11A						0.56～0.64	0.15～0.35
					50XφA	0.46～0.54	0.17～0.37
SUP12						0.51～0.59	1.20～1.60
SUP13						0.56～0.64	0.15～0.35

Chemical Composition 化学成分 (%)								Index Number 索引番号
Mn	P	S	Ni	Cr	Mo	Cu	Others	
0.70～1.00	≦0.035	≦0.035	–	0.70～1.00	–	≦0.30	B≧0.0005	E117
0.80～1.00	–	–	–	0.80～1.10	–	0.10～0.20	–	
0.60～0.90	≦0.035	≦0.035	–	0.70～0.90	0.25～0.35	–	–	E118
0.70～1.00	≦0.035	≦0.035	–	0.70～0.90	0.25～0.35	–	–	E119

6. Free-Cutting Steels 快削鋼

JIS	ASTM	BS	DIN	NF	ГОСТ		
G4804	A581 A582	970 Part 1	1651	–	–	C	Si
SUM11						0.08～0.13	by agreement
SUM12						0.08～0.13	by agreement
SUM21						≦0.13	by agreement
			9S20			≦0.13	≦0.05
SUM22						≦0.13	by agreement
		220M07	No longer listed			≦0.15	by agreement
		230M07	No longer listed			≦0.15	by agreement
			9SMn28			≦0.14	0.24～0.32
SUM22L						≦0.13	by agreement
			9Mnpb28			≦0.14	≦0.05
SUM23						≦0.09	by agreement
SUM23L						≦0.09	by agreement
SUM24L						≦0.15	by agreement
SUM25						≦0.15	by agreement
			9SMn36			≦0.14	≦0.05
SUM31						0.14～0.20	by agreement
	1117					–	–
			15S10			0.13～0.18	–
SUM31L						0.14～0.20	by agreement
SUM32						0.12～0.20	by agreement
SUM41						0.32～0.39	by agreement
	1137					–	–
SUM42						0.37～0.45	by agreement
	1141					–	–
SUM43						0.40～0.48	by agreement
	1144					–	–
		Z26M44	No longer listed			0.40～0.48	by agreement
	303					≦0.15	≦1.00
	303Se					≦0.15	≦1.00
	416					≦0.15	≦1.00
	416Se					≦0.15	≦1.00
	420F					0.03～0.40	≦1.00
	420FSe					0.20～0.40	≦1.00
	430F					≦0.12	≦1.00
	430FSe					≦0.12	≦1.00
	440F					0.95～1.20	≦1.00
	440FSe					0.95～1.20	≦1.00
	XM-1					≦0.08	≦1.00

Chemical Composition 化学成分 (%)								Index Number 索引番号
Mn	P	S	Ni	Cr	Pb	Cu	Others	
0.30～0.60	≦0.040	0.08～0.13	–	–	–	–	–	E120
0.60～0.90	≦0.040	0.08～0.13	–	–	–	–	–	E121
0.70～1.00	0.07～0.12	0.16～0.23	–	–	–	–	–	E122
0.60～1.20	≦0.100	0.18～0.25	–	–	–	–	–	
0.70～1.00	0.07～0.12	0.24～0.33	–	–	–	–	–	E123
0.90～1.30	≦0.070	0.20～0.30	–	–	–	–	Mo0.5～1.0	
0.10～1.30	≦0.070	0.20～0.30	–	–	–	–	Mo0.5～1.0	
0.9～1.3	≦0.05	≦0.10	–	–	–	–		
0.70～1.00	0.07～0.12	0.24～0.33	–	–	0.10～0.35			E124
0.90～1.30	≦0.100	0.24～0.33	–	–	0.15～0.30	–	–	
0.75～1.05	0.04～0.09	0.26～0.35	–	–	–	–	–	E125
0.75～1.05	0.04～0.09	0.26～0.35	–	–	0.10～0.35	–	–	E126
0.85～1.15	0.04～0.09	0.26～0.35	–	–	0.10～0.35	–	–	E127
0.90～1.40	0.07～0.12	0.30～0.40	–	–	–	–	–	E128
0.9～1.3	≦0.05	≦0.10	–	–	–	–	–	
1.00～1.30	≦0.040	0.08～0.13	–	–	–	–	–	E129
–	–	–	–	–	–	–	–	
1.0～1.3	≦0.05	0.08～0.13	–	–	–	–	–	
1.00～1.30	≦0.040	1.00～1.30	–	–	0.10～0.25	–	–	E130
0.60～1.10	≦0.040	0.10～0.20	–	–	–	–	–	E131
1.35～1.65	≦0.040	0.08～0.13	–	–	–	–	–	E132
–	–	–	–	–	–	–	–	
1.35～1.65	≦0.040	0.08～0.13	–	–	–	–	–	E133
–	–	–	–	–	–	–	–	
1.35～1.65	≦0.030	≦0.035	–	–	–	–	–	E134
–	–	–	–	–	–	–	–	
1.35～1.65	≦0.04	0.24～0.33	–	–	–	–	–	
≦2.00	≦2.00	≧0.15	8.00～10.00	17.00～19.00		–	–	E135
≦2.00	≦0.06	≦0.06	8.00～10.00	17.00～19.00	–	–	Se≧0.15	
≦1.25	≦0.06	≦0.15	–	12.00～14.00	–	–	–	
≦1.25	≦0.06	≦0.06	–	12.00～14.00	–	–	Se≧0.15	
≦1.25	≦0.06	≦0.15	(≦0.50)	12.00～14.00	–	–	Cu≦0.60	
≦1.25	≦0.06	≦0.06	(≦0.50)	12.00～14.00	–	–	Cu≦0.60 Se≧0.15	
≦1.25	≦0.06	≦0.15	–	16.00～18.00	–	–	–	
≦1.25	≦0.06	≦0.06	–	16.00～18.00	–	–	Se≧0.15	
≦1.25	≦0.06	≦0.15	(≦0.50)	16.00～18.00	–	–	Cu≦0.60	
≦1.25	≦0.06	≦0.06	(≦0.50)	16.00～18.00			Cu≦0.60 Se≧0.15	
5.00～6.50	≦0.04	0.18～0.35	5.00～6.50	17.00～19.00	–	–	Cu1.75～2.25	

JIS	ASTM	BS	DIN	NF	ГОСТ		
G4804	A581 A582	970 Part 1	1651	–	–	C	Si
	XM-2					≦0.15	≦1.00
	XM-5					≦0.15	≦1.00
	XM-6					≦0.15	≦1.00
	XM-34					≦0.08	≦1.00

Chemical Composition 化学成分 (%)								Index Number 索引番号
Mn	P	S	Ni	Cr	Pb	Cu	Others	
≦2.00	≦0.05	0.11～0.16	8.00～10.00	17.00～19.00	–	–	Al 0.60～1.00 Mo 0.40～0.60	E135
2.50～4.50	≦0.20	≧0.25	7.00～10.00	17.00～19.00	–	–	–	
1.50～2.50	≦0.06	≧0.15	–	12.00～14.00	–	–	–	
≦2.50	≦0.04	≧0.15	–	17.50～19.50	–	–	Mo1.50～2.50	

7. Bearing Steels　軸受鋼

JIS	ASTM	B S	DIN	N F	ГОСТ		
G4805	A295 A485 A535	970 Part 1	17230	A35 – 565	21022	C	S i
SUJ1						0.95～1.10	0.15～0.35
	51100					0.98～1.10	0.15～0.35
	50100					0.98 ～1.10	0.15～0.35
	5195					0.90 ～1.03	0.15～0.35
	K19526					0.89 ～1.01	0.15～0.35
	Gr2					0.85 ～1.00	0.50～0.80
	Gr3					0.95 ～1.10	0.15～0.35
	Gr4					0.95 ～1.10	0.15～0.35
SUJ2						0.95～1.10	0.15～0.35
	52100					0.98～1.10	0.15～0.35
		534A99	No longer listed			0.95～1.10	–
			100Cr6			0.90～1.05	0.15～0.35
				100Cr6		0.95～1.10	0.15～0.35
					WX15C	0.95～1.05	0.20～0.37
SUJ3						0.95～1.10	0.40～0.70
	Gr1					0.90～1.05	0.45～0.75
SUJ4						0.95～1.10	0.15～0.35
SUJ5						0.95～1.10	0.40～0.70
	1070M					0.65～0.75	0.15～0.35
	5160					0.56～0.64	0.15～0.35

Chemical Composition 化 学 成 分 (%)								Index Number 索引番号
Mn	P	S	Ni	Cr	Mo	Cu	Others	
≦0.50	≦0.025	≦0.025	(≦0.25)	0.90～1.20	≦0.08	≦0.25	–	E 136
0.25～0.45	≦0.025	≦0.025	≦0.25	0.90～1.15	≦0.10	≦0.35	–	
0.25～0.45	≦0.025	≦0.025	≦0.25	0.40～0.60	≦0.10	≦0.35	–	
0.75～1.00	≦0.025	≦0.025	≦0.25	0.70～0.90	≦0.10	≦0.35	–	
0.50～0.80	≦0.025	≦0.025	≦0.25	0.40～0.60	0.08～0.15	≦0.35	–	
1.40～1.70	≦0.025	≦0.025	≦0.25	1.40～1.80	≦0.10	≦0.35	–	
0.65～0.90	≦0.025	≦0.025	≦0.25	1.10～1.50	0.20～0.30	≦0.35	–	
1.05～1.35	≦0.025	≦0.025	≦0.25	1.10～1.50	0.45～0.60	≦0.35	–	
≦0.50	≦0.025	≦0.025	≦0.25	1.30～1.60	≦0.08	≦0.25	–	E 137
0.25～0.45	≦0.025	≦0.025	≦0.25	1.30～1.60	≦0.10	≦0.35	–	
0.25～0.40	≦0.015	≦0.015	–	1.30～1.60	–	–	–	
0.25～0.40	≦0.025	≦0.015	≦0.30	1.35～1.65	≦0.10	≦0.35	Al≦0.050	
0.20～0.40	≦0.030	≦0.025	–	1.35～1.60	≦0.08	≦0.35	–	
0.30～0.50	≦0.025	≦0.01	≦0.30	1.30～1.65	–	≦0.25	Cu+Ni≦0.50	
0.90～1.15	≦0.025	≦0.025	≦0.25	0.90～1.20	≦0.08	≦0.25	–	E 138
0.95～1.25	≦0.025	≦0.025	≦0.25	0.90～1.20	≦0.10	≦0.35	–	
≦0.50	≦0.025	≦0.025	≦0.25	1.30～1.60	0.10～0.25	≦0.25	–	E 139
0.90～1.15	≦0.025	≦0.025	≦0.25	0.90～1.20	0.10～0.25	≦0.25	–	E 140
0.80～1.10	≦0.025	≦0.025	≦0.25	≦0.20	≦0.10	≦0.35		
0.75～1.00	≦0.025	≦0.025	≦0.25	0.70～0.90	≦0.10	≦0.35		

8. Hollow Drill Steels 中空鋼

JIS	ASTM	BS	DIN	NF	ГОСТ			
G4410	–	–	–	–	–	C	Si	Mn
SKC3						0.70～0.85	0.15～0.35	≦0.50
SKC11						0.85～1.10	0.15～0.35	≦0.50
SKC24						0.33～0.43	0.15～0.35	0.30～1.00
SKC31						0.12～0.25	0.15～0.35	0.60～1.20

Chemical Composition 化学成分 (%)									Index Number 索引番号
P	S	Ni	Cr	Mo	W	V	Cu	Others	
≦0.030	≦0.030	(≦0.25)	(≦0.20)	–	–	(≦0.25)	(≦0.25)	Ti – (≦0.25)	E141
≦0.030	≦0.030	≦0.20	0.30~1.50	(≦0.40)	–	(≦0.25)	(≦0.25)	–	
≦0.030	≦0.030	2.50~3.50	0.30~0.70	0.15~0.40	–	–	(≦0.25)	–	
≦0.030	≦0.030	2.80~3.20	1.20~1.80	0.40~0.70	–	–	(≦0.25)	–	

9. Super Alloys 超合金鋼

JIS	ASTM	BS	DIN	NF	ГОСТ			
G4091 G4092	—	—	—	—	—	C	Si	Mn
NCF600						≦0.15	≦0.50	≦1.00
NCF601						≦0.10	≦0.50	≦1.00
NCF750						≦0.08	≦0.50	≦1.00
NCF751						≦0.10	≦0.50	≦1.00
NCF800						≦0.10	≦1.00	≦1.50
NCF800H						0.05～0.10	≦1.00	≦1.50
NCF825						≦0.05	≦0.50	≦1.00
NCF80A						0.04～0.10	≦1.00	≦1.00

Chemical Composition 化学成分 (%)									Index Number 索引番号
P	S	Ni	Cr	Fe	Al	Ti	Cu	Others	
≦0.030	≦0.015	≧72.00	14.0~17.0	6.00~1.00	–	–	≦0.50	–	E142
≦0.030	≦0.015	58.0~63.0	21.0~25.0	balance	1.00~1.70	–	≦1.00	–	
≦0.030	≦0.015	≧70.00	14.0~17.0	5.00~9.00	0.40~1.00	2.25~2.75	≦0.50	Nb+Ta 0.70~1.20	
≦0.030	≦0.015	≧70.00	14.0~17.0	5.00~9.00	0.90~1.50	2.00~2.60	≦0.50	Nb+Ta 0.70~1.20	
≦0.030	≦0.015	30.0~35.0	19.0~23.0	balance	0.15~0.60	0.15~0.60	≦0.75	–	
≦0.030	≦0.015	30.0~35.0	19.0~23.0	balance	0.15~0.60	0.15~0.60	≦0.75	–	
≦0.030	≦0.015	30.0~45.0	19.5~23.5	balance	≦0.20	0.60~1.20	1.50~3.00	Mo 2.50~3.50	
≦0.030	≦0.015	balance	18.0~21.0	≦1.50	1.00~1.80	1.00~1.80	≦0.20	–	

INDEX OF CHAPTER V

STEELS FOR SPECIAL PURPOSES

第5部　特殊用途鋼

索　　引

"Index No." stands for the "Index No." in the last column of each page.

「索引番号」は本文各頁の右端の「索引番号」の欄の数字をあらわす。

Standard No. 規格番号	Grade グレード	Index No. 索引番号
〔JIS〕		
JIS-G3320	SUS304	E035
	SUS430	E035
JIS-G4303	SUS201	E001
	SUS202	E002
	SUS301	E003
	SUS302	E004
	SUS303	E005
	SUS304	E006
	SUS304L	E006
	SUS304LN	E006
	SUS304N1	E006
	SUS304N2	E006
	SUS305	E007
	SUS309S	E008
	SUS310S	E009
	SUS316	E010
	SUS316J1	E010
	SUS316L1L	E010
	SUS316L	E010
	SUS316LN	E010
	SUS316N	E010
	SUS317	E011
	SUS317J1	E011
	SUS317L	E011
	SUS321	E012
	SUS329J1	E013
	SUS347	E014
	SUS403	E016
	SUS405	E017
	SUS410	E018
	SUS410J1	E018
	SUS410L	E018
	SUS416	E019
	SUS420F	E020
	SUS420J1	E020
	SUS420J2	E020
	SUS430	E023
	SUS430F	E023
	SUS431	E024
	SUS434	E025
	SUS440A	E027
	SUS440B	E027
	SUS440C	E027
	SUS440F	E027
	SUS447J1	E029
	SUS630	E030
	SUS631	E031
	SUSXM7	E032
	SUSXM15J1	E033
	SUSXM27	E034
JIS-G4304	SUS201	E001
	SUS202	E002
	SUS301	E003
	SUS301J1	E003
	SUS302	E004
	SUS302B	E004
	SUS304	E006
	SUS304L	E006
	SUS304LN	E006
	SUS304N1	E006
	SUS304N2	E006
	SUS305	E007
	SUS309S	E008
	SUS310S	E009
	SUS316	E010
	SUS316J1	E010
	SUS316J1L	E010
	SUS316L	E010
	SUS316LN	E010
	SUS316N	E010
	SUS317	E011
	SUS317J1	E011
	SUS317L	E011
	SUS321	E012
	SUS329J1	E013
	SUS329J2L	E013
	SUS347	E014
	SUS403	E016
	SUS405	E017
	SUS410	E018
	SUS410L	E018
	SUS410S	E018
	SUS420J1	E020
	SUS420J2	E020
	SUS429	E022
	SUS429J1	E022
	SUS430	E023
	SUS430LX	E023
	SUS434	E025
	SUS436L	E026
	SUS440L	E027
	SUS447	E029
	SUS447J1	E029
	SUS631	E031
	SUSXM15J1	E033
	SUSXM27	E034

Standard No. 規格番号	Grade グレード	Index No. 索引番号
JIS-G4305	SUS201	E001
	SUS202	E002
	SUS301	E003
	SUS301J1	E003
	SUS302	E004
	SUS302B	E004
	SUS304	E006
	SUS304L	E006
	SUS304LN	E006
	SUS304N1	E006
	SUS304N2	E006
	SUS305	E007
	SUS309S	E008
	SUS310S	E009
	SUS316	E010
	SUS316J1	E010
	SUS316J1L	E010
	SUS316L	E010
	SUS316LN	E010
	SUS316N	E010
	SUS317	E011
	SUS317J1	E011
	SUS317L	E011
	SUS321	E012
	SUS329J1	E013
	SUS329J2L	E013
	SUS347	E014
	SUS403	E016
	SUS405	E017
	SUS410	E018
	SUS410L	E018
	SUS410S	E018
	SUS420J1	E020
	SUS420J2	E020
	SUS429	E022
	SUS429J1	E022
	SUS430	E023
	SUS430LX	E023
	SUS434	E025
	SUS436L	E026
	SUS440A	E027
	SUS444	E028
	SUS447J1	E029
	SUS631	E031
	SUSXM15J1	E033
	SUSXM27	E034
JIS-G4306	SUS201	E001
	SUS202	E002
	SUS301	E003
	SUS301J1	E003
	SUS302	E004
	SUS302B	E004
	SUS304	E006
	SUS304L	E006
	SUS304LN	E006
	SUS304N1	E006
	SUS304N2	E006
	SUS305	E007
	SUS309S	E008
	SUS310S	E009
	SUS316	E010
	SUS316J1	E010
	SUS316J1L	E010
	SUS316L	E010
	SUS316LN	E010
	SUS316N	E010
	SUS317	E011
	SUS317J1	E011
	SUS317L	E011
	SUS321	E012
	SUS329J1	E013
	SUS329J2L	E013
	SUS347	E014
	SUS403	E016
	SUS405	E017
	SUS410	E018
	SUS410L	E018
	SUS410S	E018
	SUS420J1	E020
	SUS420J2	E020
	SUS429	E022
	SUS429J1	E022
	SUS430	E023
	SUS430LX	E023
	SUS434	E025
	SUS436L	E026
	SUS440A	E027
	SUS444	E028
	SUS447J1	E029
	SUS631	E031
	SUSXM15J1	E033
	SUSXM27	E034
JIS-G4307	SUS201	E001
	SUS202	E002
	SUS301	E003
	SUS301J1	E003
	SUS302	E004
	SUS302B	E004

Standard No. 規格番号	Grade グレード	Index No. 索引番号
	SUS304	E006
	SUS304L	E006
	SUS304LN	E006
	SUS304N1	E006
	SUS304N2	E006
	SUS305	E007
	SUS309S	E008
	SUS310S	E009
	SUS316	E010
	SUS316J1	E010
	SUS316J1L	E010
	SUS316L	E010
	SUS316LN	E010
	SUS316N	E010
	SUS317	E011
	SUS317J1	E011
	SUS317L	E011
	SUS321	E012
	SUS329J1	E013
	SUS329J2L	E013
	SUS347	E014
	SUS403	E016
	SUS405	E017
	SUS410	E018
	SUS410L	E018
	SUS410S	E018
	SUS420J1	E020
	SUS420J2	E020
	SUS429J1	E022
	SUS430	E023
	SUS430LX	E023
	SUS434	E025
	SUS436L	E026
	SUS440A	E027
	SUS444	E028
	SUS447J1	E029
	SUS631	E031
	SUSXM15J1	E033
	SUSXM27	E034
JIS-G4308	SUS201	E001
	SUS302	E004
	SUS303	E005
	SUS303Se	E005
	SUS304	E006
	SUS304L	E006
	SUS304N1	E006
	SUS305	E007
	SUS305J1	E007
	SUS309S	E008
	SUS310S	E009
	SUS316	E010
	SUS316L	E010
	SUS321	E012
	SUS347	E014
	SUS384	E015
	SUS410	E018
	SUS416	E019
	SUS420J1	E020
	SUS420J2	E020
	SUS420F	E020
	SUS430	E023
	SUS430F	E023
	SUS440C	E027
	SUS631J1	E035
	SUSXM7	E032
JIS-G4309	SUS201	E001
	SUS303	E005
	SUS303Se	E005
	SUS304	E006
	SUS304L	E006
	SUS304N1	E006
	SUS305	E007
	SUS305J1	E007
	SUS309S	E008
	SUS310S	E009
	SUS316	E010
	SUS316L	E010
	SUS321	E012
	SUS347	E014
	SUS410	E018
	SUS416	E019
	SUS420F	E020
	SUS420J1	E020
	SUS420J2	E020
	SUS430	E023
	SUS430F	E023
	SUS440C	E027
	SUSXM7	E032
JIS-G4311	SUH1	E051
	SUH3	E052
	SUH4	E053
	SUH11	E053
	SUH21	E047
	SUH31	E037
	SUH35	E038
	SUH36	E039
	SUH37	E040

Standard No. 規格番号	Grade グレード	Index No. 索引番号
	SUH38	E041
	SUH309	E042
	SUH309S	E042
	SUH310	E043
	SUH310S	E043
	SUH330	E044
	SUH409	E048
	SUH446	E049
	SUH600	E054
	SUH616	E055
JIS-G4312	SUH21	E047
	SUH31	E039
	SUH35	E038
	SUH36	E039
	SUH37	E040
	SUH38	E041
	SUH309	E042
	SUH309S	E042
	SUH310	E043
	SUH310S	E043
	SUH330	E044
	SUH409	E048
	SUH446	E049
	SUH600	E054
	SUH616	E055
	SUH660	E045
	SUH661	E040
JIS-G4313	SUS301	E003
	SUS304	E006
	SUS420J2	E020
	SUS631	E031
JIS-G4314	SUS302	E004
	SUS304	E006
	SUS304N1	E006
	SUS316	E010
	SUS631J1	E031
JIS-G4315	SUS304	E006
	SUS305	E007
	SUS305J1	E007
	SUS316	E010
	SUS384	E015
	SUS410	E018
	SUS430	E023
	SUS434	E025
	SUSXM7	E032
JIS-G4317	SUS302	E004
	SUS304	E006
	SUS304L	E006
	SUS316	E010
	SUS316L	E010
	SUS321	E012
	SUS347	E014
	SUS430	E023
JIS-G4318	SUS302	E004
	SUS303	E005
	SUS303Se	E005
	SUS304	E006
	SUS304L	E006
	SUS305	E007
	SUS305J1	E007
	SUS309S	E008
	SUS310S	E009
	SUS316	E010
	SUS316L	E010
	SUS321	E012
	SUS347	E014
JIS-G4320	SUS304	E006
	SUS316	E010
	SUS316L	E010
	SUS430	E023
JIS-G4401	SK1	E056
	SK2	E057
	SK3	E058
	SK4	E059
	SK5	E060
	SK6	E061
	SK7	E062
JIS-G4403	SKH2	E094
	SKH3	E096
	SKH4	E097
	SKH10	E099
	SKH51	E102
	SKH52	E103
	SKH53	E104
	SKH54	E105
	SKH55	E106
	SKH56	E107
	SKH57	E108
	SKH58	E109
	SKH59	E110

Standard No. 規 格 番 号	Grade グレード	Index No. 索引番号
JIS-G4404	SKD1	E080
	SKD4	E084
	SKD5	E085
	SKD6	E086
	SKD7	E089
	SKD8	E090
	SKD11	E081
	SKD12	E082
	SKD61	E089
	SKD62	E088
	SKS2	E064
	SKS3	E075
	SKS4	E070
	SKS5	E065
	SKS7	E066
	SKS8	E068
	SKS11	E063
	SKS21	E065
	SKS31	E076
	SKS41	E071
	SKS43	E072
	SKS44	E073
	SKS51	E066
	SKS93	E077
	SKS94	E078
	SKS95	E079
	SKT3	E091
	SKT4	E092
JIS-G4410	SKC3	E141
	SKC11	E141
	SKC24	E141
	SKC31	E141
JIS-G4801	SUP3	E112
	SUP6	E113
	SUP9	E115
	SUP9A	E115
	SUP10	E116
	SUP11A	E117
	SUP12	E118
	SUP13	E119
JIS-G4804	SUM11	E120
	SUM12	E121
	SUM21	E122
	SUM22	E123
	SUM22L	E124
	SUM23	E125
	SUM23L	E126
	SUM24L	E127
	SUM25	E128
	SUM31	E129
	SUM31L	E130
	SUM32	E131
	SUM41	E132
	SUM42	E133
	SUM43	E134
JIS-G4805	SUJ1	E136
	SUJ2	E137
	SUJ3	E138
	SUJ4	E139
	SUJ5	E140
JIS-G4901	NCF80A	E142
G4902	NCF600	E142
	NCF601	E142
	NCF750	E142
	NCF751	E142
	NCF800	E142
	NCF800H	E142
	NCF825	E142
〔ASTM〕		
ASTM-A167	9	E036
ASTM-A176	201	E036
ASTM-A240	205	E036
ASTM-A276	218	E036
ASTM-A313	302	E036
ASTM-A314	302B	E036
ASTM-A478	303	E036, E135
ASTM-A479	303Se	E036, E135
ASTM-A492	304	E036
ASTM-A493	304H	E036
ASTM-A580	304L	E036
ASTM-A581	304LN	E036
ASTM-A582	304N	E036
ASTM-A666	305	E036
	308	E036
	ER308	E036
	309	E036
	309S	E036
	309Cb	E036
	310	E036
	310MLN	E036
	310S	E036
	310Cb	E036
	310H	E036
	310Cb	E036

Standard No. 規 格 番 号	127Grade 128グレード	Index No. 索引番号	Standard No. 規 格 番 号	Grade グレード	Index No. 索引番号
	314	E036		S28200	E036
	316	E036		S30454	E036
	316Cb	E036		S30600	E036
	316H	E036		S30815	E036
	316L	E036		S31200	E036
	316LN	E036		S31254	E036
	316N	E036		S31260	E036
	316Ti	E036		S31654	E036
	317	E036		S31725	E036
	317L	E036		S31726	E036
	317LN	E036		S31803	E036
	321	E036		S32550	E036
	321H	E036		S32615	E036
	329	E036		S32950	E036
	347	E036		S41050	E036
	347H	E036		S41500	E036
	348	E036		S42010	E036
	348H	E036		S44400	E036
	384	E036		S44401	E036
	403	E036		S44625	E036
	403F	E036		S44635	E036
	405	E036		S44660	E036
	409	E036		S44700	E036
	410	E036		S44735	E036
	410S	E036		S44800	E036
	414	E036		XM-1	E036, E135
	416	E036, E135		XM-2	E036, E135
	416Se	E036, E135		XM-3	E036, E135
	420	E036		XM-5	E036, E135
	420F	E036, E135		XM-6	E036, E135
	420Se	E036		XM-7	E036
	420FSe	E036, E135		XM-8	E036
	429	E036		XM-10	E036
	430	E036		XM-11	E036
	430F	E036, E135		XM-15	E036
	430Se	E036, E135		XM-16	E036
	431	E036		XM-17	E036
	439	E036		XM-18	E036
	440A	E036		XM-19	E036
	440B	E036		XM-21	E036
	440C	E036		XM-26	E036
	440F	E036		XM-27	E036
	440Se	E036		XM-28	E036
	442	E036		XM-29	E036
	446	E036		XM-30	E036
	501	E036		XM-31	E036
	502	E036		XM-33	E036
	631	E036		XM-34	E036, E134
	S018235	E036			
	S21800	E036	ASTM-A295	51100	E136

Standard No. 規格番号	Grade グレード	Index No. 索引番号
	52100	E139
ASTM-A485	Gr. 1	E138
ASTM-535	52100	E139
ASTM-600	M1	E101
	M2 regular	E102
	m2 Highc	E102
	M3 Class1	E103
	M3 Class2	E104
	M4	E105
	M6	E111
	M7	E109
	M10 regularC	E111
	M10 high C	E111
	M30	E111
	M33	E111
	M34	E111
	M36	E111
	M41	E111
	M43	E111
	M46	E111
	M47	E111
	M48	E111
	M50	E111
	M52	E111
	M62	E111
ASTM-A681	A2	E082
	A3	E083
	A4	E083
	A5	E083
	A6	E083
	A7	E083
	A8	E083
	A9	E083
	A10	E083
	D2	E083
	D3	E083
	D4	E083
	D5	E083
	D7	E083
	H21	E093
	H22	E093
	H23	E093
	H24	E093
	H25	E093
	H26	E093
	01	E083
	02	E083
	06	E083
	07	E083
	P2	E083
	P3	E083
	P4	E083
	P5	E083
	P6	E083
	P20	E083
	P21	E083
	S1	E071
	S2	E074
	S4	E074
	S5	E074
	S6	E074
	S7	E074
ASTM-A686	W1A-8	E060
	W1A-9	E059
	W1A-9½	E058
	W1A-10	E058
	W1A-11½	E057
	W1C-8	E060
	W1C-9	E059
	W1C-9½	E058
	W1C-10	E058
	W1C-11½	E057
	W2A-9	E059
	W2A-9½	E058
	W2A-13	E056
	W2C-9	E059
	W2C-9½	E058
	W2C-13	E056
〔AISI〕		
AISI-MN06/222	201	E001
	202	E002
	301	E003
	302	E004
	302B	E004
	303	E005
	303B	E005
	304	E006
	304L	E006
	304N	E006
	305	E007
	309	E042
	309S	E008
	310	E043
	310S	E009

Standard No. 規 格 番 号	Grade グレード	Index No. 索引番号	Standard No. 規 格 番 号	Grade グレード	Index No. 索引番号
	9SMn36	E128		Z6CNU17.04	E030
	15S10	E129		Z6NC18.16	E015
DIN-17221	50CrV4	E116		Z8C17	E023
	53Cr3	E115		Z8CN19.12	E007
				Z8CNb17	E023
DIN-17224	X2CrNi1911	E006		Z8CNiAl177	E031
DIN-17440	X2CrNiMo17132	E010		Z8CD17.01	E025
DIN-17441	X5CrNi1810	E006		Z8CT17	E023
	X5CrNi1911	E007		Z10CN18.09	E005
	X5CrNiMo17122	E010		Z10CN24.13	E008
	X6Cr13	E016		Z10CNF18.09	E005
	X6CrAl13	E017		Z12C13	E018
	X6CrNb17	E023		Z12CN17.07	E003
	X6CrNiNb1810	E014		Z12CN25.20	E009
	X6CrNiNb1810	E014		Z20C13	E021
	X6CrNiTi1810	E012			
	X6CrTi17	E023	NF-A35-578	Z2NC25.15B	E045
	X6CrNiAl177	E031		Z2NC35.16	E044
	X10Cr13	E018		Z2NCN25.20	E043
	X10CrNiS189	E004, E005		Z6CT12	E048
	X20CrNi192	E024		Z6NC25.15B	E045
	X22CrNi17	E024		Z8CNS20.2	E052
				Z10C24	E049
DIN-17230	100Cr6	E139		Z15CN24.13	E042
				Z20CDNbV11	E054
DIN-17350	105WCr6	E064, E076		Z40C5D10	E051
	C 70W2	E061, E062		Z45CS9	E052, E053
	C 80W1	E060			
	C105W1	E058	NF-A35-590	32CDV28	E089
	X32CrMoV33	E089		55NCDV7	E092
	X38CrMoV51	E086		Y1 70	E061, E062
	X40CrMoV51	E087		Y1 90	E059, E060
	X210Cr12	E080		Y1 105	E058
				Y1 105V	E072
〔NF〕				Y1 120C	E057
NE-A35-565	100C6	E139		Y1 140C	E056
NF-A35-571	50CV4	E116		Y2 120C	E057
	56S7	E113		Y2 140C	E068
	61S7	E114		Z30WCV9	E084
				Z30WCV9	E085
NF-A35-573	Z2CN18.10	E006		Z35CWDV5	E088
NF-A35-574	Z2CN18.10AZ	E006		Z38CDV5	E086
NF-A35-575	Z2CND17.12	E010		Z40CDV5	E088
NF-A35-576	Z6C13	E016, E018		Z80WKCV18-04-01	E094
NF-A35-581	Z6CA13	E017		Z80WKCV18-05-04-01	E096
NF-A35-582	Z6CN18.09	E006		Z80WKCV18-10-04-02	E097
NF-A35-584	Z6CND17.11	E010		Z85WCV18-04-02	E095
NF-A35-586	Z6CND17.12	E011		Z85WDCV06-05-04-02	E102
	Z6CNNb18.10	E015		Z85WDKCV	E106
	Z6CNT18.10	E012		06-05-05-04-02	

Standard No. 規格番号	Grade グレード	Index No. 索引番号
	Z100DCWV	E109
	09-04-02-02	
	Z110DKCWV	E110
	08-08-04-02-01	
	Z120WDCV	E104
	06-05-04-03	
	Z130WDCV	E105
	06-05-04-04	
	Z130WKCDV	E108
	10-10-04-04-02	
〔ГOCT〕		
ГOCT-1435	Y7	E062
	Y8	E061
	Y8Г	E060
	Y9	E060
	Y10	E059
	Y11	E058
	Y12	E057
	Y13	E056
ГOCT-5632	03X17H4M2	E010
	03X18H11	E006
	08X13	E016
	08X17T	E023
	08X18H10	E006
	08X18H10T	E012
	08X18H12b	E014
	09X1701	E031
	10X18C0	E023
	10X23H18	E009
	12X13	E016, E018
	12X17	E023
	12X17Г9AH4	E002
	12X18H10E	E005
	14X17H2	E024
	15X25T	E034
	20X13	E027
	20X17H2	E024
	30X13	E021
ГOCT-5950	3X3M3φ	E088
	4X5Mφ1C	E088
	4X5MφC	E085
	X12	E080
	XBГ	E076
ГOCT-14959	50XφA	E117
	50XГφA	E116
	60C2	E113
	60C2Г	E114
	75	E112
	80	E112
ГOCT-21022	WX15C	E137